Phytomedicine

Phytomedicine

Research and Development

Edited by
Parimelazhagan Thangaraj
Professor
Department of Botany
Bharathiar University
Coimbatore 641046
Tamil Nadu, India

CRC Press
Taylor & Francis Group
Boca Raton London New York

CRC Press is an imprint of the
Taylor & Francis Group, an **informa** business

First edition published 2020
by CRC Press
6000 Broken Sound Parkway NW, Suite 300, Boca Raton, FL 33487-2742

and by CRC Press
2 Park Square, Milton Park, Abingdon, Oxon, OX14 4RN

CRC Press is an imprint of Taylor & Francis Group, LLC

Library of Congress Cataloging-in-Publication Data

Names: Parimelazhagan, T., editor.
Title: Phytomedicine : research and development / Parimelazhagan Thangaraj.
Description: First edition. | Boca Raton, FL : CRC Press, 2020. | Includes bibliographical references and index.
Identifiers: LCCN 2020009466 | ISBN 9780367857592 (hardback) | ISBN 9781003014898 (ebook)
Subjects: LCSH: Herbs--Therapeutic use. | Medicinal plants.
Classification: LCC RM666.H33 P4885 2020 | DDC 615.3/21--dc23
LC record available at https://lccn.loc.gov/2020009466

ISBN: 9780367857592 (hbk)
ISBN: 9781003014898 (ebk)

Typeset in Times
by Lumina Datamatics Limited

Contents

Foreword

Plants are the greatest boon to mankind. The food, shelter, clothing, and medicines that we possess are only through them. They serve in raising the economic status of the country, and meanwhile, India has a rich biodiversity that can help people survive the drastic environmental conditions. Indians use a folklore system of medicinal plants from ancient times for their curative properties. Up to recent times, there is a deficiency prevailing in the scrutiny of the scientific validation of such medicinal plants. Current estimates also prove that most of the drugs utilized by India are of plant origin. Traditional herbal medicines are getting significant attention in global health debates, playing a prominent role in the global health care. At present, there is a situation where synthetic medicines have led to various side effects, such as the growth of lifestyle related disorders, the issue of serious malnutrition, a lack of remedies for chronic disorders, and an increase in the notion against proper health maintenance. Thus, the importance of phytomedicine is being realized and research should hold the potential to achieve social value. Also, the safety and effectiveness of herbal medicines must be highly considered with reliable preclinical and clinical trials and should typically employ experimental research. This book indeed represents a decisive step forward in understanding the importance of phytomedicine. This book serves as a fascinating reference guide to develop new ideas in research and development. The excellent contribution of Prof. Dr. Parimelazhagan Thangaraj in bringing about this book is highly appreciable, opening an inspirable foundation for gathering scientific information on medicinal plants. I would like to congratulate his exceptional commitment in making this book possible. I hope this book is one of the best pieces of green literature and also a convincing tool that helps to motivate and stimulate young minds toward the improvement of modern science and technologies.

Prof. Dr. P. Kaliraj
Vice Chancellor
Bharathiar University
Coimbatore

Preface

Nutrition, medicine, and food for health care are mainly derived through plant sources, which are natural gifts. They are presently being administered in both traditional and modern medicinal systems. Research has proved their efficacy in treating various diseases. The knowledge toward pharmacology, phytomedicine, and herbalism is increasing swiftly to bring about the improvement in remedial action and therapeutics, as the plants possess numerous bioactive compounds. Though current emerging research and interpretations of phytomedicine have been published in several books, pharmacopoeia's *Phytomedicine: Research and Development* has unique principles proposed in different disciplines of phytomedicine, including ethnopharmacology, nutrition and dietetics, nutraceuticals, and phytochemistry with innovative ideas and clear understanding. The information in this book is easy to read and will help readers find cutting-edge knowledge concerning the scientific validation of medicinal plants. This book is intended to explore important medicinal plants with immense potential for treating various diseases through 26 research and review chapters written by eminent academicians and research scientists. This book is completely composed of the scientific information behind the valuable medicinal plants studied in different fields encompassing phytochemistry, molecular biology, ethnopharmaceutics, nutritional biology, pharmacognosy, pharmatherapeutics, and biomedicine.

The editor conveys his sincere gratitude to Professor P. Kaliraj, vice chancellor, and the authorities of Bharathiar University for the endless guidance and support for this fruitful outcome. The editor extends his heartfelt thanks to all the authors for publishing their research work through their exceptional contributions and consistent cooperation. The support rendered by Prof. Dr. A. Rajendran and Head, Department of Botany, Bharathiar University is also acknowledged. The editor also thanks the bioprospecting research team, Dr. Suman Thamburaj, Mr. Saikumar Sathyanarayanan, Mr. Kasipandi Muniyandi, Ms. Ashwini Lydia Manoharan, Ms. Gayathri Jagadeesan, and Ms. Gayathri Nataraj for their involvement in the compilation of the manuscripts in an elegant manner. The editor also extends his warm gratitude to the publishers and their team members for all the technical support and continuous back-up for bringing out this book in time successfully.

Prof. Dr. Parimelazhagan Thangaraj
Coimbatore, India

Editor

Prof. Dr. Parimelazhagan Thangaraj completed his education from St. Joseph's College (autonomous), Trichy, and PhD from Bharathiar University, Coimbatore, India. His areas of specialization include bioprospecting of medicinal plants. Dr. Parimelazhagan has worked as a scientist in the Defence Research & Development Organization, Ministry of Defence, Government of India. He has also worked as a scientist in the Rubber Research Institute of India, Ministry of Commerce, Government of India, Tura, Meghalaya. He has 20 years of research and 12 years of teaching experience. He has been awarded the Raman Fellowship for Post-Doctoral Research for Indian Scholars in the United States by the University Grants Commission, Government of India, and he undertook research for one year. In addition to these credits, he has authored several books and book chapters in his field of specialization. He has filed for four patents and has two patents granted. He has more than 80 research publications in international and national peer-reviewed journals. He has received the National Science Day Medal 2005 and Laboratory Scientist of the Year award 2005 from DRDO, New Delhi. Currently, he is engaged in the investigation of phytomedicine to discover novel herbal drugs and nutraceuticals products to treat some of the most common human diseases like diabetes and cancer.

Contributors

S. Abiraami Valli
Department of Plant Biology and Plant Biotechnology
Ethiraj College for Women (Autonomous)
Chennai, India

Heidi Abrahamse
Laser Research Centre
Faculty of Health Sciences
University of Johannesburg
Doornfontein, South Africa

A. S. Afolabi
Biotechnology Advanced Research Center
Sheda Science and Technology Complex
FCT-Abuja, Nigeria

R. Akshaya
Heavenly Fuel Pvt Ltd
Chennai, India

E. C. Aniogo
Laser Research Centre
Faculty of Health Sciences
University of Johannesburg
Doornfontein, South Africa

Umrao Ankita
School of Biotechnology and Bioinformatics
D.Y. Patil University
Navi Mumbai, India

Adriano A. S. Araújo
Laboratory of Pharmaceutical Assays and Toxicity (LeFT)
Department of Pharmacy
Federal University of Sergipe
São Cristóvão, Brazil

Y. Aron Santhosh Kumar
Botanical Survey of India
Southern Regional Centre
Coimbatore, India

K. Arun Prasath
Biomedical Research Laboratories
Department of Botany
Bharathiar University
Coimbatore, India

K. Arunachalam
Department of Basic Sciences in Health, Area Pharmacology
Faculty of Medicine
Federal University of Mato Grosso (UFMT)
Cuiabá, Brazil

Palavesam Arunachalam
Department of Animal Science
Manonmanium Sundaranar University
Tirunelveli, India

M. Ashwini Lydia
Bioprospecting Laboratory
Department of Botany
Bharathiar University
Coimbatore, India

Muniappan Ayyanar
Department of Botany
A. Veeriya Vandayar Memorial Sri Pushpam College (Autonomous)
Thanjavur, India

Spoorthy N. Babu
Centre for Bio-Separation Technology
Vellore Institute of Technology
Vellore, India

V. Bhanu Deepthi
Heavenly Fuel Pvt Ltd
Chennai, India

L. S. Bilbis
Department of Biochemistry
Usmanu Danfodiyo University
Sokoto, Nigeria

Ajitabh Bora
Defence Research Laboratory (DRDO)
Tezpur, Assam-India

Rahul Chandran
Laser Research Centre
Faculty of Health Sciences
University of Johannesburg
Doornfontein, South Africa

Pronobesh Chattopadhyay
Defence Research Laboratory (DRDO)
Tezpur, Assam-India

Vijayakumar Dhaarani
Department of Botany
Bharathiar University
Coimbatore, India

Sanjai K. Dwivedi
Defence Research Laboratory (DRDO)
Tezpur, Assam-India

S. Gangadevi
PG and Research Department of Botany
Government Arts College (Autonomous)
Coimbatore, India

Blassan P. George
Laser Research Centre
Faculty of Health Sciences
University of Johannesburg
Doornfontein, South Africa

Elizabeth George
Department of Botany
Mar Thoma College
Chungathara, India

Baikuntha J. Gogoi
Defence Research Laboratory (DRDO)
Tezpur, Assam-India

S. Govindarajan
Centre for Bio-Separation Technology
Vellore Institute of Technology
Vellore, India

D. Inbakandan
Centre for Ocean Research
Sathyabama Institute of Science and Technology
Chennai, India

G. Iyappan
Genes, Repair, and Regeneration of Ophthalmic Workstation
Narayana Nethralaya Foundation
Bommasandra, India

Gayathri Jagadeesan
Bioprospecting Laboratory
Department of Botany
Bharathiar University
Coimbatore, India

Subramaniam Kalidass
Department of Animal Science
Manonmanium Sundaranar University
Tirunelveli, India

and

Institute of Food Security and Sustainable Agriculture (IFSSA)
Universiti of Malaysia Kelantan
Jeli, Kelantan, Malaysia

R. Kalidoss
Biomedical Research Laboratories
Department of Botany
Bharathiar University
Coimbatore, India

K. Kalimuthu
PG and Research Department of Botany
Government Arts College (Autonomous)
Coimbatore, India

M. C. Kamaraj
Heavenly Fuel Pvt Ltd
Chennai, India

A. V. P. Karthikeyan
PG and Research Department of Botany
Government Arts College
Thanthonimalai, India

Karuppasamy Kaviarasan
Department of Animal Science
Manonmanium Sundaranar University
Tirunelveli, India

R. Krishnamurthy
C. G. Bhakta Institute of Biotechnology
Uka Tarsadia University
Surat, India

N. Lawal
Department of Biochemistry & Molecular Biology
Federal University Dutsinma
Dutsin-Ma, Nigeria

B. D. Macham
Chemistry Advanced Research Centre
Sheda Science and Technology Complex
FCT-Abuja, Nigeria

S. Mahadevan
Department of Botany
Government Arts College (Autonomous)
Coimbatore, India

Ramasamy Manikandan
Botanical Survey of India
Southern Regional Centre, TNAU Campus
Coimbatore, India

M. Mariraj
Biomedical Research Lab
Department of Botany
Bharathiar University
Coimbatore, India

Amarnath Mathan Babu
Department of Animal Science
Manonmanium Sundaranar University
Tirunelveli, India

Mehala Devi Ramasamy
Botanical Survey of India
Southern Regional Centre, TNAU Campus
Coimbatore, India

J. Merlin Seles
Biomedical Research Laboratories
Department of Botany
Bharathiar University
Coimbatore, India

Kasipandi Muniyandi
Bioprospecting Laboratory
Department of Botany
Bharathiar University
Coimbatore, India

Chithambaram Murugan
Botanical Survey of India
Southern Regional Centre
Coimbatore, India

S. Murugan
Department of Biology
School of Nature Science
Madawalabu University
Bale, Ethiopia

Khonamai S. Nakhuru
Defence Research Laboratory (DRDO)
Tezpur, India

A. Narayanasamy
Department of Botany, Government Arts College (Autonomous)
Coimbatore, India

Gayathri Nataraj
Bioprospecting Laboratory
Department of Botany
Bharathiar University
Coimbatore, India

Nithya Sivagami Ponnurangam
Botanical Survey of India
Southern Regional Centre
Coimbatore, India

Ayesha Noor
Centre for Bio-Separation Technology
Vellore Institute of Technology
Vellore, India

Olivier Nzamwita
Department of Environmental Sciences
Bharathiar University
Coimbatore, India

R. O. Omosimua
Biotechnology Advanced Research Centre
Sheda Science and Technology Complex
Abuja, Nigeria

A. Onanuga
Department of Pharmaceutical Microbiology & Biotechnology, Faculty of Pharmacy
Niger Delta University
Amassama, Nigeria

M. Palanisamy
Botanical Survey of India
Southern Regional Centre
Coimbatore, India

P. Ponmurugan
Biomedical Research Laboratories
Department of Botany
Bharathiar University
Coimbatore, India

P. Premasudha
Department of Nanoscience and Technology
Bharathiar University
Coimbatore, India

Jullyana S. S. Quintans
Department of Physiology, Laboratory of Neuroscience and Pharmacological Assays (LANEF)
Federal University of Sergipe
São Cristóvão, Brazil

Lucindo J. Quintans-Júnior
Department of Physiology, Laboratory of Neuroscience and Pharmacological Assays (LANEF)
Federal University of Sergipe
São Cristóvão, Brazil

Vasantharaja Raguraman
Centre for Ocean Research
Sathyabama Institute of Science and Technology
Chennai, India

N. Rajaprabu
Biomedical Research Lab
Department of Botany
Bharathiar University
Coimbatore, India

Arumugam Rajendran
Department of Botany
Bharathiar University
Coimbatore, India

S. Ramalingam
Plant Genetic Engineering Laboratory
Department of Biotechnology
Bharathiar University
Coimbatore, India

S. G. A. Rao
Sri Raghavendra Biotechnologies Pvt Ltd
Bangalore, India

A. Rao Gururaj
International Stem Cell Services Ltd
Bangalore, India

A. Rao Jyothsna
Sri Raghavendra Biotechnologies Pvt Ltd
Bangalore, India

Lakshmanan Ranjith
Marine Biodiversity Division
Tuticorin Research Centre
ICAR–Central Marine Fisheries Research Institute
Thoothukudi, India

A. A. Sabir
Department of Medicine
Usmanu Danfodiyo University Teaching Hospital
Sokoto, Nigeria

Saikumar Sathyanarayanan
Bioprospecting Laboratory
Department of Botany
Bharathiar University
Coimbatore, India

B. Sathya Priya
Department of Environmental Sciences
Bharathiar University
Coimbatore, India

Luciana Scotti
Department of Chemistry
Federal University of Paraíba
JoaoPessoa, Brazil

Marcus T. Scotti
Department of Chemistry
Federal University of Paraíba
Joao Pessoa, Brazil

Giftson J. Senapathy
C. G. Bhakta Institute of Biotechnology
Uka Tarsadia University
Surat, India

T. Shanmuganathan
Graduate Teacher (BT Assistant)
Panchayat Union Middle School
Kattur, Karur, India

Saravanan Shanmugam
Laboratory of Pharmaceutical Assays and Toxicity (LeFT)
Department of Pharmacy
Federal University of Sergipe
São Cristóvão, Brazil

M. Shenbagam
Biomedical Research Laboratories
Department of Botany
Bharathiar University
Coimbatore, India

T. Stalin
Faculty Research & Development Unit
Molecular Biology Division
Karur, India

L. Stanley Abraham
Centre for Ocean Research
Sathyabama Institute of Science and Technology
Chennai, India

K. M. Smitha
Centre for Ocean Research
Sathyabama Institute of Science and Technology
Chennai, India

P. S. Sreeja
Bioprospecting Laboratory
Department of Botany
Bharathiar University
Coimbatore, India

M. M. Sudheer Mohammed
Department of Botany
Government Arts College (Autonomous)
Coimbatore, India

Jeyalatchagan Sureshkumar
Department of Botany
A. Veeriya Vandayar Memorial Sri Pushpam College (Autonomous)
Thanjavur, India

Suman Thamburaj
Bioprospecting Laboratory
Department of Botany
Bharathiar University
Coimbatore, India

Parimelazhagan Thangaraj
Bioprospecting Laboratory
Department of Botany
Bharathiar University
Coimbatore, India

Sajeesh Thankarajan
Department of Botany
Mar Ivaneos College
Thiruvananthapuram, India

B. T. Thomas
Department of Microbiology
Olabisi Onabanjo University
Ago-Are, Nigeria

Sunday Asuquo Thomas
Chemistry Advanced Research Centre
Sheda Science and Technology Complex
FCT-Abuja, Nigeria

R. Thirugnanasambandam
Centre for Ocean Research
Sathyabama Institute of Science and Technology
Chennai, India.

C. O. Ugbomor
Biotechnology Advanced Research Centre, Sheda Science and Technology Complex
FCT-Abuja, Nigeria

S. Uma Gowrie
Department of Plant Biology and Plant Biotechnology
Ethiraj College for Women (Autonomous)
Chennai, India

P. Umadevi
Department of Biology
School of Nature Science
Madawalabu University
Bale, Ethiopia

R. A. Umar
Department of Biochemistry
Usmanu Danfodiyo University
Sokoto, Nigeria

P. Viswanathan
PG and Research Department of Botany, Government Arts College (Autonomous), Coimbatore, India

S. Vivek
Botanical Survey of India, Southern Regional Centre, TNAU Campus
Coimbatore, India

1 Ethnic Uses and Commercial Applications of *Capsicum assamicum* (*Bhut jolokia*)

Ajitabh Bora, Khonamai S. Nakhuru, Baikuntha J. Gogoi, Pronobesh Chattopadhyay, and Sanjai K. Dwivedi

CONTENTS

1.1 INTRODUCTION

Capsicum assamicum, an extremely pungent chilli, is endemic to northeast (NE) India, particularly in the states of Assam, Nagaland, and Manipur. The NE region, lying between 20.59° latitude and 78.96° longitude, has a unique environment with a high degree of humidity, which has led to the development of extreme pungency in this chilli. It is known by different names and *Bhut jolokia* is the name as it is called in Assam, while it is called Naga chilli in Nagaland.

The identity of this species has been a point of controversy since its discovery, regarding its phylogenetic position. Mathur et al. (2000) reported the *Bhut jolokia* to be a variety of *Capsicum frutescens* L., while in the *Guinness Book of World Records* (2006), *Bhut jolokia* has been recorded as belonging to the species of *C. chinense* (Purkayastha et al. 2012b). On the other hand, Bosland and Baral (2007) reported *Bhut jolokia* to be a natural hybrid of *C. frutescens* and *C. chinense* based on Random Amplified Polymorphic DNA (RAPD) analysis. But the RAPD technique has several limitations, such as non-reproducibility of results, competition between different DNA fragments for amplification (Williams et al. 1993; Halldén et al. 1996), and identical mobility of fragments with different sequences (Xu 2006), which limits its uses for interpopulation or interspecies comparisons. Purkayastha et al. (2012b) undertook molecular studies of *Bhut jolokia* to determine the distinctness of this species from *C. frutescens* and *C. chinenese* through sequencing of the ribosomal RNA gene-internal transcribed region along with its phylogenetic analysis. The phylogenetic analysis of the internal transcribed 1, 5.8S, and internal transcribed 2 sequences lead to a distinct clade for all the accessions of *Bhut jolokia*, while *C. frutescens* and *C. chinenese* occupied discrete lineages. Further, a unique 13 base pair deletion was observed in all the accessions of *Bhut jolokia*, making it distinct from all other chilli species. The morphological features of *Bhut jolokia*, such as profusely branched habitat, yellow green corolla, pale blue anther, orange red color, and sub-conical to conical fruits with rough, uneven, dented skin, and anatomical details of the stem make its distinct from *C. frutescens* and *C. chinense* (Purkayastha et al. 2012a). Moreover, phylogeny, based on median joining network analysis, revealed clear distinction of the Naga King Chilli from its closest relatives (*C. chinense* and *C. frutescens*) and the absence of a star-like network of haplotypes indicates the expansion of an ancient population of this chilli species. Besides, the differential proteomic analysis of *Bhut jolokia* revealed 22 differential protein spots, which appeared only in this species, but not in *C. frutescens* and *C. chinense*. Thus, *Bhut jolokia* has been established as a distinct species within the genus *Capsicum* and christened as *C. assamicum* Jubilee Purkayastha & L. Singh sp. nov. (Purkayastha et al. 2012a). This was further validated by Kehie et al. (2016), who also observed distinct 13 base deletions in the 5.8S gene in all the accessions of the Naga King Chilli, unlike other *Capsicum* species. These works clearly establish *Bhut jolokia* or Naga King Chilli as a distinct species of chilli endemic to northeast India. This chilli has been patented under the Geographical Indication Registry by the Government of Nagaland (Kehie et al. 2014).

1.2 BOTANY, DISTRIBUTION, AND DIVERSITY

Capsicum assamicum, belonging to the family Solanaceae, is a self-pollinated crop, though cross-pollination occurs (10%) when the insect population is high. Under optimum conditions, the plant behaves as perennials and grows to a height of 50–100 cm or sometimes even taller. The plant is the most similar to *C. frutescens* and *C. chinense*, but can be distinguished by its profusely branched habitat, yellow green corolla, pale blue anther, orange red color, and sub-conical to conical fruits with rough, uneven, dented skin, and anatomical details of the stem.

The leaves are ovate in shape and the size ranges from 5.5–7.5 cm in width and 10–14 cm in length. Flowers are pendant shaped, having creamy white corollas. The fruit is usually oblong in shape with a rough texture and is bright red or orange in color, which varies with soil and climate. A single fruit bears about 25–35 slightly wrinkled seeds and weighs around 5 gram (g) on average. A single plant produces, on average, around 15–20 full-sized fruit and 10–14 smaller ones (Borgohain and Devi 2007) (Figures 1.1 and 1.2).

Bhut jolokia is widely cultivated in northeast India, mainly in the states of Assam, Manipur, and Nagaland. This chilli is known by different names in the regions. It is popularly called *Bhut jolokia* in Assam. The term *Bhut* means ghost and *jolokia* means chilli in the Assamese language, which may be due to its ghostly bite or due to its introduction by the Bhutias from Bhutan according to another legend. It is also called *Bih jolokia* or *Borbih Jolokia* (*Bih* meaning poison, indicating high hotness of the chilli fruit) and by other names, such as *Nagahari, Naga Morich, Naga Moresh, Naga Jolokia* (named after the once ferocious "Naga" warrior tribe of northeast India), *Raja Mirchi*, and *Dorset Naga* (its derivative from Bangladesh).

FIGURE 1.1 *Bhut jolokia* plant.

FIGURE 1.2 *Bhut jolokia* fruits.

In Nagaland, it is known as Naga King Chilli and in the local Angami dialect of Nagaland as "Kedi Chusi," meaning the "King of Chillies" (Kehie et al. 2016). In Manipur, it is known as *Omorok* (Oo meaning tree and morok meaning chilli in the Manipuri language, probably due to the perennial growing habit of this chilli like a tree) (Talukdar et al. 2015). Besides, it is also known by several other names as *Saga jolokia*, Indian mystery chilli, India rough chilli (due to the rough texture of the fruits), and ghost pepper by the Western media (Meghvansi et al. 2010). Regardless of its name, it is cultivated and consumed in different states of northeast India as a spice or eaten raw with staple food.

Diversity of this chilli is seen in terms of fruit size, color, and bearing capacity, which may be termed as land races. Three colors of matured fruit can be found *viz.* orange, light to dark red, and dark chocolate. Two types of plants can be classified based on the immature fruit color which bears light and dark green. The color of the immature fruit is green in color, which turns to orange or red or chocolate on maturity depending on the landraces (Figures 1.3 and 1.4).

FIGURE 1.3 Different types of *Bhut jolokia* fruits.

FIGURE 1.4 Unripe fruits of *Bhut jolokia*.

1.3 ETHNIC USES

Traditionally, *Bhut jolokia* is used as a spice or consumed raw due to its pleasant and palatable aroma that is unique to this chilli. The chilli is an integral part of the local cuisine of most of the tribes of the NE region (Table 1.1). Since time immemorial, this chilli has been used in treating a myriad of human ailments such as asthma, gastrointestinal abnormalities, toothache, muscle pain, arthritis, and removal of puss from boils as seen in Table 1.1 (Bhagowati and Changkija 2009). Besides its use in treating various human ailments, the chilli is smeared on fences or used in smoke bombs to ward off elephants since the smell of chilli is deterrent to elephants. The elephants get repelled as this chilli is highly pungent, it is quite effective in repelling the elephants.

1.4 CHEMISTRY OF HIGH HOTNESS

The hotness and pungency of a chilli is due to the presence of capsaicinoids, a group of lipophilic alkaloids produced by the plants as a self-defense mechanism against the invasion of predators and herbivores. Among the capsaicinoids, capsaicin and dihydrocapsaicin are responsible for 90% of the pungency in most of the chilli species (Suzuki et al. 1980). These compounds are found only in the members of the genus *Capsicum* and are mainly localized in the placenta and flesh of the chilli fruit. The gland on the placenta of the fruit produces capsaicinoids (Bosland 1996). One fruit of *Bhut jolokia* can sustain intense heat sensation in the mouth for up to 30 minutes before subsiding (Purkayastha et al. 2012b).

Conventionally, an organoleptic test is being used to measure the hotness of chillies. This test was first developed by Wilber Scoville in 1912 and, hence, the unit of measure of hotness was given as Scoville Heat Unit (SHU). However, in modern times, with the advent of sophisticated instruments, the capsaicinoid content is being estimated more accurately with the help of High Performance

TABLE 1.1
Few Ethnic Uses of *Capsicum assamicum* in Treating Human Aliments

Sl. No.	Ailments	Mode of Application	Mode of Action
1.	Asthma	Regular consumption of raw fruits in low quantities	Dilation of blood vessels by capsaicin, thereby relieving chronic congestion
2.	Gastrointestinal abnormalities	Regular consumption of raw fruits in low quantities	Stimulation of saliva and gastric juice by capsaicin as well as protection of mucous membrane
3.	Toothache/muscle pain	Hot infusions of fruits. But it should not be applied on injured tissues	Alleviation of pain by capsaicin
4.	Removal of puss from boils/arthritis/headache	Paste of leaves applied locally	Removal of puss from boils/pain alleviation

Liquid Chromatography (HPLC) and spectrophotometer. Pure capsaicin measures 15,000,000–16,000,000 SHUs. The higher the capsaicinoid content, the higher is the hotness (Borgohain and Devi 2007). Unlike other chilli species, cultivated in the Indian sub-continent, which contain around 1% capsaicin, *Bhut jolokia* has around 3%–5% capsaicin (Mathur et al. 2000; Sanatombi and Sharma 2008). When Mathur et al. (2000) reported hotness of *Bhut jolokia* to be 855,000 SHUs, it superseded the Red Savina habanero of Mexico whose pungency was 577,000 SHUs.

Bhut jolokia was officially recognized as the world's hottest chilli in the year 2006, measuring 1,001,304 SHUs after testing its hotness by the New Mexico State University, thereby replacing the Red Savina, which had a SHU of 248,566 (*Guinness Book of World Records* 2006; Bosland and Baral 2007). However, after holding its glory for 5 years, *Bhut jolokia* has been superseded by other chilli species in subsequent years, as in 2011 by the Infinity chilli, in 2012 by the Naga Viper, in 2012 by the Trinidad Moruga Scorpion, and in 2013 by the Carolina Reaper (Kehie et al. 2016) until now. Presently, *Bhut jolokia* holds seventh position in the list of world's hottest chilli—2018.

The pungency in a *Bhut jolokia* is affected by the climatic conditions and the time of cultivation. High rainfall and humidity with the moderate climate of northeast India favors the development of its extreme pungency. When this chilli was cultivated elsewhere, lower levels of pungency were recorded. For example, when it was grown in Gwalior, there was a reduction of capsaicin and dihyrocapsaicin by 50% (Tiwari et al. 2005), and when cultivated in Pithoragarh, in Uttarakhand, the pungency level came down to 254,896 SHUs (Pandey et al. 2009).

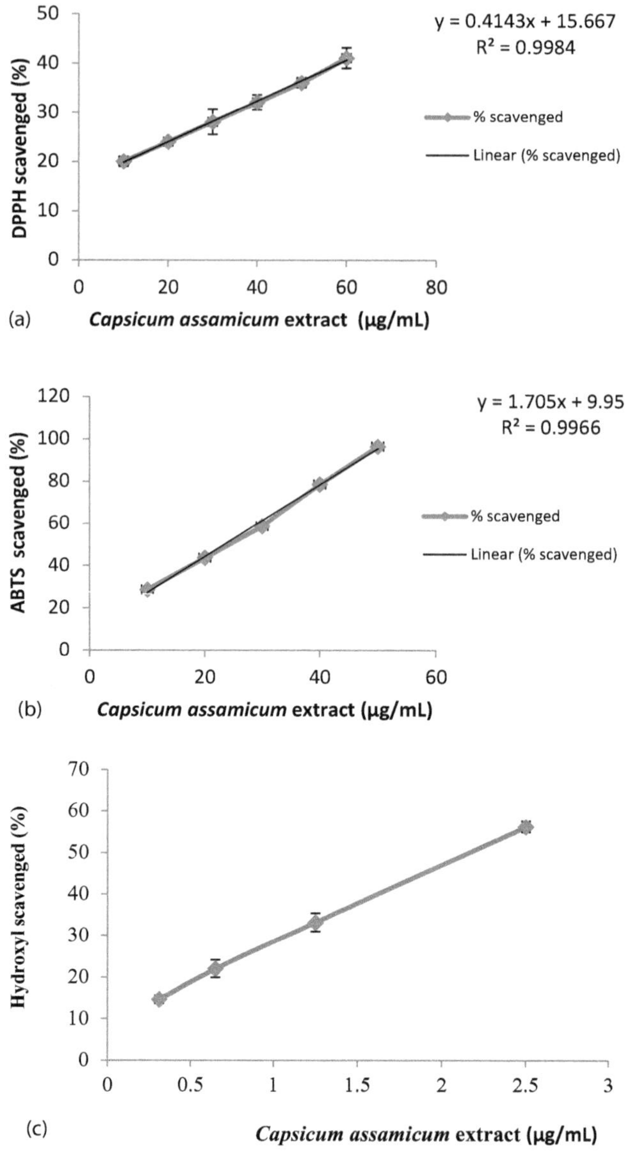

FIGURE 1.5 Anti-oxidant and radical scavenging activities of aqueous methanol extract of *Bhut jolokia* fruit in different concentrations. Each value represents mean±SD. (a) DPPH, (b) ABTS, and (c) hydroxyl.

1.5 NUTRITIVE VALUE

The anti-oxidant and anti-radical potentials of *Bhut jolokia* were investigated in vitro to validate the claims of its uses by the local folks. It was found to have immense potential as a natural source of anti-oxidant and anti-radical agents, as given in Figure 1.5 with half maximum inhibitory concentration (IC_{50}) values of 74.42±2.31, 25.00±0.31, and 2.15±0.04 against 2, 2-Diphenyl-1-picrylhydrazyl (DPPH), 2,2′-Azino-bis-3-ethylbenzothiazoline-6-sulfonic acid (ABTS), and hydroxyl, respectively (Nakhuru et al. 2018). Therefore, this chilli is a potential candidate for exploitation as a source of natural anti-oxidant for dietary supplements and for pharmaceutical/cosmetic uses.

1.6 COMMERCIAL APPLICATIONS

Bhut jolokia has tremendous potential for its use as a value-added product in the form of dry chilli, as whole, powdered, or flakes, paste, distilled oil for flavor, pickles, sauce, etc., which needs to be explored. This chilli is gaining popularity in the Western world and is a popular ingredient of several recipes of restaurants worldwide. In recent times, *Bhut jolokia* sauce, prepared from the chilli paste, is gaining much popularity for its off-the-shelf uses in various cuisines apart from the traditional recipes. The sauce is available in online marketing platforms, which has given it the much-needed hype and popularity. The sauce is being produced by Naga Spices & Herbs in Senapati, Manipur under the brand name *Poumai* and marketed by Lunar Agro Chemicals. Another start-up firm, Zizira, based in Meghalaya, is marketing a couple of products based on *Bhut jolokia*, while another firm, East by Northeast (ENE), has launched three variants of the *Bhut jolokia* sauce—sweet, hot, and extra hot to cater to different degrees of endurance of the taste buds. Frontech Agritech Pvt. Ltd. in Jorhat, an agri-business company, is involved in the exporting of *Bhut jolokia* to markets in Europe and the United States (Figure 1.6).

FIGURE 1.6 Dried fruits of *Bhut jolokia.*

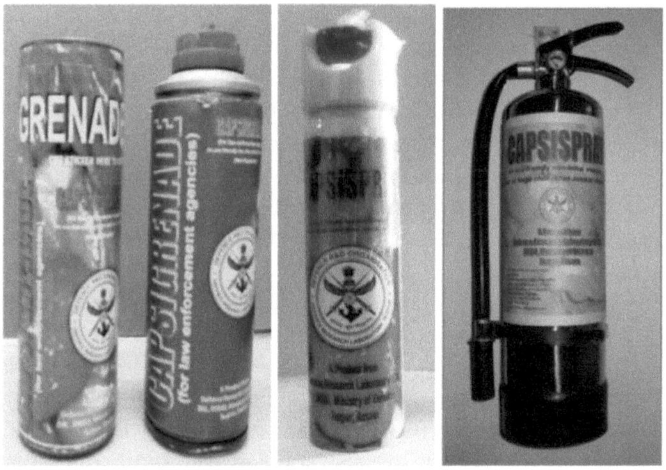

(a) (b) (c)

FIGURE 1.7 Capsigrenade (a), capsispray for personal protection (b), and mob dispersal (c).

Due to the high capsaicin content of *Bhut jolokia*, it has immense potential for commercial extraction of the oleoresin, which has wide applications in pharmaceutical as well as in defense sectors. Oleoresin contains capsaicin, which is in great demand both in domestic as well in international markets. The capsaicin compound has several pharmacological activities, such as hepatoprotective, anti-inflammatory, analgesic, anti-obesity, anti-cancer, anti-oxidants, anti-microbial, anti-pyretic, anti-diabetic, etc. (Meghvansi et al. 2010; Roy 2016). Defence Research Laboratory, DRDO, Tezpur (Assam) has found unusual applications of this chilli for mob dispersal and self-defense. The oleoresin extracted from *Bhut jolokia* is being used in chilli grenades and chilli spray in different versions and forms. The chilli grenade is mainly meant for use by law enforcement agencies as a non-lethal mob dispersal agent in lieu of tear gas, which is harmful to human eyes. The chilli grenade has been tested by state police of different states and has been inducted in their armory. The Indian Army has also used it for expelling terrorists from their hideouts in Jammu & Kashmir, which proved to be very useful in such circumstances. The chilli spray is meant for use by ladies and senior citizens for their self-defense, which they can carry along with them and can spray in the eyes of assaulters (Figure 1.7).

1.7 AGRO-TECHNOLOGY

The climatic condition of northeast India favors the natural growth of *Bhut jolokia* and the subsequent development of its unique extreme hotness without much care. Traditionally, it was grown as a sporadic intercrop with paddy under *Jhum* cultivation in the hills, as well as in homestead gardens. In the NE region, there are two planting seasons for this chilli, *Kharif* and *Rabi*. In the hilly states, *Kharif* cultivation is followed where planting is done in the month of February/March and harvesting starts from May/June onward, while *Rabi* is followed in the plains where planting is done in August/September and harvesting from November/December. In the homestead garden, it is cultivated under partial shade and the plants are kept for 2–3 years, beyond which, there is a reduction in fruit size and yield. The plants grow to a height of 12–13 feet at the age of 3 years. The plants are reported to be infested by aphid and mealy bugs in traditional fields. The farmers usually sprinkle sieved ash on the leaves and the base of the plant as an indigenous crop management practice (Borgohain and Devi 2007; Bhagowati and Changkija 2009). Though no package of practices for *Bhut jolokia* could be found, Singh (2009) reported highest cost: benefit ratio (1:8.7) on sowing the crop in September 15 and planting at a spacing of 75 cm × 75 cm followed by October 15 sowing (1:7.3).

The average yield of fresh fruits of *Bhut jolokia* under rainfed conditions is estimated to be around 80–100 q/ha, which after dehydration comes to around 10–12 q/ha (Borgohain and Devi 2007). The price per kilogram of the fruit of *Bhut jolokia* in local markets is Rs. 100–150, as reported by Borgohain and Devi (2007), while the current price in the Tezpur region ranges from Rs. 400– 600 depending upon the season. Since the cultivation of *Bhut jolokia* is an unorganized sector spread out in different pockets of northeast India, there is no documented data available on area, production, and yield loss due to diseases. As per a rough estimate, 1000 tons of *Bhut jolokia* is produced per year in different regions of northeast India, which may increase up to five times, if environmental factors are conducive (Meghvansi et al. 2010).

1.8 CONCLUSION

Bhut jolokia or Naga King Chilli is a chilli species of northeast India with a high potential for commercial applications. This chilli crop brought fame and glory to this region by being the world's hottest chilli and, if explored judiciously with proper post-harvest handling and good market network, can bring about an economic upsurge for the farmers of this region. With the increase in popularity and demand for this native chilli of NE India, the first King Chilli Festival was organized in the Tamenglong District of Manipur in 2018 to attract tourists as well as entrepreneurs for setting up post-harvest processing units for export of this chilli in the form of value-added products.

Due to the rising demand of this chilli, both in the domestic and international markets, this chilli is being cultivated on a wide scale in different regions of northeast India. The chilli is being sold in the open market at the price of Rs. 300–400 per kg in Assam depending upon the season and size of the fruit. They are also sold loose at 3–4 fruits for ten rupees. This reflects the high demand of this chilli for domestic consumption, since the supply is not adequate as the cultivation is restricted to only a few areas owing to the suitability of the agro-climate. Since the availability of the chilli is restricted to only a few months of the year, the demand is high in the market. Moreover, extraction of capsaicin is limited to the time of availability of the fruit, for which, R&D efforts are being made by the Defence Research Laboratory, Tezpur for in vitro production of capsaicin through cell suspension culture. Breakthroughs have been made at laboratory scale, which need to be scaled-up, for which further studies are being carried out.

REFERENCES

Bhagowati, R.R., and Changkija, S. 2009. Genetic variability and traditional practices in Naga King Chilli landraces of Nagaland. *Asian Agri-History* **13**: 171–180.

Borgohain, R., and Devi, J. 2007. The hottest chilli: A new horizon in agri-entrepreneurship. *Entrepreneur* August 2007. www.assamagribusiness.nic.in

Bosland, P.W. 1996. Capsicums: Innovative uses of an ancient crop. In Janick, J., (Ed.), *Progress in New Crops.* ASHS Press, Arlington, VA, 479–487.

Bosland, P.W., and Baral, J.B. 2007. 'Bhut jolokia'—The world's hottest known chille pepper is a putative naturally occurring interspecific hybrid. *HortScience* **42**: 222–224.

Guinness Book of World Records. 2006. http://en.wikipeia.org/wiki/Bhut_jolokia.

Halldén, C., Hansen, M., Nilsson, N.O., Hjerdin, A., and Sall, T. 1996. Competition as a source of errors in RAPD analysis. *Theoretical and Applied Genetics* **93**: 1185–1192.

Kehie, M., Kumaria, S., Devi, K.S., and Tandon, P. 2016. Genetic diversity and molecular evolution of Naga King Chilli inferred from internal transcribed spacer sequence of nuclear ribosomal DNA. *Meta Gene* **7**: 56–63.

Kehie, M., Kumaria, S. and Tandon, P. 2014. Manipulation of culture strategies to enhance capsicin biosynthesis in cell cultures of Capsisum Chinense Jacq. cv Naga King Chilli. *Bioprocess Biosyst. Bioprocess Eng.* 37: 1055–1063.

Mathur, R., Dangi, R.S., Das, S.C., and Malhotra, R.C. 2000. The hottest chilli variety in India. *Current Science* **79**: 287–288.

Meghvansi, M.K., Sidiqui, S., Khan, H., Gupta, V.K., Vairale, M.G., Gogoi, H.K., and Singh, L. 2010. Naga chilli: A potential source of capsaicinoids with broad-spectrum ethnopharmacological applications. *Journal of Ethnopharmacology* **132**: 1–14.

Pandey, V., Pandey, H.K., Dayal, D., Joshi, U.C., Pant, T., and Ahmed, Z. 2009. Studies on antioxidant constituents of some domesticated capsicums in the middle hill conditions of western Himalayas. *Horticultural Science* **36**: 26–30.

Purkayastha, J., Alam, S.I., Gogoi, H.K., and Singh, L. 2012a. *Capsicum assamicum* sp. Nov. (Solanaceae) from Assam, Northeastern India. *Ozean Journal of Applied Science* **5**: 55–66.

Purkayastha, J., Alam, S.I., Gogoi, H.K., Singh, L., and Veer, V. 2012b. Molecular characterization of 'Bhut jolokia' the hottest chilli. *Journal of Biosciences* **37**: 757–768.

Roy, A. 2016. *Bhut jolokia* (*Capsicum chinense* Jacq.): A review. *International Journal of Pharmaceutical Sciences and Research* **7**: 882–889.

Sanatombi, K., and Sharma, G.J. 2008. Capsaicin content and pungency of different *Capsicum* spp. Cultivars. *Notulae Botanicae Horti. Agrobotanici Cluj-Napoca* **36**: 89–99.

Singh, M.S. 2009. Standardization of sowing time and plant population of *Bhut Jolokia* (*Capsicum chinense*). MSc Thesis. Assam Agricultural University.

Suzuki, T., Fujiyake, H., and Iwai, K. 1980. Intracellular localization of capsaicin and its analogues, capsaicinoid in *Capsicum* fruit 1. Microscopic investigation of the structure of the placenta of *Capsicum annuum* cv. Karayatsubusa. *Plant and Cell Physiology* **21**: 839–853.

Talukdar, J., Saikia, A., and Borah, P. 2015. Survey and detection of the diseases of Bhut jolokia (*Capsicum chinense* Jacq.) in Assam. *Journal of Crop and Weed* **11**: 186–192.

Tiwari, A., Kaushik, M.P., Pandey, K.S., and Dangi, R.S. 2005. Adaptability and production of hottest chilli variety under Gwalior agro-climatic conditions. *Current Science* **88**: 1545–1546.

Williams, J.G.K., Hanafey, M.K., Rafalski, J.A., and Tingey, S.V. 1993. Genetic analysis using random amplified polymorphic DNA markers. *Methods in Enzymology* **218**: 705–740.

Xu, J. 2006. Fundamentals of fungal molecular population genetic analyses. *Current Issues in Molecular Biology* **8**: 75–90.

2 Plant-Based Compounds as Alternative Adjuvant Therapy for Multidrug-Resistant Cancer

E. C. Aniogo, Blassan P. George, and Heidi Abrahamse

CONTENTS

2.1 INTRODUCTION

Cancer is a term used to describe a collection of diseases widely acknowledged as the leading cause of death worldwide (Kapinova et al. 2018). Breast cancer is one of the most common cancers diagnosed in women. This cancer is associated with many risk factors, like: family history, age, genetic factors, exposure to radiation, menstrual and menopausal history, and other lifestyle changes, among others (Bak et al. 2016). Advances made in cancer research to date have identified chemotherapy as a promising treatment option. Other treatment options include: hormonal therapy, radiation therapy, or a combination of both. Although these therapies have improved the living standards of patients, cancer recurrence still persists because some cells evade the treatment effect and survive, or metastasis to other tissue, thus becoming resistant and regrow (Mansoori et al. 2017). Multidrug cancer resistance is the ability of cancerous cells to tolerate a given drug treatment and, thus, become insensitive to them. Approximately 50% of the tumors treated with chemotherapy become resistant to the same or to any structurally related chemotherapeutic agent, which is a serious problem in cancer treatment (Wang et al. 2018). Evidence have shown that cancer drug resistance is a complex process that involves many factors, such as variation in individual genetic composition especially in somatic cells, apoptotic cell inhibition, enhanced DNA repair, and altering of drug metabolism and targets (Mansoori et al. 2017). The prevalence of these drug-resistant cancers necessitates further research to develop better treatment modalities. Abundant evidences have shown that plant-derived bioactive compounds like: flavonoids, carotenoids, phenolic acids, and organosulfur compounds possess beneficial disease-preventive properties. They possess anticarcinogenic and antimutagenic effects, which can lower various types of neoplasia (Gonzalez-Vallinas et al. 2013; Dandawate et al. 2016). This review summarizes the current knowledge on mechanisms that promote multidrug resistance and outlines some of the bioactive compounds used toward multidrug resistance in cancer treatment.

2.2 MULTIDRUG RESISTANCE MECHANISMS

The mechanism of drug resistance is a critical issue that continues to evolve and disseminate among many cancer types, like: breast, gastrointestinal, lung, and ovarian. These cancer types become tolerant to chemotherapeutic drugs with the same structural and mechanism of action similarities, leading to the phenomenon called multidrug resistance (Eid et al. 2015). One explanation to this may be due to the heterogeneity, stemness, and other changes within the tumor microenvironment (Carli et al. 2013). Several hypotheses with experimental evidence have tried to explain this phenomenon and the mechanisms, such as: altered drug transport across cell membrane, genetic and drug target alterations, and increased DNA repair mechanisms have been proposed to account for cancer drug resistance (Chorawala et al. 2012).

One of the most widely known resistance mechanisms is through altered membrane transport by adenosine triphosphate (ATP)-binding cassette (ABC) transporters. The ABC transporters belong to the ubiquitous superfamily of transmembrane proteins that modulate drug and other biomolecules absorption, distribution, and excretion across cell membrane. Today, there are 48 identified human ABC genes grouped into seven subclasses (A–G) based on their sequence of homology and genomic organization (Chorawala et al. 2012; Eid et al. 2015). In clinical transport-associated multidrug resistance, the MDR1 gene, which encodes for P-glycoprotein (P-gp; MDR1, ABCB1), is the widely studied. The P-gp has two highly hydrophobic integral membrane and nucleotide-binding domains that make up its structure. This structure enables them to efficiently efflux cytotoxic drugs through a "pump" and "flippase" model of transport from the inner leaflet to the outer leaflet of the lipid bilayer into the extracellular space (Sharom 2014). Other ABC transporters: the multidrug resistance-associated protein 1 (MRP-1, ABCC1) and the breast cancer-resistant protein (BCRP, ABCG2) have been implicated as major efflux transporters responsible for cancer resistance in chemotherapy (Mao and Unadkat 2015). The MRP1 is similar in structure with the P-gp/MDR1 and contains an added amino terminal in the domains attached to the core that can recognize both neutral and anionic hydrophobic natural products and transports glutathione (Eid et al. 2015). The ABCG2 gene located at 4q22 encodes human BCRP. Structurally, the BCRP contains an ATP-binding domain and six transmembrane segments in a homodimer of two-half transporters. The BCRP is highly expressed and widely distributed in the gastrointestinal tract, excretory tissue, and blood-tissue barriers (Fletcher et al. 2016).

Genetic responses that lead to changes in cellular processes have been identified as another drug resistance mechanism. Alterations in cell cycle, increased DNA damage repair, reduced apoptosis, and altered drug metabolism are among the cellular changes that affect the ability of cytotoxic drugs to kill cancerous cells (Cree and Charlton 2017). Rapid downregulation of a target gene expression and mutations are common in cancer cells, which enables them to escape upon DNA damage, and become susceptible to increased DNA repair mechanisms for the replication process to continue. Thus, producing abnormal cells with increased survival proteins to protect themselves and inhibit cell death. These, in addition to reduced apoptosis and other forms of cell death, have underpinned drug resistance. Other factors like tumor microenvironments, where cancer cells reside, and intratumoral genetic heterogeneity with proximity to stromal cells can function in paracrine manner to support the survival of cancer cells. For example, various tumor landscapes create numerous microenvironmental niches, and each niche with genetically identical cancer cells will respond differentially to the same treatment due to different biophysical properties, such as: vascularity, Potential of Hydrogen (PH), hypoxia, and extracellular matrix organization of the niche (Rybinski and Yun 2016). Tumor-initiating cells otherwise known as cancer stem cells may also play a role in tumor resistance. These cells utilize the expression of ABC transporters, altered DNA damage response, and expression of pro-survival proteins to evade and better tolerate drug exposure. The cancer stem cells are thus refractory to drug treatment and often contribute to cancer relapse, recurrence, and resistance (Abdullah and Chow 2013).

2.3 CLASSIFICATIONS OF PLANT-DERIVED COMPOUNDS

Phytochemicals are bioactive compounds found in plants. Many of these compounds are in synergy with several nutrients, vitamins, and minerals found in foods to possess disease preventive effects (Kapinova et al. 2018). These effects are expected to be integrated into the human diet, thus, a growing emphasis has been placed on dietary plant products as an alternative approach toward the prevention and treatment of cancer, especially breast cancer (Bak et al. 2016). Based on the chemical structure and biological properties, phytochemicals can be classified as phenolics, carotenoids, alkaloids, nitrogen-containing, and organosulfur compounds, as discussed in Figure 2.1 (Kapinova et al. 2018).

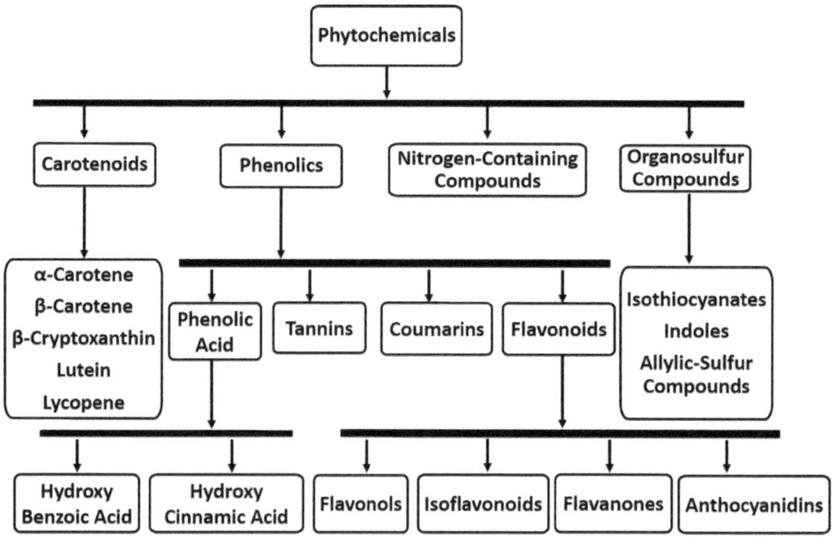

FIGURE 2.1 Basic classification of phytochemicals.

2.3.1 Carotenoids

They are lipid-soluble compounds that play an important role together with chlorophylls in photosynthesis and photoprotection (Kapinova et al. 2018). This molecule has a long conjugated double bond series in its center, which gives them their sharp, chemical reactivity and light-absorbing properties. Carotenoids inhibit the oxidative effects of reactive and singlet oxygen species that are produced from a plant's exposure to light and air. They react with radical molecules and delocalize the unpaired electron, thus inhibiting the oxidative effects of the radicals. This antioxidant activity makes them an excellent free radical scavenger (Ramel et al. 2012). Carotenoids are widely found in nature and can be used to prevent lipid oxidation and other related oxidative stress. Fruits including carrots, watermelons, apricots, pink guavas, tomatoes, pumpkin, mango, and sweet potatoes are rich sources of carotenoids (Jaswir et al. 2011). Generally, carotenoids are classified into α-carotene, β-carotene, β-cryptoxanthin, lutin, and lycopene (Liu 2004). Many epidemiological studies have found beneficial effects of carotenoid-rich foods in the reduction of cancer progression. As a chemoprevention strategy against certain human cancers, carotenoids and other antioxidants could be used, although there appears to be a detrimental interaction between β-carotene, smoke, and alcohol (Tanaka et al. 2012).

2.3.2 Phenolics

These are aromatic compounds with one or more hydroxyl groups that are produced from secondary plant metabolism that provide essential functions in plants growth and reproduction. They are richly found in fruits and vegetables, like cranberry, apple, red grape, pineapple, lemon, spinach, red pepper, carrot, lettuce, cucumber, etc., and are categorized as phenolics acids, flavonoids, stilbenes, coumarins, and tannins. Phenolic compounds act as a defense mechanism for plants to pathogens, predators, parasites, etc. (Kapinova et al. 2018). Researches have shown that regular consumption of fruits rich in phenolics compounds can help improve some chronic disease conditions (Liu 2004). The protective attributes of some phenolics compounds against cancer have been linked to their alterations in various signaling pathways like angiogenesis, cell cycle, and apoptosis (Garvin et al. 2006; Ma et al. 2008; Zhang et al. 2013).

2.3.3 Nitrogen-Containing Compounds

These are groups of molecules unique to fruits from the genus *Capsicum* (chilli peppers). They are purported to have antimicrobial and analgesic effects. Although there are reports linking their consumption with cancer chemoprevention, the exact mechanism of action is not fully understood. Hence, more research is expected to support its therapeutic efficacy against cancer, especially the drug-resistant type.

2.3.4 Organosulfur Compounds

Foods such as cruciferous and *Allium* genus vegetables like onions, garlic, etc. are high in sulfur compounds. These compounds have shown to exert diverse biological effects, like: free radical scavenging and detoxification activity, inhibition of DNA and tumor cell proliferation, cell cycle arrest, and the ultimate cell death (Moriarty et al. 2007). Evidence supports the protective effects of garlic in stomach, colorectal, and breast cancer in humans that appear to be associated with the presence of organosulfur compounds. The predominantly allyl derivatives in these compounds have also shown inhibitory effects against esophagus, colon, mammary gland, and lung carcinogenesis in experimental animals (Omar and Al-Wabel 2010).

2.4 BIOACTIVE PLANT-DERIVED COMPOUNDS WITH ANTITUMOR ACTIVITY AGAINST MULTIDRUG-RESISTANT CELLS

One of the oldest human practices is the use of medicinal plants as a herbal remedy in the treatment of various minor ailments through regular consumption of fruits and vegetables. These earliest practices, though still much practiced, recently have gained popularity in their association with a reduced risk of developing chronic diseases like cancer and cardiovascular diseases (Liu 2004). Although chemically derived drugs have been developed and used for cancer treatment, the search for antitumor drugs from plants continues to dominate due to the major limitations of high toxicity, affordability, and often the resistance of these synthesized drugs (Greenwell and Rahman 2015). These limitations and awareness of the perception that many allopathic drugs currently in use are based on natural products necessitate a further search of a safe and more efficient product without adverse effects to improve treatment efficacy (Song et al. 2014). The availability and abundance of plant compounds make them the desired product to study to find new drugs against tumors. In addition, these natural products, aside from their anti-inflammatory, antifungal, and antiviral properties, also possess antitumor activities. The widely exhibited antitumor properties of phytochemicals thus reported have been classified according to the mechanism of action, namely: DNA damage or antioxidant drugs, mitotic disruptors, histone deacetylase, and methyltransferase inhibitors (Greenwell and Rahman 2015). Based on this, phytochemicals have been proposed to be useful in the treatment and complete elimination of cancer. Furthermore, it has been suggested and showed that naturally occurring phytochemicals have the potential to modulate and reverse the development of MDR in cancer due to their multitargeting properties (Dandawate et al. 2016; Muthusamy et al. 2016; Budisan et al. 2017).

TABLE 2.1
Summary of the Modulatory and Multidrug Resistance Reversal Activities of Selected Bioactive Compounds

Bioactive Compounds	Classification	Mechanisms Against MDR	MDR Cells Used	References
Curcumin	Polyphenol flavonoid	Inhibition of P-gp ATPase activity function. Decrease proliferation and increase cytotoxicity.	Human cervical carcinoma KB-V-1 Human breast cancer MCF-7/MDR	Limtrakul et al. (2007)
		Interaction with the substrate-binding sites. Stimulation of basal ATPase activity and ATP hydrolysis of MRP1.	HEK 293	Chearwae et al. (2006)
		Regulation of NF-kB activity and expression of ABC transporter function.	L1210/Adr cells	Xue et al. (2013); Choi et al. (2008)
Ferulic acid	Phenolic acid	Induction of cell cycle arrest at G2/M phase and apoptotic signaling effect. Inhibition of P-gp transport function. Downregulation of P-gp expression.	MDR KB Ch^R 8-5 cells HEK 293/ABCB1 cells	Muthusamy et al. (2016)
Quercetin	Flavonoid	Reversal of gene-encoding P-gp and inhibition of its expression.	KB/VCR oral cancer-resistant cells	Yuan et al. (2015)
		Modulation and inhibition of FZD7/β-catenin pathway. Downregulation of ABCB1, ABCC1, and ABCG2 protein expression. Increased intracellular accumulation of rhodamine 123 and Adriamycin.	MDR human hepatocellular carcinoma BEL/5-Fu cells	Chen et al. (2018)
Retinoid	Carotenoids	Competitive inhibitor of P-gp. Modulation of pregnan-X-receptors, which regulates P-gp expression.	MDR colon and leukemia cells	Eid et al. (2012)
Xanthophylls	Carotenoids	Inhibition of P-gp activity and enhancement of antiproliferative function of anticancer drugs.	MDR lymphoma and breast cancer cells	Gyemant et al. (2006); Molnar et al. (2006)
Phenethyl Isothiocyanate	Organosulfur compounds	Suppress expression and activation of ABCG2 and pluripotency-associated transcription factors such as Octamer-binding transcription factor (Oct_4), sex-determining region Y-box (Sox-2), and Nanog. Inhibition of self-renewal, clonogenicity, and tumor growth of cancer stem cells.	EpCAM-positive NCCIT human embryonic carcinoma cells	Yun et al. (2017)

The compounds being discussed are represented in Table 2.1 with their origin, resistant cell line used and anticancer activity.

Curcumin polyphenol has been extensively studied in cancer for its ability to reverse MDR. This compound was able to restore drug sensitivity in MDR cancer cells through direct inhibition of P-gp and MRP1 transport pumps (Chearwae et al. 2006; Limtrakul et al. 2007). Similarly, Choi and colleagues (2008) reported that curcumin not only inhibited *mdr1b* gene promoter activity, but also prohibits the constitutive transcription factor nuclear factor $_k$B (NF-$_k$B) transactivation, thereby regulating the antiapoptotic proteins including: bcl-2, bcl-xL. Thus, curcumin can reverse cancer-resistant phenotypes through the suppression of MDR protein expression and inhibition of P13K/Akt/NF-kB signaling pathways (Xue et al. 2013). Ferulic acid, another phenolic acid richly found in wheat and barley leaves has demonstrated the potential of reversing P-gp-mediated paclitaxel-resistant cells. Muthusamy et al. (2016) reported that ferulic acid inhibits P-gp transport function and significantly downregulates its expression in resistant KB Ch^R 8–5 cell lines. Their report further stated that the compound could also be used to arrest the cell cycle and induce apoptosis in the resistant cell line. These effects were attributed to the chemosensitizing potential of the compound. Quercetin is a flavonoid found in leafy greens, tomatoes, berries, and broccoli. It is considered as one of the most abundant antioxidants in the human diet. Apart from its antioxidant properties, it has also exhibited anti-inflammatory, immunomodulatory, and vasodilating activities (Chen et al. 2018). Hashemzaei and colleagues (2017) studied the putative anticancer potential of quercetin in eight different cancer cell lines, namely, colon, prostrate, pheochromocytoma, breast, ovarian, leukemia, myeloma, and lymphoid cancer and found that quercetin at optimal concentrations was able to induce apoptosis in all the tested cell lines. Importantly, quercetin exerts its tumor suppressive effect through mechanisms like: inhibition of DNA topoisomerase I/II, release of cytochrome c, activation of caspase 3, and heat shock protein 27 (Badziul et al. 2014). Furthermore, Yuan and colleagues (2014) reported that quercetin is a potent modulator of P-gp, and their study showed that quercetin from 25 to 100 µmol/L reversed gene-encoded P-gp-mediated MDR in KB/VCR oral cancer resistant cells by inhibiting its expression. Similarly, Chen et al. (2018) confirmed the potential role of quercetin and ABC transporter-mediated MDR in BEL/5-FU hepatocellular carcinoma cells. Evidently, their study showed that quercetin increased the sensitivity of a resistant cell line to chemotherapeutic drugs with a significant increase in rhodamine and Adriamycin drug accumulation.

The biological activity of carotenoids as reported by previous studies indicated that they exhibit anticancer properties through cell cycle arrest, growth factor inhibition transduction, stimulation of anti-oncogene proteins, and apoptotic mechanism induction (Garattini et al. 2007; Yu et al. 2011). The carotenes, xanthophylls, and retinoids class of carotenoids have a high affinity to a P-gp transporter and are thus used as a competitive inhibitor for the transporter (Gyemant et al. 2006). This was apparent in the findings of Eid et al. (2012) using MDR-resistant colon and leukemia cells. Eid and colleagues confirmed that the five carotenoids (β-carotene, crocin, retinoic acid, canthaxanthin, and fucoxanthin) studied competitively inhibit the P-gp activity in both the colon and leukemic cell line. These compounds increased drug accumulation and efficacy in the resistant cells and reversed the MDR effect of the chemotherapeutic drugs in a synergistic fashion. Likewise, many naturally occurring isothiocyanates display anticarcinogenic activity because they reduce the activation of carcinogens and increase their detoxification (Wu et al. 2009). Recent studies show that phenethyl isothiocyanate, a natural compound found in cruciferous vegetables, can suppress cancer stem cell properties in vitro and in a xenograft model. Yun and colleagues (2017) revealed that ABCG2 as well as pluripotent-associated transcription factors in epithelial cellular adhesion molecule (EpCAM)-positive NCCIT, a pluripotent stem cell line established by Shinichi Teshima (National Cancer Institute, Tokyo, Japan), human embryonic carcinoma cells, such as octamer-binding transcription (Oct4), sex-determining region Y-box (Sox-2), and NANOG were downregulated by phenethyl isothiocyanate. Previous studies have reported that cancer stem cells exhibit drug resistance through the expression of ABCG2, which closely correlates with the expression of pluripotency factors. Thus, isocyanate compounds are useful in targeting cancer stem cells, which are a source of tumor initiating cells and contribute in drug resistance.

2.5 CONCLUSION AND FUTURE PROSPECTS

The ability of cancer cells to develop resistance to chemotherapeutic agents seems to be a continuous process with a substantial effect on therapeutic outcome and limited efficacy. These shortfalls make natural products to be in the forefront of becoming a very important alternative source of treatment for resistant cancer cells, generally, because of their lower cost, toxicity, and natural abundance within our environment. Many publications and reports have documented the antiproliferative effect of plant extracts against breast, lung, colon cancer, etc., while others have also shown the important pharmacological applications of these extracts against several human diseases. From all these reports, it is evident that medicinal plants can be a rich source in potential anticancer compounds. Hence, it is necessary to consider and look toward the use of new molecules from nature in drug development with an unexploited mode of action to fight against cancer drug resistance.

ACKNOWLEDGMENTS

The South African Research Chairs Initiative of the Department of Science and Technology and National Research Foundation of South Africa (Grant No 98337) supports this work.

REFERENCES

Abdullah, L.N., and Chow, E.K. 2013. Mechanisms of chemoresistance in cancer stem cells. *Clinical and Translational Medicine* 2: 3.

Badziul, D., Jakubowicz-Gil, J., Langner, E., Rzeski, W., Glowniak, K., and Gawron, A. 2014. The effect of quercetin and imperatorin on programmed cell death induction in T98G cells *in vitro*. *Pharmacological Reports* 66(22): 292–300.

Bak, M.J., Gupta, S.D., Wahler, J., and Suh, N. 2016. Role of dietary bioactive natural products in estrogen receptor-positive breast cancer. *Seminars in Cancer Biology* 40(41): 170–191.

Budisan, L., Gulei, D., Zanoaga, O.M., Irimie, A.I., Chira, S., Braicu, C., Gherman, C.D., and Berindan-Neagoe, I. 2017. Dietary intervention by phytochemicals and their role in modulating coding and non-coding genes in cancer. *International Journal of Molecular Sciences* 18(6): 1178.

Carli, C.B.A., Quilles, M.B., and Carlos, I.Z. 2013. Natural products with activity against multidrug resistance tumor cells. In *Fighting Multidrug Resistance with Herbal Extracts, Essential Oils and Their Components.* (Ed.), Rai M., and Kon K. Academic Press, Burlington, VT, pp. 237–244.

Chearwae, W., Wu, C.P., Chu, H.Y., Lee, T.R., Ambudkar, S.V., and Limtrakul, P. 2006. Curcuminoids purified from turmeric powder modulate the function of human multidrug resistance protein 1 (ABCC1). *Cancer Chemotherapy and Pharmacology* 57(3): 376–388.

Chen, Z., Huang, C., Ma, T., Jiang, L., Tang, L., Shi, T., Zhang, S. et al. 2018. Reversal effect of quercetin on multidrug resistance via FZD7/β-catenin pathway in hepatocellular carcinoma cells. *Phytomedicine* 43: 37–45.

Choi, B.H., Kim, C.G., Lim, Y., Shin, S.Y., and Lee, Y.H. 2008. Curcumin down-regulates the multidrug-resistance mdr1b gene by inhibiting the P13K/Akt/NF kappa B pathway. *Cancer Letters* 259(1): 111–118.

Chorawala, M.R., Oza, P.M., and Shah, G.B. 2012. Mechanisms of anticancer drugs resistance: An overview. *International Journal of Pharmaceutical Sciences and Drug Research* 4(1): 1–9.

Cree, I.A., and Charlton, P. 2017. Molecular chess? Hallmarks of anticancer drug resistance. *BMC Cancer* 17: 10.

Dandawate, P.R., Subramaniam, D., Jensen, R.A., and Anant, S. 2016. Targeting cancer stem cells and signaling pathways by phytochemicals: Novel approach for breast cancer therapy. *Seminars in Cancer Biology* 40(41): 192–208.

Eid, S.Y., El-Readi, M.Z., Fatani, S.H., Eldin, E.E.M.N., and Wink, M. 2015. Natural products modulate the multifactorial multidrug resistance of cancer. *Pharmacology and Pharmacy* 6: 146–176.

Eid, S.Y., El-Readi, M.Z., and Wink, M. 2012. Carotenoids reverse multidrug resistance in cancer cells by interfering with ABC-transporters. *Phytomedicine* 19(11): 977–987.

Fletcher, J.I., Williams, R.T., Henderson, M.J., Norris, M.D., and Haber, M. 2016. ABC transporters as mediators of drug resistance and contributors to cancer cell biology. *Drug Resistance Updates* 26: 1–9.

Garattini, E., Gianni, M., and Terao, M. 2007. Retinoids as differentiating agents in oncology: A network of interactions with intracellular pathways as the basis for rational therapeutic combinations. *Current Pharmaceutical Design* 13(13): 1375–1400.

Garvin, S., Ollinger, K., and Dabrosin, C. 2006. Resveratrol induces apoptosis and inhibits angiogenesis in human breast cancer xenografts in vivo. *Cancer Letters* 231(1): 113–122.

Gonzalez-Vallinas, M., Gonzalez-Castejon, M., Rodriguez-Casado, A., and Ramirez de Molina, A. 2013. Dietary phytochemicals in cancer prevention and therapy: A complementary approach with promising perspectives. *Nutrition Reviews* 71(9): 585–599.

Greenwell, M., and Rahman, P.K.S.M. 2015. Medicinal plants: Their use in anticancer treatment. *International Journal of Pharmaceutical Science Research* 6(10): 4103–4112.

Gyemant, N., Tanaka, M., Molnar, P., Deli, J., Mandoky, L., and Molnar, J. 2006. Reversal of multidrug resistance of cancer cells *in vitro:* Modification of drug resistance by selected carotenoids. *Anticancer Research* 26(1A): 367–374.

Hashemzaei, M., Far, A.D., Yari, A., Heravi, R.E., Tabrizian, K., Taghdisi, S.M., Sadegh, S.E. et al. 2017. Anticancer and apoptosis-inducing effects of quercetin *in vitro* and *in vivo. Oncology Reports* 38(3): 819–828.

Jaswir, I., Noviendri, D., Hasrini, R.F., and Octavianti, F. 2011. Carotenoids: Sources, medicinal properties and their application in food and nutraceutical industry. *Journal of Medicinal Plants Research* 5(33): 7119–7131.

Kapinova, A., Kubatka, P., Golubnitschaja, O., Kello, M., Zubor, P., Solar, P., and Pec, M. 2018. Dietary phytochemicals in breast cancer research: Anticancer effects and potential utility for effective chemoprevention. *Environmental Health and Preventive Medicine* 23: 36.

Limtrakul, P., Chearwae, W., Shukla, S., Phisalphong, C., and Ambudkar, S.V. 2007. Modulation of function of three ABC drug transporters, P-glycoprotein (ABCB1), mitoxantrone resistance protein (ABCG2) and multidrug resistance protein 1 (ABCC1) by tetrahydrocurcumin, a major metabolite of curcumin. *Molecular and Cellular Biochemistry* 296(1–2): 85–95.

Liu, R.H. 2004. Potential synergy of phytochemicals in cancer prevention: Mechanism of action. *Journal of Nutrition* 134(12): 3479S–3485S.

Ma, H., Lai, F., Xie, H., Wang, J., and Wang, H. 2008. Involvement of the Bcl-2 family members in *Pinus massoniana* bark extract induced apoptosis in HeLa cells. *Phytotheray Research* 22(11): 1472–1476.

Mansoori, B., Mohammadi, A., Davudian, S., Shirjang, S., and Baradaran, B. 2017. The different mechanisms of cancer drug resistance: A brief review. *Advanced Pharmaceutical Bulletin* 7(3): 339–348.

Mao, Q., and Unadkat, J.D. 2015. Role of the breast cancer resistance protein (BCRP/ABCG2) in drug transport: An update. *The AAPS Journal* 17(1): 65–82.

Molnar, J., Gyemant, N., Tanaka, M., Hohmann, J., Bergmann-Leitner, E., Molnar, P., Deli, J., Didiziapetris, R., and Ferreira, M.J. 2006. Inhibition of multidrug resistance of cancer cells by natural diterpenes, triterpenes and carotenoids. *Current Pharmaceutical Design* 12(3): 287–311.

Moriarty, R.M., Naithani, R., and Surve, B. 2007. Organosulfur compounds in cancer chemoprevention. *Mini Reviews in Medicinal Chemistry* 7(8): 827–838.

Muthusamy, G., Balupillai, A., Ramasamy, K., Shanmugam, M., Gunaseelan, S., Mary, B., and Prasad, N.R. 2016. Ferulic acid reverses ABCB1-mediated paclitaxel resistance in MDR cell lines. *European Journal of Pharmacology* 786: 194–203.

Omar, S.H., and Al-Wabel, N.A. 2010. Organosulfur compounds and possible mechanism of garlic in cancer. *Saudi Pharmaceutical Journal* 18: 51–58.

Ramel, F., Birtic, S., Cuine, S., Triantaphylides, C., Ravanat, J., and Havaux, M. 2012. Chemical quenching of singlet oxygen by carotenoids in plants. *Plant Physiology* 158: 1267–1278.

Rybinski, B., and Yun, K. 2016. Addressing intra-tumoral heterogeneity and therapy resistance. *Oncotarget* 7(44): 72322–72342.

Sharom, F.J. 2014. Complex interplay between the P-glycoprotein multidrug efflux pump and the membrane: Its role in modulating protein function. *Frontiers in Oncology* 4: 41.

Song, Y., Sun, H., Zhang, A., Yan, G., Han, Y., and Wang, X. 2014. Plant-derived natural products as leads to anticancer drugs. *Journal of Medicinal Plant and Herbal Therapy Research.* 2: 6–15.

Tanaka, T., Shnimizu, M., and Moriwaki, H. 2012. Cancer chemoprevention by carotenoids. *Molecules* 17: 3202–3242.

Wang, M., Liu, Z.F., Tang, H., and Chen, B.A. 2018. Application of alkaloids in reversing multidrug resistance in human cancers. *Chinese Journal of Natural Medicines* 16(8): 561–571.

Wu, X., Zhou, Q., and Xu, K. 2009. Are isothiocyanates potential anti-cancer drugs? *Acta Pharmacologica Sinica* 30(5): 501–512.

Xue, X., Yu, J., Sun, D., Zou, W., Kong, F., Wu, J., Liu, H., Qu, X., and Wang, R. 2013. Curcumin as a multidrug resistance modulator—A quick review. *Biomedicine and Preventive Nutrition* 3: 173–176.

Yu, R.X., Hu, X.M., Xu, S.Q., Jiang, Z.J., and Yang, W. 2011. Effects of fucoxanthin on proliferation and apoptosis in human gastric adenocarcinoma MGC-803 cells via JAK/STAT signal pathway. *European Journal of Pharmacology* 657(1–3): 10–19.

Yuan, Z., Wang, H., Hu, Z., Huang, Y., Yao, F., Sun, S., and Wu, B. 2015. Quercetin inhibits proliferation and drug resistance in KB/VCR oral cancer cells and enhances its sensitivity to vincristine. *Nutrition and Cancer* 67(1): 126–136.

Yun, J.H., Kim, K., Yoo, G., Kim, S.Y., Shin, J.M., Kim, J.H., Jung, S.H., Kim, J., and Nho, C.W. 2017. Phenethyl isothiocyanate suppresses cancer stem cell properties *in vitro* and in a xenograft model. *Phytomedicine* 30: 42–49.

Zhang, W., Chen, H., Liu, D.L., Luo, J., Zhang, J.H., Li, Y., Chen, K.J., Tong, H.F., and Lin, S.Z. 2013. Emodin sensitizes the gemcitabine-resistant cell line Bxpc-3/Gem to gemcitabine via downregulation of NF-kB and its regulated targets. *International Journal of Oncology* 42(4): 1189–1196.

3 Ethnomedicinally Important Pteridophytes of Kolli Hills from Eastern Ghats, India

Muniappan Ayyanar and Jeyalatchagan Sureshkumar

CONTENTS

3.1 INTRODUCTION

Natural resources give vital needs like food, clothing, fuel, shelter, and medications to treat a wide range of diseases, and documentation of folk knowledge is essential for the utilization of medicinal plants to avoid and cure ailments (Adhikari et al. 2018). Traditional knowledge has been described as a cumulative body of knowledge, practice, and belief, evolving through adaptive processes, and transferred over generations by cultural transmission (Ayyanar and Ignacimuthu 2011; Ayyanar 2015). The widespread use of ethnic medicine could be attributed to cultural acceptability, economic affordability, and efficacy against various diseases compared to modern medicines (Ayyanar and Ignacimuthu 2005). Mostly, native plants are named by the local people in their local languages and are important resources for their livelihood and primary health care needs.

The traditional system of medicine plays an important role in the health care of rural and tribal people in treating almost all types of ailments (Albuquerque and Hanazaki 2006). Even nowadays, traditional medicine is widely practiced in many developing countries. The World Health Organization (WHO) estimated that about 80% of populations in developing countries depend on traditional medicine for their primary health care needs (Mukherjee 2002). Most of the traditional medicine practices, Chinese medicine and India's Ayurveda and Siddha, are based on folk knowledge systems. Folk knowledge can be conserved as a part of cultural-ecological systems, helping to maintain local knowledge and practices, and reinforcing links between people and the environment (Silambarasan et al. 2017). Due to the easy accessibility and affordability of herbal medicines, they are playing a major role in primary health care services, and their use is an integral part of the cultural heritage among ethnic minorities in India.

Pteridophytes have been considered as a source of medicine and their medicinal importance is known to humans for more than 2000 years in treating various illnesses (Rout et al. 2009). Pteridophytes have been frequently used in Ayurvedic, Unani, homeopathic, and other systems of medicine. Ayurvedic system of medicine referred by Sushruta and Charaka (100 CE) cited the uses of *Marsilea minuta* and *Adiantum capillus-veneris* in their famous book on *Samhitas*. It was reported that pteridophytes have been considered as an excellent source of medicine and remain underexplored in ethnobotanical aspects when compared to flowering plants, and they are not often indicated as medicinally important in various ethnobotanical surveys (Moerman 1979, 1991; Ankli et al. 1999; Di-Stasi et al. 2002; Leonti et al. 2003; Amiguet et al. 2006). Pteridophytes are also used for different purposes of livelihood other than medicinal, viz., ornamental, decoration, growing media for epiphytic orchids, bio-fertilizers, etc. (Manickam and Irudayaraj 1992). Studies on pteridophytes in India have been carried out by several investigators majorly during the last few decades, and most of these studies are focused on the taxonomic aspects with a lacuna in ethnobotany.

In the Kolli Hills of the Eastern Ghats, the medicinal value of pteridophytes has received less attention with very few studies in the last decade (Perumal 2010; Gowrisankar and Chandrasekaran 2011; Karthik et al. 2011; Revathi et al. 2013). Scientific reports on pharmacological evidences of ethnomedicinally important pteridophytes were also scarce in literature. So, documentation of folk knowledge through ethnobotanical studies is important for conservation and utilization of biological resources. In this connection, we made a detailed report on previous studies about medicinally important pteridophytes of India, and this was deliberated in our recent review paper (Sureshkumar et al. 2018).

The present study was designed with the objective of documenting ethnomedicinal uses of pteridophytes as practiced by the *Malayali* people in the study area with detailed information on the preparation and usage of herbal medicines to treat various diseases.

3.2 METHODS

The present study was carried out in the Kolli Hills of the southern area of the Eastern Ghats with extensive field surveys. The Eastern Ghats constitutes an important biogeographic region as one of the nine floristic zones in India. It forms a broken chain of mountains spreading along the states of Orissa, Andhra Pradesh, Karnataka, and Tamil Nadu. The Eastern Ghats lies between the Rivers Mahanadhi in Orissa and Vaigai in Tamil Nadu. The southern part of the Eastern Ghats in Tamil Nadu covers six major hill complexes, such as Bodamalai, Chitteri Hills, Kalrayan Hills, Pachamalai, Kolli Hills, and Shervarayan Hills. The Kolli Hills represent a low hilly area, meeting the high mountain ranges of the Western Ghats in the Nilgiris.

The Kolli Hills are one of the six hill ranges in the southern part of Eastern Ghats, and are situated in the Namakkal District of Tamil Nadu and lie between 11° 11′ to 11° 30′ N and 78° 15′ to 78° 30′ E with a total geographical area of 485 km^2 (Figure 3.1). The nearest city is Namakkal, which is 45 km away from foothills of the Kolli Hills. Karavalli (foothills) to Semmedu (top hills) could be reached by a road with a distance of 26 km, which has 70 hairpin bends. Semmedu is the headquarters for the Kolli Hills and is connected to various cities (Rasipuram, Salem, Namakkal, and Thammampatti) by road. Parts of the Ghats road are mostly covered by forest plantation, vegetation in the valley region, scrubland, and built-up area (Anbazhagan et al. 2010). The frequent use of the Ghats's road in the hills leads to soil erosion and deforestation.

The *Malayalis* (*Malai* = hills, *ali* = dwells or inhabits), popularly called *Kolli Malayalis*, were believed to be descendants of the Proto Australoid group who lived here before the Harappan period (Rajendran and Manian 2011). The *Malayali* people also inhabit other parts of the Eastern Ghats of Tamil Nadu, such as Yercaud, Javadhu Hills, and Bodemalai (Thurston and Rangachari 1909). The mountain is named as "Kolli Hills" because the *Malayali* people believe the goddess "Kollipaavai or Ettukkai Amman Temple" protect the hills from evil.

The *Malayalis* are one of the major ethnic people out of 36 known ethnic groups in Tamil Nadu with around 54% of the

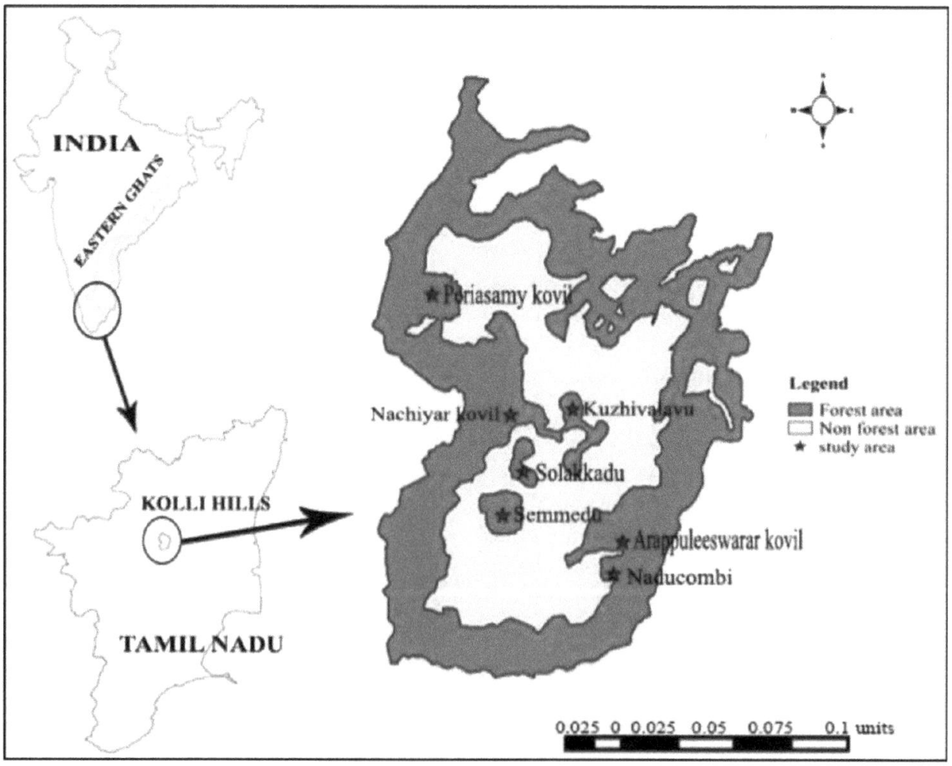

FIGURE 3.1 Location map of Kolli hills showing study areas in Eastern Ghats.

total ethnic population of the state (Jayasree 2002). *Malayalis* are generally illiterate and speak the local language, Tamil. Physically, they resemble the Semong of Malaya. The main occupation of *Malayalis* is cattle rearing, the collection of honey, bee's wax, minor forest products, cultivation of crops like paddy, cholam (jowar), cumbu (bajra), tapioca, thinai (foxtail millet), varagu (kodo millet), beans, black pepper, and some fruit crops like jackfruit, pineapple, citrus, guava, mango, and banana (Baskaran et al. 2015).

The geology of the area is characterized by charnockite associated with gneisses and metamorphic rocks (Ramesh and Anbazhagan 2015). December and January are the coldest months (15°C), whereas a higher temperature (35°C) is noticed during March to June. The mean annual temperature in the hills ranges from 15°C to 35°C during March to June and the average between 18°C and 25°C during October to January. The annual rainfall is between 300 and 750 mm. The water holding capacity of the hills is 40%, and the soil type varies between red to black clay. It has a mild climate, and the location is known for the cultivation of cash crops like banana, jackfruit, pineapple, etc. The Kolli Hills are frequently visited by nature lovers, trekking fans, tourists, and meditation practitioners (Bharath and Raj 2016). The famous Lord Siva Arappaleeswarar Temple, horticulture farm, herbal farm, Agaya gangai waterfalls (300 ft), boathouse, Ettukai Amman Temple, Periya Swamy Temple, pineapple farms, viewpoints, and telescope house are the places attracted by the tourists.

An extensive field investigation was carried out in different seasons from October 2017 to September 2018 throughout the Kolli Hills of Tamil Nadu. The frequency of field trips was varied to ensure the maximum coverage of the study area. Field explorations were made twice in a month, and most of the field surveys were focused after the rainy season. In each field study, we spend at least 4 to 6 days in the study area. The pteridophyte sample specimens were collected in a scientific manner without harming other species (Jain and Rao 1977). Normally 2–3 specimens of each species in fertile and sterile conditions were collected for identification and preparation of herbarium. Voucher specimens of documented plants were deposited in the herbarium of A.V.V.M. Sri Pushpam College, Poondi (SPCH), Thanjavur, India for future reference.

The documentation of medicinal uses of pteridophytes by the *Malayalis* has been done in accordance with the consensus statement on ethnopharmacological field studies (Heinrich et al. 1998). Experienced resource persons/informants were identified based on their knowledge of folk medicinal practices. The surveys were conducted through semi-structured and open-ended interviews (Martin 1995). The purpose of the fieldwork was explained to the informants with the importance of documenting their traditional knowledge on medicinal practices. The help acquired from the ethnic people of the study area for documentation of medicinal uses was duly acknowledged. A questionnaire was designed to collect information about the plants used in ethnobotanical practice, viz., a local name of the plant used, useful plant part(s), preparation method, and medicinal uses of the specific plant.

3.3 RESULTS

3.3.1 Informants of the Study Area

In total, 30 informants were interviewed during the study period, aged between 30 and 69 years, with 66% male and 34% female. Mostly, the female groups are not much interested in practicing herbal medicines due to the difficulties in collecting plant materials from forest regions. The traditional knowledge held by elders in the study area becomes eroded among the younger generations, because of the usage of modern medicines.

3.3.2 Diversity of Pteridophytes Species with Medicinal Uses

A total of 46 species of pteridophytes are recorded during the study period representing 28 families and 36 genera. A family which contributed the highest number of medicinally important species was Sinopteridaceae with four species followed by Polypodiaceae, Dryopteridaceae, Adiantaceae, and Thelypteridaceae with three species each. This indicated that the species of Sinopteridaceae were recognized as the most utilized plants by *Malayalis*, since species richness of this family is comparatively higher than other families in the study area and the easy accessibility of these species in the Kolli Hills.

Most of the recorded species of ethnomedicinally important pteridophytes are grown in terrestrial habitats (62%) followed by terrestrial/lithophytes (13%), epiphytes/lithophytes (13%), epiphytes (7%), lithophytes (3%), and aquatic (2%). Major terrestrial growing pteridophytic genera with ethnomedicinal importance recorded in the Kolli Hills are *Adiantum, Asplenium, Cheilanthes, Cyathea, Dryopteris, Equisetum, Hemionitis, Ophioglossum, Pteris, Selaginella*, and *Tectaria*, in which some species are also grown as lithophytes. Epiphytic forms of pteridophytes like *A. indicum, A. plantagineum, A. lanciniatum, D. quercifolia, H. phlegmaria, M. punctatum, N. auriculata, P. nudum, P. lanceolata, P. porosa*, and *V. elongata*. *Cheilanthes farinose* and *Hymenophyllum javanicum* are grown as lithophytes in rock crevices and among rock boulders along water channels.

Fronds, rhizomes, whole plant parts, and stem/rachis of pteridophytes are used in preparation of medicine by *Malayalis* in the study area to treat various ailments. In which, fronds (29 reports, 37%) are frequently used in parts for preparation of medicine followed by whole plant (24 reports, 31%), rhizome (19 reports, 25%), and stem (5 reports, 7%). The herbal practitioners are preparing herbal medicines six ways, such as powder (61 reports, 31%) as a predominantly used method followed by decoction (50 reports, 26%), paste (45 reports, 23%), juice (30 reports, 16%), cooked (5 reports, 3%), and raw (2 reports, 1%).

The dependence of herbal medicines by local people is attributed to the fact that the study area is far away from hospitals and, hence, people rely on the resources in their environs to cure various illnesses. Folk practices are strongly influenced by cultures, traditions, and their own beliefs of *Malayalis* since their folk knowledge was transmitted from their forefathers orally from generation to generation. There are four administration methods employed by *Malayalis* for healing practices, in which oral intake (143 reports, 74%) is predominantly used to treat various illnesses, followed by topical application (45 reports, 24%), bathing (1%), and smoking (1%).

3.3.3 Enumeration of Ethnomedicinally Important Pteridophytes

The documented information about the medicinal uses of pteridophytes are summarized as given below with botanical name of plant, family, common names, vernacular names, and medicinal uses, which includes parts used for medicinal preparation, method of preparation, and mode of application. Previously published literature and other bibliographic databases with regard to the present study were thoroughly searched to find out the novelty of the present study. Hence, most of the documented medicinal uses of studied plants were not reported or scarcely reported earlier in ethnobotanical literature across and outside the country.

Actiniopteris radiata (Sw.) Link (Actiniopteridaceae); Peacock's tail, fan fern; Vernacular names: Saava chedi, Korai panai. The juice extracted from young fronds is taken orally twice a day to cure a fever. The powdered fronds are mixed with boiled water and taken orally to cure uterine problems, female fertility, sperm production, and dysentery. The fronds were made into a paste and applied with water to cure rheumatism. The whole plant is powdered and taken orally with hot water to cure epilepsy. The decoction obtained from the whole plant is taken orally to treat asthma and piles. The dried whole plant is ground with water and taken orally with honey twice a day to cure intestinal worms. The whole plant is made into a paste with aloe oil and applied topically to heal cuts and wounds.

Adiantum incisum Forssk. (Adiantaceae); Maiden-hair fern; Vernacular name: Sentharai. The powdered fronds are mixed with butter and used as a cooling agent. The decoction obtained from the whole plant is taken orally to cure diabetes and bronchitis. The juice obtained from fronds is mixed with turmeric and hot water and taken orally to treat fever. The powdered fronds are taken orally with honey to cure skin diseases. The decoction obtained from the whole plant is taken orally twice a day to cure menstrual problems.

Adiantum lunulatum Burm. f. (Adiantaceae); Walking maiden hair fan; Vernacular name: Roadu keerai. The powdered rhizome is taken orally to cure a snakebite. The whole plant is ground into a paste with turmeric and applied over the affected places to cure pimples and wounds.

Adiantum raddianum C. Presl (Adiantaceae); Maidenhair fern; Vernacular name: Parapadappai. The juice obtained from the whole plant is taken orally twice a day to get relief from a cough and fever. The decoction obtained from the whole plant is taken orally along with honey to cure asthma. The whole plant is ground into juice and taken orally to cure leprosy. The whole plant is ground into a paste and applied topically on the head for relief from hair fall.

Anemia wightiana Gardner (Schizaeaceae); Flowering fern; Vernacular name: Kalkakadai. The whole plant is ground into a paste with water and turmeric and applied topically to treat rheumatism.

Angiopteris evecta (G. Forst.) Hoffm. (Angiopteridaceae); Giant fern; Vernacular name: Yanaivanagi. The decoction obtained from fronds is taken orally with lemon juice to treat an intestinal ulcer, leprosy, and a stomachache. The young fronds are made into a paste and mixed with coconut oil and applied over the affected places to get relief from body pain and skin diseases. The juice obtained from young fronds is taken orally to treat dysentery. The juice obtained from a young stem in taken orally with honey for body strength. The scales removed from rhizomes are made into a paste and applied topically to treat bone fractures and piles. The rhizome is ground into juice and taken orally to treat heart problems.

Asplenium inaequilaterale Willd. (Aspleniaceae); Spleenwort; Vernacular name: Erathalai. The fronds are made into a paste and applied topically over the affected places before bathing for 30 minutes to get relief from body pain.

Asplenium lanciniatum D. Don (Aspleniaceae); Spleenwort. The juice obtained from the rhizome is mixed with hot water and taken orally to cure leucorrhea.

Blechnum orientale L. (Blechnaceae); Centipede fern; Vernacular name: Palsai. The young fronds are ground with hot water and taken orally with honey to cure urinary problems. The decoction of fronds and rhizomes is taken orally twice a day to treat an abscess and typhoid fever. The young fronds are made into a paste with turmeric and applied over affected places to cure skin diseases. The rhizomes are ground into a paste and applied topically to treat sores, cuts, and wounds.

Botrychium daucifolium Wall. ex Hook. & Grev. (Ophioglossaceae); Moonwort; Vernacular name: Nandukuddhi. The juice obtained from the whole plant is taken orally to treat body pain and dysentery.

Ceratopteris thalictroides (L.) Brongn. (Parkeriaceae); Water fern; Vernacular name: Thalatharai. The young fronds are ground into a paste with turmeric and applied over the affected places to cure skin diseases.

Cheilanthes farinosa (Forssk.) Kaulf. (Sinopteridaceae). The decoction obtained from the fronds is taken orally to cure menstrual diseases and fever. The rhizome is ground into a paste with hot water and applied topically over the body before bathing for 30 minutes to get relief from skin irritation. The powdered rhizome is mixed with lemon and water and taken orally to cure epilepsy and stomachaches. The dried whole plant is ground into a paste and taken orally with honey for body strength and chest pain.

Cheilanthes mysurensis Wall. ex Hook. (Sinopteridaceae); Vernacular name: Pachai. The juice obtained from the fronds is mixed with hot water and taken orally with honey to get relief from throat pain.

Cheilanthes tenuifolia (Burm. f.) Sw. (Sinopteridaceae); Vernacular name: Parai Chedi. The fronds are made into a paste and applied topically to treat an abscess. The rhizomes are made into a paste and applied topically to heal cuts and wounds. The powdered rhizomes are mixed with honey and taken orally for body strength.

Christella dentata (Forssk.) Brownsey & Jermy (Thelypteridaceae); Vernacular name: Korai panai. The young fronds are made into a paste, mixed with hot water, and applied topically to treat swellings.

Christella parasitica H. Lev. (Thelypteridaceae); Vernacular name: Panai chedi. The decoction obtained from the fronds is taken orally to increase sperm production and to treat snakebites. The powdered fronds are taken orally to get relief from rheumatism and swellings. The fronds are made into a paste with hot water and applied topically to heal cuts and wounds.

Cyathea gigantea (Wall. ex Hook.) Holttum (Cyatheaceae); Tree fern. The dried fronds are ground with cold water and taken orally to get relief of body pain. The decoction obtained from the rhizome is taken orally twice a day to treat dysentery and diabetes.

Cyathea nilgirensis Holttum (Cyatheaceae); Tree fern. The decoction obtained from the rhizomes is taken orally to cure diabetes.

Dicranopteris linearis (Burm. f.) Underw. (Glaicheniaceae); Thicket fern; Vernacular name: Padachi. The juice obtained from the young fronds is taken orally to treat asthma. The juice obtained from the fronds is mixed with honey and taken orally to treat throat pain. The powdered fronds are made into a paste with water and applied topically to the affected place to cure wounds. The powdered stem is mixed with hot water and taken orally to cure chest pain. The juice obtained from the whole plant is taken orally to treat indigestion.

Diplazium esculentum (Retz.) Sw. (Athyriaceae); Vegetable fern; Vernacular name: Pasali. The juice obtained from the young fronds is taken orally twice a day to get relief from bronchitis.

Doryopteris concolor (Langsd. &Fisch.) Kuhn (Sinopteridaceae); The decoction obtained from the fronds and rhizomes is taken orally to treat chronic disorders.

Drynaria quercifolia (L.) J. Sm. (Polypodiaceae); Oak leaf fern; Vernacular name: Aattukal-kilangu. The powdered young fronds are taken orally to treat swellings and a throat infection. The powdered rhizome is mixed with pepper, cumin seeds, onion, and garlic, and taken orally to get relief from heart problems, blood clotting, intestinal disorders, dysentery, typhoid fever, indigestion, and a cough. The rhizome is made into a paste and applied topically to treat rheumatism, cuts, and wounds.

Dryopteris cochleata (D. Don) C. Chr. (Dryopteridaceae); Vernacular name: Kalkarai. The decoction obtained from the stem is taken orally to treat an ulcer. The powdered stem is mixed with cold water and taken orally to treat swelling, cuts, and wounds. The powdered rhizome is taken orally to treat epilepsy. The decoction obtained from the rhizome is taken orally to cure rheumatism and dysentery. The juice obtained from the rhizome is taken orally to cure leprosy and intestinal worms. The dried whole plant is ground with water and taken orally twice day to treat snakebites.

Equisetum ramosissimum Desf. (Equisetaceae); Common name: Horsetail. The dried aerial stem is ground with water and taken orally to get relief from stone formations, stomachaches, and rheumatism. The dried whole plant is ground with water and taken orally with honey to cure skin diseases and male sterility. The dried whole plant is made into a paste and applied topically to treat bone fractures. The juice obtained from the whole plant is taken orally to get relief from back pain and muscle pain.

Hemionitis arifolia (Burm. f.) T. Moore (Hemionitidaceae); Rabbit ear's fern; Vernacular name: Kambithamarai. The juice obtained from the rhizome is taken orally twice a day to cure hypertension. The rhizome is made into a paste with turmeric and applied over the affected places to heal wounds. The decoction obtained from the rhizome is taken orally to cure diabetes. The powdered rhizome is mixed with cumin seeds, pepper, and garlic, and taken orally to treat cancer. The dried whole plant is ground with water and taken orally to cure female sterility and body strength. The powdered whole plant is mixed with hot water and taken orally to cure rheumatism.

Huperzia phlegmaria (L.) Rothm. (Lycopodiaceae); Vernacular name: Easwarisadai. The dried whole plant is made into a paste and applied topically to heal wounds.

Hymenophyllum javanicum Spreng. (Hymenophyllaceae); Filmy fern; Vernacular name: Chakarathai. The dried whole plant is mixed with garlic and onion, and the mixture is applied topically over the affected place to cure headaches.

Leucostegia immerse Wall. ex C. Presl (Davalliaceae); Vernacular name: Thesachai. The rhizome is ground into a paste with turmeric and applied over the affected places to get relief from swellings.

Lycopodiella cernua (L.) Pic. Serm. (Lycopodiaceae); Club mosses. The dried whole plant is made into a paste and applied topically to heal wounds. The powdered whole plant is mixed with water and taken orally to treat bronchitis.

Lygodium microphyllum (Cav.) R. Br. (Schizaeaceae); Climbing fern; Vernacular name: Mullu kodi. The decoction obtained from the fronds is taken orally twice a day to cure dysentery and swellings.

Macrothelypteris torresiana (Gaudich.) Ching (Thelypteridaceae); The powdered whole plant is ground into a paste with water and applied topically over the head before a bath to cure hair fall.

Marsilea minuta L. (Marsileaceae); Water fern; Vernacular name: Thanniyarilai. The decoction obtained from the fronds is taken orally along with ginger to cure bronchitis and epilepsy. The powdered fronds are mixed with water and taken orally to cure dysentery, fevers, and stomachaches. The juice obtained from the fronds is taken orally to treat leprosy and mental disorders. Small pieces of the rhizome are soaked in bathing water before a bath to cure skin irritation.

Microsorum punctatum (L.) Copel. (Polypodiaceae); Vernacular name: Nundukuddhi. The decoction obtained from the fronds is taken orally to cure snakebites. The powdered fronds are mixed with water and taken orally twice a day to cure dysentery. The decoction obtained from the fronds is taken orally with honey to cure urinary problems. The fronds are made into a paste and applied topically over the affected places twice a day with hot water to heal wounds and swellings.

Nephrolepis auriculata Trimen (Oleandraceae); Vernacular name: Pasalithazhai. The powdered fronds are mixed with water and honey and taken orally twice a day to cure a cough. The fronds are ground into a paste and topically applied over the wounds. The powdered stem is mixed with hot water and taken orally to cure urinary problems and stomachaches.

Odontosoria chinensis (L.) J. Sm. (Lindsaceae); Farsley fern; Vernacular name: Ochalai. The powdered fronds are mixed with hot water and taken orally to cure urinary and gastric problems. The fronds are ground into a paste with water and applied topically over the affected places to cure swellings and sprains. The powdered whole plant is taken orally along with water to heal cuts and wounds.

Ophioglossum reticulatum L. (Ophioglossaceae); Adder's tongue fern; Vernacular name: Kalkurunjal. The fronds are ground into a paste with hot water and applied topically to treat swellings, wounds, and headaches.

Pityrogramma calomelanos (L.) Link (Hemionitidaceae); Golden fern; Vernacular name: Aaswarisadai. The decoction obtained from the whole plant is taken orally early in the morning to treat kidney problems and hypertension. The powdered whole plant is mixed with honey and water and is taken orally twice a day to cure fevers and coughs.

Psilotum nudum (L.) P. Beauv. (Psilotaceae); Whisk fern; Vernacular name: Tharathani. The decoction obtained from the whole plant is taken orally twice a day to treat dysentery. The whole plant is ground into a paste and mixed with turmeric and applied topically over the affected places to heal wounds.

Pteridium revolutum (Blume) Nakai (Dennstaedtiaceae); Bracken fern; Vernacular name: Thennarai. The juice obtained from the rhizome is taken orally with turmeric to cure diabetes and gastric problems. The powdered whole plant is taken orally with honey, pepper, and hot water to treat intestinal worms, stomachache, and urinary problems.

Pteris quadriaurita Retz. (Pteridaceae); Vernacular name: Kuddhi. The fronds and stems are ground into a paste and applied topically to heal cuts and wounds. The rhizome is ground into a paste and applied topically over the affected places to get relief from body pain. The decoction obtained from the rhizome is taken orally to cure chronic disorders.

Pteris vittata L. (Pteridaceae); Vernacular name: Pannel chedi. The fronds are ground into a paste and applied topically over the affected places to treat wounds and skin diseases. The whole plant is ground into a paste and mixed with honey and onion, and the mixture is taken orally to cure fever, hypertension, body strength, and bronchitis.

Pyrrosia lanceolata (L.) Farw. (Polypodiaceae); Vernacular name: Kalsendai. The fronds are made into a paste with pepper and taken orally to treat throat infections.

Selaginella tenera Spring (Selaginellaceae); Little club mosses; Vernacular name: Sanjeevani. The decoction obtained from the whole plant is taken orally to cure urinary problems. The powdered whole plant is taken orally along with water to cure gonorrhea.

Tectaria paradoxa (Fee) Sledge (Dryopteridaceae); The decoction obtained from the rhizome is taken orally along with hot water to cure stomachaches, intestinal disorders, and bronchitis. The powdered whole plant is taken orally along with hot water to cure asthma and dysentery.

Tectaria wightii Ching. (Dryopteridaceae); Vernacular name: Thatharai. The decoction obtained from the whole plant is taken orally twice a day to cure asthma.

Vittaria elongata Sw. (Vittariaceae); Vernacular name: Chamattai. The fronds are ground into a paste and applied topically over the affected places to treat rheumatism.

3.4 DISCUSSION

In the present study, male informants dominates over female informants in practicing herbal medicine. Interestingly, women informants play a major role in practicing herbal medicines and take part in nutrition-rich food preparations with plants available in their environs by the local population in northeast Algeria (Weckerle et al. 2018). Sinopteridaceae, Polypodiaceae, Dryopteridaceae, Adiantaceae, and Thelypteridaceae are recorded with more number of ethnomedicinally important pteridophytes in the study area. Sathiyaraj et al. (2015) reported that 50 species of pteridophytes belonging to 35 genera in 19 families are used traditionally for primary health care by tribal people of the Palani Hills, in which Adiantaceae and Woodsiaceae are recorded with 7 species followed by Polypodiaceae, Thelypteridaceae, Dryopteridaceae, Selaginellaceae, and Aspleniaceae.

The results of the present study indicated that species of Sinopteridaceae, Polypodiaceae, Dryopteridaceae, and Adiantaceae are recognized as the most utilized plants by *Malayalis*, since the species richness of these families is comparatively higher than other families in the Kolli Hills, as well as the easy accessibility of these species in the Kolli Hills. These results are in accordance with other recent reports in different parts of the country (Benniamin 2011; Singh and Upadhyay 2014; Sathiyaraj et al. 2015). As with the present study, Shil and Choudhury (2009) reported 16 species of pteridophytes belonging to 14 genera in 10 families which are frequently used to cure various diseases belonging to the genera *Cyathea*, *Drynaria*, *Pteris*, and *Cheilanthes* collected from Reang tribe of Tripura, northeast India.

Mannan et al. (2008) recorded medicinal uses of 54 pteridophytic species by various ethnic peoples from different regions of India, including, *Actinopteris radiata* (anthelmintic and stypic), *Cheilanthes farinosa* (eczema and stomachache), *Drynaria quercifolia* (fever, dyspepsia, cough, and phthisis), *Equisetum ramosissimum* (gonorrhea), *Lycopodium cernum* (coughs), *Pityrogramma calomelanos* (kidney troubles), and *Psilotum nudum* (diarrhea). The present study reports that similar species are medicinally used by *Malayalis* ethnic peoples, such as *Actinopteris radiata* (uterine problem, epilepsy, and piles), *Cheilanthes farinosa* (menstrual diseases, skin irritation, body strength, and chest pain), *Drynaria quercifolia* (throat infection, dysentery, coughs, cuts, and wounds), *Equisetum ramosissimum* (stone formation), *Lycopodium cernum* (healing wounds and bronchitis), *Pityrogramma calomelanos* (kidney trouble and fever), and *Psilotum nudum* (dysentery and healing wounds) were recorded as frequently used pteridophytes. In the present study too, these plants are used by the *Malayalis* of the Kolli Hills in preparation of herbal medicines in treating various diseases which are prevalent in the area.

Singh and Upadhyay (2012) documented the ethnobotanical importance of 31 pteridophytic species belonging to 16 families which are used by the local tribe of Pachmarhi, and the frequently used species by them are *Actiniopteris radiata*, *Adiantum incisum*, *Blechnum orientale*, *Dicranopteris linearis*, *Diplazium esculentum*, *Drynaria quercifolia*, *Equisetum ramosissimum*, *Lycopodiella cernua*, *Marsilea minuta*, and *Ophioglossum reticulatum*. *Adiantum incisum*, *Blechnum orientale*, *Dicranopteris linearis*, *Drynaria quercifolia*, *Equisetum ramosissimum*, and *Marsilea minuta* are recorded in the present study and considered as highly utilized plants for the preparation of herbal medicines by the *Malayalis*.

In the present study, the fronds are the most used plant part for the preparation of medicine, followed by whole plants, rhizomes, stems/rachis, and spores. Most of the previously reported studies on the ethnobotany of pteridophytes also stated fronds as a commonly used part for the preparation of medicine in treating various diseases. Likewise, tribal people in southern Nigeria (Nwosu 2002) and Amerindian ethnic groups in Amazonian Bolivia and Ecuador (Macia 2004) regularly use fronds for herbal preparations, which may be due to the presence of more bioactive compounds than other plant parts, easy to collect, and in a conservation point of view, fronds are a renewable resource in an area of study (Giday et al. 2003).

In the present study, *Drynaria quercifolia* has been reported as a highly cited species with 48 uses reported followed by *Actiniopteris radiata* (39), *Equisetum ramosissimum* (34), and *Blechnum orientale* (30). Similarly, Karthik et al. (2011) reported a high number of pteridophytic species from the Kolli Hills, like *Adiantum lunulatum*, *Botrychium lanuginosum*, *Drynaria quercifolia*, *Hemionitis arifolia*, *Lycopodium cernum*, *Marsilea quadrifolia*, and *Pityrogramma calomelanos*.

Kidney problems and stone formation had the highest agreement among the studied informants in the study area, followed by hair care, ear, nose, throat problems, common fever and typhoid fever, respiratory system disorders including asthma, bronchitis, chest pain and cough, gastro-intestinal ailments including dysentery, gastric complaints, indigestion, intestinal disorders, intestinal ulcer, intestinal worms, stomachaches, and ulcers. As with the present study, Revathi et al. (2013) also reported *Equisetum ramosissimum* and *Pityrogramma calomelanos* as good remedies to treat kidney problems from the Kolli Hills.

Out of the 46 species of pteridophytes recorded as medicinal in the present study in treating 57 types of ailments, the largest number of species are used to treat dermatological infections and diseases (*D. linearis* as a frequently used species), followed by skeleto-muscular system disorders (*D. quercifolia* as a frequently used species), gastro-intestinal ailments (*C. gigantea* as a frequently used species), and respiratory system disorders (*D. esculentum* as a frequently used species). Ethnomedicinal documentations among different ethnic people in India revealed that a greater number of plants is employed in treating various kinds of diseases with *C. thalictroides* as a chiefly referred to species (Rout et al. 2009; Singh and Upadhyay 2014). In the present study also, three traditional practitioners use the fronds of *C. thalictroides* in treating various kinds of diseases. It was also reported that the fresh young leaves of *C. thalictroides* are fried and used as food by the ethnic communities in China (Liu et al. 2012).

3.5 CONCLUSION

The present study revealed that traditional medicine is still in common use by the *Malayali* people. Thus, the study ascertains the value of a great number of plants which are used in traditional medicine especially with pteridophytes, one of the unexplored groups of vascular plants in an ethnobotanical point of view. There is obviously still much unknown information about plants to treat various ailments. So far, very few studies have been carried out on a particular compound which confers the biological or pharmacological activity of the pteridophytic species. The efficacy of the documented species in traditional medicine revealed that there is a possibility of bioactive compounds. A comprehensive evaluation of the plants with pharmacological activity on the basis of traditional medicine may possibly give new compounds that could be used as prominent drugs in modern day diseases. This study evidently points out that instead of trying to identify the active components of herbs through a massive collection of plants from natural sources, it is better to start investigating the efficacy of the natural product from the traditional use. Based on the results of this study, it was suggested that scientific steps should be undertaken to pharmacologically validate the pteridophytes which have more citations in experimental animals to ensure the sustainability of their uses. Experimental proof of these herbal medicines may help in developing novel drugs which can be used to cure inevitable diseases which are prevalent in the study area.

ACKNOWLEDGMENTS

The authors are grateful to the Science and Engineering Research Board (SERB), Government of India, New Delhi for providing financial support through the EMR Scheme (Ref. No. EMR/2016/007164). The authors wholeheartedly acknowledge all informants in the study area who shared their knowledge on medicinal plants and their co-operation throughout the field survey, without which, this study would not have been achievable with fruitful results. We also thank the Forest Ranger, District Forest Officer, Namakkal Division, and the Principal Chief Conservator of Forests, Tamil Nadu Forest Department, Tamil Nadu for granting permission to undertake fieldwork in the Kolli Hills.

REFERENCES

Adhikari, Y., Fischer, A., Fischer, H.S., and M.B. Rokaya. 2017. Diversity, composition and host-species relationships of epiphytic orchids and ferns in two forests in Nepal. *Journal of Mountain Science* 14 (6): 1–11.

Albuquerque, U.P.D., and N. Hanazaki. 2006. Ethnodirected research in the discovery of new drugs of medical and pharmaceutical interest: Flaws and perspectives. *Revista Brasileira de Farmacognosia* 16: 678–689.

Amiguet, V.T., Arnason, J.T., Maquin, P., Cal, V., Sanchez-Vindas, P., and L. Poveda Alvarez. 2006. A regression analysis of Q'eqchi' maya medicinal plants from Southern Belize. *Economic Botany* 60: 24–38.

Anbazhagan, S., Shanmugavalli, G., and V. Ramesh. 2010. Geomorphology and rainfall data analysis for landslide susceptibility mapping in Kolli Hills: A preliminary study. *National Conference on Geomatics in Disaster Management 9th–10th of April*, JTCDM, Mumbai: TISS.

Ankli, A., Sticher, O., and M. Heinrich. 1999. Medical ethnobotany of the Yucatec Maya: Healers' consensus as a quantitative criterion. *Economic Botany* 53: 144–160.

Ayyanar, M. 2015. Medicinal plants with active principles in the treatment of diabetes mellitus. In *Modern Methods in Phytomedicine*, ed. T. Parimelazhagan, pp. 77–100. New Delhi, India: Daya Publishing House.

Ayyanar, M., and S. Ignacimuthu. 2011. Role of traditional medicinal plants in drug discovery: Ethnobotanical approaches. In *Herbal Perspectives: Present and Future*, eds. T. Parimelazhagan et al., pp. 1–23. New Delhi, India: Satish Serial Publishing House.

Ayyanar, M., and S. Ignacimuthu. 2005. Traditional knowledge of Kanitribals in Kouthalai of Tirunelveli hills, Tamil Nadu, India. *Journal of Ethnopharmacology* 102: 246–255.

Baskaran, N., Ilanthirayan, A., and K. Venkatachalam. 2015. Educational status in Kolli Hill, using geographical information system (GIS). *International Journal of Innovative Science, Engineering and Technology* 2(9): 487–496.

Benniamin, A. 2011. Medicinal ferns of north eastern India with special reference to Arunachal Pradesh. *Indian Journal of Traditional Knowledge* 10(3): 516–522.

Bharath, S., and J.M. Raj. 2016. A study on farm products marketing with special reference to Kolli Hills in Tamil Nadu state. *International Journal of Applied Research* 2(5): 1017–1019.

Di-Stasi, L.C., Oliveira, G.P., Carvalhaes, M.A., Queiroz, M., Tien, O.S., Kakinami, S.H., and Reis, M.S. et al. 2002. Medicinal plants popularly used in the Brazilian Tropical Atlantic forest. *Fitoterapia* 73: 69–91.

Giday, M., Asfaw, Z., Elmqvist, T., and Z. Woldu. 2003. An ethnobotanical study of medicinal plants used by the Zay people in Ethiopia. *Journal of Ethnopharmacology* 85: 43–52.

Gowrisankar, K., and R. Chandrasekaran. 2011. Ethnomedicinal use of Pteridophyte from Kolli Hills Eastern Ghats, Namakkal Dist., Tamil Nadu, India. *Indian Journal of Natural Science* 1: 316–323.

Heinrich, M., Ankli, A., Frei, B., Weimann, C., and O. Sticher. 1998. Medicinal plants in Mexico: Healers' consensus and cultural importance. *Social Science and Medicine* 47: 91–112.

Jain, S.K., and R.R. Rao. 1977. *A Handbook of Field and Herbarium Methods*, New Delhi, India: Today & Tomorrow's Printers and Publishers.

Jayasree, G. 2002. Impact of development programmes on the Malayali of Javadhu Hills, PhD Thesis University of Madras Library

Karthik, V., Raju, K., Ayyanar, M., Gowrishankar, K., and T. Sekar. 2011. Ethnomedicinal uses of pteridophytes in Kolli Hills, Eastern Ghats of Tamil Nadu, India. *Journal of Natural Products and Plant Resources* 1(2): 50–55.

Leonti, M., Ramirez, R.F, Sticher, O., and M. Heinrich. 2003. Medicinal flora of the Popoluca, Mexico: A botanico-systematical perspective. *Economic Botany* 57: 218–230.

Liu, Y., Wujisguleng, W., and C. Long. 2012. Food uses of ferns in China: A review. *Acta Societalis Botanicorum Poloniae* 81(4): 263–270.

Macia, M.J. 2004. A comparison of useful pteridophytes between two Amerindian groups from Amazonian Bolivia and Ecuador. *American Fern Journal* 94 (1): 39–46.

Manickam, V.S., and V. Irudayaraj. 1992. Pteridophyte flora of the Western Ghats (South India). New Delhi, India: B.I. Publications.

Mannan, M., Maridass, M., and B. Victor. 2008. A review on the potential uses of ferns. *Ethnobotanical Leaflets* 12: 281–285.

Martin, G.J. 1995. *Ethnobotany—A Methods Manual*. Dordrecht, the Netherlands: Springer Science, Business media.

Moerman, D.E. 1979. Symbols and selectivity: A statistical analysis of native American medical ethnobotany. *Journal of Ethnopharmacology* 1: 111–119.

Moerman, D.E. 1991. The medicinal Flora of native North America: An analysis. *Journal of Ethnopharmacology* 31: 1–42.

Mukherjee, P.W. 2002. *Quality Control of Herbal Drugs: An Approach to Evaluation of Botanicals*. New Delhi, India: Business Horizons Publishers.

Nwosu, M.O. 2002. Ethnobotanical studies on some pteridophytes of Southern Nigeria. *Economic Botany* 56(3): 255–259.

Perumal, G. 2010. Ethnomedicinal use of Pteridophyte from Kolli Hills, Namakkal District, Tamil Nadu, India. *Ethnobotanical Leaflets* 14: 161–172.

Rajendran, A., and S. Manian. 2011. Herbal remedies for diabetes from the Kolli Hills, Eastern Ghats, India. *Indian Journal of Natural Products and Resources* 2(3): 383–386.

Ramesh, V., and S. Anbazhagan. 2015. Landslide susceptibility mapping along Kolli hills Ghat road section (India) using frequency ratio, relative effect and fuzzy logic models. *Environmental Earth Sciences* 73(12): 8009–8021.

Revathi, R., Muthuraja, R., Thomas, B., and K. Raju. 2013. Ethno medicinal fern and fern allies used by tribal *Malayalis* of Kolli hills, Eastern Ghats, Tamil Nadu. *Pteridological Research* 2: 1–10.

Rout, S.D., Panda, T., and N. Mishra. 2009. Ethnomedicinal studies on some pteridophytes of Similipal biosphere reserve, Orissa, India. *International Journal of Medical Sciences* 1(5): 192–197.

Sathiyaraj, G., Muthukumar, T., and K.C. Ravindran. 2015. Ethnomedicinal importance of fern and fern allies traditionally used by tribal people of Palani hills (Kodaikanal), Western Ghats, South India. *Journal of Medicinal Herbs and Ethnomedicine* 1(1): 4–9.

Shil, S., and M.D. Choudhury. 2009. Ethnomedicinal importance of pteridophytes used by Reang tribe of Tripura, North East India. *Ethnobotanical Leaflets* 13: 634–643.

Silambarasan, R., Sureshkumar, J., and M. Ayyanar. 2017. Traditional herbal medicines practiced by the ethnic people in Sathyamangalam forests of Western Ghats, India. *European Journal of Integrative Medicine* 16: 61–72.

Singh, B.P., and R. Upadhyay. 2012. Ethno-botanical importance of Pteridophytes used by the tribe of Pachmarhi, Central India. *Journal of Medicinal Plants Research* 6(1): 14–18.

Singh, B.P., and R. Upadhyay. 2014. Medicinal pteridophytes of Madhya Pradesh. *Journal of Medicinal Plants Studies* 2(4): 65–68.

Sureshkumar, J., Silambarasan, R., Bharati, K.A., Krupa, J., Amalraj, S., and M. Ayyanar. 2018. A review on ethnomedicinally important pteridophytes of India. *Journal of Ethnopharmacology* 219: 269–287.

Thurston, F., and K. Rangachari. 1909. *Caste and Tribes of Southern India*, Vol, IV: Government Press (Reprinted 1975. Cosmo Publication, New Delhi), India.

Weckerle, C.S., de Boer, H.J., Puri, R.K., van Andel, T., Bussmann, R.W., and M. Leonti. 2018. Recommended standards for conducting and reporting ethnopharmacological field studies. *Journal of Ethnopharmacology* 210: 125–132.

4 Bioprospecting of Antimicrobial *Carica papaya* Leaf Extract for Finishing of Fabrics in Health Care Industries

B. Sathya Priya, T. Stalin, and Olivier Nzamwita

CONTENTS

4.1 INTRODUCTION

Natural fibers like wool, cotton, and jute are easily attacked by microbes because of their water retention capacity and easy breakdown of polymer linkages by microbial enzymes (Gupta and Bhaumik 2007). The cotton fabrics, in contact with the human body, are prone to an ideal environment for the growth of microorganisms, which results in allergic reactions and the degradation of fabrics. It is a necessity that the textile products should be finished with antimicrobial agents for good hygiene, but the majority of the products are finished with synthetic agents that are not safe to skin and the environment. The textiles finished with disinfectants that lead to bacterial resistance were reported by Russell (2004), and those finished with metals cause many disorders.

The production of natural antimicrobial health care textiles is very important in the health care industries to prevent infections and diseases that are caused by the pathogenic microbes. The natural antimicrobial finishing is an emerging technology in the production of medical textiles. The different parts of the plants have significant antimicrobial properties and are used for the finishing of textiles. The natural antimicrobial textiles have more advantages, such as safe to the skin, eco-friendly, inhibit the infections caused by pathogenic microbes, prevention of bad odor, and deterioration of fabrics. The cotton fabrics that were finished with the different parts of the plant extracts were reported by many researchers, which are as follows: neem and Mexican daisy extracts (Thilagavathi and Bala 2007); green tea leaf extracts (Syamili et al. 2012); *Ocimum sanctum* and rind of *Punica granatum* (Sathianarayanan et al. 2010); aloe vera gel and neem leaves extract (Khurshid et al. 2015); *E. globulus, T. cordifolia*, and *T. procumbens* leaf extracts (Vastrad and Byadgi 2018); neem, henna, konrai, and papaya leaves extracts (Ganesan and Vardhini 2015); *Aerva lanata* extract (Ganesan et al. 2013); and *Vitex negundo* leaf extract (Mohanraj et al. 2012).

Papaya (*Carica papaya* Linn) belongs to the family Caricaceae. The papaya leaves have a broad spectrum of antimicrobial activity. So, the present study aimed for the finishing of cotton fabrics with *Carica papaya* leaf extract and analyzed the antimicrobial activity of leaf extract against the known test pathogens. The antimicrobial efficiency of the washed fabrics was also investigated.

4.2 MATERIALS AND METHODS

4.2.1 Collection and Extraction of Papaya Leaves

Carica papaya fresh leaves were collected from Coimbatore, Tamil Nadu, India. The leaves were washed and shadow dried at room temperature. The fine powder obtained after grinding and sieving was used for the extraction process. The solvent extracts of the leaf samples were prepared by mixing 10 g of papaya leaf powder with every 100 mL of 80% methanol, ethanol, and acetone in an airtight conical flask and kept at room temperature. The centrifugation was done at 8000 rpm for 10 minutes after 24 hours of extraction. It was followed by filtration and evaporation.

4.2.2 Finishing of Cotton Fabrics Using Papaya Leaf Extract

The washed, sterilized cotton fabrics were immersed in different extracts of *Carica papaya* leaves with 6% of citric acid as a binding agent for 10 minutes, and the excess solution was removed by the pad dry method. Later, the fabrics were dried at 80°C for 5 minutes and cured for 3 minutes at 150°C (Sathianarayanan et al. 2010).

4.2.3 Agar Diffusion Method

The treated fabrics were then tested for antibacterial activity. The plates were prepared by pouring the sterilized agar media into sterile Petri plates. The plates were allowed to solidify, and the 24-hour inoculum of gram-positive (*Staphylococcus aureus*) and gram-negative bacteria (*Escherichia coli*) were swabbed using a sterile cotton swab. The finished fabric of 15 mm was placed over the inoculated bacterial plates and was kept for incubation at 37°C for 24 hours. At the end of incubation, the zone of inhibition formed around the fabric was measured.

4.2.4 Wash Durability of Fabrics

The finished fabrics were allowed 15 wash cycles using a standard detergent of 0.2% concentration at 40°C, and the antibacterial activity of the fabrics on the test organisms was assessed.

4.3 RESULTS

The cotton samples were treated with methanol, ethanol, and acetone leaf extracts of *Carica papaya* and assessed for their antimicrobial activity on gram-positive *Staphylococcus aureus* and gram-negative *Escherichia coli*. The evaluation was made by using the zone of inhibition that formed around the fabric and was measured in mm. Figure 4.1 represents the antimicrobial activity of the finished fabrics on the test organisms. The ethanol leaf extract showed the maximum zone of inhibition for both the gram-positive (33 mm) and gram-negative (32 mm) test organisms.

Figures 4.2 and 4.3 show the antimicrobial efficiency of washed fabrics on *S. aureus* and *E. coli*. The increased wash cycles resulted in decreasing the zone of inhibition. It was observed that nearly 50% of the antimicrobial activity was retained by all the finished fabrics after ten wash cycles.

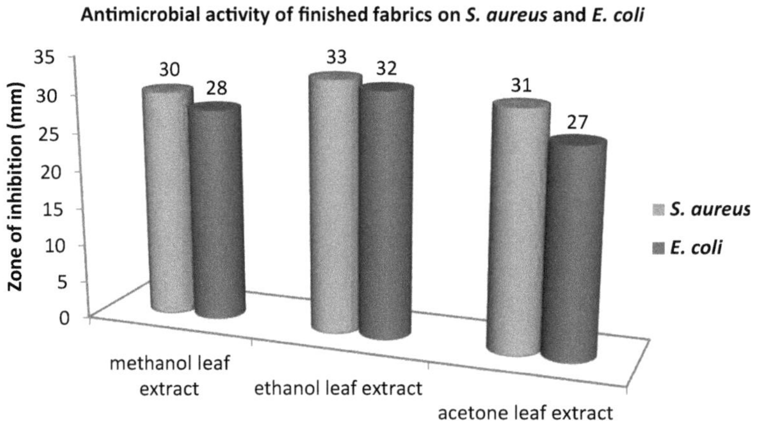

FIGURE 4.1 Antimicrobial activity of *Carica papaya* leaf extract finished fabrics on *Staphylococcus aureus* and *Escherichia coli*.

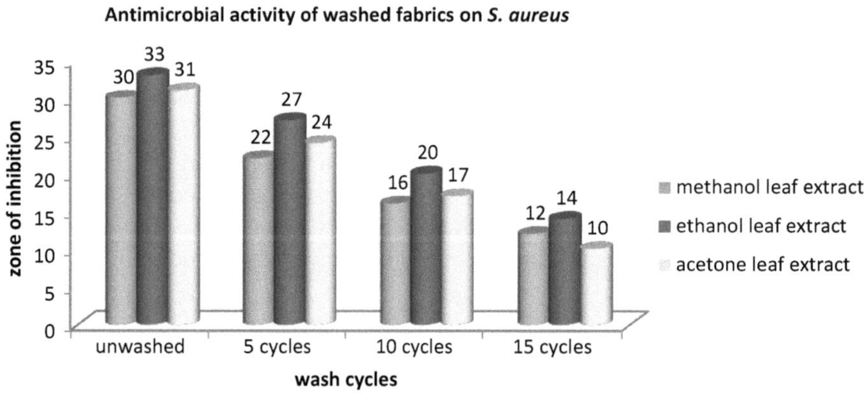

FIGURE 4.2 Antimicrobial activity of washed fabric samples on *Staphylococcus aureus*.

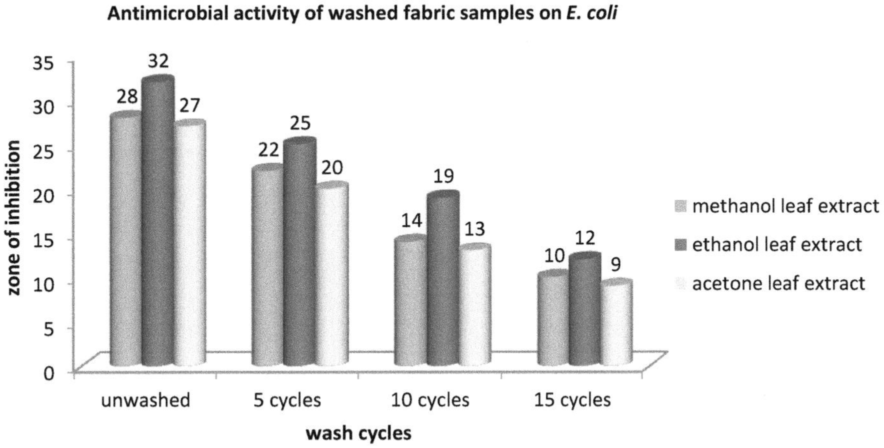

FIGURE 4.3 Antimicrobial activity of washed fabric samples on *Escherichia coli.*

4.4 DISCUSSION

The present investigation showed the maximum zone of inhibition for *S. aureus* and *E. coli.* It revealed the broad-spectrum activity of leaf extracts against both the gram-positive and gram-negative bacteria. It was observed that the finished fabrics showed more antibacterial efficiency on gram-positive bacteria than gram-negative bacteria, which may be due to the difference in thickness of the cell wall and mode of action of the active principle in the leaf extracts. Similarly, the results were in accordance with Thilagavathi and Bala (2007); Mohan raj et al. (2012); and Vastrad and Byadgi (2018). The phytochemical analysis of *Carica papaya* leaf extracts showed the presence of alkaloids, flavonoids, glycosides, tannins, saponins, phenols, and steroids (Vijay et al. 2014; Anjum et al. 2013). It was proposed that flavonoids may inhibit the cell wall integrity, saponins caused the leakage of vital constituents and tannins to coagulate the cell wall proteins (Alabi et al. 2012; Hassan et al. 2007; Onwuliri and Wonang et al. 2005). The ethanol extract showed higher antibacterial activity than the methanol and acetone extracts. It may be due to the nature of the solvent which is more potent to dissolve the antimicrobial agents of the leaf. Similar results were obtained by Alabi et al. (2012). The methanol extract showed a 30 mm zone of inhibition for *S. aureus* and 28 mm for *E. coli.* Similarly, Ganesan and Vardhini (2015) observed the higher zone of inhibition (28 mm) in the methanol extract of a papaya leaf against *S. aureus*. It was observed that the zones of inhibition for both the test organisms were gradually decreased due to the increase in wash cycles. It may be due to the hydrogen bonds and weak van der Waals forces. The *Carica papaya* leaf extract finished cotton fabric samples retained 50% of the antibacterial activity in up to 10 wash cycles. Similar results were obtained by Khurshid et al. (2015).

4.5 CONCLUSION

The natural fabrics may be widely used in medical and surgical applications, but the larger surface area and the retention of moisture made the textile fabrics more prone to bacterial growth. To prevent the infection by pathogens and transmission of diseases, it is very important to finish these textile materials with antimicrobial herbal extracts. Hence, it is concluded that *Carica papaya* leaf extract finished cotton fabric samples could be used as an effective alternative to synthetic antimicrobial agents, disinfectants, and heavy metals in health care industries.

REFERENCES

Alabi, O.A., Haruna, M.T., Anokwuru, C.P., Jegede, T., Abia, H., Okegbe, V.U., and Esan, B.E. 2012. Comparative studies on antimicrobial properties of extracts of fresh and dried leaves of *Carica papaya* (L) on clinical bacterial and fungal isolates. *Advances in Applied Science Research* 3(5): 3107–3114.

Anjum, V., Ansari, S.H., Naquvi, K.J., Arora, P., and Ahmad, A. 2013. Development of quality standards of *Carica papaya* Linn. leaves. *Scholars Research Library* 5(2): 370–376.

Ganesan, P., and Vardhini, K.J. 2015. Herbal treated microbial resistant fabrics for health care textiles. *Indian Journal of Fibre & Textile Research* 6: 227–230.

Ganesan, P., Ramachandran, T., Karthik, T., Anand, V.P., and Gowthaman, T. 2013. Process optimization of *Aerva lanata* extract treated textile material for microbial resistance in healthcare textiles. *Fibers and Polymers* 14(10): 1663–1673.

Gupta, D., and Bhaumik, S. 2007. Antimicrobial treatments for textiles, *Indian Journal of Fibre & Textile Research* 32: 254–263.

Hassan, S.W., Umar, R.A., Ladan, M.J., Nyemike, P., Wasagu, R.S.U., Lawal, M., and Ebbo, A.A. 2007. Nutritive value, phytochemical and antifungal properties of *Pergularia tomentosa* L. (Asclepiadaceae). *International Journal of Pharmacology* 3(4): 334–340.

Khurshid, M.F., Ayyoob, M., Asad, M., and Shah, S.N.H. 2015. Assessment of eco-friendly natural antimicrobial textile finish extracted from aloe vera and neem plants. *Fibres and Textiles in Eastern Europe* 23: 120–123.

Mohanraj, S., Vanathi, P., Sowbarniga, N., and Saravanan, D. 2012. Antimicrobial effectiveness of *Vitex negundo* leaf extracts. *Indian Journal of Fibre & Textile Research* 37: 389–392.

Onwuliri, F.C., and Wonang, D.L. 2005. Studies on the combined antibacterial action of ginger (*Zingiber officinale* L.) and garlic (*Allium sativum* L.) on some bacteria. *Nigerian Journal of Botany* 18: 224–228.

Russell, A.D. 2004. Bacterial adaptation and resistance to antiseptics, disinfectants and preservatives is not a new phenomenon. *Journal of Hospital Infection* 57(2): 97–104.

Sathianarayanan, M.P., Bhat, N.V., Kokate, S.S., and Walunj, V.E. 2010. Antibacterial finish for cotton fabric from herbal products. *Indian Journal of Fibre & Textile Research* 35: 50–58.

Syamili, E., Elayarajah, B., Rajendran, R., Venkatrajah, B., and Kumar, P.A. 2012. Antibacterial cotton finish using green tea leaf extracts interacted with copper. *Asian Journal of Textile* 2(1): 6–16.

Thilagavathi, G., and Bala, S.K. 2007. Microencapsulation of herbal extracts for microbial resistance in healthcare textiles. *Indian Journal of Fibre & Textile Research* 32: 351–354.

Vastrad, J.V., and Byadgi, S.A. 2018. Eco-friendly antimicrobial finishing of cotton fabric using plant extracts. *International Journal of Current Microbiology and Applied Sciences* 7 (2): 284–292.

Vijay, Y., Goyal, P.K., Chauhan, C.S., Anju, G., and Bhupendra, V. 2014. *Carica papaya* Linn: An overview. *International Journal of Herbal Medicine*, 2(5 Part A): 1–8.

5 Analysis of Discrete and Combined Effect of Solvent, Extraction Time, and Extraction Temperature on Polyphenol Compounds Extraction from Roxburgh Fig (*Ficus auriculata* Lour.) Fruit Using Response Surface Methodology

Gayathri Jagadeesan, Kasipandi Muniyandi, M. Ashwini Lydia, Gayathri Nataraj, Suman Thamburaj, Saikumar Sathyanarayanan, and Parimelazhagan Thangaraj

CONTENTS

5.1 INTRODUCTION

The advancement in the custom of using medicinal plant-based bioactive compounds influenced the international market for further exploration of important ethnopharmacological plants (Mendonça-Filho 2006). Despite all the nutrition from different diets, plant-based products are rich in antioxidants and therapeutic potentials. Amidst countless bioactive compounds, the plant secondary metabolites are utilized in pharmacological research for drug development and also used as therapeutic agents against oxidative stress, metabolic disorder, and apoptosis (Chandran et al. 2015; George et al. 2016; Muniyandi et al. 2017; Sathyanarayanan et al. 2017; Suman et al. 2018; Manoharan et al. 2019). The phenolic compounds are the natural antioxidants with dietary elements, which play an outstanding role against oxidative stress by defending the endogenous reactive oxygen species (Sivaraj et al. 2018; Sreeja et al. 2018). Therefore, the screening and extraction of phytocompounds have become decisive in the development of nutraceuticals, dietary ingredients, and pharmaceutical components.

The family Moraceae, generally known as the fig family, and the fig fruits are being utilized for therapeutic effects against liver diseases, menstrual disorders, diabetes, and respiratory problems in the Indian traditional systems (Upadhyay et al. 2007; Arunachalam and Parimelazhagan 2013). *F. auriculata*, commonly distributed all over India, and the fruits are edible either raw or in the form of jam, roasted fruits, and juices also cooked as curries for the medication of diarrhea and dysentery. Ethnobotanically, *F. auriculata* fruits were also used for wound healing, mumps, cholera, and jaundice (Gairola and Biswas 2008). The fruits were reported to contain sterols, terpenes, flavonoids, coumarins, laxative substances, sugars, and vitamins A and C with significant antioxidant, antihepatotoxic, antimicrobial, anticancer, antidiabetic, antipyretic, and anti-inflammatory activities (Gairola and Biswas 2008). The phenolic compounds from *F. auriculata* fruit have significant therapeutic properties; therefore, the voluminous extraction of phenolic compounds would be effective for the pharmaceutical industries for the advancement of drug and nutraceutical development.

The phytochemical extraction may be affected by many factors, such as solvent composition, extracting time, extraction temperature, and solvent to sample ratio (Jagadeesan et al. 2019), therefore optimization should be emphasized to improve the extraction process by considering all the factors. The response surface methodology technique certainly influenced the field of extraction of phytochemicals from medicinal plants (Cacace and Mazza 2003; Rajha et al. 2014; Jagadeesan et al. 2019). Response surface methodology has an extensive application in the design, development, and formulation of new products by optimizing. The technique improves the performance of the systems and increases the yield of the process without increasing the cost of the study (Bezerra et al. 2008). Hence, with all the above detail, this study was designed using response surface methodology to determine the discrete and combined effect of solvent, extraction temperature, and extraction time on the polyphenol extraction from *F. auriculata* fruits.

5.2 MATERIALS AND METHODS

5.2.1 Chemicals and Solvents

The analytical grade chemicals and reagents used in this study were purchased from HiMedia Laboratories (Mumbai, India); Sisco Research Laboratories (Mumbai, India); Sigma-Aldrich (St. Louis, Missouri, USA); and Merck (Bengaluru, India). The assays were carried out using water obtained from a Milli-Q water system (Merck Millipore, Billerica, MA, USA).

5.2.2 Effect of Ethanol Concentration, Extracting Temperature, and Extraction Time on the Polyphenol Extraction

The main aim for the present investigation was to determine the discrete and combined effects of solvent, extraction temperature, and extracting time on the polyphenolic content extraction from *F. auriculata* fruits. The fruits were collected from Western Ghats, Coimbatore, Tamil Nadu, India, and identified by the Botanical Survey of India, Southern Regional Centre, Coimbatore, Tamil Nadu, India. Before shade drying, the sample was rinsed with distilled water, after shade drying, the fruits were finely grounded and used for extraction. The central composite design was used to determine the effects of three parameters, such as ethanol concentration, extraction temperature, and extracting time, and the design parameters were given in our previously published paper (Jagadeesan et al. 2019). The total phenolic content, total flavonoid content, and antioxidant activity were selected as responses. The central composite design for the experiment parameters was presented in Table 5.1. *F. auriculata* fruit extracts were prepared by adding the fruit powder with various combinations of parameters as tabulated in Table 5.1. The prepared extracts were centrifuged at 3000 x g for 10 minutes, and then the supernatant was collected and dried at room

TABLE 5.1
Summary of Response Surface Methodology Design for Optimizing the *F. auriculata* Fruit Polyphenol Extracting Conditions

Design Expert—Trial Version 11.0.3.0 **Study: Response Surface Method (Central Composite Design)**

Model: Quadratic Model **Total Experiments: 16**

	Coded Variables Levels		
Variables	**−1**	**1**	**Responses/Units**
X–Ethanol concentration (%)	30	50	R1–Total phenolic content (mg gallic acid equivalents/g fruit extract)
Y–Extraction temperature (°C)	30	55	R2–Total flavonoid content (mg rutin equivalents/g fruit extract)
Z–Extraction time (minutes)	40	120	R3–2, 2-diphenyl-1–picryl hydrazyl radical scavenging activity (% inhibition)

temperature. After some time, the dried extract was collected and used for the proposed studies. The optimal conditions for the extraction of the fruit polyphenolic contents were determined by comparing the experimental and predicted values of three responses from the following regression equations.

$$Y = \beta_0 + \sum \beta_i X_i + \sum \beta_{ii} X_i^2 + \sum \beta_{ij} X_i X_j, \quad (5.1)$$

where Y—predicted response, β_0—central point response, β_i, β_{ii}, and β_{ij}—linear, quadratic, and interaction coefficients, and X_i and X_j—variables levels.

5.2.3 Estimation of *F. auriculata* Fruit Extracts Total Phenolics Contents

The extracted phenolics of *F. auriculata* fruit run order extracts were estimated by Thangaraj (2016). About 100 µL of the fruit's extract was taken from the sample stock solution, and the total volume made up to 1 mL using distilled water with 1 N Folin–Ciocalteu phenol reagent of 0.5 mL added. After 5 minutes of dark incubation, 2.5 mL of 5% sodium carbonate was added and mixed well. After 40 minutes of dark incubation, the formed blue color was read at 725 nm against the reagent blank. The total phenolic content present in the fruit samples was quantified using a gallic acid calibration curve and expressed in milligram gallic acid equivalents per gram of fruit extract.

5.2.4 Estimation of *F. auriculata* Fruit Extracts Total Flavonoid Contents

The flavonoid contents of the *F. auriculata* fruit run order extracts were determined by using the method described in Thangaraj (2016). The fruit extracts (0.5 mL) were taken in triplicates and made up to an equal volume with distilled water and kept for 6 minutes of incubation after adding 0.15 mL of 5% sodium nitrite. After the incubation, 0.15 mL of 10% aluminum chloride was added including the reagent blank. Then 2 mL of 4% sodium hydroxide was added, and the final volume made up to 5 mL using distilled water and incubated for 15 minutes at room temperature. After the incubation, the developed pink color was read at 510 nm, and the results were expressed in milligrams of rutin equivalents in grams of fruit extract using a rutin calibration curve.

5.2.5 Analysis of 2, 2-Diphenyl-1–Picryl Hydrazyl Radical Scavenging Activity

The 2, 2-diphenyl-1–picryl hydrazyl radical scavenging ability of prepared *F. auriculata* fruit experimental extracts were analyzed by the method described by Thangaraj (2016). The desired series of concentration of the extract was taken from the prepared stock, and then the final tube volume was made to 100 µL by adding methanol. Then the 0.1 mM 2, 2-diphenyl-1–picryl hydrazyl methanol solution of 3 mL was mixed with the extracts, after the incubation period, the optical density was read at 517 nm against the reagent blank. The tube with 3 mL 2, 2-diphenyl-1–picryl hydrazyl methanolic solution alone served as the negative control. The radical inhibition was expressed in terms of inhibition percentage.

$$\text{Inhibition percentage} = \left(1 - \left(\text{Sample Abs} / \text{negative control Abs}\right)\right) \times 100 \quad (5.2)$$

5.2.6 Statistical Analyses

The in vitro studies calculations were carried out using Microsoft Excel 2007 (Microsoft Corporation, Washington, DC, USA). The Response Surface Methodology (RSM) experiments were designed by Design-Expert, Trial Version 11.0.3.0 (Stat-Ease Inc., Minneapolis, MN, USA). One-way Analysis of Variance (ANOVA) was used to produce lack of fit, coefficient of determination (R^2), and F-test, and it was used to evaluate the adequacy of the model and response surface plots with multiple linear regressions with three responses to fit the model.

5.3 RESULTS AND DISCUSSION

5.3.1 Validation of the Model

For the optimization of phenolic compound extraction from plants, the second-order polynomial model is often used with the variables such as extraction time, extraction temperature, and solvent composition, and so on (Cacace and Mazza 2003; Rajha et al. 2014; Jagadeesan et al. 2019). And the proposed model is significantly fitted for the phenolic compounds extraction from *F. auriculata* fruit, and the ANOVA results were presented in Table 5.2. The model was validated by insignificant lack of fit ($P > 0.05$), which indicated the precision of the proposed model and the accurate prediction of the variables, which effects the polyphenol extraction with the experimental data. Also, the coefficient of determination (R^2) was ranging from 0.8 to 0.9 for the studied responses (Table 5.2), indicating the adequacy of this model. The Folin–Ciocalteu reagent was used to quantify phenolics using a gallic acid calibration graph ($y = 0.0276x - 0.011$; $R^2 = 0.9913$). The total phenolic content of *F. auriculata* showed significant P (0.0010) and F (10.20) values and the lack of fit was found to be an insignificant P (0.4916) value, which proved that the model was accurate (Table 5.2). The P and F values of the parameters and the linearity between the predicted and actual values showed that the model has a significant effect on the phenolic extraction with the slight variations in the graph (Figure 5.1a). The flavonoid content in *F. auriculata* fruit was determined by the rutin calibration curve, and the results were expressed in milligram

TABLE 5.2
Analysis of Variance for Response Surface Quadratic Model for the Extraction of Phenolic Compounds from *F. auriculata*

R^2 of Total Phenolic Content = 0.9107

R^2 of Total Flavonoid Content = 0.9073

R^2 of 2, 2-Diphenyl-1–Picryl Hydrazyl Radical Scavenging Activity = 0.8449

	Sum of Square			*df*	Mean Square			F Value			P Value		
	R1	R2	R3		R1	R2	R3	R1	R2	R3	R1	R2	R3
Model	859.85	9223.88	482.46	9	95.54	1024.88	53.61	10.20	9.78	5.45	0.0010***	0.0011***	0.0094***
X	142.65	0.6033	13.00	1	142.65	0.6033	13.00	15.23	0.006	1.32	0.0036***	0.9412	0.2800
Y	4.45	1330.44	104.68	1	4.45	1330.44	104.68	0.474	12.70	10.64	0.5082	0.0061***	0.0098***
Z	56.59	259.54	2.83	1	56.59	259.54	2.83	6.04	2.48	0.287	0.0363*	0.1499	0.6046
XY	130.98	853.81	30.55	1	130.98	853.81	4.40	13.99	8.15	3.10	0.0046***	0.0189	0.1119
XZ	72.37	1019.79	24.33	1	72.37	1019.79	7.89	7.73	9.74	2.47	0.0214*	0.0123*	0.1503
YZ	328.73	2366.58	115.91	1	328.73	2366.58	39.70	35.11	22.59	11.78	0.0002***	0.0010***	0.0075***
Residual	84.28	942.69	88.58	9	9.36	104.74	9.84						
Lack of Fit	48.01	448.26	25.32	5	9.60	89.65	5.06	1.06	0.725	0.320	0.4916[a]	0.6396[a]	0.8783[a]
Pure Error	36.27	494.43	63.25	4	9.07	123.61	15.81						
Cor Total	944.12	10166.57	571.04	18									

Notes: R^2—coefficient of determination; X, Y, Z—selected variables for the present study; X—ethanol concentration; Y—extraction temperature; Z—extraction time; X, Y, Z—the linear effects of selected variables on polyphenol extraction; XY, XZ, YZ—combined effects of selected variables on polyphenol extraction; R1, R2, R3—selected responses for the present study; R1—total phenolic content; R2—total flavonoid content; R3—2, 2-diphenyl-1–picryl hydrazyl radical scavenging activity.

[a] Not significant. Statistically significant at $*P \leq 0.05$; $**P \leq 0.01$; and $***P \leq 0.001$.

rutin equivalents per gram of extract ($y = 0.0022x - 0.0089$; $R^2 = 0.9926$). The results of the quadratic polynomial model (Table 5.2) suggested that this model has high response for the flavonoid extraction from fig fruits. The ANOVA of the model showed F and P values of 9.78 and 0.0011, respectively (Table 5.2), which explained the significance of the model with the R^2 value of 0.9073 in total flavonoid content extraction. The 2, 2-diphenyl-1–picryl hydrazyl radical scavenging activity of the *F. auriculata* fruit extract experimental values were given in Table 5.3, and the insignificant lack of fit P value 0.8783 showed that the model was fit. So, these antioxidant phenolic and flavonoid compounds could be obtained by the enhanced solvent composition with temperatures under various extracting times (Cacace and Mazza 2003; Rajha et al. 2014; Jagadeesan et al. 2019). The relationship between the significant independent variables and responses was described by equations 5.3 through 5.5.

$$\text{Total phenolic content} = 36.24 - 3.23 \times X + 0.5105 \times Y + 2.04 \times Z - 4.05 \times XY - 3.01 \times XZ + 6.41 \times YZ + 0.0057 \times X^2 + 2.97 \times Y^2 + 0.1413 \times Z^2 \quad (5.3)$$

$$\text{Total flavonoid content} = 77.27 - 0.2102 \times X + 9.87 \times Y + 4.36 \times Z + 10.33 \times XY - 11.29 \times XZ + 17.20 \times YZ + 0.2328 \times X^2 + 14.20 \times Y^2 - 4.62 \times Z^2 \quad (5.4)$$

$$\text{2, 2-diphenyl-1–picryl hydrazyl radical scavenging activity} = 35.05 - 0.9758 \times X + 2.77 \times Y + 0.4555 \times Z + 1.95 \times XY - 1.74 \times XZ + 3.81 \times YZ + 1.24 \times X^2 + 2.48 \times Y^2 - 2.13 \times Z^2 \quad (5.5)$$

From equations 5.3 through 5.5, X—discrete effect of ethanol concentration; Y—discrete effect of extraction temperature; Z—discrete effect of extraction time; XY—combined effect of ethanol concentration and extraction temperature; XZ—combined effect of ethanol concentration and extraction time; YZ—combined effect of extraction temperature and

TABLE 5.3
Central Composite Design Data and Triplicate Determination Results of Total Phenolic Content, Total Flavonoid Content, and 2, 2-Diphenyl-1–Picryl Hydrazyl Radical Scavenging Activity from the Obtained Experimental Data

Experiment	X	Y	Z	R1	R2	R3
1	50	30	120	31.20 ± 2.18	46.30 ± 3.97	26.39 ± 4.91
2	40	42.5	147	41.83 ± 2.38	64.73 ± 2.64	28.06 ± 3.00
3	40	42.5	13	29.47 ± 0.84	51.78 ± 5.50	28.57 ± 0.83
4	23	42.5	80	40.70 ± 10.60	71.55 ± 0.70	39.64 ± 2.71
5	40	63.5	80	42.75 ± 3.82	123.49 ± 24.17	43.91 ± 3.38
6	57	42.5	80	29.83 ± 2.41	72.38 ± 4.90	36.03 ± 1.76
7	50	30	40	47.83 ± 6.77	90.97 ± 2.64	34.86 ± 0.63
8	30	55	40	38.16 ± 9.57	58.64 ± 1.75	32.88 ± 1.04
9	40	21	80	44.57 ± 2.54	99.45 ± 3.69	38.81 ± 0.21
10	40	42.5	80	39.67 ± 2.54	82.68 ± 1.85	35.51 ± 1.17
11	30	55	120	59.21 ± 5.44	127.94 ± 7.35	46.61 ± 0.38
12	50	55	120	38.18 ± 4.01	122.08 ± 4.38	44.36 ± 0.28
13	30	30	40	40.64 ± 14.19	92.99 ± 6.13	37.96 ± 1.53
14	30	30	120	35.13 ± 3.70	87.74 ± 11.00	34.74 ± 1.42
15	50	55	40	30.08 ± 9.89	103.70 ± 10.65	39.32 ± 3.59
16	40	42.5	80	36.23 ± 2.38	78.64 ± 4.86	36.37 ± 1.67

Notes: X, Y, Z—selected variables for the present study; X—ethanol concentration (%); Y—extraction temperature (°C); Z–extraction time (minutes); R1, R2, R3—selected responses for the present study; R1—total phenolic content (mg gallic acid equivalents/g fruit extract); R2—total flavonoid content (mg rutin equivalents/g fruit extract); R3—2, 2-diphenyl-1-picryl hydrazyl radical scavenging activity (% inhibition); values are mean of triplicate determination (n = 3) ± standard deviation.

extraction time; X^2—integrative effect of ethanol concentration; Y^2—integrative effect of extraction temperature; and Z^2—integrative effect of extraction time.

5.3.2 Discrete Effect of Solvent, Extraction Time and Extraction Temperature on Polyphenol Extraction

The preparation of extracts is the important process for separating phytochemicals, and there are various factors which affect the extraction conditions, such as the sample nature, solvent, extraction temperature, and extraction time (Bezerra et al. 2008). In the present work, *F. auriculata* fruit extracts were obtained by using different concentrations of aqueous ethanol ranging from 30% to 50% for effective extraction (Jagadeesan et al. 2019). The obtained results indicated that the increasing ethanol concentration increases the extraction of phenolic and flavonoid content. The total phenolic content of the run order extracts was calculated using the Folin–Ciocalteu method and the values range from 29.47 ± 0.84 to 59.21 ± 5.44 mg gallic acid equivalents/g sample (Table 5.3). In the case of flavonoid, the estimated amount ranged from 51.78 ± 5.50 to 127.94 ± 7.35 mg rutin equivalents/g extract, whereas in the case of free radical scavenging activity, the scavenging percentage varies from 26.39% ± 4.91% to 46.61% ± 0.38% (Table 5.3). From the ANOVA results, the variable ethanol concentration contributed significantly to the better extraction of the phenolic content from fruit samples with the P value of 0.0036 and F value of 15.23 (Table 5.2). But in the case of flavonoid, the ethanol concentration revealed moderate significant values (F = 0.006; P = 0.9412), and the antioxidant results supported that the solvent effect was significant with higher radical inhibition (F = 1.32; P = 0.2800) (Table 5.2). The extractions of phenolic and flavonoid compounds from a fruit are highly associated with the concentration of the solvent concentration (Zhang et al. 2007; Do et al. 2014; Vajić et al. 2015), this conclusion is supported by the obtained results of this study that were the ethanolic concentrations greatly influenced the phenolic and flavonoid content extraction. These results were highly correlated with our earlier result (Jagadeesan et al. 2019), where the phenolic and flavonoid contents were higher at moderate ethanol concentration. The statistical analysis showed that the extracting temperatures highly influenced the flavonoids extraction (P = 0.0061; F = 12.70), as well as the antioxidant activity (P = 0.0098; F = 10.64) (Table 5.2). The significant extraction of the flavonoids could be responsible for the higher antioxidant activity of the fruit extract. The higher temperature increases the extracting ability of the solvent by enhancing polyphenol diffusion and the solubility (Al-Farsi and

Lee 2008). The previous reports (Wang et al. 2007; Zhang et al. 2007; Al-Farsi and Lee 2008; Do et al. 2014; Vajić et al. 2015; Jagadeesan et al. 2019) also supported the present results that the extraction temperature could favor the breakdown of bound polyphenol in the sample. The extraction time is also important in polyphenol extraction (Wang et al. 2007; Zhang et al. 2007; Al-Farsi and Lee 2008; Do et al. 2014; Vajić et al. 2015; Jagadeesan et al. 2019), in the ANOVA analysis, the extraction time showed significant values for phenolic ($P = 0.0363$; $F = 6.04$) and flavonoid ($P = 0.1499$; $F = 2.48$) content (Table 5.2).

5.3.3 Combined Effect of Solvent, Extraction Time, and Extraction Temperature on Polyphenol Extraction

The extraction combined effect of solvent, extraction time, and extraction temperature on the phenolic and flavonoid content was given in Table 5.2 and Figures 5.1 through 5.3. All of the selected variables significantly contributed toward the extraction of polyphenol from *F. auriculata* fruits. Based on the ANOVA results, solvent and temperature (XY) were combined and influenced the extracting ability ($P = 0.0046$; $F = 13.99$) (Table 5.2). Also, the interactive effect (YZ) of extraction temperature and extraction time highly established the phenolic extraction condition with the P value of 0.0002 and F value of 35.11. Whereas, the flavonoid ($P = 0.0010$; $F = 22.59$) and antioxidant activity ($P = 0.0075$; $F = 11.78$) were significantly influenced by the combined effect of extraction temperature and extraction time (Table 5.2). Moreover, the ethanol concentration and extraction time also slightly influenced the phenolic and flavonoid content extraction with the significant value of $P \leq 0.05$ (Table 5.2). Therefore, the respective and combined parameters of the ethanol concentration, extraction temperature, and extraction time were highly supported for

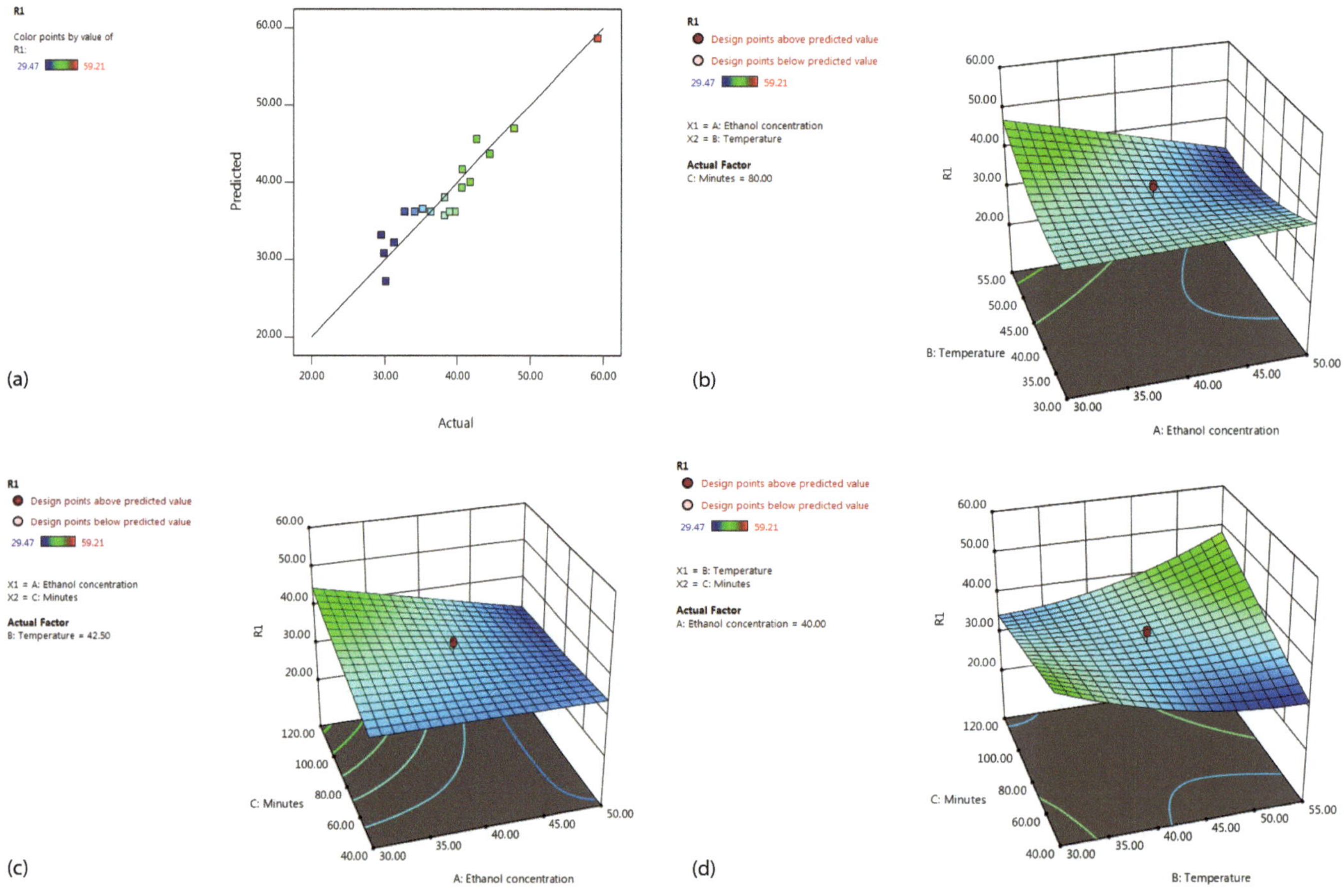

FIGURE 5.1 Response surface plots for the combined effect of variables on total phenolic content of *F. auriculata*. (a) Correlation between experimental and predicted value, (b) effect of ethanol concentration and extraction temperature, (c) effect of ethanol concentration and incubation time, and (d) effect of extraction temperature and incubation time.

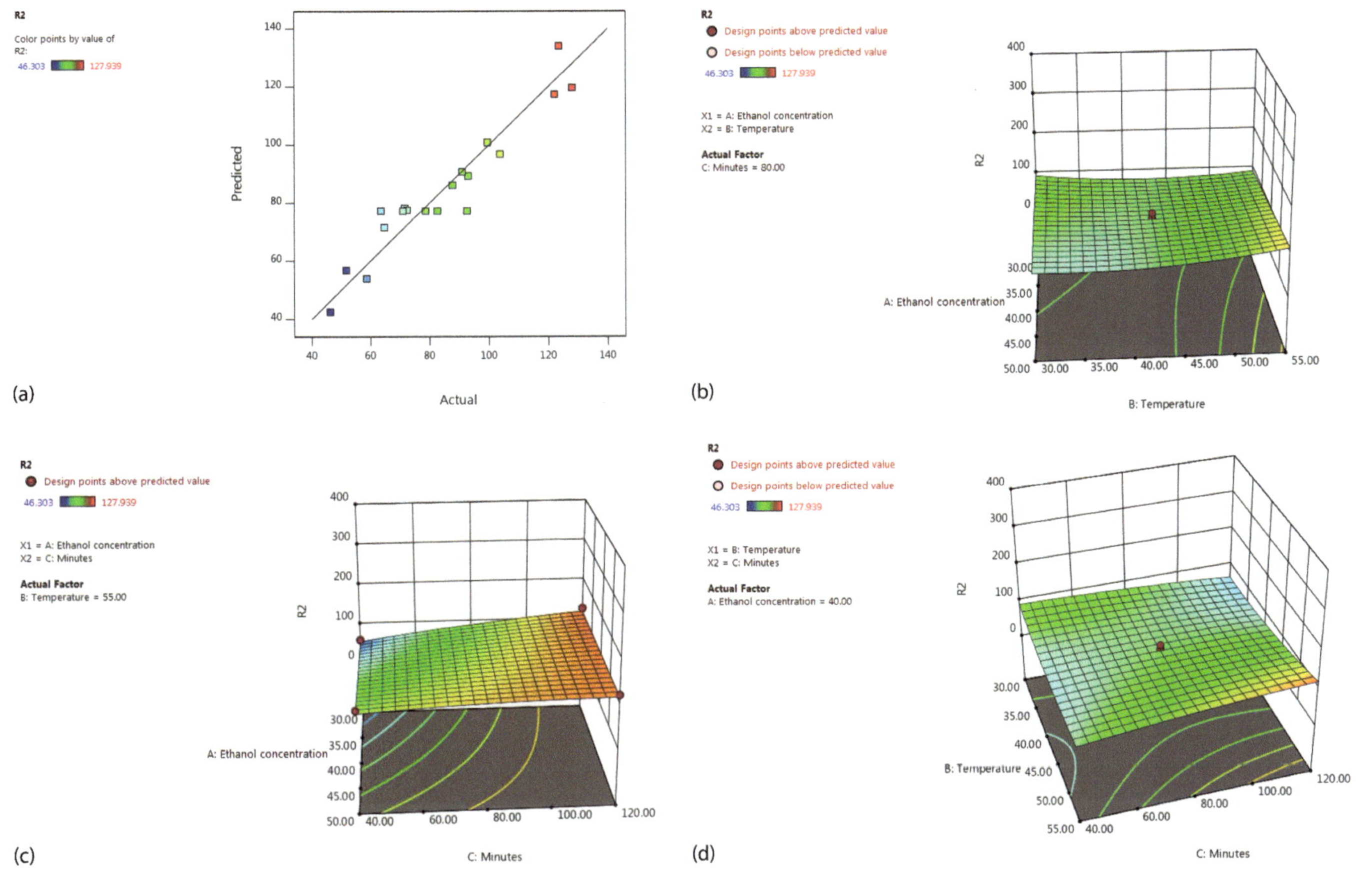

FIGURE 5.2 Response surface plots for the combined effect of variables on total flavonoid content of *F. auriculata.* (a) Correlation between experimental and predicted value, (b) effect of ethanol concentration and extraction temperature, (c) effect of ethanol concentration and incubation time, and (d) effect of extraction temperature and incubation time.

the extraction of phenolic content (Figure 5.1). From this developed quadratic model, the higher flavonoid content was achieved at the temperature of 45°C–50°C at the extracting time of 80 minutes with the ethanol concentration of 45%–50% (Figure 5.2) and the antioxidant activity was increased with the ethanol concentration of 30%–40%, with the temperature of 55°C–60°C, and at the extraction time of 120 minutes (Figure 5.3). The obtained results were highly correlated with the previous reports on the phenolic extraction through optimized conditions (Wang et al. 2007; Zhang et al. 2007; Al-Farsi and Lee 2008; Hossain et al. 2012; Gomes et al. 2013; Do et al. 2014; Vajić et al. 2015; Jagadeesan et al. 2019). The obtained results indicated that the moderate ethanol concentration and extraction temperature with a higher extraction time were highly favored for the extraction of phenolic and flavonoids compounds. Therefore, the phenolic and flavonoid content was independently affected by ethanol concentration, extracting temperature, and extraction time. Finally, the results concluded that 40% of ethanol concentration with 80 minutes of extraction time at 42.50°C could be optimum for the *F. auriculata* fruit phenolic compounds extraction.

5.3.4 Confirmation of the Proposed Model

In conclusion, the 40% ethanol, 80 minutes, and 42.50°C was suggested for the efficient extraction of phenolic compounds from *F. auriculata* fruits. The prepared extracts from these optimized conditions resulted in 59.21 ± 5.44 mg gallic acid equivalents/g sample of phenolic content, 127.94 ± 7.35 mg rutin equivalents/g sample of flavonoid, and 46.61% ± 0.38% of DPPH radical scavenging activity (Table 5.4).

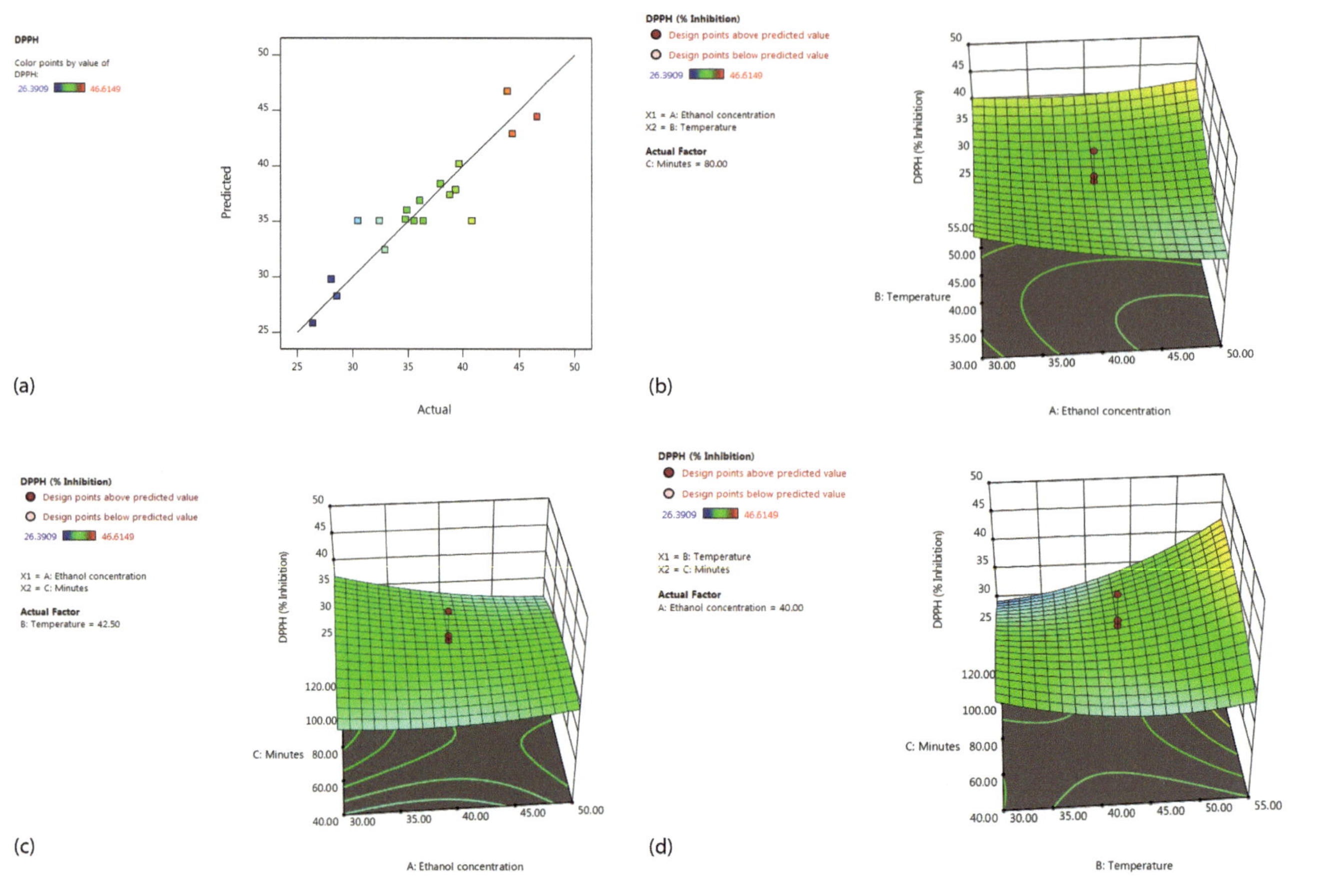

FIGURE 5.3 Response surface plots for the combined effect of variables on 2, 2-diphenyl-1–picryl hydrazyl radical scavenging activity of *F. auriculata*. (a) Correlation between experimental and predicted antioxidant activity, (b) effect of ethanol concentration and extraction temperature, (c) effect of ethanol concentration and incubation time, and (d) effect of extraction temperature and incubation time.

TABLE 5.4

Analysis of Total Phenolic Content, Total Flavonoid Content, and 2, 2-diphenyl-1–picryl Hydrazyl Radical Scavenging Activity of *F. auriculata* Fruit Extract Prepared from Optimized Conditions

Total Phenolic Content (mg Gallic Acid Equivalents/g Fruit Extract)	Total Flavonoid Content (mg Rutin Equivalents/g Fruit Extract)	2, 2-diphenyl-1–picryl Hydrazyl Radical Scavenging Activity (% Inhibition)
59.21 ± 5.44	127.94 ± 7.35	46.61 ± 0.38

Notes: Values are mean of triplicate determination (n = 3) ± standard deviation.

5.4 CONCLUSION

Therefore, this study resolutely showed the optimizing condition for better extraction of phenolic compounds from *F. auriculata* fruit using response surface methodology. From the results, it was concluded that three variables significantly affected the extraction of polyphenols from *F. auriculata* fruit. Particularly, the ethanol concentration and extraction temperature highly influenced the fig polyphenols content extraction. The proposed model was found to be highly significant for better extraction, and the studied values were fitted well. Hence, this method would be better for the extraction of natural phenolics from *F. auriculata* fruits with optimized conditions.

CONFLICT OF INTEREST

The authors declare no conflict of Interest.

REFERENCES

Al-Farsi, M.A., and Lee, C.Y. 2008. Optimization of phenolics and dietary fibre extraction from date seeds. *Food Chemistry* 108(3): 977–985.

Arunachalam, K., and Parimelazhagan, T. 2013. Antidiabetic activity of *Ficus amplissima* Smith. bark extract in streptozotocin induced diabetic rats. *Journal of Ethnopharmacology* 147(2): 302–310.

Bezerra, M.A., Santelli, R.E., Oliveira, E.P., Villar, L.S., and Escaleira, L.A. 2008. Response surface methodology (RSM) as a tool for optimization in analytical chemistry. *Talanta* 76(5): 965–977.

Cacace, J.E., and Mazza, G. 2003. Optimization of extraction of anthocyanins from black currants with aqueous ethanol. *Journal of Food Science* 68(1): 240–248.

Chandran, R., Sathyanarayanan, S., Rajan, M., Kasipandi, M., and Parimelazhagan, T. 2015. Anti-oxidant, hypoglycemic and anti-hyperglycemic properties of *Syzygium calophyllifolium*. *Bangladesh Journal of Pharmacology* 10(3): 672–680.

Do, Q.D., Angkawijaya, A.E., Tran-Nguyen, P.L., Huynh, L.H., Soetaredjo, F.E., Ismadji, S., and Ju, Y.H. 2014. Effect of extraction solvent on total phenol content, total flavonoid content, and antioxidant activity of *Limnophila aromatica*. *Journal of Food Drug Analytical* 22(3): 296–302.

Gairola, Y., and Biswas, S. 2008. Bioprospecting in Garhwal Himalayas, Uttarakhand. *Current Science* 94(9): 1139–1143.

George, E., Kasipandi, M., Venkataramana, M., Kumar, K.N., Allen, J.A., Parimelazhagan, T., and Gopalan, N. 2016. In vitro antioxidant and cytotoxic analysis of *Pogostemon mollis* Benth. *Bangladesh Journal of Pharmacology* 11: 148–158.

Gomes, T., Delgado, T., Ferreira, A., Pereira, J.A., Baptista, P., Casal, S., and Ramalhosa, E. 2013. Application of response surface methodology for obtaining lettuce (*Lactuca sativa* L.) by-products extracts with high antioxidative properties. *Industrial Crop Production* 44: 622–629.

Hossain, M.B., Brunton, N.P., Patras, A., Tiwari, B., O'Donnell, C.P., Martin-Diana, A.B., and Barry-Ryan, C. 2012. Optimization of ultrasound assisted extraction of antioxidant compounds from marjoram (*Origanum majorana* L.) using response surface methodology. *Ultrasonics Sonochemistry* 19(3): 582–590.

Jagadeesan, G., Muniyandi, K., Manoharan, A.L., Thamburaj, S., Sathyanarayanan, S., and Thangaraj, P. 2019. Optimization of phenolic compounds extracting conditions from *Ficus racemosa* L. fruit using response surface method. *Journal of Food Measurement and Characterization* 13(1): 312–320.

Manoharan, A.L., Thamburaj, S., Muniyandi, K., Jagadeesan, G., Sathyanarayanan, S., Nataraj, G., and Thangaraj, P. 2019. Antioxidant and antimicrobial investigations of *Elaeocarpus tectorius* (Lour.) Poir. fruits against urinary tract infection pathogens. *Biocatalysis and Agricultural Biotechnology* 20: 101–260.

Mendonça-Filho, R.R. 2006. Bioactive phytocompounds—New approaches in the phytosciences. In Iqbal, A., Farrukh, A., Mohammad, O. (Eds.), *Modern Phytomedicine: Turning Medicinal Plants into Drugs*, Wiley, New York. pp. 1–25.

Muniyandi, K., George, E., Mudili, V., Kalagatur, N.K., Anthuvan, A.J., Krishna, K., Thangaraj, P., and Natarajan, G. 2017. Antioxidant and anticancer activities of *Plectranthus stocksii* Hook. f. leaf and stem extracts. *Agriculture and Natural Resources* 51(2): 63–73.

Rajha, H.N., El Darra, N., Hobaika, Z., Boussetta, N., Vorobiev, E., Maroun, R.G., and Louka, N. 2014. Extraction of total phenolic compounds, flavonoids, anthocyanins and tannins from grape by products by response surface methodology. Influence of solid-liquid ratio, particle size, time, temperature and solvent mixtures on the optimization process. *Journal of Food Science and Nutrition* 5(4): 397.

Sathyanarayanan, S., Muniyandi, K., George, E., Sivaraj, D., Sasidharan, S.P., and Thangara, P. 2017. Chemical profiling of *Pterolobium hexapetalum* leaves by HPLC analysis and its productive wound healing activities in rats. *Biomedicine and Pharmacotherapy* 95: 287–297.

Sivaraj, D., Shanmugam, S., Rajan, M., Sasidharan, S.P., Sathyanarayanan, S., Muniyandi, K., Thangaraj, P., and de Souza Araújo, A.A. 2018. Evaluation of *Aristolochia indica* L. and *Piper nigrum* L. methanol extract against centipede *Scolopendra moristans* L. using Wistar albino rats and screening of bioactive compounds by high pressure liquid chromatography: A polyherbal formulation. *Biomedicine and Pharmacotherapy* 97: 1603–1612.

Sreeja, P.S., Arunachalam, K., de Oliveira Martins, D.T., da Silva Lima, J.C., Balogun, S.O., Pavan, E., Saikumar, S., Dhivya, S., Kasipandi, M., and Parimelazhagan, T. 2018. *Sphenodesme involucrata* var. *paniculata* (CB Clarke) Munir.: Chemical characterization, anti-nociceptive and anti-inflammatory activities of methanol extract of leaves. *Journal of Ethnopharmacology* 225: 71–80.

Suman, T., Elangomathavan, R., Kasipandi, M., Chakkaravarthi, K., Tamilvendan, D., and Parimelazhagan, T. 2018. Diphyllin: An effective anticandidal agent isolated from *Cleistanthus collinus* leaf extract. *Egyptian Journal of Basic and Applied Sciences* 5(2): 130–137.

Thangaraj, P. 2016. *Pharmacological Assays of Plant-Based Natural Products*. Springer, Switzerland.

Upadhyay, B., Roy, S., and Kumar, A. 2007. Traditional uses of medicinal plants among the rural communities of Churu district in the Thar Desert, India. *Journal of Ethnopharmacology* 113(3): 387–399.

Vajić, U.J., Grujić-Milanović, J., Živković, J., Šavikin, K., Gođevac, D., Miloradović, Z., and Mihailović-Stanojević, N. 2015. Optimization of extraction of stinging nettle leaf phenolic compounds using response surface methodology. *Industrial Crops and Products* 74: 912–917.

Wang, J., Sun, B., Cao, Y., Tian, Y., and Li, X. 2007. Optimisation of ultrasound-assisted extraction of phenolic compounds from wheat bran. *Food Chemistry* 106(2): 804–810.

Zhang, Z.S., Li, D., Wang, L.J., Ozkan, N., Chen, X.D., Mao, Z.H., and Yang, H.Z. 2007. Optimization of ethanol–water extraction of lignans from flaxseed. *Separation and Purification Technology* 57(1): 17–24.

6 GC-MS Analysis of Methanolic Extract of *Rubus ellipticus*

A Wild Edible Fruit

Elizabeth George, Blassan P. George, Sajeesh Thankarajan, Parimelazhagan Thangaraj, Kasipandi Muniyandi, and Saikumar Sathyanarayanan

CONTENTS

6.1 INTRODUCTION

India is blessed with a great wealth of plants, especially the wild edible plants. There are large numbers of edible plants around us, and, unfortunately, we often fail to utilize them. The wild edible and underutilized fruits are sometimes more nutritious than the widely consumed ones (Nazarudeen 2010; Saravanan and Parimelazhagan 2014). Fruits are found to be rich in carbohydrates, amino acids, proteins, fats, vitamins, etc. (Sundriyal and Sundriyal 2001; Agbaire and Emoyan 2012). Berries are well known for their comparable levels of phenolic compounds like anthocyanins, which are good for the body and provide better health. They encompass therapeutic effects like anti-carcinogenic, anti-inflammatory, anti-mutagenic, chemo-protection, protection against radiation, etc. (Sreeram 2008; Zang et al. 2010; Saklani et al. 2012).

Free radicals are generated in the body mainly by various metabolic activities and environmental effects. These can cause various health disorders and diseases including cancer, heart diseases, Alzheimer's disease, neurodegenerative diseases, etc. (Lako et al. 2007). Anti-oxidants play a key role in terminating the direct attack of the reactive species and are effective in the treatment of various ailments. Anti-oxidants prevent damage against various components of the cell (Hussin et al. 2004; Patel et al. 2010). Gas chromatography is used as a confirmatory test, especially for volatile compounds. Although it is very difficult to detect anti-oxidant compounds such as phenolics and flavonoids using gas chromatography-mass spectroscopy (GC-MS) analysis, the chemical profile of essential oils of herbs and plant extracts having pharmacological activities is detected in many cases (Tsao and Deng 2004).

Rubus is one of the largest genera in the family Rosaceae, and the members are well known as wild edibles with high nutritional and curative values (George et al. 2013; Srivastava et al. 2010). *R. ellipticus* is used mainly among tribals for gynecological problems, stomach problems, diarrhea, curing bone fracture, relieving stomach worms in children, anti-fertility activity in males, and for edible purposes (Maity et al. 2004; Gupta and Sharma 2006; Pradhan and Badola 2008).

The present work is carried out using the fruit extract of the *R. ellipticus* to find out the anti-oxidant activity of the extract and detect, using GCMS, the phytochemicals present in the extract showing the best anti-oxidant activity.

6.2 MATERIALS AND METHODS

6.2.1 Collection and Preparation of Fruit Samples

The fruits of the *Rubus ellipticus* were collected from the Shola Forest of Marayoor, Kerala, India. The taxonomic identity of the plant was confirmed from the Botanical Survey of India, Southern Circle, Coimbatore, Tamil Nadu. The collected fruits were washed under running tap water and air dried under shade. Then they were homogenized into fine powder using a mixer grinder and used for further studies.

6.2.2 Chemicals Required

Methanol, ethyl acetate, petroleum ether, 2, 4, 6-Tris(2-pyridyl)-s-triazine, ferric chloride, acetate buffer, etc. were used for the study, and all of the chemicals used were of analytical grade.

6.2.3 Extraction of Plant Material

The powdered fruits were packed in a small thimble and extracted successively with different organic solvents—petroleum ether, ethyl acetate, and methanol in the increasing order of polarity using Soxhlet apparatus. Then the material was macerated using hot water with constant stirring for 48 hours, and the water extract was filtered using Whatman No. 1 filter paper. The extracts were concentrated by rotary vacuum evaporator and then air dried.

6.2.4 Ferric Reducing Anti-oxidant Power Assay

The anti-oxidant capacities of the samples were estimated according to the procedure described by Pulido et al. (2000). The ferric reducing anti-oxidant power (FRAP) reagent (900 μL), prepared freshly and incubated at 37°C for 30 minutes, was mixed with 90 μL of distilled water and 30 μL of the test sample or methanol (for the reagent blank). The test samples and reagent blank were incubated at 37°C for 30 minutes in a water bath. The final dilution of the test sample in the reaction mixture was 1/34. At the end of incubation, the absorbance readings were taken immediately at 593 nm against the reagent blank using a spectrophotometer. Methanolic solutions of known Fe (II) concentration, ranging from 100 to 2000 μM ($FeSO_4.7H_2O$), were used for the preparation of the calibration curve. The parameter equivalent concentration was defined as the concentration of the anti-oxidant having a ferric-tripyridyltriazine (TPTZ) reducing ability equivalent to that of 1 mM $FeSO_4{\cdot}7H_2O$.

6.2.5 Gas Chromatography-Mass Spectroscopy Analysis

GC-MS analysis was carried out using a Clarus 600 gas chromatograph system equipped with a Clarus 600 C mass spectrometer (PerkinElmer Precisely, USA). An Elite-5MS fused silica capillary column coated with a 5% diphenyl/95% dimethyl polysiloxane stationary phase was used for the GC/MS. The injector temperature was kept at 200°C, whereas the oven temperature was programmed from 70°C to 300°C for a total run time of 35 minutes. Helium was used as the carrier gas at a flow rate of 1.0 mL/min, and the solvent delay was fixed to 4 minutes after running a blank. The following parameters were set for the MS analysis: electron ionization mode with ionization energy of 70 eV; ion source temperature of 200°C; GC interface temperature of 240°C; scan interval of 0.2 second; and fragments range from 50 to 600 m/z. About 1 μL of the methanol extract of the *R. ellipticus* fruit was injected manually in a splitless mode. The mass spectra of the respective peaks obtained in the GC/MS were compared with the mass fragmentation patterns of standards in the National Institute of Standards and Technology library.

6.3 RESULTS AND DISCUSSIONS

6.3.1 Ferric Reducing Anti-oxidant Power Assay

The anti-oxidant potential of the fruit extracts of the *R. ellipticus* were estimated from their ability to reduce a TPTZ-Fe (III) complex to TPTZ-Fe (II). The results clearly show that the methanol extract of the *R. ellipticus* had the best activity in the FRAP assay with 57.05 mM Fe (II)/mg extract. The hot water and ethyl acetate extracts of the *R. ellipticus* fruits showed anti-oxidant potential to reduce the TPTZ-Fe (III) complex to TPTZ-Fe (II) with 24.35 and 23.05 mM Fe (II)/mg extract. The lowest activity was shown by petroleum ether extracts of the *R. ellipticus* (0.44 mM Fe (II)/mg extract). The results of the assay are given in Table 6.1. The FRAP assay depends upon the conversion of Fe (III)-TPTZ complex

TABLE 6.1
Result of FRAP Assay

***R. ellipticus* Samples/Standards**	**FRAP Assay (mM Fe (II)/mg extract)**
Petroleum ether	0.44 ± 0.03
Ethyl acetate	23.05 ± 1.35
Methanol	57.05 ± 0.71[a]
Hot water	24.35 ± 1.72
Rutin	99.19 ± 14.90
Quercetin	307.47 ± 82.70
BHT	112.56 ± 9.58
BHA	133.68 ± 3.93

[a] Values are mean of triplicate determination (n = 3) ± standard deviation.

to the Fe (II)-TPTZ by a reductant at low pH. Fe (II)-TPTZ has an intensive blue color and can be monitored at 593 nm. This in vitro assay is highly effective in the measurement of total anti-oxidant power of the fresh biological fluids, such as plant extracts which are pharmacologically important (Gohari et al. 2011).

The ferric reducing anti-oxidant power assay measures the anti-oxidant effect of any substance in the reaction medium for its reducing ability. Polyphenols are the most important bioactive compounds universally present in plants and are largely being used in different health related activities due to their established role in absorbing and neutralizing the free radicals. Phenols are also known to exert a modifying influence on the secretion of anti-carcinogenic enzymes or inhibit the induction of cancer-promoting enzymes (Badhani et al. 2012). Thus, the ferric reducing power of the different extracts of *Rubus* fruits reveals that there are compounds that may be of polyphenolic contents in the methanolic and hot water extracts, which have a high affinity to the ferrous ions and thereby quench or scavenge them through redox reactions. It can be these phytoactive biomolecules which offer the pharmacological characters for the fruits.

6.3.2 GC-MS Analysis of *R. ellipticus* Methanol Extract

The result of FRAP analysis, which is an in vitro anti-oxidant assay, suggests that the *R. ellipticus* methanol extract to be the best among all the others. Therefore, the extract was analyzed through GC-MS, and the chromatogram is shown is Figure 6.1. Around 29 compounds were separated, among which, 17 compounds were identified by searching in the National Institute of Standards and Technology library. The details of the compounds detected are presented in Table 6.2.

The compounds detected from GC-MS analysis have showed many therapeutical uses. The compound 1,3-propanediamine, N′-[3-(dimethylamino) propyl]-NN dimethyl is found to be used as a curing agent for epoxy resins and polyurethanes and as a skin sensitizer. It is also used as a chemical intermediate in the production of surface-active agents for water treatment, motor oil additives, fabric softeners, etc. Some new derivatives of 2-aminomethyl-N-ethylpyrrolidine and of the 3-amino-N-ethylpiperidine possess medicinal properties like anti-emetic and anti-ulcer activity (Orzalesi et al. 1976).

The compound 3-piperidinecarboxamide, N,N-diethyl, inhibits the secretion of prolactin and growth hormones and reverses levodopa-induced dyskinesias in Parkinsonian monkeys (Cabergoline Basic information 2008). Moreover, anti-teratogenic effects of tumor inhibitors and analgesic properties were shown by [2-(4-hydroxy-phenyl)-ethyl]-carbamic acid ethyl ester. (The mass spectra of compounds identified through GC-MS are provided in Figure 6.2). However, there are many other unknown compounds in the chromatogram which need to be further identified.

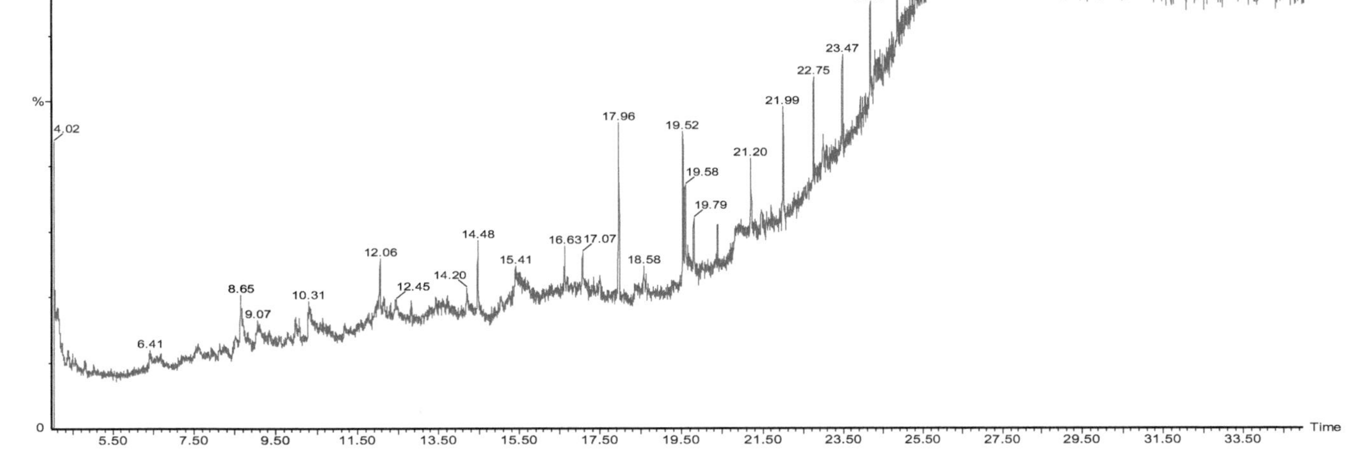

FIGURE 6.1 Chromatogram of GC-MS analysis of *R. ellipticus* methanol extract of fruit.

TABLE 6.2
Chemical Compounds Identified Through GC-MS Analysis

S No.	Name	R.T.	Mol. Wt.	Structure
1	3,3-diethoxypropylamine	8.65	147	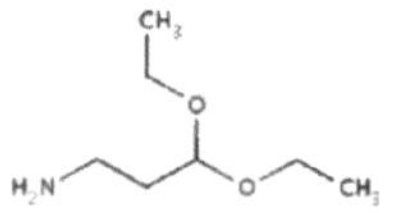
2	[2-(4-hydroxy-phenyl)-ethyl]-carbamic acid ethyl ester	9.07	105	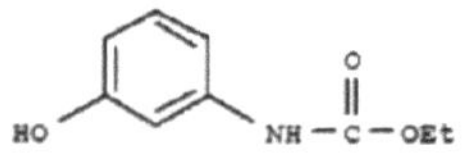
3	7,9-dimethyl-1,4-dioxa-7,9-diazacycloundecan 8one	10.31	202	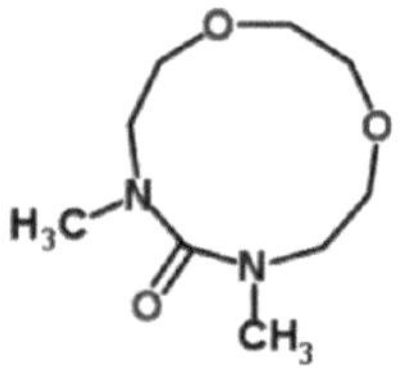
4	2-bromopropionic acid, tridecyl ester	12.06	334	
5	Glucurolactone	13.72	176	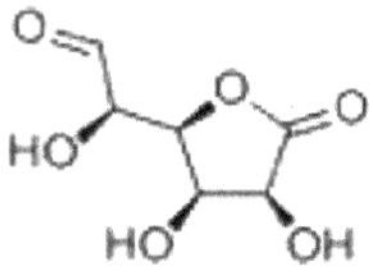
6	1(diethylamino)ethylidenimino]sulfur pentafluride	14.48	240	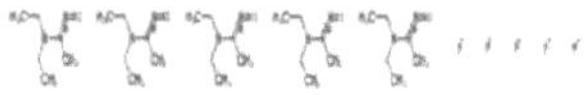
7	Carbamic acid, hydroxy-, ethyl ester	15.41	105	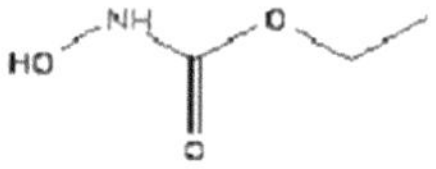
8	4-fluoro-n-[2-(4-methyl-piperazine-1-carbonyl phenyl benzamide	16.63	341	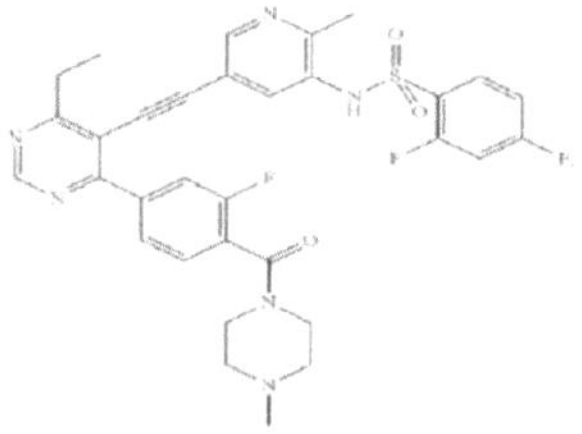
9	Acetamide, N-[3-(3-dimethylaminopropylamino)propyl]-2-hydroxyimino-2-phenyl	18.58	306	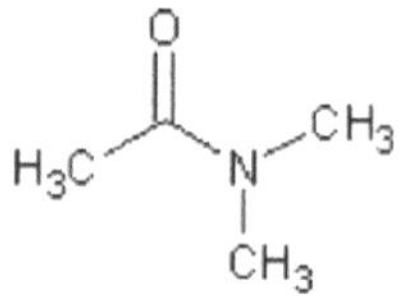
10	(e)-9,11-dodecadien-1-ol	19.52	182	
11	1,3-propanediamine, N′-[3-(dimethylamino) propyl]-NN dimethyl	20.38	187	
12	3-piperidinamine, 1-ethyl-	21.20	128	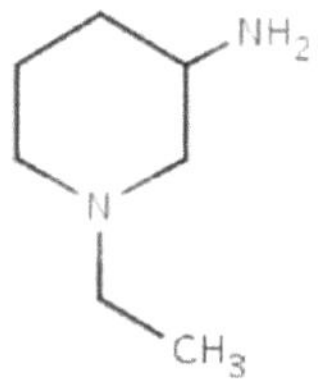

(Continued)

TABLE 6.2 (*Continued*)
Chemical Compounds Identified Through GC-MS Analysis

S No.	Name	R.T.	Mol. Wt.	Structure
13	2-propanamine, N-methyl-1-[4-[2-(1-piperidyl)ethoxyphenyl]	21.99	276	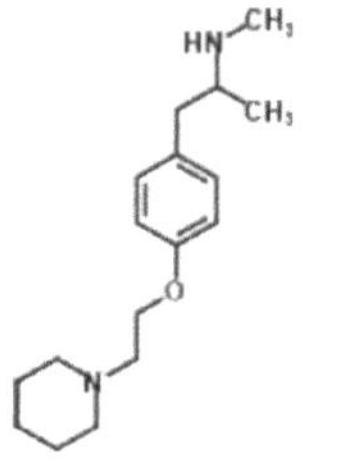
14	1,1-(diethylcarbamoyl)succinimide	22.75	198	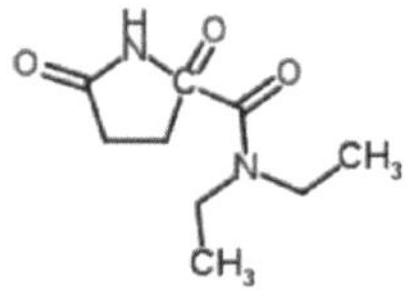
15	4-dimethylamino-2,2,6,6-tetramethylpiperidinde	23.47	184	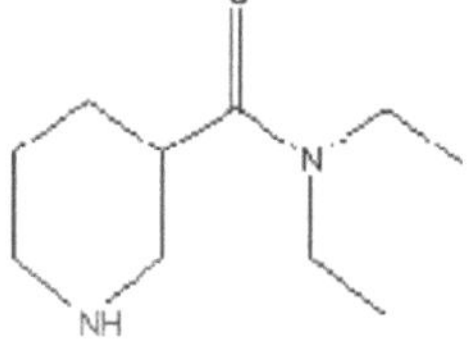
16	3-piperidinecarboxamide, N, N-diethyl-	24.17	184	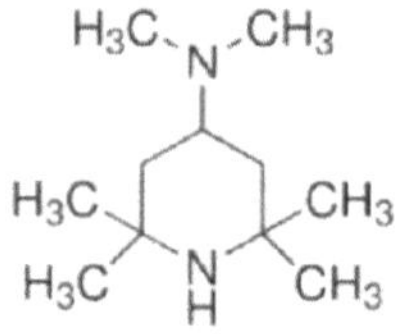
17	4(equat)-N-butyl-1,2(axial)-dimethyl-trans-decahydroquinol-4-ol	25.51	239	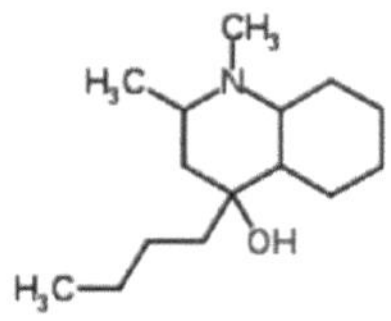

R.T.—Retention time

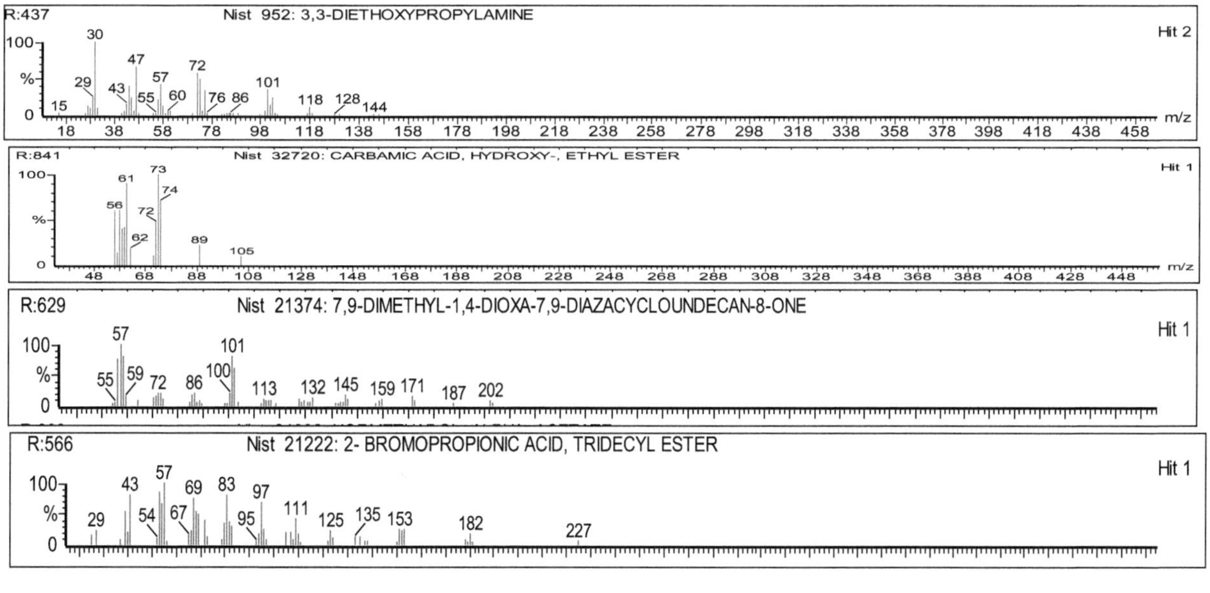

FIGURE 6.2 Mass spectra of compounds identified through GC-MS. (*Continued*)

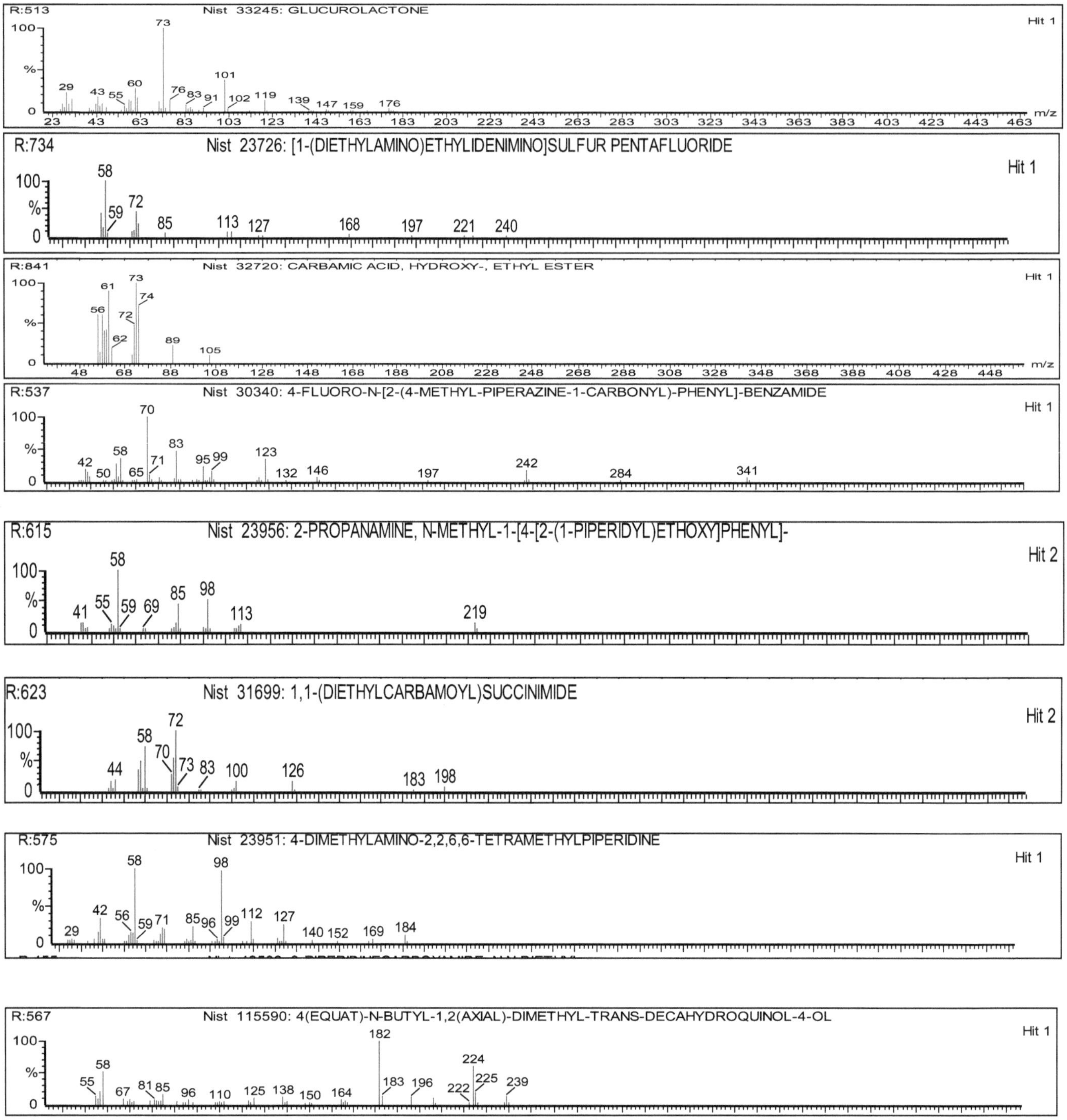

FIGURE 6.2 (Continued) Mass spectra of compounds identified through GC-MS.

6.4 CONCLUSION

This study is an attempt to bring such a wild edible and underutilized fruit into the mainstream. Even though *Rubus* plants possess a wide range of nutritionally and therapeutically important phytochemicals, limited reports are available revealing the phytochemical and pharmacological properties of their fruits. The present study thus focuses on the antioxidant and GC-MS profiling of *R. ellipticus* fruit extracts. The dried fruit powders were used to prepare the successive Soxhlet extracts using various solvents, and finally the sample was macerated with hot water. In the ferric reducing antioxidant power assay performed, the methanolic extract of the *R. ellipticus* fruit exhibited significantly more activity than the other extracts with a value of 57.05 mM Fe (II)/mg extract. The GC-MS analysis of the methanolic extract of the *R. ellipticus* revealed the presence of 17 known compounds that have different therapeutic effects. Further, the identification of other unknown compounds in the chromatogram will pave the way to explore the potential of therapeutic uses of the methanolic extract of the *R. ellipticus*. This qualifies the fruits of the *R. ellipticus* to be a potential candidate for nutiritonal and pharmacological consumption.

ACKNOWLEDGMENTS

The authors are grateful to Dr. S. Manian, former professor and head, Department of Botany, Bharathiar University, Coimbatore for rendering the permission and necessary facilities to carry out the lab work in the department during this project. The authors also acknowledge the authorities of DRDO-Bharathiar University Centre for Life Sciences, for providing the necessary facilities to carry out the GC-MS analysis in their center.

REFERENCES

Agbaire, P.O., and Emoyan, O.O. 2012. Nutritional and antinutritional levels of some local vegetables from Delta State, Nigeria. *African Journal of Food Science* 6(1): 8–11.

Badhani S., Jasrotia, N., Sharma, I., Parashar, B., and Gupta R. 2012. A review on some Indian medicinal plants for antiulcer activity. *Journal of Scientific Research in Pharmacy* 1(2): 6–9.

Cabergoline Basic information, Chemical Book. 2008. http://www.chemicalbook.com/ProductChemicalPropertiesCB0282495_EN.html.

George, B.P., Parimelazhagan, T., Saravanan, S., and Chandran, R. 2013. Anti-inflammatory, analgesic and antipyretic properties of *Rubus niveus* thunb. root acetone extract. *Pharmacologia* 4: 228–235.

Gohari, A.R., Hajimehdipoor, H., Saeidnia, S., Ajani, Y., and Hadjiakhoondi, A. 2011. Antioxidant activity of some medicinal species using FRAP assay. *Journal of Medicinal Plants* 1: 54–60.

Gupta, R.S., and Sharma, R. 2006. A review on medicinal plants exhibiting antifertility activity in males. *Natural Product Radiance* 5: 389–410.

Hussin, M., Abdul-Hamid, A., Mohamad, S., Saari, N., Ismail, M., and Bejo, M. H. 2004. Protective effect of *Centella asiatica* extract and powder on oxidative stress in rats. *Food Chemistry* 100: 535–545.

Lako, J., Trenerry, V.C., Wahlqvist, M., Wattanapenpaiboon, N., Sotheeswaran, S., and Premier, R. 2007. Phytochemical flavonols, carotenoids and the antioxidant properties of a wide selection of Fijian fruit, vegetables and other readily available foods. *Food Chemistry* 101(4): 1727–1741.

Maity, D., Pradhan, N., and Chauhan, A.S. 2004. Folk uses of some medicinal plants from North Sikkim. *Indian Journal of Traditional Knowledge* 3: 66–71.

Nazarudeen, A. 2010. Nutritional composition of some lesser-known fruits used by ethnic communities and local folks of Kerala. *Indian Journal of Traditional Knowledge* 9: 398–402.

Orzalesi, G., Selleri, R., and Volpato, I. 1976. Pyrrolidine derivative, anti-ulcer drug, and antibacterial drug. *Bollettino chimico Farmaceutico* 115: 317–328.

Patel, V.R., Patel, P.R., and Kajal, S.S. 2010. Antioxidant activity of some selected medicinal plants in western region of India. *Advances in Biological Research* 4(1): 23–26.

Pradhan, B.K., and Badola, H.K. 2008. Ethnomedicinal plant use by Lepcha tribe of Dzongu valley, bordering Khangchendzonga Biosphere Reserve, in north Sikkim, India. *Journal of Ethnobiology and Ethnomedicine* 4(1): 22.

Pulido, R., Bravo, L., and Saura-Calixto, F. 2000. Antioxidant activity of dietary polyphenols as determined by a modified ferric reducing/antioxidant power assay. *Journal of Agricultural and Food Chemistry* 48(8): 3396–3402.

Saklani, S., Chandra, S., Badoni, P.P., and Dogra, S. 2012. Antimicrobial activity, nutritional profile and phytochemical screening of wild edible fruit of *Rubus ellipticus*. *International Journal of Medicinal and Aromatic Plants* 2: 269–274.

Saravanan, S., and Parimelazhagan, T. 2014. *In vitro* antioxidant, antimicrobial and anti-diabetic properties of polyphenols of *Passiflora ligularis* Juss. Fruit pulp. *Food Science and Human Wellness* 3: 56–64.

Sreeram, N.P. 2008. Berry fruits: Compositional elements, biochemical activities, and the impact of their intake on human health, performance, and disease. *Journal of Agriculture and Food Chemistry* 56: 627–629.

Srivastava, R.C., Singh, R.K., and Mukherjee, T.K. 2010. Indigenous biodiversity of Apatani plateau: Learning on biocultural knowledge of Apatani tribe of Arunachal Pradesh for sustainable livelihoods. *Indian Journal of Traditional Knowledge* 9: 432–442.

Sundriyal, M., and Sundriyal, D.C. 2001. Wild edible plants of the Sikkim Himalaya: Nutritive values of selected species. *Economic Botany* 55(3): 377–390.

Tsao, R., and Deng, Z. 2004. Separation procedures for naturally occurring antioxidant phytochemicals. *Journal of Chromatography B* 812(1–2): 85–99.

Zang, L., Li, J., Hogan, S., Chung, H., Welbaum, G.E., and Zhou, K. 2010. Inhibitory effect of raspberries on starch digestive enzyme and their antioxidant properties and phenolic composition. *Food Chemistry* 119(2): 592–599.

7 Chemoprotective and Immunomodulating Effects of Ferulic Acid on Cisplatin Chemotherapy in Dalton's Lymphoma Xenografted Mice

Giftson J. Senapathy, R. Krishnamurthy, Heidi Abrahamse, P. Umadevi, and S. Murugan

CONTENTS

7.1 INTRODUCTION

Scientific investigations into the treatment of cancer are increasingly anomalous with the aim of controlling the disease or increasing the life span of the affected individuals. Chemotherapy is a common treatment for cancer usually used when surgery or radiotherapy is not effective. The drugs used in chemotherapy are chemically designed to target cells which are dividing and growing rapidly. Besides targeting abnormally proliferating cancer cells, they also affect normal cells. Thus, chemotherapy is still a major challenge to the cancer patients because the highly potent drugs employed in therapy can be toxic and less than 1% of the injected drug molecules can reach their target cells, whereas the rest may damage healthy cells and tissue especially bone marrow, epithelial tissues, reticuloendothelial system, and gonads (Kathiriya et al. 2010). The adverse effect of this therapy is the suppression of the immune system (Devasagayam and Sainis 2002).

Cisplatin (*cis*-Diaminedichloroplatinum(II)) is well established as a potent chemotherapeutic drug and used effectively against various solid tumors either singly or in combination with other drugs, radiotherapy, and/or surgery (Prasad and Giri 1994; Go and Adjei 1999). However, its therapeutic efficacy is limited due to the development of acquired drug resistance by the cancer cells and various side effects in the hosts, such as nephrotoxicity, ototoxicity, neurotoxicity, hepatotoxicity, hematotoxicity, mutagenicity, and gastrointestinal toxicity (Prasad et al. 2010).

These associated toxicities often defeat the purpose of providing relief from the disease, as cancer patients suffer more from the side effects of the chemotherapeutic agents rather than the disease itself. Thus, in spite of the therapeutic efficacy of these drugs, toxicity to normal cells will continue to limit the chemotherapy of cancer, unless new treatment paradigms are developed (Gibbs 2000). So, to end these problems, there is a search for a therapy that might minimize the associated toxic effects and destroy tumor cells preferentially.

The interest has increased in the search of modulating agents from plant origin that are capable of minimizing the toxicity induced by chemotherapy to normal cells without compromising its therapeutic effect. Nowadays, naturally occurring plant polyphenols are emerging as excellent chemosensitizers, which can sensitize the cancer cells which are resistant to chemotherapy. Besides sensitizing the cancer cells, they also can augment the potency of a cancer drug and decrease the side effects caused by it (Garg et al. 2005). Phenolic acids are a group of polyphenols that are widely distributed in the diet, mostly in whole grains, fruits, vegetables, and beverage (Martin and Appel 2010). These phenolic compounds exhibit good anti-oxidant and chemoprotective properties which may have beneficial effects in vivo. They are also believed to modulate the immune functions of the host (Zhou and Raffoul 2012). Ferulic acid (4-hydroxy, 3-methoxy cinnamic acid) (FA) is a ubiquitous phenolic compound of plant tissues and thus constitutes a bioactive ingredient of many foods. It arises from the metabolism of phenylalanine and tyrosine by shikimate pathway in plants. Ferulic acid (Figure 7.1) is rich in rice bran, whole grain foods, citrus fruits, banana, beet root, cabbage, spinach, and broccoli. In certain countries, ferulic acid has been approved as a food additive to prevent lipid peroxidation. FA exhibited anti-oxidant, anti-hyperlipidemic, hypotensive, anti-microbial, anti-carcinogenic, and radioprotective properties (Nyaradzo et al. 2009). It exhibits beneficial effects against various diseases like cancer, diabetes, and cardiovascular and neurodegenerative disorders. It was reported that ferulic acid protected against carbon-tetrachloride-induced hepatotoxicity in an experimental animal model. FA suppressed carcinogenesis in the forestomach, lungs, skin, tongue, colon, oral, skin, and mammary carcinogenesis in experimental animal models (Mori et al. 1999; Huang et al. 1988; Jayaprakasam et al. 2006; Baskaran et al. 2010).

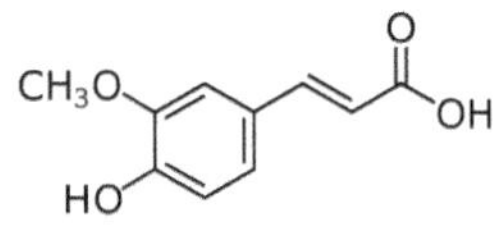

FIGURE 7.1 Chemical structure of ferulic acid (4-hydroxy, 3-methoxy cinnamic acid).

Even though many reports suggest the chemopreventive role of FA, its chemotherapeutic action in combination with a cytotoxic drug on tumor xenografted mice was not yet documented. So, the present study was designed to investigate the role of FA in cancer chemotherapy with a special focus on its immunostimulating and chemoprotective effects on lymphoma xenografted mice.

7.2 MATERIALS AND METHODS

7.2.1 Chemicals

FA was procured from Sigma-Aldrich Chemicals Co. (St. Louis, MO, USA). All other chemicals used were of high grade unless otherwise specified.

7.2.2 Animal Care

Inbred Swiss albino mice in the age group of 11–12 weeks, weighing 25–30 g, were purchased from Small Animal Breeding Station, Agricultural University (Mannuthy, Thrissur, Kerala, India). They were housed individually in polypropylene cages, maintained under normal laboratory conditions (temperature 24°C–28°C, relative humidity 60%–70%, and 1:1 dark and light cycle), and fed with commercially available food pellets (Hindustan Unilever Ltd., Mumbai, India) and tap water ad libitum. All animal experiments were carried out according to the guidelines and approval of the Institutional Animal Ethics Committee, Government of India (Reg. No. KMCRET/Ph.D./2011).

7.2.3 Tumor Induction

Dalton's lymphoma, a spontaneous T-cell lymphoma, is characterized by its highly invasive and malignant nature. Ascites Dalton's lymphoma is being maintained in vivo by intraperitoneal serial transplantation of approximately 1×10^7 viable tumor cells in 0.25 mL phosphate buffered saline per animal at pH 7.4. The tumor-transplanted mice usually survived for 19–21 days (Prasad et al. 2010).

7.2.4 Drugs Treatment

7.2.4.1 Dosage and Preparation of Treatment Drugs

FA was solubilized in water just before treatment and was administered orally thrice a week on alternative days at a dose of 40 mg/kg body weight (Alias et al. 2009) from the next day of the tumor inoculation for 15 days.

Cisplatin was dissolved in 0.9% NaCl (saline) for administration. The therapeutic dose of cisplatin against a malignant tumor has been established to be 8–10 mg/kg body weight (Prasad et al. 2010). A single dose of cisplatin (8 mg/kg body weight) was administered intraperitoneally in tumor-bearing mice on the 10th day post-tumor transplantation when the tumor was in the log phase of growth.

7.2.4.2 Grouping of Animals

The animals were randomly divided into five groups of six animals each. Group 1 served as the control (untreated animals). The animals in groups 2–5 were induced with Dalton's lymphoma ascites (DLA) cells on day 0 as described above. Group 2 served as the DLA tumor control; and group 3 and 5 animals received FA on alternate days for 14 days. In groups 4 and 5, the DLA-induced mice were treated with a single dose of cisplatin on the 10th day alone as described above. All of the treatments were continued until day 14. The experiment was terminated at the 15th day.

7.2.4.3 Euthanasia of Experimental Animals

At the end of the experimental period, all the mice were deprived of food overnight and euthanized by cervical dislocation under ketamine-hydrochloride anesthesia. The ascitic fluid, blood samples, and liver and kidney tissues were collected carefully for various assays.

7.2.4.4 Morphological Data and Body Weight Index

Throughout the experimental period, the body weight of the animals was recorded on the recording chart, and the percentage increase in the body weight of the animals of each was calculated according to the formula, % increase in body weight = [(Final body weight – Initial body weight)/Initial body weight] × 100.

7.2.4.5 Hematological Parameters

From the blood samples of the animals collected after sacrifice, hematological parameters, such as the total White Blood Cells (WBC), differential count (using Leishman's stain), platelets count, and packed cell volume were assessed.

7.2.4.6 Bone Marrow Cells

From the femur part, the bone marrow cells were collected and made into a single cell suspension, and the number of cells was determined using a hemocytometer.

7.2.4.7 Marker Enzymes in Liver and Kidney

Immediately after sacrifice, the liver and kidney tissues washed with saline were cut into fragments, homogenized with 3 volumes (w/v) of the appropriate buffer using a homogenizer, and centrifuged at 1200 × g for 20 minutes at 4°C. The supernatant was used for the estimation of liver marker enzymes like aspartate amino transferase (AST), alanine amino transferase (ALT), alkaline phosphatase (ALP) by using Ecoline Kits.

7.2.4.8 Total Protein

The total protein content was determined by the method of Lowry et al. (1951) using bovine serum albumin as the standard at 660 nm.

7.2.5 Statistical Analysis

The data were analyzed using one-way analysis of variance, and a significant difference among treatment groups was evaluated by Duncan's multiple range test. The results were considered statistically significant at $P < 0.05$. All statistical analyses were made using the SPSS 12.0 software package (SPSS, Tokyo, Japan).

7.3 RESULTS

The stimulating effect on the immune system and the protective effect over the host organs by the polyphenol FA during the cisplatin treatment on Dalton's lymphoma bearing mice are depicted in the following tables and figures.

7.3.1 Effect of FA and Cisplatin on the Morphology of the Mice

Figure 7.2 shows the morphological variation of the tumor control and drug-treated mice. The size of the DLA (group

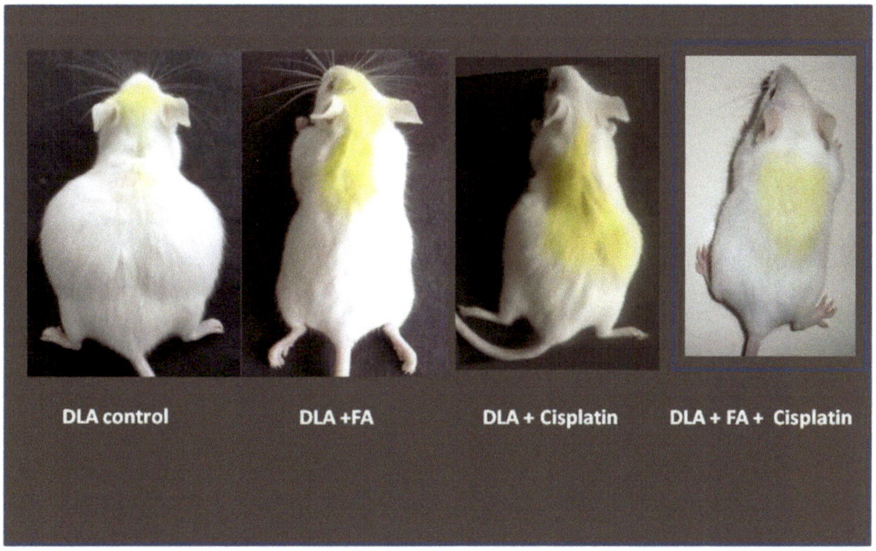

FIGURE 7.2 Morphological variations due to the effect of FA and cisplatin on control and experimental mice.

2) control was found to be increased with an increase in the peritoneal size compared to the normal mice. In group 3 (FA-treated) and group 4 (cisplatin-treated), the size of the peritoneum was decreased due to treatment. A further decrease in the size of the animal was seen in group 5, which was treated with a combination of FA and cisplatin.

7.3.2 Effect of FA and Cisplatin on the Body Weight of Animals

Table 7.1 summarizes the effect of FA on the body weight of DLA-induced mice on treatment with the chemotherapeutic drug cisplatin. It was found that the % increase in body weight of the DLA alone-induced mice (DLA control) was increased significantly ($P < 0.05$) when compared to the control mice. The cisplatin and FA treatment decreases the body weight significantly when compared to the DLA control mice, but a more pronounced effect on body weight drop was seen in the cisplatin-treated DLA mice treated with FA, which was close to that of the normal group.

7.3.3 Effect of FA and Cisplatin on Total WBC, Differential Count, Platelet Counts, and Packed Cell Volume of Control and Experimental Mice

The immunostimulating effect of FA on the total WBC (on days 0, 5, 10, and 15) and WBC differential count of cisplatin-treated DLA-induced mice were shown in Figure 7.3 and Table 7.2. In Figure 7.3, the total WBC counts of the DLA control group were significantly increased on days 5, 10, and 15 when compared to the control group on these days. The treatment with FA significantly decreased the WBC levels from days 10 and 15 when compared to the DLA control. Whereas on the cisplatin-treated group, though there was a decrease in the total WBC counts when compared to

TABLE 7.1
Effect of FA and Cisplatin on the Body Weights of Control and Experimental Animals

Groups	Increase in Body Weight (%)[a]
Control	18.83 ± 0.57^{b}
DAL Control	58.98 ± 1.82^{c}
DAL + FA	31.88 ± 0.97^{d}
DAL + Cisplatin	$21.05 \pm 1.04^{b,e}$
DAL + Cisplatin + FA	17.99 ± 0.59^{b}

[a] (Final body weight – Initial body weight)/Initial body weight × 100; Final body weight—on day 15; Initial body weight—on day 0.
Data are presented as the means ± SD; n = 6
[b–e] $P < 0.05$, the values not sharing a common superscript letter are significantly different.

TABLE 7.2
Effect of FA and Cisplatin on WBC Differential Count in Dalton's Lymphoma Xenografted Mice

Groups	Lymphocytes (%)	Neutrophils (%)	Eosinophils (%)
Control	69.44 ± 2.39^{a}	35.63 ± 2.99^{a}	6.51 ± 0.40^{a}
DLA Control	28.37 ± 9.56^{b}	57.23 ± 3.40^{b}	3.42 ± 0.28^{b}
DLA + FA	66.72 ± 4.67^{a}	31.74 ± 3.62^{c}	6.00 ± 0.53^{a}
DLA + Cisplatin	43.00 ± 5.66^{c}	48.24 ± 1.90^{d}	2.37 ± 0.43^{c}
DLA + Cisplatin + FA	65.83 ± 6.66^{a}	36.90 ± 2.35^{a}	6.51 ± 0.36^{a}

Data are presented as the means ± SD; n = 6; on day 15.
[a–d] $P < 0.05$, the values not sharing a common superscript letter are significantly different.

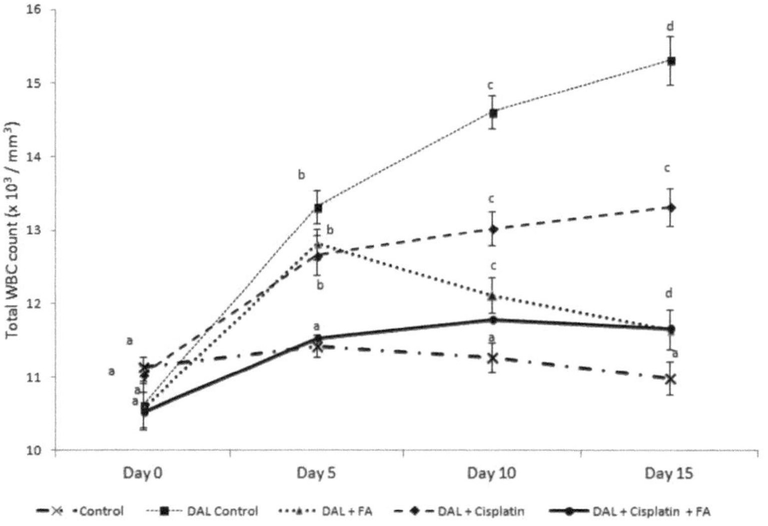

FIGURE 7.3 Effect of FA and cisplatin on total WBC counts in Dalton's lymphoma xenografted mice. Data are presented as the means ± SD; n = 6; as on day 15. [a–d] $P < 0.05$, the values not sharing a common superscript letter are significantly different.

the DLA control group, it was still considered to be higher in comparison with the normal and other treatment groups. When the cisplatin-treated DLA-induced mice were further treated with FA, the increased WBC counts by DLA induction and cisplatin treatment were significantly lowered, close toward normal levels.

In the DLA alone-induced mice, lymphocytes and eosinophil counts were decreased, whereas the neutrophil count was increased significantly ($P < 0.05$). On treatment with cisplatin alone, the condition was reversed slightly, but when treated with both FA and cisplatin, the levels of lymphocytes, eosinophils, and neutrophils were significantly altered to that of the normal values. Moreover, treatment with FA alone showed more enhanced effects than treatment with cisplatin alone in Table 7.2.

In Table 7.3, the platelets count was significantly ($P < 0.05$) decreased and the packed cell volume (PCV) was significantly ($P < 0.05$) increased in the DLA control mice when compared to the control animals, which clearly indicated the diseased condition. In the cisplatin-alone treatment, the platelet counts were slightly increased. But in the presence of FA, the platelet counts were significantly ($P < 0.05$) increased further in the cisplatin-treated DLA mice, close to that of the normal values. A similar effect was seen in FA-alone-treated DLA-induced mice. This pattern of changes was reversed for PCV.

TABLE 7.3
Effect of FA and Cisplatin on Platelets and PCV in Dalton's Lymphoma Xenografted Mice

Groups	Platelets (10^3/mm^3)	Packed Cell Volume (mL)
Control	180.13 ± 13.62[a]	0.25 ± 0.01[a]
DLA Control	132.00 ± 11.11[b]	0.92 ± 0.07[b]
DLA + FA	183.04 ± 18.72[a]	0.50 ± 0.03[d]
DLA + Cisplatin	151.67 ± 14.60[c]	0.65 ± 0.05[c]
DLA + Cisplatin + FA	186.22 ± 16.74[a]	0.33 ± 0.02[e]

Data are presented as the means ± SD; n = 6; on day 15.
[a–e] $P < 0.05$ the values not sharing a common superscript letter are significantly different.

7.3.4 Effect of FA and Cisplatin on Bone Marrow Cells

The effect of FA and cisplatin on the bone marrow cells of DLA-induced mice is shown in Figure 7.4. The DLA-alone-induced mice had a significantly ($P < 0.05$) decreased number of bone marrow cells. FA treatment alone significantly ($P < 0.05$) enhanced the bone marrow cellularity when compared with DLA-alone-induced mice. Due to toxicity, cisplatin-alone treatment caused a greater decrease in the bone marrow cells than FA-alone treatment. The DLA-induced mice treated with a combination of cisplatin and FA had significantly ($P < 0.05$) increased bone marrow cellularity when compared with the cisplatin-alone-treated group.

7.3.5 Effect of FA and Cisplatin on Hepatic Marker Enzymes in Liver and Kidney Tissues

The induction of Dalton's lymphoma to the animals caused a significant increase in the level of AST, ALT, and ALP when compared to the normal animals. The treatment with a combination of FA and cisplatin significantly reversed the changes toward the normal values when compared to the cisplatin-alone-treated mice, which is shown in Table 7.4. The results were found to be significantly different.

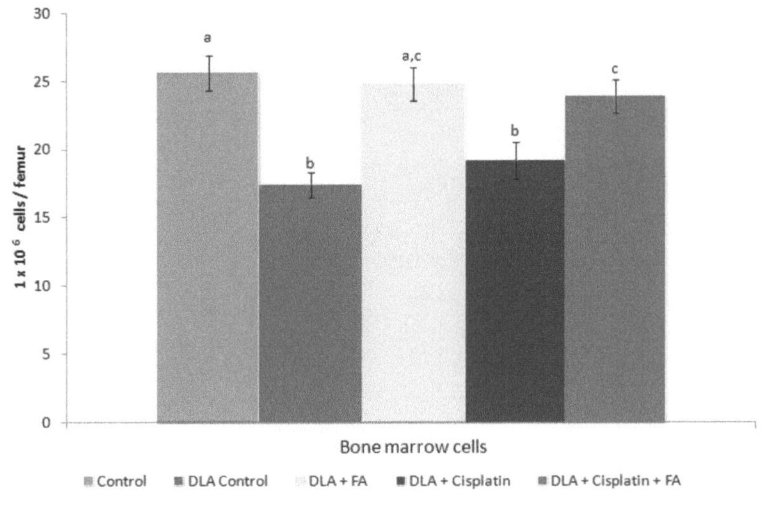

FIGURE 7.4 Effect of FA and cisplatin on bone marrow cells in Dalton's lymphoma xenografted mice. Data are presented as the means ± SD; n = 6; as on day 15. [a–c] $P < 0.05$, the values not sharing a common superscript letter are significantly different.

TABLE 7.4
Effect of FA and Cisplatin AST, ALT, and ALP on Liver and Kidney of Control and Experimental Mice

Groups	Liver (U/l)	Kidney (U/l)
	AST	
Control	57.04 ± 4.15[a]	63.89 ± 5.14[a]
DAL Control	96.02 ± 7.98[b]	165.63 ± 14.42[b]
DAL + FA	74.32 ± 5.66[c]	96.10 ± 8.12[c]
DAL + Cisplatin	55.81 ± 4.77[a]	60.31 ± 4.94[a]
DAL + Cisplatin + FA	60.09 ± 5.90[a]	65.67 ± 4.68[a]
	ALT	
Control	103.81 ± 8.76[a]	73.27 ± 6.77[a]
DAL Control	154.92 ± 11.46[b]	136.37 ± 11.26[b]
DAL + FA	121.64 ± 11.57[c]	117.46 ± 10.83[c]
DAL + Cisplatin	111.18 ± 10.70[a,c]	106.27 ± 10.94[c]
DAL + Cisplatin + FA	99.38 ± 8.77[a]	86.02 ± 7.33[d]
	ALP	
Control	166.13 ± 13.46[a]	153.76 ± 13.99[a]
DAL Control	215.67 ± 19.76[b]	212.35 ± 21.98[b]
DAL + FA	174.32 ± 17.62[a]	112.72 ± 11.42[c,d]
DAL + Cisplatin	93.46 ± 7.59[c]	122.84 ± 11.46[d]
DAL + Cisplatin + FA	107.64 ± 8.77[c]	99.40 ± 8.88[c]

Data are presented as the means ± SD; n = 6; on day 15.
[a–c] P < 0.05, the values not sharing a common superscript letter are significantly different.

7.3.6 Effect of FA and Cisplatin on Tissue Total Protein of DLA-Induced Mice

The effect of FA and cisplatin on the total protein content of liver and kidney tissues in DLA-induced mice is shown in Table 7.5. The total protein content in the tissues of DLA-alone-induced mice was significantly ($P < 0.05$) decreased as compared to the mice of the control group. Treatment with FA alone or treatment with FA along with cisplatin significantly ($P < 0.05$) increased the protein value when compared to the cisplatin alone treatment.

TABLE 7.5
Effect of FA and Cisplatin on Total Protein in Dalton's Lymphoma Xenografted Mice

Groups	Liver (g/dL)	Kidney (g/dL)
Control	16.42 ± 1.20[a]	15.47 ± 1.09[a]
DAL Control	10.27 ± 1.08[b]	11.37 ± 1.06[b]
DAL + FA	13.56 ± 1.14[c]	14.92 ± 1.08[a]
DAL + Cisplatin	14.68 ± 1.21[c]	14.37 ± 1.15[a]
DAL + Cisplatin + FA	14.87 ± 1.09[c]	15.02 ± 1.32[a]

Data are presented as the means ± SD; n = 6; on day 15.
[a–c] P < 0.05, the values not sharing a common superscript letter are significantly different.

7.4 DISCUSSION

Cisplatin is one of the most active and important chemotherapeutic drugs and is known for its cytotoxic action in the treatment of cancer. The primary action of cisplatin in a cell is its interaction with the DNA, forming DNA adducts (Kartalou and Essigmann 2001). Besides its anti-cancer action by the interaction with DNA in a cell, its clinical usefulness is limited due to immunotoxicity, nephrotoxicity, hepatotoxicity, neurotoxicity, genotoxicity, and ototoxicity to the host on continuous administration (Collins and Kao 1989). Thus, selective targeting of cancer cells and low toxicity for normal host tissues are, therefore, fundamental requisites for ideal chemotherapeutic drugs since a non-selective mechanism of action may lead to potentially severe immune toxicity.

FA is receiving greater attention nowadays in the research world because of its chemopreventive effects over a wide spectrum of cancer in experimental animals and cells. The results of the present study clearly demonstrate that FA treatment markedly reduced cisplatin-induced immunotoxicity and systemic toxicity as confirmed by the biochemical assays. It also played a significant role in stimulating the immune system.

In the present study, a significant gain in the percentage increase in body weight in lymphoma bearing mice was observed, which may be due to the increase in the ascitic fluid in the peritoneal cavity (Gupta et al. 2004). Body weight is directly proportional to ascitic fluid volume. On treatment with cisplatin, the body weight was significantly reduced, but the morphology of the animals revealed signs of hair loss ("alopecia"). This may be due to the effect of cisplatin actively dividing hair follicles (Yáñez et al. 2003). Treatment with FA on cisplatin-treated mice not only reduced weight gain further, but also reduced the hair loss caused by cisplatin treatment and produced a normal hair texture as seen in Figure 7.2. This suggested the protective effect of FA over actively diving normal cells of the host.

Most commonly, chemotherapy acts by killing cells that divide rapidly which is an important property of the cancer cells. These drugs are also harmful to cells that divide rapidly under normal circumstances, such as cells in the bone marrow, digestive tract, and hair follicles, resulting in side effects such as myelosuppression (decreased production of blood cells) and immunosuppression (Rüther et al. 2000). It is known that tumor development causes immunosuppression in the host. Most anti-cancer drugs are anti-proliferative and will also affect actively dividing normal cells. Furthermore, the side effects of chemotherapeutics include toxicity to bone marrow. Bone marrow depression occurs after cisplatin administration, which is indicated by a fall in circulating leukocytes and platelets. This may lead to bleeding disorders, increased susceptibility to infection, and bone marrow aplasia. FA restores back the bone marrow cells, which were decreased by cisplatin treatment.

The decreased level of WBCs observed after cisplatin treatment may indicate the emergence of complications like myelosuppression and thrombocytopenia, which are usually associated during cancer chemotherapy (Hoagland 1982). The presence of a tumor alters the body's hematopoietic system, which may be reversed by any cytotoxic drug like cisplatin. In the present study, alteration in the levels of lymphocytes, neutrophils, eosinophils, packed cell volume, and platelets by cisplatin treatment was brought back to normalcy by FA treatment.

FA could restore immunodepletion due to a tumor burden, and as a result of this lymphocyte or the bone marrow progenitor cell numbers were restored to normal levels. These results suggested that FA might stimulate the body's immune system during cancer therapy. Reports showed that our body's immune system can be benefitted from FA (Srinivasan et al. 2007). From earlier studies, we found that FA itself is nontoxic to the host, rather it could ameliorate tumor-induced systemic toxicity.

It was reported that the presence of a tumor in humans or experimental animals is known to affect many functions of the vital organs, especially in the liver and kidney, even when the site of the tumor does not interfere directly with organ functions (Natesan et al. 2007). In the present study, AST, ALT, and ALP levels in the liver and kidney are elevated, indicating tumor-induced general toxicity to the host. On cisplatin treatment, these levels were reduced, but still the levels were higher than that of the normal values, indicating hepatic and renal dysfunction. Liver and kidney damage by cisplatin were already recorded in reports (Collins and Kao 1989).

The reduction in the levels of these liver marker enzymes toward their respective normal values in liver and kidney tissues by FA is an indication of the stabilization of the plasma membrane, as well as the repair of hepatic and renal tissue damage caused by the chemotherapeutic drug cisplatin. Moreover, FA is shown to preserve physiological integrity of the cells and organs exposed to various stress (Srinivasan et al. 2007). FA showed a protective role over normal cells in many cancers (Sudheer et al. 2007). All these findings indicate that by restoring the immune system and host cells to normalcy, one can combat cancer, and ferulic acid may have the potency for that, either when used alone or with a chemotherapeutic drug.

7.5 CONCLUSION

To conclude, FA, when introduced in chemotherapy, could play a significant role by stimulating the immune system to fight against cancer and protect the host from therapy-induced damage. The research findings of the present work may provide a significant contribution for cancer treatment in clinical trials in the near future.

DECLARATION OF INTEREST

The authors declare no conflict of interest.

ACKNOWLEDGMENTS

The authors are thankful to Dr. P.R. Padma, professor (Department of Biotechnology, Avinashilingam Deemed University for Women, Coimbatore), and Mr. Ariharasivakumar, professor (Department of Pharmacology, KMCH College of Pharmacy), for their suggestions and help in in-vivo studies.

REFERENCES

Alias, L. M., Manoharan, S., Vellaichamy, L., Balakrishnan, S., and Ramachandran, C. R. 2009. Protective effect of ferulic acid on 7, 12-dimethylbenz [a] anthracene-induced skin carcinogenesis in Swiss albino mice. *Experimental and Toxicologic Pathology* 61(3): 205–214.

Baskaran, N., Manoharan, S., Balakrishnan, S., and Pugalendhi, P. 2010. Chemopreventive potential of ferulic acid in 7, 12-dimethylbenz [a] anthracene-induced mammary carcinogenesis in Sprague–Dawley rats. *European Journal of Pharmacology* 637(1–3): 22–29.

Collins, J. L., and Kao, M. S. 1989. The anticancer drug, cisplatin, increases the naturally occurring cell-mediated lysis of tumor cells. *Cancer Immunology, Immunotherapy* 29(1): 17–22.

Devasagayam, T. P. A., and Sainis, K. B. 2002. Immune system and antioxidants, especially those derived from Indian medicinal plants. *Indian Journal of Experimental Biology* 40: 639–655.

Garg, A. K., Buchholz, T. A., and Aggarwal, B. B. 2005. Chemosensitization and radiosensitization of tumors by plant polyphenols. *Antioxidants and Redox Signaling* 7(11–12): 1630–1647.

Gibbs, J. B. 2000. Mechanism-based target identification and drug discovery in cancer research. *Science 287*(5460): 1969–1973.

Go, R. S., and Adjei, A. A. 1999. Review of the comparative pharmacology and clinical activity of cisplatin and carboplatin. *Journal of Clinical Oncology* 17(1): 409–409.

Gupta, M., Mazumder, U. K., Kumar, R. S., Sivakumar, T., and Vamsi, M. L. M. 2004. Antitumor activity and antioxidant status of *Caesalpinia bonducella* against Ehrlich ascites carcinoma in Swiss albino mice. *Journal of Pharmacological Sciences* 94(2): 177–184.

Hoagland, H. C. 1982. Hematologic complications of cancer chemotherapy. *Seminars in Oncology* 9(1): 95–102.

Huang, M. T., Smart, R. C., Wong, C. Q., and Conney, A. H. 1988. Inhibitory effect of curcumin, chlorogenic acid, caffeic acid, and ferulic acid on tumor promotion in mouse skin by 12-O-tetradecanoylphorbol-13-acetate. *Cancer Research* 48(21): 5941–5946.

Jayaprakasam, B., Vanisree, M., Zhang, Y., Dewitt, D. L., and Nair, M. G. 2006. Impact of alkyl esters of caffeic and ferulic acids on tumor cell proliferation, cyclooxygenase enzyme, and lipid peroxidation. *Journal of Agricultural and Food Chemistry* 54(15): 5375–5381.

Kartalou, M., and Essigmann, J. M. 2001. Recognition of cisplatin adducts by cellular proteins. *Mutation Research/Fundamental and Molecular Mechanisms of Mutagenesis* 478(1–2): 1–21.

Kathiriya, A., Das, K., Kumar, E. P., and Mathai, K. B. 2010. Evaluation of antitumor and antioxidant activity of *Oxalis corniculata* Linn. against Ehrlich ascites carcinoma on mice. *Iranian Journal of Cancer Prevention* 3(4): 157–165.

Lowry, O. H., Rosebrough, N. J., Farr, A. L., and Randall, R. J. 1951. Protein measurement with the Folin phenol reagent. *Journal of Biological Chemistry* 193: 265–275.

Martin, K. R., and Appel, C. L. 2010. Polyphenols as dietary supplements: A double-edged sword. *Nutrition and Dietary Supplements* 2(1): 12.

Mori, H., Kawabata, K., Yoshimi, N., Tanaka, T., Murakami, T., Okada, T., and Murai, H. 1999. Chemopreventive effects of ferulic acid on oral and rice germ on large bowel carcinogenesis. *Anticancer Research* 19(5A): 3775–3778.

Natesan, S., Badami, S., Dongre, S. H., and Godavarthi, A. 2007. Antitumor activity and antioxidant status of the methanol extract of *Careya arborea* bark against Dalton's lymphoma ascites-induced ascitic and solid tumor in mice. *Journal of Pharmacological Sciences* 103(1): 12–23.

Nyaradzo, T. L., Chingorimbo-Murefu, A., Riva, S., and Burton, S. G. 2009. Lipase-catalysed synthesis of esters of ferulic acid with natural compounds and evaluation of their antioxidant properties. *Journal of Molecular Catalysis B: Enzymatic* 56: 277–282.

Prasad, S. B., and Giri, A. 1994. Antitumor effect of cisplatin against murine ascites Dalton's lymphoma. *Indian Journal of Experimental Biology* 32(3), 155–162.

Prasad, S. B., Rosangkima, G., and Kharbangar, A. 2010. Structural and biochemical changes in mitochondria after cisplatin treatment of Dalton's lymphoma-bearing mice. *Mitochondrion* 10(1): 38–45.

Rüther, U., Nunnensiek, C., and Schmoll, H. J. (Eds.). 2000. *Secondary Neoplasias Following Chemotherapy, Radiotherapy and Immunosuppression* (Vol. 55). Basel, Switzerland: Karger Medical and Scientific Publishers.

Srinivasan, M., Sudheer, A. R., and Menon, V. P. 2007. Ferulic acid: Therapeutic potential through its antioxidant property. *Journal of Clinical Biochemistry and Nutrition* 40(2): 92–100.

Sudheer, A. R., Muthukumaran, S., Kalpana, C., Srinivasan, M., and Menon, V. P. 2007. Protective effect of ferulic acid on nicotine-induced DNA damage and cellular changes in cultured rat peripheral blood lymphocytes: A comparison with N-acetylcysteine. *Toxicology in Vitro* 21(4): 576–585.

Yáñez, J. A., Teng, X. W., Roupe, K. A., Fariss, M. W., and Davies, N. M. 2003. Chemotherapy induced gastrointestinal toxicity in rats: Involvement of mitochondrial DNA, gastrointestinal permeability and cyclooxygenase-2. *Journal of Pharmacy and Pharmaceutical Sciences* 6(3): 308–314.

Zhou, K., and Raffoul, J. J. 2012. Potential anticancer properties of grape antioxidants. *Journal of Oncology* https://doi.org/10.1155/2012/803294.

8 Anti-microbial and Anti-oxidant Properties of Solvent Extract of Lichen Species Collected from Kodaikanal Hills, Western Ghats of Tamil Nadu

R. Kalidoss, M. Mariraj, M. Shenbagam, J. Merlin Seles, K. Arun Prasath, N. Rajaprabu, and P. Ponmurugan

CONTENTS

8.1 INTRODUCTION

Fungi in association with algae form an autotrophic, one single thallus with two different organisms, or lichens. They are able to grow on rock (saxicolous), soil (terricolous), and bark (corticolous). Because of its small size, the contribution of lichen to pharmaceutical industry has been ignored for several years. A total population of about 22,000 species are widespread in the world. It was estimated that about 2714 species occur in India. Five hundred and fifty-seven species of lichen were recognized as endemic in India. Lichens are known for unique secondary metabolites of about 1050 total compounds. Over 550 compounds are unique to lichens. The derivatives of lichen compounds are fatty acids, macrolytic lactones, zeorins, pulvic acid derivatives, cumarone derivatives, dibenzofurans, depsides, depsidones, terpenoids, anthroquinone derivatives, steroids, carotenoids, and diphenyl ethers. All lichen compounds have generally poor solubility in water. The amount of lichen substances in the lichen species are quite variable. The thallus of *Pamotrema tinctorum* had 23.5% lecanoric acid. The thallus of wolf lichen showed 10.5% vulpinic acid. Atranorin, parietin, usnic acid, and fungal melanins are some of the lichen substances found in the cortex. The lichen extracts with protocetratic acid, physodalic acid, and physodic acid are organized in the medullary layer. The lichen substances are identified by spot test and thin layer chromatography or high pressure liquid chromatography. Lichens are used to cure dyspepsia, bleeding piles, bronchitis, scabies, stomach disorders, and other disorders of the blood and heart. Cetrarin accelerated peristalsis and the substance was recommended for anemia and lack of appetite, and the usnic acid extracted from lichens rendered wide applications as analgesics and anti-pyretics to treat psoriasis and parasitic infections and skin diseases in the pharmaceutical industry.

In recent years, the research has unveiled that the ecosystems worldwide are contaminated with drugs that exceed safety limits. The misuse and widespread contamination of antibiotics contributed to the cause of microbial drug resistance. Feeding animals food containing antibiotics is the key to the wide distribution of the drug tolerant microbial strains. As an outcome of the development of the microbial drug resistance, even the common cold and recurrent fever are challenging the world in recent years. Microbial drug resistant strains take away the lives of about 23,000 people each year.

Lichen compounds rendered several biological activities, and they are a promising anti-microbial agent alternative to antibiotics. Over 50% of the known lichen can have medicinal value (Vartia 1973). Depending on the compounds enclosed in lichen, which encompass specific properties that are documented in the pharmacopoeias, the lichen compounds exert a wide variety of biological actions including anti-microbial, anti-viral, anti-inflammatory, anti-oxidant, analgesic, anti-pyretic, anti-proliferative and cytotoxic effects, besides lichens also have anti-snake venom activity (Tanas et al. 2010; Shrestha et al. 2015). The secondary metabolites are inevitable chemical substances, which assist lichen identification as well.

8.2 MATERIALS AND METHODS

8.2.1 Experimental Site

The lichen biodiversity was noticed in Kodaikanal Hills. It is located in Western Ghats of Tamil Nadu, India. The height of the hills is about 2130 meters. Lichen samples were collected from the Madhiketan cholai, Paneparpuram village, Guna Cave, Mother Teresa University, and Silent Valley. The specimens were collected from different substratum including the bark (corticolous) of a palm tree, rock (saxicolous), and soil (terricolous) samples. The species *Parmotrema reticulatam (terriclous)* was scraped out of the soil substratum, while *P. tinctorum* (corticolous) and *Usnea stigmatoides* (corticolous) lichens were collected from the bark of the tree, and the species *Parmotrema austrosinense* (saxicolous) was collected along with the rock using a chisel. The lichen field survey was supported by a hill guard, tribals, and forest conservation department guards.

The general objective of the present study was to examine the anti-microbial activities of different lichen extracts against clinical test pathogens and to evaluate the efficacy of lichen substances for their in vitro anti-oxidant properties. In addition, the study analyzed the total phenolic content of lichen extracts. Ecological data used in the present study include statistical data.

8.2.2 Collection of Lichen

The lichens were collected from thalli attached to soil, a rock, tree bark, and moss plants. The lichen specimen was collected to the measurable height from the bark of a tree and sliced using a chisel. The lichen thallus attached to the rock was cut using a chisel and a hammer from the base. However, care was taken that the intact saxicolous lichen thallus was not damaged. The terricolous lichen was picked up from the base of the soil and transferred to a specimen pocket.

8.2.3 Lichen Identification

Identification of the lichen was carried out at the Biomedical Research Lab, Bharathiar University. The individual lichen species were confirmed by comparing the representative specimen with the herbarium preserved in National Botanical Research Institute (NBRI), Lucknow. The identification procedures involved are morphological characterization, microscopic examination, chemical tests (spot test), and the investigation of a secondary metabolite using thin layer chromatography. The morphological and anatomical identification of the thalli were carried out using compound light microscopy. The macro lichen were identified using Awasthi's identification key manual (1991, 2007).

8.2.4 Spot Test (Chemical Test)

The spot test was carried out by scraping the cortex using a sharp blade to expose the medulla. The color reaction was observed by applying the reagent in the medulla. The chemical reagents used are potassium hydroxide (K), aqueous solution of calcium hypochlorite or bleaching powder (C), and para phenylene diamine (Culberson 1969). The color appears on the thallus as red or yellow. By introducing a drop of test solution on the cortex or medulla using a capillary tube or glass rod, the reaction was indicated by changes in color. The reactions positive to the tests were marked as "+" and negative as "–". The KC test was carried out by applying K solution first, followed by C solution on the cortex or medulla to witness the color changes. The iodine test was also applied on the cortex to note the presence or absence of the polysaccharides.

8.2.5 Extraction of Lichen Materials

The crude extract was collected by a solvent extraction procedure. A known measure of 5 g of a lichen sample was packed in a fat free thimble and introduced into the extraction tube. The receiving beaker was transferred with 100–150 mL of petroleum ether and fitted into the apparatus. Water was drained inside the condenser to start the extraction. After siphoning for 6 hours, the power supply was disconnected. The ether extract was transferred to the clean glass dish and concentrated by placing it in a water bath at 103°C for 30 minutes. The concentrate was cooled in a desiccator. The acetone and methanolic extracts of the lichen were collected using the same procedure.

8.2.6 Thin Layer Chromatography

Standard thin layer chromatography (TLC) protocol was used for the TLC experiment (Orange et al. 2001). Retention factor (Rf) class was compared with the standard chart.

8.2.7 Screening for Anti-microbial Activity

An in vitro anti-microbial assay was carried out using the methanolic extracts of the lichen samples by following the method of the disc diffusion test according to Kirby et al. (1956); Bauer et al. (1959); Stokes and Ridgway (1980); and Tibuhwa (2008). The following were the microbial test strains used, such as a yeast *Candida albicans*, a Gram-positive bacterium *Staphylococcus aureus*, and clinical test strains of Gram negative bacteria included *Proteus vulgaris, Kelbsiella pneumonia, Salmonella typhii, Escherichia coli*, and *Pseudomonas aeruginosa*, and they were obtained from multispecialty clinic in Coimbatore. Overnight bacterial cultures were grown in Mueller-Hinton broth and incubated at 37°C for 24 hrs. A known concentration of 250 µg/mL of the methanolic extract was prepared using 20% sterile dimethyl sulfoxide (DMSO) as a diluent. In a conical flask, about 100 mL of the overnight microbial cultures were adjusted to an Optical Density (OD) value of 0.1 at 600 nm using a UV spectrophotometer. The cell population was equivalent to 10^6 cells/mL McFarland standard. This culture suspension was used for swabbing on the Mueller-Hinton agar surface.

Microbial cultures were seeded into the Mueller-Hinton agar for bacterial suspension and potato dextrose agar plates for fungal suspension. Hard binding cardboard discs measuring 5 mm were made using a punching machine. Each sterile disc was flooded with 20 µL of the lichen extract. The plates were kept inside the refrigerator at 4°C for 3 hrs to allow the proper diffusion of the extracts into the medium. Extracts with activities at this concentration were considered as having anti-microbial potencies, while others with no potencies at this concentration were considered the extracts that failed the test (Hirasawa et al. 1999). The complete procedure was done in triplicates, while an evaporated methanol impregnated disc served as the negative control and the chloramphenicol (30 mg/mL) impregnated disc served as the positive control. The plates were incubated further at 37°C for 24 hrs. The hollow zone around the discs were measured using a ruler in millimeters (mm). The substance impregnated discs that showed no hollow zone were presumed as no activity for anti-microbial test.

8.2.8 Determination of MIC

The minimum inhibitory concentration (MIC) test was conducted using the 96-wells microtiter plate method by following the National Committee for Clinical Laboratory Standards, USA guidelines. The clinical test strains of the overnight culture adjusted to the turbidity equivalent of 0.5 McFarland standard were seeded by dispensing 100 microliters to each well that already contained 100 µL of sterile Mueller-Hinton broth medium. Prior to seeding the overnight culture, the compounds were separated using Soxhlet apparatus that measured 80 mg/mL in 20% sterile DMSO as stock and 100 µL stock was transferred to the first well. The two-fold dilution was made by transferring each 100 µL Mueller-Hinton broth mixed sample to adjacent wells to get 0.4 mg, 0.2 mg, and up to 6.25 µg/mL. The titer plate was further incubated overnight at 37°C. A resazurin indicator 0.05% was prepared in sterile distilled water and filter-sterilized to transfer 30 µL in each well on the next day. The titer plate was incubated again for 4 hours to develop pink color. All experiments were conducted in triplicates. The titer plate with the Mueller-Hinton broth alone served as the negative control, and the titer plate with bacterial culture and 100 µL served as the positive control.

8.2.9 Determination of DPPH Scavenging Assay

The methanol extracts of lichen substances were evaluated for 1,1-diphenyl-2-picrylhydrazyl (DPPH) radical-scavenging activity. The substance in the extract that reacts by donating a hydrogen atom and DPPH reduced. The purple color changed to slightly yellow and was measured at 517 nm using a UV visible light spectrophotometer. The solution of DPPH in methanol 0.2 mM was prepared freshly. One milliliter of this solution was mixed with various solvent extracts at different concentrations (25 mg, 50 mg, 75 mg, and 100 µg/mL). The set up was left undisturbed in the dark for 15 minutes at room temperature and the absorbance decrease was measured. The entire procedure was performed in triplicates. Radical-scavenging activity was calculated by the following formula:

$$\text{DPPH Scavenging Activity}\,(\%) = 100 - (\text{Ac} - \text{As}/\text{AC}) \times 100$$

A_C = control is the absorbance and A_S = Extract is the absorbance of reaction mixture (in the presence of extract).

8.2.10 Determination of Total Phenolic Compounds

A total phenolic compound estimation was performed by the Folin-Ciocalteu method. A known volume of 2 mL methanolic extract in different concentrations such as 25, 50, 75, and 100 µL was made up to 500 µL and methanol from the final concentration of 1 mg/mL of the lichen compound was added to 2.5 mL of the Folin's reagent along with 2 mL of sodium carbonate solution. The solution was incubated at 45°C for 15 minutes. The absorbance was read against a gallic acid calibration curve spectrophotometrically at 765 nm (Slinkard and Slingleton 1977).

8.3 RESULTS AND DISCUSSION

The lichens used for the present study are presented in Figure 8.1. Thin layer chromatographic identification test results are presented in Figure 8.2.

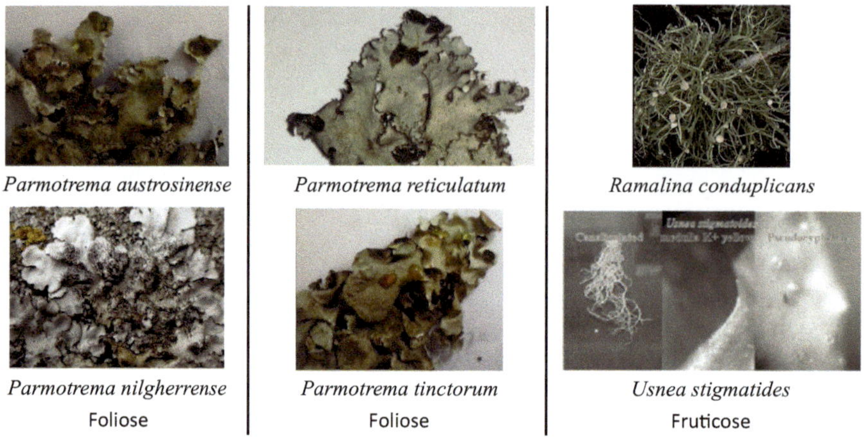

FIGURE 8.1 The macrolichen samples used for the present study.

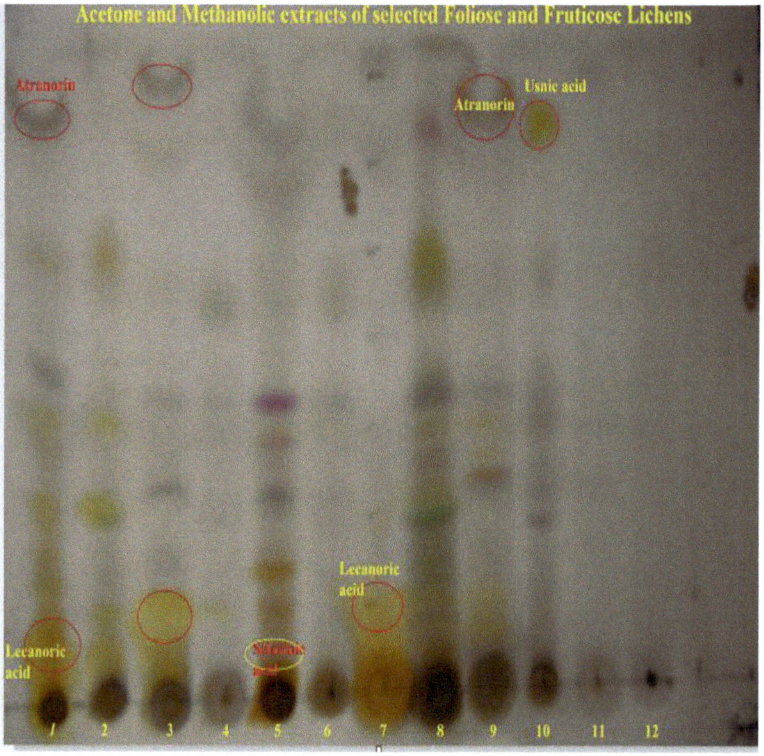

FIGURE 8.2 Thin layer chromatography separation of lichen species in solvent system. 1, Acetone extract of *P. austrosinense* with lecanoric acid and Atranorin. 2, Methanolic extract of *P. austrosinense* with lecanoric acid. 3, Acetone extract of *P. nilgherrense* with Atranorin. 4, Methanolic extract of *P. nilgherrense* with unknown compounds. 5, Acetone extract of *P. reticulatum* with salazinic acid. 6, Methanolic extract of *P. reticulatum* with unknown compounds. 7, Acetone extract of *P. tinctorum* with lecanoric acid. 8, Methanolic extract of *P. tinctorum* with unknown compounds. 9, Acetone extract of *Ramalina conduplicans* with Atranorin. 10, Methanolic extract of *Ramalina conduplicans* with usnic acid. 11, Acetone. 12, methanolic extract of *Usnea stigmatoides* with stictic acid.

8.3.1 Determination of Anti-microbial Activity of Lichen Compound

The methanolic extract of *Parmotrema austrosinense* showed a higher zone of inhibition against *Candida albicans* and *Proteus* species with the diameter of above 8.33 ± 0.58 mm at a concentration of 250 μg. *Parmotrema nilgherrense* and *Ramalina conduplicans* showed the least significant activity with the value of 6.0 mm against *Candida albicans.* The methanolic extract of *Parmotrema tinctorum* showed potential anti-bacterial activity at 8.33 ± 0.58 mm against *Candida albicans.* A better pathogenic control was observed in acetone extracts of *P. nilgherrense* with a 9.8 ± 0.28 mm zone against *Staphylococcus aureus* (Table 8.1; Figure 8.3).

TABLE 8.1
Anti-microbial Assay of Methanolic Extracts

Sl. No.	Phytochemical Tests	*Parmotrema austrosinense*	*Parmotrema nilgherrense*	*Parmotrema reticulatum*	*Parmotrema tinctorum*	*Ramalina conduplicans*	*Usnea stigmatoides*	Standard
		250 mg	**250 mg**	**250 mg**	**250 mg**	**250 mg**	**250 mg**	**30 mg/mL**
1.	*Candida albicans*	8.0 ± 0.58	6.0 ± 0.58	6.17 ± 0.28	8.33 ± 0.58	6.00 ± 0.58	ND	15.00 ± 0.58
2.	*Escherichia coli*	7.00 ± 0.58	1.70 ± 2.88	6.06 ± 0.120	6.0 ± 0.58	1.83 ± 3.17	ND	20.33 ± 0.58
3.	*Kelbsiella pnemuoniae*	4.00 ± 3.45	2.00 ± 3.47	1.70 ± 2.88	1.83 ± 3.17	1.73 ± 3.00	ND	6.66 ± 0.58
4.	*Proteus vulgaris*	2.0 ± 3.45	1.83 ± 3.17	2.00 ± 3.46	3.7 ± 3.17	1.67 ± 2.89	ND	15.33 ± 0.58
5.	*Pseudomonas aeruginosa*	8.0 ± 0.58	7.33 ± 0.58	4.00 ± 3.45	7.5 ± 0.50	1.67 ± 2.89	ND	17.33 ± 0.58
6.	*Salmonella typhii*	7.3 ± 0.58	2.0 ± 3.45	6.67 ± 0.58	7.0 ± 1.0	1.83 ± 3.17	ND	19.80 ± 0.06
7.	*Staphylococcus aureus*	8.33 ± 0.58	9.8 ± 0.28	6.83 ± 0.28	6.5 ± 0.50	7.00 ± 0.58	ND	20.67 ± 0.58

Note: The value represented shows mean ± standard error of mean, where n = 3.

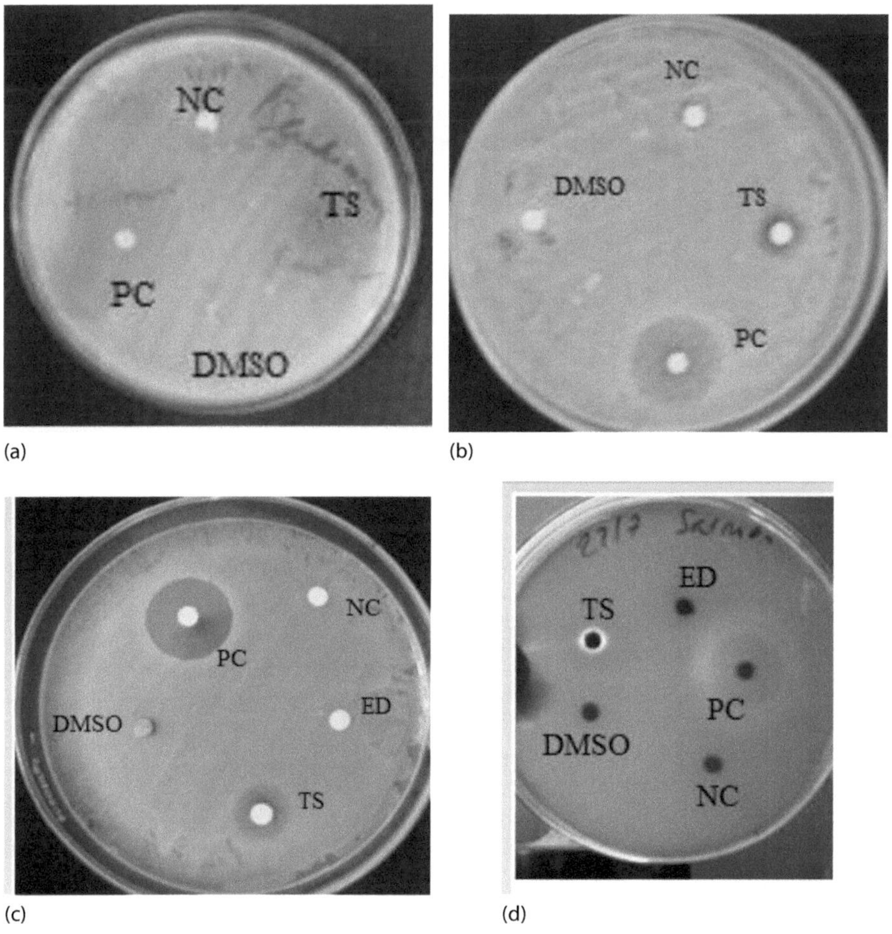

FIGURE 8.3 Anti-microbial activity of lichen extracts. (a) *Parmotrema austrosinense*; methanolic extract *against Pseudomonas aeruginosa*; (b) *Ramalina conduplicans methanolic extract against Candida albicans—Salmonella typhii*; (c) *Parmotrema reticulatum* methanolic extract against *Staphylococcus aureus*; (d) *Parmotrema reticulatum* methanolic extract against *Salmonella typhii*;

(*Continued*)

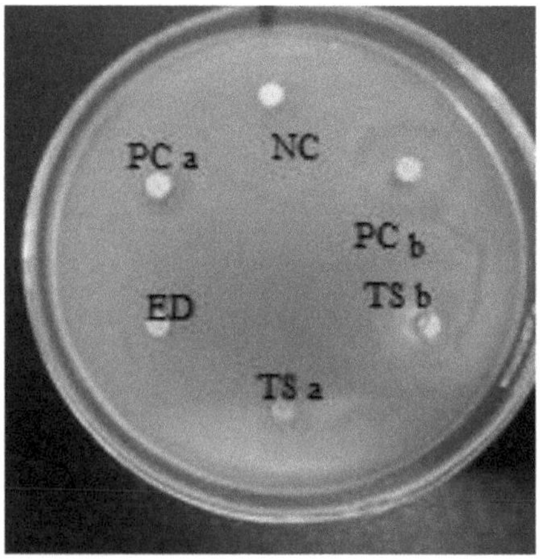

(e)

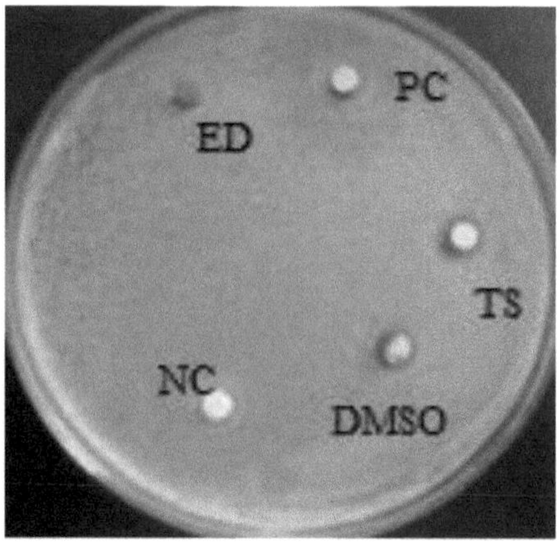

(f)

FIGURE 8.3 (Continued) (e) *Parmotrema nilgherrense* methanolic extract against *E. coli*; and (f) *P. tinctorum* against *Candida albicans*. (NC)—negative control—Disc with water; (PC) positive control with corresponding antibiotic (PC a)—positive control—with antibiotic 10 μL; (PC b) positive control with antibiotic 20 μL; (TS) test sample; (TS a) test sample lichen extract with 10 μL; (TS b) test sample containing the lichen extract with the 20 μL; and (ED) empty sterile disc.

TABLE 8.2
Minimum Inhibitory Concentration (μg/mL) of Lichen Extracts

Sl. No.	Clinical Test Strains	*Parmotrema austrosinense*	*Parmotrema nilgherrense*	*Parmotrema reticulatum*	*Parmotrema tinctorum*	*Ramalina conduplicans*	*Usnea stigmatoides*
1.	*Candida albicans*	500	500	500	500	500	ND
2.	*Escherichia coli*	500	250	1000	1000	500	ND
3.	*Klebsiella pneumoniae*	1000	2000	2000	1000	500	ND
4.	*Proteus vulgaris*	1000	1000	2000	500	500	ND
5.	*Pseudomonas aeruginosa*	500	1000	2000	500	500	ND
6.	*Salmonella typhii*	500	1000	2000	500	500	ND
7.	*Staphylococcus aureus*	500	1000	2000	500	500	ND

Note: ND represents not detected.

8.3.2 Determination of MIC

The MIC values of the lichen extract against a selected pathogen showed the concentration ranges between 2000 and 250 μg/mL. The results are presented in Table 8.2. The greater the microbial load, the faster the disappearance of resazurin color was observed. The purple color indicated the inhibitory activity of the lichen compound (Figure 8.4). The lichen extract of *Parmotrema austrosinense* showed significant activity against all tested pathogens.

8.3.3 Determination of Anti-oxidant

The methanolic extracts of the lichen samples were used to investigate the anti-oxidant property using DPPH. Lower absorbance is an indication of higher DPPH scavenging activity. The methanolic extract of *Usnea stigmatoides* exhibited strong anti-oxidant activity with an IC_{50} of 50.2275. Half the inhibitory concentration the IC_{50} value of the lichen extract was close to standard ascorbic acid (Figure 8.5).

8.3.4 Determination of Total Phenolic Content

The methanolic extracts of the lichen samples were assessed for total phenolic content using the Folin-Ciocalteu method. The highest phenolic content was found in *P. nilgherrense*, *P. reticulatum*, and *Usnea stigmatoides* with the concentrations of 38.49, 32.80, and 32.088 mg gallic acid equivalent/g, respectively (Figure 8.6).

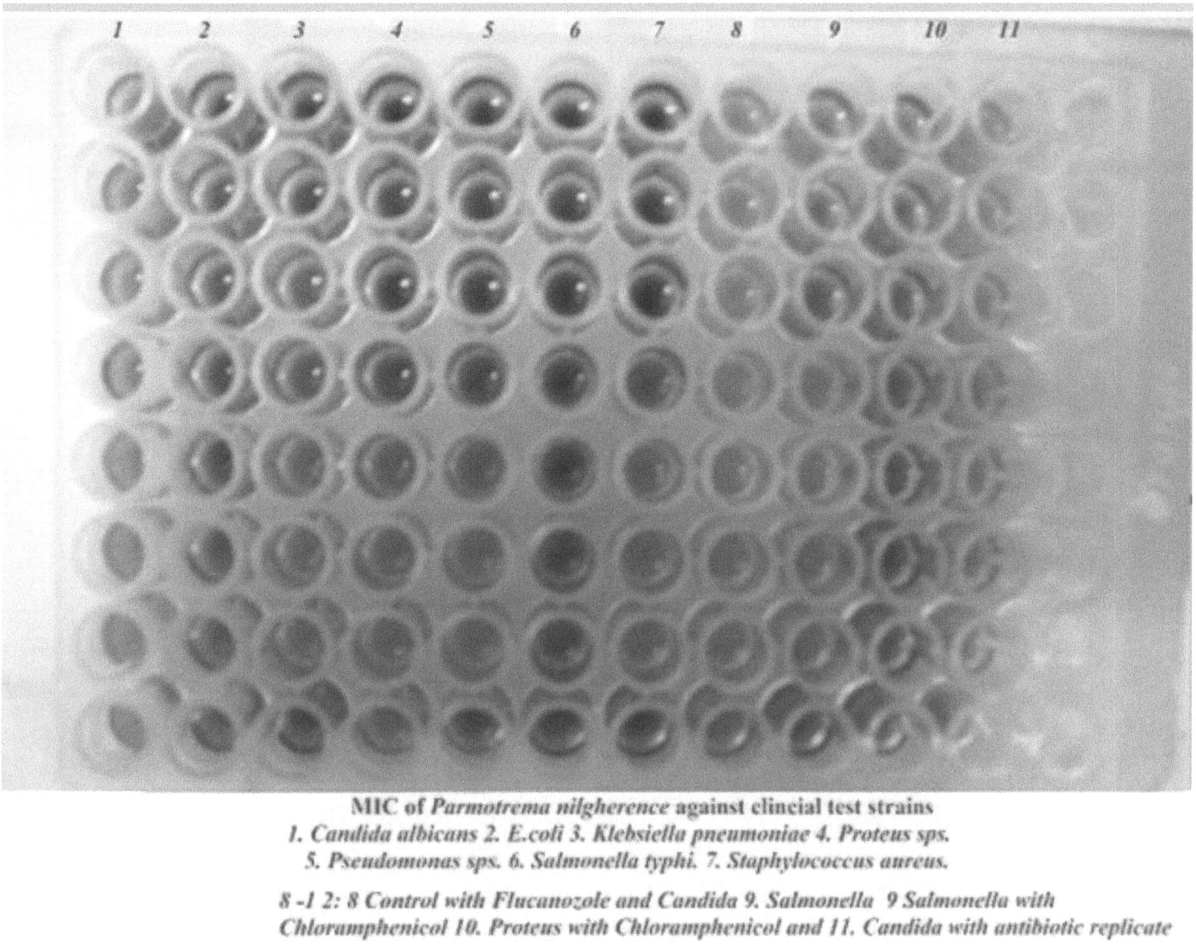

FIGURE 8.4 MIC of *Parmotrema nilgherrense* against clinical test strains.

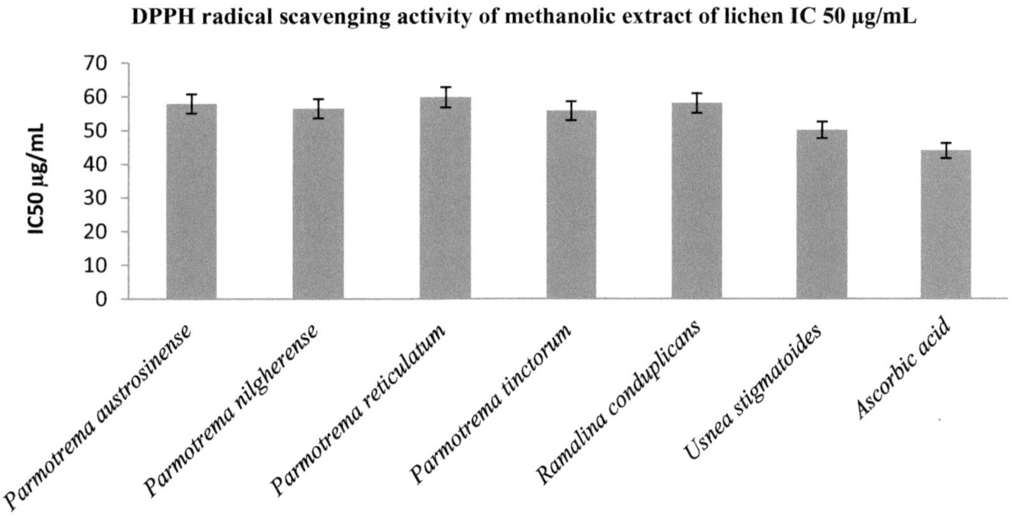

FIGURE 8.5 Anti-oxidant activity—DPPH radical scavenging activity.

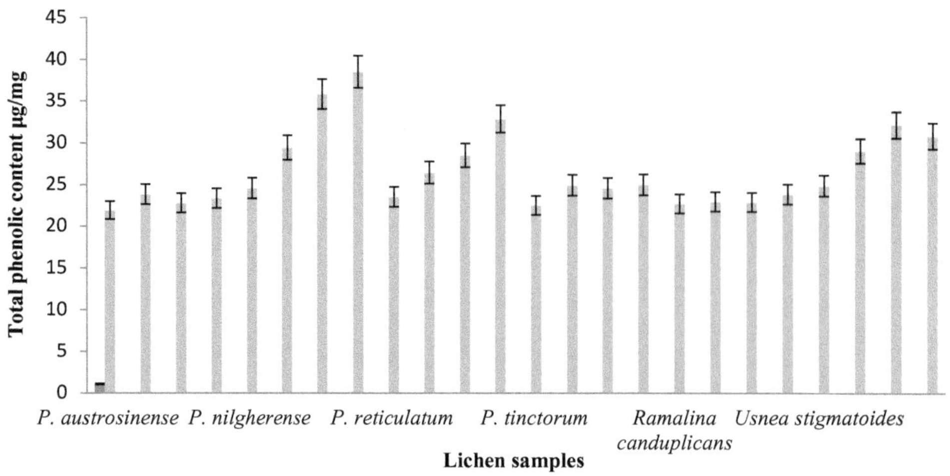

FIGURE 8.6 Total phenolic content of lichen samples.

8.4 DISCUSSION

The results of anti-microbial and anti-oxidant assays are quite eye opening. The results are in accordance with the findings of Burkholder et al. (1944) and Silva et al. (1986), who supported that the lichens inhibit mostly Gram-positive bacteria. On the contrary, the petroleum ether extracts did not show inhibition activity against human pathogens. The methanolic extracts of *Ramalina conduplicans* of fruticose lichen exhibited better anti-bacterial activity against *Staphylococcus aureus*. The extracts of *Parmotrema grayanam* and *P. tinctorum* exhibited strong anti-bacterial activity against *Staphylococcus aureus* (Poornima et al. 2017), with the inhibition values of 10.83 $\pm$ 1.26 mm and 10.23 $\pm$ 0.75 mm, respectively. Variations in the results are probably due to the following reasons, such as the extracts of lichen and concentrations of the extracts used for the test.

As far as the MIC is concerned, the work conducted by Srivastava et al. (2013) reported that the MIC values of the extract against bacterial strains varied between 25 and 3.125 µg/mL in the case of Gram-positive bacteria. Our results are not in accordance with the findings of Srivastava et al. (2013). The high MIC value in our present study was evidently due to a high accumulation of a lichen compound in the acetone extract and a low in the methanolic extract as noticed in a TLC chromatographic profile (Figure 8.2).

The in vitro radical scavenging activity conducted by Poornima et al. (2017) showed that the concentration of a lichen extract at 50% inhibition in a methanol extract of *P. austrosinense* had exhibited significant DPPH scavenging activity with the IC_{50} value of 26.39 µg/mL. Singleton et al. (1999) reasoned that the number of phenolic groups in the native structure of the compound is responsible for its anti-oxidant activity.

The amount of phenolic content in the methanolic extract of our present study showed that the value ranges between 20 and 40 mg GA/g in the methanolic extract (Figure 8.5). Ganesan et al. (2015) reported that the benzene extract of lichen exhibited high phenolic content with the value of 154.2 mg gallic acid equivalent/g of the lichen extracts. Therefore, this study has indicated that the particular solvent extract could affect the biological activity of a compound to a great extent.

8.5 CONCLUSION

A bioprospection study was carried out in four species of foliose and two fruticose lichens. With an exception of *Usnea stigmatoides*, the methanolic extract of all lichens exhibited moderate anti-bacterial activity against clinical pathogens. The *Parmotrema nilgherrense* lichen extract revealed higher phenolic content. The methanolic extract of *Usnea stigmatoides* showed good anti-oxidant properties. The lichen substances identified in the present study were lecanoric acid, salazinic acid, and usnic acid. Drug resistance of all known microorganisms to all existing anti-microbial agents is fully established. It is necessary to identify the new anti-microbial agents and to determine which of the potential compound is more effective against unidentified microorganisms. Therefore, this work will provide the baseline information for the future pharmacological studies.

ACKNOWLEDGMENTS

The authors wish to acknowledge DST FIST for the support. The authors are immensely grateful to Dr. K.S. Vinayaka, principal, Kumadvathi First Grade College, Shikaripura, Karnataka, and Dr. G. Ayyappadasan KSR College of Technology, Tiruchengode, Tamil Nadu, for their extensive support in lichen identification. The authors wish to extend their sincere gratitude to Dr. Sanjeeva Nayaka, National Botanical Research Institute, Lucknow, for his noble guidance in identification of the lichen. The authors express their gratitude to Dr. A. Rajendran, professor and head, Department of Botany, Bharathiar University, Coimbatore, for providing excellent help throughout this research work.

REFERENCES

Awasthi, D. D. 1991. A key to the microlichens of India, Nepal and Sri Lanka. *Lichenologica* 40: 112–337.

Awasthi, D. D. 2007. *Compendium of the Macrolichens from India, Nepal and Sri Lanka*. Bishen Singh Mahendra Pal Singh: Dehradun, 112–580.

Bauer, A. W., Perry, D. M., and Kirby, W. M. 1959. Single-disk antibiotic-sensitivity testing of staphylococci: An analysis of technique and results. *AMA Archives of Internal Medicine* 104(2): 208–216.

Burkholder, P. R., Evans, A. W., McVeigh, I., and Thornton, H. K. 1944. Antibiotic activity of lichens. *Proceedings of the National Academy of Sciences of the United States of America* 30(9): 250.

Culberson, C. F. 1969. *Chemical and Botanical Guide to Lichen Products*. University North Carolina Press: Chapel Hill, NC, 85–93.

Ganesan, A., Thangapandian, M., Ponnusamy, P., Sundararaj, J. P., and Nayaka, S. 2015. Antioxidant and antibacterial activity of parmeliod lichens from Shevaroy hills of Eastern Ghats, India. *International Journal of Pharm Tech Research* 8(9): 13–23.

Hirasawa, M., Shouji, N., Neta, T., Fukushima, K., and Takada, K. 1999. Three kinds of antibacterial substances from *Lentinus edodes* (Berk.) Sing. (Shiitake, an edible mushroom). *International Journal of Antimicrobial Agents* 11(2): 151–157.

Kirby, W. M., Yoshihara, G. M., Sundsted, K. S., and Warren, J. H. 1956. Clinical usefulness of a single disc method for antibiotic sensitivity testing. *Antibiotics Annual* 1: 892.

Orange, A., James, P. W., and White, F. J. 2001. *Microchemical Methods for the Identification of Lichens*. British Lichen Society, Natural History Museum: London, UK.

Poornima, S., Ponmurugan, P., Gnanamangai, B. M., Nagarjun, N., and Ayyappadasan, G. 2017. Evaluation of antimicrobial, antioxidant and anticancer activities of few macrolichens collected from Eastern Ghats of Tamil Nadu, India. *International Research Journal of Pharmacy* 8(3): 39–43.

Shrestha, G., El-Naggar, A. M., St. Clair, L. L., and O'Neill, K. L. 2015. Anticancer activities of selected species of North American lichen extracts. *Phytotherapy Research* 29(1): 100–107.

Silva, D. A., Oliveira, J., Maileite, J. E., Paulo, M. Q., and Filho, L. X. 1986. Antimicrobial activity of Brazilian lichens. *Biological Society Broteriana* 59: 87–96.

Singleton, V. L., Orthofer, R., and Lamuela-Raventós, R. M. 1999. Analysis of total phenols and other oxidation substrates and antioxidants by means of Folin-Ciocalteu reagent. *Methods in Enzymology* 299: 152–178.

Slinkard, K., and Singleton, V. L. 1977. Total phenol analysis: Automation and comparison with manual methods. *American Journal of Enology and Viticulture* 28(1): 49–55.

Srivastava, P., Upreti, D. K., Dhole, T. N., Srivastava, A. K., and Nayak, M. T. 2013. Antimicrobial property of extracts of Indian lichen against human pathogenic bacteria. *Interdisciplinary Perspectives on Infectious Diseases* 10: 1155–1161.

Stokes E. J., and Ridgway G. I. 1980. *Clinical Bacteriology*. 5th ed. Edward Arnold: London, 209–226.

Tanas, S., Odabasoglu, F., Halici, Z., Cakir, A., Aygun, H., Aslan, A., and Suleyman, H. 2010. Evaluation of anti-inflammatory and antioxidant activities of *Peltigera rufescens* lichen species in acute and chronic inflammation models. *Journal of Natural Medicines* 64(1): 42.

Tibuhwa, D. D. 2008. Taxonomic studies on *Cantharellus* (Mushrooms) from the Miombo woodlands of Tanzania. PhD Thesis, University of Dar es Salaam.

Vartia, K. O. 1973. Antibiotics in lichens. In *The Lichens*. Academic Press, 547–561.

Walker, F. J., and James, P. W. 1980. *A Revised Guide to Microchemical Techniques for the Identification of Lichen Products*. British Lichen Society.

9 Preliminary Phytochemical Screening and Identification of Bioactive Compounds from *Banana Inflorescence* and to Find the Interactions on Molecular Docking for PCOS

M. C. Kamaraj, Suman Thamburaj, R. Akshaya, and V. Bhanu Deepthi

CONTENTS

9.1 INTRODUCTION

Polycystic ovarian syndrome (PCOS) is a common endocrine disorder affecting 5%–10% of women in the reproductive age group. According to the survey of diseases across the world, PCOS was found to affect most of the younger women (Azziz et al. 2005). PCOS is a syndrome characterized by a hormonal imbalance resulting in enlarged ovaries with a large number of cysts filled with fluid. PCOS manifests due to inappropriate secretory levels of the follicle stimulating hormone and luteinizing hormone (LH). If left untreated, it may result in infertility, which is troublesome. The menstrual cycle depends on hormone regulation. During this cycle, the follicle stimulating hormone is produced from the pituitary gland in the brain, which is responsible for the production of the watery sac which surrounds the egg. After the formation of the egg, the pituitary gland secretes the LH, which induces the follicle to release the egg that settles in the fallopian tube for ovulation. Upon releasing the egg, the follicle slowly disappears. After ovulation, the ovaries produce estrogen and progesterone, which make the uterus stronger for pregnancy. At this time, the ovary secretes a small amount of androgen, which slowly gets converted into estrogen. In the patient with PCOS, the follicle fails to mature and the ovulation does not occur. Hence, the follicle remains and forms into a cyst which settles in the ovary. This cyst may result in a hormonal imbalance that might contain elevated levels of testosterone, which may cause abnormal hair growth in the face and body. The imbalance of hormonal level leads to infertility, hirsutism (Ramanand et al. 2013). The various signs of PCOS start at an early age, when a girl is maturing into a young woman. During this pubertal transition, several features may be in evolution, and many findings may be transitory, which stabilize later during adolescence. It is important to make an early diagnosis in order to prevent early and late growth of the syndrome. The symptoms of PCOS include diabetes mellitus, irregular menstrual bleeding, increased levels of male hormones, male pattern baldness or thinning hair, excess facial and body hair growth, acne and abnormal BMI, oily skin or dandruff, dark colored patches of skin specially on neck, groin, underarms, chronic pelvic pain, increased weight or obesity, lipid abnormalities, and high blood pressure. In this study, we have examined the effect of banana inflorescence on PCOS. Banana inflorescence of *Musa acuminata* from the genus *Musa*, which belongs to the family (Musaceae), is studied. It is also referred as "monthan" in Tamil. Various species of *Musa* are reported to possess diverse medicinal properties. The banana inflorescence has been reported to control excessive bleeding and infertility. It is also being traditionally used as part of the diet to stop excessive bleeding and painful menstruation. However, there is no scientific evidence on its action on excessive bleeding, painful menstruation, and infertility to date. Understanding the science behind the association of banana inflorescence over infertility will be very useful, as it can be used for treating bleeding disorders. Disorders of menstruation can lead to an increased risk of infertility. The World Health Organization reported that about 80% of people use traditional medicine for primary health care. In India, banana inflorescence is used to treat heart pain, asthma, and diabetes. The use of banana inflorescence also helps to treat diarrhea and stomach cramps. For women, it helps in reducing painful menstruation and menopausal bleeding. Banana inflorescence contains many nutrients and vitamins that are also used for infantile malnutrition. Therefore, this study was conducted to scientifically validate the effect of *Musa acuminata* on PCOS using *Yuvati* as a dietary supplement. We have prepared *Yuvati* by extracting the florets of the inflorescence into a capsule. This study includes phytochemical screening and identification of the bioactive compounds of *Yuvati* using molecular docking studies to understand its interactions on PCOS.

9.2 MATERIALS AND METHODS

The florets of a banana were collected from agricultural lands in and around Papanasam, Thanjavur District, Tamil Nadu, India. The florets were cut into small pieces and soaked with water. After drying, the aqueous extracts were ground into a fine powder using a dry grinder and filled into capsule.

9.2.1 Preliminary Phytochemical Screening

9.2.1.1 Qualitative and Quantitative Determination of Phytoconstituents of *Banana Inflorescence*

The detection of alkaloids was carried out by Dragendorff's test according to the method previously described by Waldi (1965), while phenol, flavonoids, tannins, terpenoids, glycosides, saponin, quinine, and carbohydrates were detected by Evans (1989). The banana florets' extract was also subjected to quantitative determination of phenol and flavonoids. The total phenol content was determined by the methodology previously described by Singleton and Rossi (1965). Flavonoids were estimated according to Jain et al. (2010), using quercitin as the reference standard.

9.2.1.2 Test for Alkaloids

Dragendorff's test: One milliliter of the banana florets' extract was added and 2–3 drops of Dragendorff's reagent was added. A yellow color precipitate was not formed. This indicates the absence of alkaloids.

9.2.1.3 Test for Phenols

Lead acetate test: One milliliter of the banana florets' extract was dissolved in 3 mL of water and 10% lead acetate solution was added. A bulky white precipitate was formed. This indicates the presence of phenols.

9.2.1.4 Test for Flavonoids

Alkaline reagent test: Two milliliters of banana florets' extract were taken and a few drops of ammonium hydroxide were added. The formation of yellow fluorescence indicated the presence of flavonoids.

9.2.1.5 Test for Tannins

Lead acetate test: One milliliter of banana florets' extract was taken and a few drops of 1% solution of lead acetate was added. The formation of white color precipitate was observed. This indicates the presence of tannins.

9.2.1.6 Test for Saponins

Foam test: One milliliter of banana florets' extract was diluted with 2 mL of water and agitated for 15 minutes. A 1 cm layer of foam was observed. This indicates the presence of saponins.

9.2.1.7 Test for Carbohydrates

Fehling's test: An equal volume of Fehling's solution A (copper sulphate in distilled water) and Fehling's solution B (potassium tartarate and sodium hydroxide in distilled water) reagents were mixed with a few drops of extract and boiled, a brick red precipitate was formed. This indicates the presence of carbohydrates.

- **Fehling's solution**
  - **Solution A:** Copper sulphate (34.66 g) was dissolved in distilled water and made up to 500 mL distilled water.
  - **Solution B:** Potassium sodium tartarate (173 g) and sodium hydroxide (50 g) were dissolved in water and made up to 500 mL.

9.2.1.8 Test for Glycosides

One milliliter of banana florets' extract was dissolved in 1 mL of water and a few drops of aqueous sodium hydroxide were added, and a yellow color was formed. This indicates the presence of glycosides.

9.2.1.9 Test for Steroids

Salkowski's test: One milliliter of banana florets' extract was dissolved in 2 mL of chloroform and a few drops of sulphuric acid was added carefully. A reddish brown color ring was formed in the interface. This indicates the presence of steroids.

9.2.2 Determination of Total Phenolic Content

The total phenols were estimated by the method of Singleton and Rossi (1965).

Phenols react with phosphomolybdic acid in a Folin-Ciocalteu reagent in an alkaline medium and produce a blue colored complex (molybdenum blue).

Reagents Required

- Folin-Ciocalteu reagent
- Na_2CO_3–20%
- Standard: Gallic acid, 100 mg, was dissolved in 100 mL of water. The working standard was prepared by diluting 1–10 mL.

Procedure: The amount of phenol in the test sample of banana florets' extract was determined with a Folin-Ciocalteu reagent using this method. Two hundred microliters of the diluted sample was added to 1 mL of 1:10 diluted Folin-Ciocalteu reagent. After 4 minutes, 800 μL of saturated sodium carbonate (75 g/L) was added. After 2 hours of incubation at room temperature, the absorbance at 750 nm was measured. Gallic acid (0–500 mg/L) was used as the standard calibration curve. The absorbance was read at 750 nm spectrometrically. The sample was analyzed in triplicate. The Folin-Ciocalteu reagent being sensitive to reducing compounds including polyphenol, produced a blue color upon reaction which was measured spectrophotometrically.

9.2.3 Determination of Total Flavonoid Content

The total flavonoid content was measured by Jain et al. (2010).

Flavonoid reacts with aluminum chloride to produce a colored product, which was measured spectrophotometrically.

Reagents Required

- 2% methanolic aluminum chloride
- Quircetin standard (100 mg/mL)

Procedure: The aluminum chloride method was used for the determination of the total flavonoid content of the sample. One milliliter of banana florets' extract was mixed with 1 mL of 2% methanolic aluminum chloride. The mixer was shaking well, kept at room temperature for 10 minutes, and absorbance was measured at 450 nm. A blank was prepared with water instead of the sample. The results of the flavonoids are expressed in terms of quercetin in mg/mL of the extract. The total flavonoid content was determined as milligrams of the quercetin equivalent per gram using the equation obtained from a calibration curve.

9.2.4 Molecular Docking

A computational tool offers the advantage of delivering new drug candidates more quickly and at a lower cost. The present work used a computational approach for the following software manipulation of drugs, using the molecular docking server online web service for the calculation of drug likeness. The identified compounds from the *banana inflorescence* were used to interact with the polycystic ovarian syndrome protein (CYP17A) retrieved from the Protein Data Bank (PDB).

9.2.4.1 Protein Data Bank

The PDB is the database that contains information about the 3D structure of biomacromolecules and their complexes as determined by X-ray crystallography, NMR spectroscopy, and includes more than a few Nobel Prize winning structures. Polycystic ovarian syndrome protein (CYP17A) was retrieved from the protein data bank with the specific resolution, and the PDB ID is 3RUK.

9.2.4.2 Protein Sequences

The polycystic ovarian syndrome protein sequence was retrieved from the online database of SWISSPROT (Arnold et al. 2006). The input keyword of CYP17A was used to search in the entire database. A number of CYP17A sequences were shown. Among which, human CYP17A was accessed in a FASTA format, and it was used for the further computational analysis.

9.2.4.3 Primary Structure Analysis

The ProtParam (http://us.expasy.org/tools/protparam.html) prediction server was used to determine the primary structure and their physiochemical parameters, such as amino acid composition, theoretical isoelectric point (PI), molecular weight, total number of positive and negative residues, extinction coefficient (Gill and Von Hippel 1989), half-life (Bachmair et al. 1986; Gonad et al. 1989; Tobias et al. 1991; Ciechanover and Schwartz 1989), instability index (Ikai 1980), and grand average hydropathy (GRAVY) (Kyte and Doolittle 1982).

9.2.4.4 Secondary Structure Prediction

The secondary structure of this protein was predicted by using the FASTA sequences of the protease and predicted using SOPMA and SOPM (Combet et al. 2000).

9.2.4.5 Transmembrane Region Identification

The identification of the transmembrane region of CYP17A was performed by using SOSUI (Hirokawa et al. 1998). The evaluated transmembrane region was analyzed and visualized by pepwheel using the EMBOSS 2.7 suite (Ramachandran and Sasisekharan 1968).

9.2.4.6 Homology Modeling and Validation

The protein sequence was subjected for comparative homology modeling via Swiss-model (Kiefer et al. 2008) and evaluated by the RAMPAGE online server (Lovell et al. 2003). The protein was validated by using the online server PROCHECK (Laskowski et al. 1996). The Swiss-model performs the sequence alignments and searches for the putative template protein for generating the 3D model.

9.2.4.7 Sequence Subjected for Modeling

>target
MWELVALLLLTLAYLFWPKRRCPGAKYPKSLLSLPLV
GSLPFLPRHGHMHNNFFKLQKKYGPIYSVRMGTKTT
VIVGHHQLAKEVLIKKGKDFSGRPQMATLDIASNNRK
GIAFADSGAHWQLHRRLAMATFALFKDGDQKLEKIIC
QEISTLCDMLATHNGQSIDISFPVFVAVTNVISLICFNT
SYKNGDPELNVIQNYNEGIIDNLSKDSLVDLVPWLKIF
PNKTLEKLKSHVKIRNDLLNKILENYKEKFRSDSITNM
LDTLMQAKMNSDNGNAGPDQDSELLSDNHILTTIGDI
FGAGVETTTSVVKWTLAFLLHNPQVKKKLYEEIDQN
VGFSRTPTISDRNRLLLLEATIREVLRLRPVAPMLIPH
KANVDSSIGEFAVDKGTEVIINLWALHHNEKEWHQP
DQFMPERFLNPAGTQLISPSVSYLPFGAGPRSCIGEILA
RQELFLIMAWLLQRFDLEVPDDGQLPSLEGIPKVVFL
IDSFKVKIKVRQAWREAQAEGST

9.2.4.8 GC-MS Analysis

Banana florets' extract was subjected to GC-MS analysis, and the distinguished compounds were identified.

9.2.4.9 Ligand Retrieved

The identified compounds were retrieved from the PubChem compound database and were used for further studies.

9.2.4.10 Receptor Retrieved

The receptor of the CPY17A protein was downloaded from the PDB, and the PDB ID is 3RUK.

9.2.4.11 Computational Molecular Docking Studies

The X-ray crystallographic structures of the protein 3RUK were retrieved from the RCSB database with PDB ID 3RUK. Computational analysis was done to compute the ligand protein binding affinity of the compound. Docking calculations were carried out using DockingServer (Bikadi and Hazai 2009). The MMFF94 force field was used for the energy minimization of the ligand using DockingServer (Halgren 1996). Gasteiger partial charges were added to the ligand atoms. The non-polar hydrogen atoms were merged and the rotatable bonds were defined. The docking calculations were carried out on a rennin-angiotensin II protein model. The essential hydrogen atoms, Kollman united atom type charges, and salvation parameters were added with the aid of AutoDock tools. Affinity (grid) maps of 20 × 20 × 20 Å grid points and 0.375 Å spacing were generated using the AutoGrid program (Morris et al. 1998). AutoDock parameters set and distance—dependent dielectric functions were used in the calculation of the van der Waals and the electrostatic terms, respectively. Docking simulations were performed using the Lamarckian genetic algorithm and the Solis and Wets local search method (Solis and Wets 1981). The initial position, orientation, and torsions of the ligand molecules were set randomly. Each docking experiment was derived from ten different runs that were set to terminate after a maximum of 250,000 energy evaluations. The population size was set to 150. During the search, a translational step of 0.2 Å and quaternion and torsion steps of 5 were applied (Kuldeep and Satpal 2013).

9.3 RESULTS AND DISCUSSION

9.3.1 Qualitative Analysis of *Banana Inflorescence*

The banana inflorescence is being used as traditional medicine not only in different parts of India, but also throughout the world for time immemorial. The results of the phytochemical screening of the banana florets' extracts showed the presence of various phytoconstituents like phenols, flavonoids, carbohydrates, tannins, saponins, glycosides, and steroids as shown in Table 9.1. The determination of total phenolic

TABLE 9.1
Qualitative Phytochemical Screening of Banana Florets' Extract

Qualitative Phytochemical Analysis	
Phytoconstituents	**Present (+)/absent (−)**
Alkaloids	−
Phenols	+
Flavonoids	+
Tannins	+
Saponins	+
Carbohydrates	+
Glycosides	+
Steroids	+

content was done by the Folin-Ciocalteu method. The results showed that the banana florets' extract contains a polyphenolic compound (9.1% ± 0.36%). The total flavonoid content was estimated by the aluminum chloride method. The flavonoid content was found to be high in the banana florets' extract (5.48% ± 0.32%).

9.3.2 Primary Structure Analysis

The human gene has the name CYP17A (P05093), the term polycystic ovarian syndrome was used as the protein, and it was retrieved from the SWISSPROT in the FASTA format. The primary physiochemical parameters were performed, and the amino acid composition was identified, and the details are given in Table 9.2. The results represented that the CYP17A protein is composed of 22 amino acids with different ratios. Among them, the leucine content was found to be more (12.6%), and that indicated the hydrophobic nature of the protein because it has an aliphatic isobutyl side chain and also essential amino acids. The molecular weight of the protein was found to be 57,370.52, and the protein has 8.72 isoelectric points that represent the protein is alkaline in nature. The extinction coefficient was 63,160 at 280 nm; it might be possible to avoid the interference of other substances. The evaluated value was used to determine the quantification of protein-protein or protein-ligand interactions. The quantitative measurement of the dynamic equilibrium was based on the half-life time. The CYP17A has 30 hours in mammalian reticulocytes, in yeast, it has 20, and 10 hours in *E. coli*, respectively. The stability of the protein was determined by using the instability index (33.18). The aliphatic index represented that the volume of the protein occupied by aliphatic chains (alanine, valine, isoleucine, and leucine), CYP17A, had a value of 99.80 that makes it more stable in high thermal conditions. Grand average hydropathicity denoted the hydrophobicity of the amino acid residues. Here, CYP17A with a value of −0.154 had a reasonable interaction with a water molecule. The protein molecule has four different atoms, such as C, H, N, O, and S, with the molecular formula of $C_{2598}H_{4116}N_{698}O_{731}S_{17}$.

TABLE 9.2
Amino Acid Composition (%) of CYP17A Computed in ProtParam

Amino Acids	Numbers	Percentage (%)
Ala (A)	30	5.9
Arg (R)	22	4.3
Asn (N)	28	5.5
Asp (D)	28	5.5
Cys (C)	5	1.0
Gln (Q)	20	3.9
Glu (E)	26	5.1
Gly (G)	29	5.7
His (H)	15	3.0
Ile (I)	36	7.1
Leu (L)	64	12.6
Lys (K)	37	7.3
Met (M)	12	2.4
Phe (F)	24	4.7
Pro (P)	27	5.3
Ser (S)	32	6.3
Thr (T)	25	4.9
Trp (W)	9	1.8
Tyr (Y)	9	1.8
Val (V)	30	5.9
Pyl (O)	0	0.0
Sec (U)	0	0.0

9.3.3 Secondary Structure of Protein

The secondary structure of CYP17A (Table 9.3) was predicted using SOPMA and SOPM (Table 9.4). The protein was α helix with other structures, such as an extended stand, β turn, and random coil present in the comparative analysis of SOPMA and SOPM. From which it is clear, that the random coil is predominantly present when the structure was predicted both by SOPMA and SOPM, followed by an extended strand and alpha helix. So this protein is stable in nature.

Mol. wt.—Molecular Weight; PI—Isoelectric point; −R—number of negatively charged residues; +R—number of positively charged residues; EC—Extinction coefficient at 280 nm; II—Instability Index; AI—Aliphatic Index; GRAVY—grand average of hydropathicity.

9.3.4 Protein Structure Validation

9.3.4.1 Ramachandran Plot

The predicted protein structure was validated by using the Ramachandran plot using PROCHECK software that shows the protein molecule contains 1874 in that 1425 amino acid most favored region, 231 amino acid additionally allowed region, and 8 generally allowed and 0 amino acid disallowed. The results are shown in Figure 9.1.

9.3.4.2 Docking

Polycystic ovarian syndrome (Figure 9.2) and ligands (isolated compounds of banana inflorescence, Clomifene, and

TABLE 9.3
Parameters Computed Using ExPASy's ProtParam Tool

Name	Accession Number	Seq. Length	Mol. Wt	PI	−R	+R	EC	II	AI	GRAVY
CYP17A	P05093	508	57370.52	8.72	54	59	63160	33.18	99.80	−0.154

TABLE 9.4
Secondary Structure of CYP17A Protein by SOPMA and SOPM

Secondary Structure	SOPMA	SOPM
Alpha helix (Hh)	227 is 44.69%	205 is 40.35%
310 helix (Gg)	0 is 0.00%	0 is 0.00%
Pi helix (Ii)	0 is 0.00%	0 is 0.00%
Beta bridge (Bb)	0 is 0.00%	0 is 0.00%
Extended strand (Ee)	59 is 11.61%	85 is 16.73%
Beta turn (Tt)	21 is 4.13%	48 is 9.45%
Bend region (Ss)	0 is 0.00%	0 is 0.00%
Random coil (Cc)	201 is 39.57%	170 is 33.46%
Ambiguous states (?)	0 is 0.00%	0 is 0.00%
Other states	0 is 0.00%	0 is 0.00%

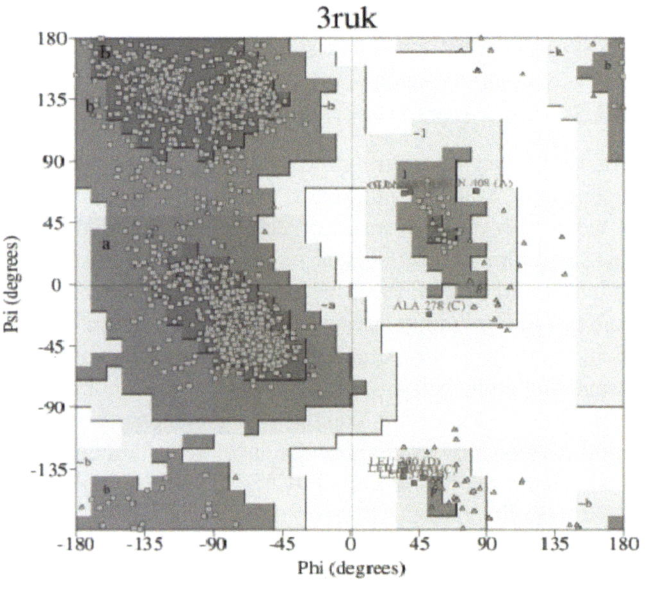

FIGURE 9.1 Ramachandran plot.

Tranexamic acid) (Tables 9.5 and 9.6) were subjected to docking studies using online AutoDock software. The software ran ten docking programs that are shown in Table 9.7. The 3D structure of the PCOS protein (PDB ID: 3RUK) was optimized to achieve minimal potential energy using the molecular docking server. The minimization values are summarized. The docking simulation of ten runs of plant compound

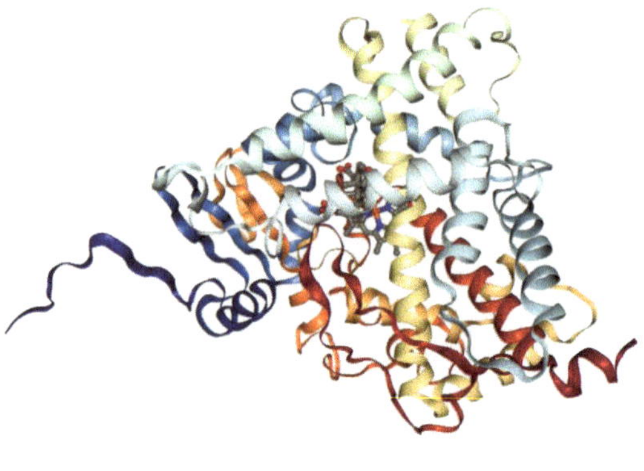

FIGURE 9.2 PCOS protein CYP17A.

vitamin E, gallocatechin, epicatechin, β-tocopherol, estragole, and phytol was performed for a set of catalytic active sites of the PCOS protein. The best docked conformation was selected based on the lowest docking energy and binding free energy. Other features present in the compound in this study, such as binding energy, interaction score and hydrogen bonding with residues, drug-like behavior of the compound, and binding capacity contribute to the docking score.

The size of the active site was determined by generating the AutoDock receptor. The most feasible orientation of the ligands in the binding pocket was predicted, and the strength of the interactions in the particular orientation was quantified from a scoring function. The AutoDock was preferred over the standard mode mainly to determine the better correlation between good poses and good scores. The docking analysis was performed between ligands and the target protein CYP17A using the AutoDock server, and the docked images are illustrated in Figure 9.3. The more negative the value on the scoring function indicates the better the docking, and the docking of the ligand and protein were ranked according to the corresponding free energy of binding. The scoring function of the AutoDock program was exhibited in the form of the docking score. The docking score indicates the binding ability of a ligand to the specific conformation of the receptor protein. The accuracy of a docking procedure is determined from the object scoring function, which was predicted even in the lowest energy pose (binding conformation). In this study, the validation of the AutoDock procedure was done by eliminating the inhibitor compound with the protein, CYP17A, that was evaluated from the binding energy, interaction score, and hydrogen bonding with residues. Six bioactive compounds were identified through ethanolic extract from banana florets' of *Musa*

TABLE 9.5
Plant Compounds Identified by GC-MS

S. No.	Name of the Compound	Mol. Formula	Mol. Weight g/mol	Mol. Structure
1.	β-tocopherol	$C_{28}H_{48}O_2$	416.68	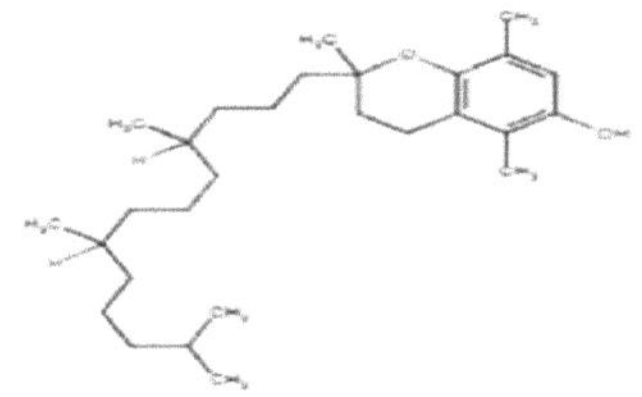
2.	Phytol	$C_{20}H_{40}O$	128.17	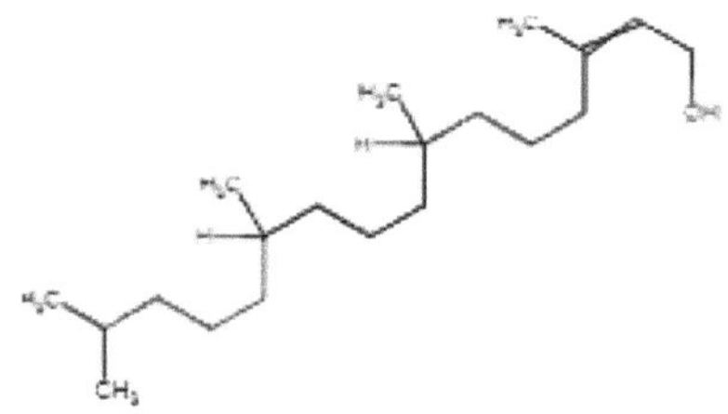
3.	Epicatechin	$C_{15}H_{14}O_6$	290.271	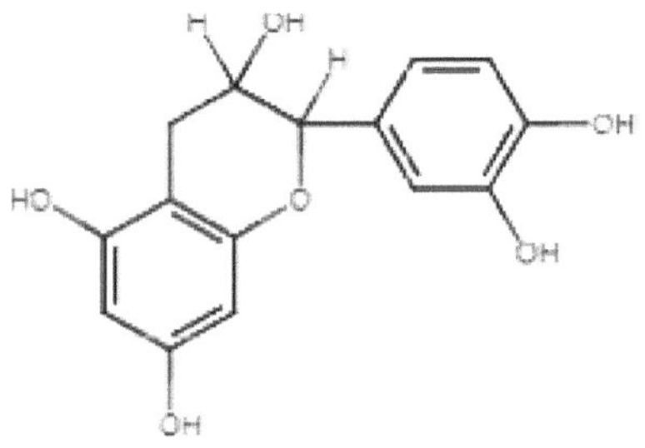
4.	Gallocatechin	$C_{15}H_{14}O_7$	306.267	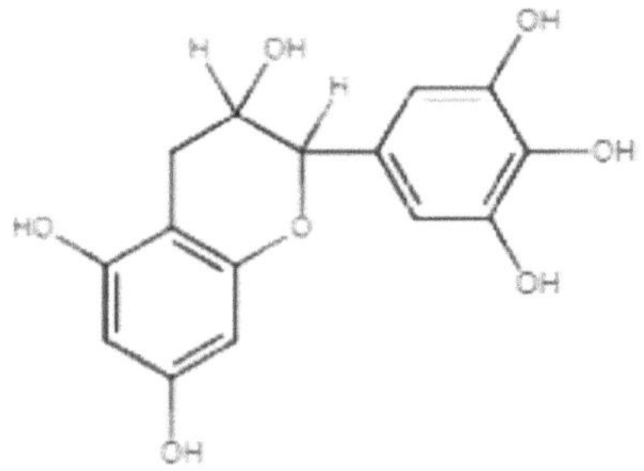
5.	Estragole	$C_{10}H_{12}O$	148.2	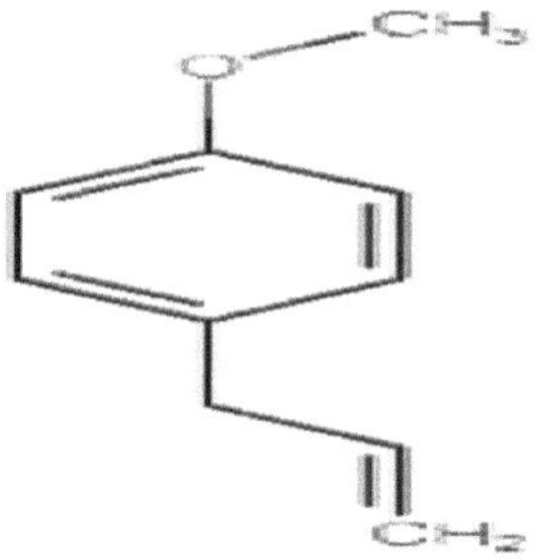
6.	Vitamin E	$C_{29}H_{50}O_2$	430.717	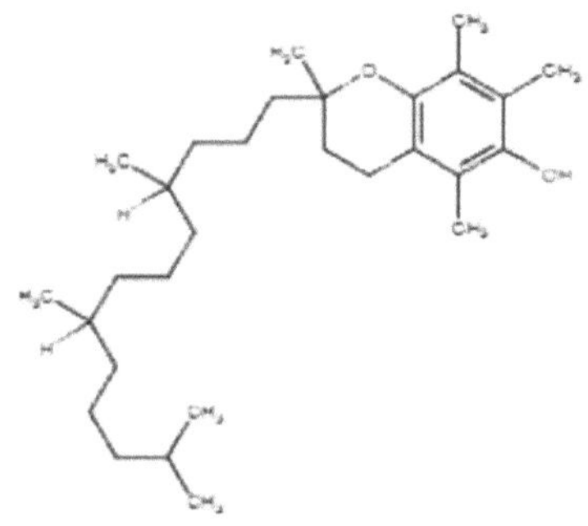

TABLE 9.6
Standard Drug

S. No.	Name	Molecular Formula	Molecular Weight	Molecular Structure
1.	Clomifene	$C_{26}H_{28}ClNO$	405.966	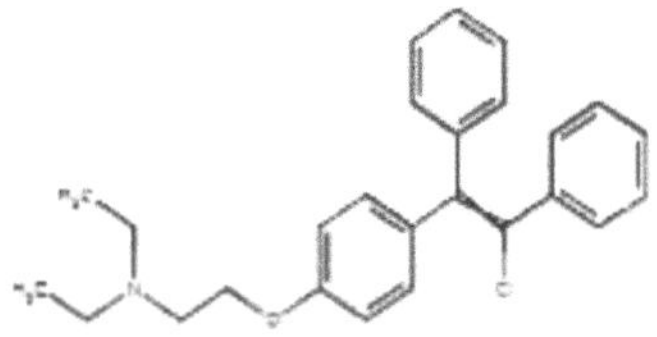
2.	Tranexamic acid	$C_8H_{15}NO_2$	157.21	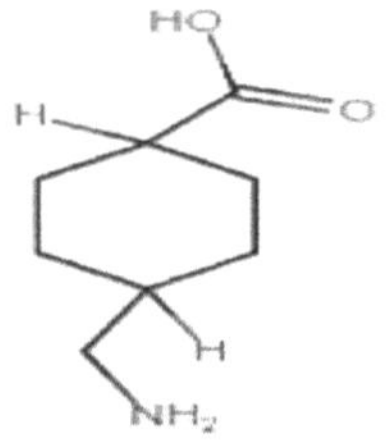

TABLE 9.7
Interacting Residues Responsible for Docking

Docking Result	Est. Free Energy of Binding	Est. Inhibition Constant, Ki	vdW + Hbond + desolv Energy	Electrostatic Energy	Total Intermolecs. Energy	Frequency (%)	Interact. Surface
Beta tocopherol with 3RUK	−8.36 kcal/mol	748.08 nM	−11.73 kcal/mol	−0.03 kcal/mol	−11.76 kcal/mol	40	781.487
Phytol with 3RUK	−6.90 kcal/mol	8.80 uM	−10.63 kcal/mol	−0.03 kcal/mol	−10.66 kcal/mol	30	588.665
Epicatechin with 3RUK	−6.44 kcal/mol	19.02 uM	−7.09 kcal/mol	−0.39	−7.48 kcal/mol	50	494.309
Gallocatechin with 3RUK	−5.36 kcal/mol	118.03 uM	−6.25 kcal/mol	−0.23 kcal/mol	−6.48 kcal/mol	100	509.658
Estragole with 3RUK	−4.91 kcal/mol	250.42 uM	−5.75 kcal/mol	−0.05 kcal/mol	−5.80 kcal/mol	100	376.952
Vitamin E	−4.61 kcal/mol	420.62 uM	−11.13 kcal/mol	−0.12 kcal/mol	−11.25 kcal/mol	40	727.251
Clomifene with 3RUK	−9.20 kcal/mol	180.80 nM	−10.73 kcal/mol	−0.29 kcal/mol	−11.01 kcal/mol	80	728.889
Tranexamic acid with 3RUK	−6.08 kcal/mol	35.18 uM	−6.57 kcal/mol	−0.39 kcal/mol	−6.96 kcal/mol	80	352.25

acumulata using GC-MS analysis. The molecular interaction and binding affinity of the ligand with protein, CYP17A, were studied by docking every ligand into the active site of the protein, and the docking results of the ligands were tabulated. For each minimized complex, the different interaction energies, such as van der Waals energy, intermolecular hydrogen bonding, and electrostatic energy were considered. The free energy of six ligands using the AutoDock were in the range of −4.6 to −8.3 against CYP17A. The free energy of the known fertility and excessive bleeding drugs, namely, Clomifene and Tranexamic acid docked against CYP17A were found to be −9.20 and −6.08 kcal/mol, respectively. Compounds from the *Musa acuminata*, such as β-tocopherol and phytol have a docking score of −8.36 and −6.90 kcal/mol, respectively. Hence, β-tocopherol and phytol from the *Musa acuminata* exhibit similar activity as that of the already established drug. β-tocopherol and phytol compounds have the highest binding ability upon docking against the crystal structure of the protein. The residues ASN 108 and SER 488 present in the protein CYP17A have a significant role in the receptor binding activity with ligands, which was revealed from the conformational analysis of different docked complexes. Docking studies using the AutoDock server have proved that the above inhibitors fit into the binding pocket of the protein CYP17A. In the results, it was observed that successful docking has been correlated with good intermolecular hydrogen bonding and liphophilic interactions between the ligand and the receptor. From this study, it can be inferred that the phytoconstituents that exist in the *Musa acumulata* could be promising alternatives for the development of polycystic ovarian syndrome.

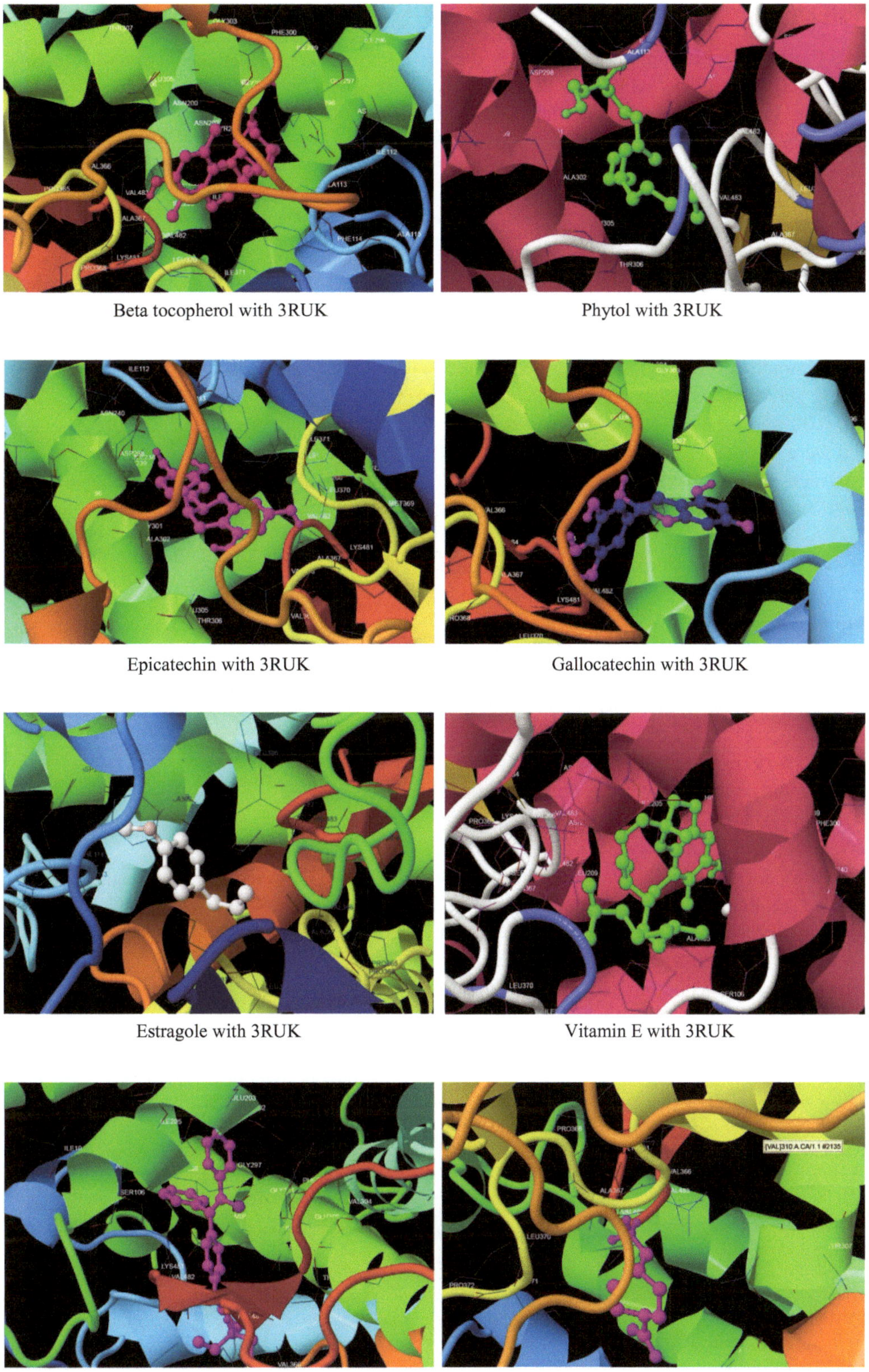

Beta tocopherol with 3RUK

Phytol with 3RUK

Epicatechin with 3RUK

Gallocatechin with 3RUK

Estragole with 3RUK

Vitamin E with 3RUK

Clomifene with 3RUK

Tranexamic acid with 3RUK

FIGURE 9.3 Docking results.

9.4 CONCLUSION

Phytochemicals are the core of phytomedicines, which therapeutic efficiency directly correlates with the presence of various phytochemicals. In this study, we have identified and quantified the major phytochemicals present in the banana inflorescence. The banana inflorescence (*Yuvati*) is being used as a traditional food supplement. The result of the phytochemical screening of the banana florets' extract shows that it exhibits as an ethnomedicine. We detected the presence of various secondary metabolites like phenols, flavonoids, tannins, saponins, carbohydrates, glycosides, and steroids. Alkaloids were absent in the extract. The qualitative phytochemical analysis of banana florets' extract was performed. The total phenol and flavonoid content of the extract were found to be 1.90 ± 0.02 and 0.665 ± 0.001 mg, respectively. The standards used were gallic acid and quercetin equivalent/mL of the flower extract. This showed that the banana florets' extract (*Musa acumulata*) contained more amounts of phenols and flavonoids.

In addition, *in silico* molecular docking studies revealed that the bioactive compounds from the banana inflorescence exhibit good interactions with the PCOS protein (CYP17A) namely, vitamin E, epicatechin, gallocatechin, β-tocopherol, estragole, and phytol. During drug discovery, the effective screening procedures should be applied to reduce the cost and time. In this study, we have utilized molecular docking by the *AutoDock server* to analyze the binding ability between six compounds with CYP17A. The compounds such as β-tocopherol, phytol, epicatechin, and gallocatechin were found to have good binding affinity with the protein by analyzing its AutoDock. Among these, *beta tocopherol* was found to be the lead compound with maximum binding affinity. This study shows the presence of active compounds from using the *Musa* florets' extract as a docking platform. Further, *in vitro* and *in vivo* studies are to be carried out in order to establish the effect of *Yuvati* on polycystic ovarian syndrome.

REFERENCES

Arnold, K., Bordoli, L., Kopp, J., and Schwede, T. 2006. The Swiss-model workspace: A web-based environment for protein structure homology modelling. *Bioinformatics* 22(2): 195–201.

Azziz, R., Marin, C., Hoq, L., Badamgarav, E., and Song, P. 2005. Health care-related economic burden of the polycystic ovary syndrome during the reproductive life span. *The Journal of Clinical Endocrinology and Metabolism* 90(8): 4650–4658.

Bachmair, A., Finley, D., and Varshavsky, A. 1986. In vivo half-life of a protein is a function of its amino-terminal residue. *Science* 234(4773): 179–186.

Bikadi, Z., and Hazai, E. 2009. Application of the PM6 semi-empirical method to modeling proteins enhances docking accuracy of AutoDock. *Journal of Cheminformatics* 1(1): 15.

Ciechanover, A., and Schwartz, A. L. 1989. How are substrates recognized by the ubiquitin-mediated proteolytic system. *Trends in Biochemical Sciences* 14(12): 483–488.

Combet, C., Blanchet, C., Geourjon, C., and Deleage, G. 2000. NPS@: Network protein sequence analysis. *Trends in Biochemical Sciences* 25(3): 147–150.

Evans, W. C. 1989. *Trease and Evan's Textbook of Pharmacognosy.* 13th ed., Cambridge University Press, London, p. 546.

Gill, S. C., and Von Hippel, P. H. 1989. Calculation of protein extinction coefficients from amino acid sequence data. *Analytical Biochemistry* 182(2): 319–326.

Gonad, D. K., Bachmir, A., Wuning, I., Tobias, J. W., Lane, W. S., and Varshavsky, A. J. 1989. University and structure of the N-end rule. *Journal of Biological Chemistry* 264: 16700–16712.

Halgren, T. A. 1996. Merck molecular force field. I. Basis, form, scope, parameterization, and performance of MMFF94. *Journal of Computational Chemistry* 17(5–6): 490–519.

Hirokawa, T., Boon-Chieng, S., and Mitaku, S. 1998. SOSUI: Classification and secondary structure prediction system for membrane proteins. *Bioinformatics (Oxford, England)* 14(4): 378–379.

Ikai, A. 1980. Thermostability and aliphatic index of globular proteins. *The Journal of Biochemistry* 88(6): 1895–1898.

Jain, A., Sharma, P., and Sunil, K. 2010. Evaluation of phenolic & flavonoid profile and screening of antioxidant activity of plant Croton sparsiflorus by bio-autographic method. *Journal of Pharmacy Research* 3: 1146–1148.

Kiefer, F., Arnold, K., Künzli, M., Bordoli, L., and Schwede, T. 2008. The SWISS-MODEL repository and associated resources. *Nucleic Acids Research* 37(suppl_1): D387–D392.

Kuldeep, S., and Satpal, B. S. 2013. Virtual screening approach of drug design for Parkinson's disease. *International Journal of Pharma and Bio Sciences* 4(2): 821–830.

Kyte, J., and Doolittle, R. F. 1982. A simple method for displaying the hydropathic character of a protein. *Journal of Molecular Biology* 157(1): 105–132.

Laskowski, R. A., Rullmann, J. A. C., MacArthur, M. W., Kaptein, R., and Thornton, J. M. 1996. AQUA and PROCHECK-NMR: Programs for checking the quality of protein structures solved by NMR. *Journal of Biomolecular NMR* 8(4): 477–486.

Lovell, S. C., Davis, I. W., Arendall III, W. B., De Bakker, P. I., Word, J. M., Prisant, M. G., and Richardson, D. C. 2003. Structure validation by Cα geometry: ϕ, ψ and Cβ deviation. *Proteins: Structure, Function, and Bioinformatics* 50(3): 437–450.

Morris, G. M., Goodsell, D. S., Halliday, R. S., Huey, R., Hart, W. E., Belew, R. K., and Olson, A. J. 1998. Automated docking using a Lamarckian genetic algorithm and an empirical binding free energy function. *Journal of Computational Chemistry* 19(14): 1639–1662.

Ramachandran, G. T., and Sasisekharan, V. 1968. Conformation of polypeptides and proteins. In Mathematical Philosophy Group, *Advances in Protein Chemistry*, 23, 283–437. Academic Press, Bangalore.

Ramanand, S. J., Ghongane, B. B., Ramanand, J. B., Patwardhan, M. H., Ghanghas, R. R., and Jain, S. S. 2013. Clinical characteristics of polycystic ovary syndrome in Indian women. *Indian Journal of Endocrinology and Metabolism* 17(1): 138.

Singleton, V. L., and Rossi, J. A. 1965. Colorimetry of total phenolics with phosphomolybdic-phosphotungstic acid reagents. *American Journal of Enology and Viticulture* 16(3): 144–158.

Solis, F. J., and Wets, R. J. B. 1981. Minimization by random search techniques. *Mathematics of Operations Research* 6(1): 19–30.

Tobias, J. W., Shrader, T. E., Rocap, G., and Varshavsky, A. 1991. The N-end rule in bacteria. *Science* 254(5036): 1374–1377.

Waldi, D. 1965. Spray reagents for thin-layer chromatography. In *Thin-Layer Chromatography*. Springer, Berlin, Germany, 483–502.

10 Effect of Anti-oxidant-Rich Nutraceutical on Serum Glucose, Lipid Profile, and Oxidative Stress Markers of Salt-Induced Metabolic Syndrome in Rats

N. Lawal, L. S. Bilbis, R. A. Umar, and A. A. Sabir

CONTENTS

10.1 INTRODUCTION

Metabolic syndrome is a cluster of metabolic abnormalities involving diabetes, abdominal obesity, high cholesterol, and high blood pressure (Ford 2004). The incidence of metabolic syndrome, the most threatening epidemic in industrialized countries, is rapidly rising (Lakka et al. 2002; Park et al. 2003). High blood pressure is one of the key features of metabolic syndrome (Zavaroni et al. 1992; Hayman 2000; Ford et al. 2002). Individuals with metabolic syndrome are more prone to diabetes mellitus than the hypertensive individuals without this syndrome (Franklin 2005). Insulin resistance is a key factor to developing these components of metabolic syndrome (Ogihara 2004). Based on the recent progress of research on adipocytes, visceral obesity plays a critical role in the development of insulin resistance (Sironi et al. 2004). Adipokines such as angiotensinogen, tumor necrosis factor-α and non-esterified fatty acid, might contributed to the development of insulin resistance in the muscle and adipose tissues (Rahmouni et al. 2004). In contrast, lack of insulin resistance in the kidney increases tubular sodium reabsorption by hyperinsulinemia, leading to sodium retention in the body, and resultant salt-sensitive hypertension (Zheng et al. 2005). Therefore, there is an intimate relationship between insulin resistance and salt-sensitive hypertension in obese hypertensive patients with metabolic syndrome (Suzuki et al. 2000). Reactive oxygen species are a key factor in the onset and progression in the pathogenesis of metabolic syndrome, such as dyslipidemia, insulin resistance, hyperlipidemia, and hypertension (Hopps et al. 2010). Metabolic syndrome (MS)-involved clusters of cardiovascular risk factors closely linked to insulin resistance, hyperglycemia, and dyslipidemia have a high prevalence and are rapidly rising in the Western population (Haffner and Taegtmeyer 2003). In the category of over 50 years of age, MS affects more than 40% of the population in the United States and nearly 30% in Europe (Kjeldsen et al. 2008). From the studies by Isezuo et al. (2003); Isezuo and Ezunu (2005); and Umar et al. (2010), the incidence of full blown metabolic syndrome in Nigeria was 20.5% in 52 patients, about 72.4% were dyslipidemic and 54.3% were hypertensive. Concurrent hypertension and dyslipidemia, obesity and dyslipidemia, and hypertension and obesity occurred in 44.4%, 42.5%, and 33.1% of type 2 diabetes patients, respectively. Nutraceuticals range from isolated nutrients, herbal products, dietary supplements, and diets to genetically engineered foods and beverages (Dillard and German 2000). Many of these products possess pertinent physiological functions and valuable biological activities (Andlauer and Furst 2002). With the passage of the Dietary Supplements Health and Education Act of 1994, in the United States, the definition of nutraceuticals has been expanded to include vitamins and minerals that have anti-oxidant activities in scavenging free radicals from biological systems (Roberfroid 2002). Nutraceutical-rich anti-oxidants are believed to delay the onset and development of metabolic syndrome. The process can be divided into major classes based on their mode of action, anti-oxidants enzymes, and non-enzymatic anti-oxidants (scavengers). Enzymatic anti-oxidants consist of superoxide dismutase (SOD), catalase, and glutathione peroxidase. Whereas non-enzymatic anti-oxidant comprises vitamins (A, C, E), glutathione, phenols, flavonoids, albumin, and ubiquinone (Teoh and Davies 2002). Selenium, manganese, copper, zinc, and chromium are referred to as anti-oxidants minerals, have no anti-oxidant activity themselves, but are required for the activity of some anti-oxidant enzymes (Papas 1999).

10.2 MATERIALS AND METHODS

10.2.1 Chemicals and Reagents

Analytical grade laboratory chemicals and reagents were used for this study.

10.2.2 Experimental Design

10.2.2.1 Experimental Animals

Wistar albino rats of both sexes weighing between 150 and 220 g were used for the study. The animals were purchased from the Department of Biological Science, Usmanu Danfodiyo University, Sokoto, Nigeria, and were allowed to acclimatize for 2 weeks before the commencement of the experiment. The animals were grouped into six groups of eight rats each and were fed with pelletized growers' feed (Vital Feed, Jos, Nigeria) and allowed access to water ad libitum before and during the experimental period.

10.2.2.2 Grouping of Animals and Treatments

The animals were randomly divided into six groups of eight rats each.

- Group I Salt-loaded, treated with 250 mg/kg of nutraceutical
- Group II Salt-loaded, treated with 500 mg/kg of nutraceutical
- Group III Salt-loaded, treated with 10 mg/kg of nifedipine
- Group IV Salt-loaded, treated with 500 mg/kg of nutraceutical + 10 mg/kg of nifedipine
- Group V Salt-loaded, untreated
- Group VI Normal, control

10.2.2.3 Induction of Metabolic Syndrome in Rats

The rats were placed on 8% w/w salt diet (Bilbis et al. 2012), except the control group, for 6 weeks and treatment with nutraceuticals for an additional 4 weeks.

10.2.2.4 Measurement of Blood Pressure

The baseline blood pressure was measured by the tail-cuff method using the non-invasive Ugo Basile, series 58500 blood pressure recorder. The average of three readings was taken for each rat, and the blood pressure of the rats was monitored throughout the experimental period. The mean arterial blood pressure was calculated according to the following equation: DP + 1/3 (SP − DP), where SP and DP are systolic and diastolic pressures, respectively. Subsequent measurements were done every week.

10.2.2.5 Preparation of Anti-oxidant Micronutrients Supplements

Anti-oxidant micronutrients supplements prepared from locally available foodstuff based on the recommended daily allowances, vitamin A (1000 µg/70 kg), vitamin C (60 mg/70 kg), vitamin E (10 mg/70 kg), copper (3 mg/70 kg), chromium (200 µg/70 kg), and manganese (15 mg/70 kg). The recommended daily allowances of rats and humans are similar (Norlin et al. 1995).

The sources of micronutrients were: garlic, ginger, watermelon seed, onion, lemon, palm oil, and tomato. They were purchased from the Sokoto Central Market. The above listed foodstuffs were identified in the herbarium of the department of Biological Sciences, Usmanu Danfodiyo University, Sokoto, Nigeria.

The anti-oxidant vitamins and minerals were determined using Higher Performance Liquid Chromatography (HPLC) and Atomic Absorption Spectrophotometry (AAS) methods of analysis, respectively, from onion, tomato, ginger, garlic, lemon, watermelon seed, and palm oil and are presented in Table 10.1. The nutraceutical was prepared from onions, garlic, tomatoes, ginger, watermelon seeds, lemon, and palm oil in a ratio of 4:4:4:4:2:1:1. This was done by mixing 20 g of onions; 20 g of garlic, 20 g of tomatoes, and 20 g of ginger in 100 mL distilled water and blended using an electric blender. Ten grams of ground watermelon seeds were then added and blended once again. To this, 5 g of lemon juice and 5 g of palm oil were added and mixed. The nutraceutical solution was packaged into clean, dry containers and stored frozen at −20°C until required.

TABLE 10.1
Composition of the Prepared Nutraceutical

Food Stuff	β-Carotene (mg/dL)	α-Tocopherol (mg/dL)	Vit. C (mg/dL)	Mn (mg/dL)	Cr (mg/dL)	Cu (mg/dL)	Zn (mg/dL)	Flavonoid (%)
A	0.030 ± 0.00	0.011 ± 0.00	6.300 ± 1.00	0.100 ± 0.01	–	–	0.400 ± 0.01	–
B	0.040 ± 0.00	0.013 ± 0.00	10.4 ± 3.00	4.400 ± 0.05	0.500 ± 0.01	1.400 ± 0.01	0.300 ± 0.01	39.000 ± 2.70
L	0.090 ± 0.01	0.003 ± 0.00	2.200 ± 0.05	0.500 ± 0.01	–	0.100 ± 0.03	0.500 ± 0.01	46.000 ± 3.60
D	0.012 ± 0.00	0.008 ± 0.00	1.100 ± 0.01	0.100 ± 0.01	–	0.800 ± 0.01	0.200 ± 0.01	59.300 ± 1.50
E	–	0.001 ± 0.00	1.600 ± 0.05	–	–	0.800 ± 0.01	0.400 ± 0.01	11.010 ± 1.00
F	2.400 ± 0.10	0.002 ± 0.00	0.017 ± 0.00	0.100 ± 0.03	0.300 ± 0.02	0.300 ± 0.02	0.100 ± 0.01	15.000 ± 3.00
G	0.040 ± 0.00	0.097 ± 0.01	1.800 ± 0.06	0.400 ± 0.01	–	0.900 ± 0.01	0.300 ± 0.01	35.000 ± 1.00
N	3.100 ± 0.50	0.150 ± 0.02	27.800 ± 2.30	11.600 ± 0.20	–	0.500 ± 0.02	3.200 ± 0.02	137.000 ± 1.00

Note: A: 20 g of garlic, B: 20 g of ginger, L: 10 g of watermelon seeds, D: 20 g of onion, E: 5 g of lemon, F: 5 g of palm oil, G: 20 g of tomato, and N: 100 g of constituted nutraceutical. The results were expressed as mean ± standard deviation.

The appropriate dosages of the nutraceutical and drug were administered to the animals orally once daily by intubation using an intravenous cannula tube for 4 weeks, weight changes of the rats were monitored throughout the experimental period.

10.2.2.6 Preparation of Serum and Tissue Homogenate

Twenty-four hours after the last treatment, the animals were anesthetized with chloroform vapor, and blood samples were collected through a cardiac puncture into labeled tubes for biochemical analyses. The animals were forced to fast for overnight before sacrifice, and the venous blood was collected after sacrificed, the blood samples collected were allowed to clot and then centrifuged at 4000 x g for 10 minutes, and the serum obtained was pipeted into labeled tubes for the estimation of serum glucose, lipid profiles, vitamin E and C, insulin, superoxide dismutase, catalase, glutathione peroxidase, and malondialdehyde (MDA). The animals were sacrificed, and the liver of each rat was dissected out and rinsed with ice-cold saline to remove the blood. Ten percent of the homogenate was prepared in an ice-cold 0.1 M Tris buffer, pH 7.4 using a homogenizer. The homogenate was centrifuged at 4000 x g for 15 minutes. The supernatants were used for the estimation of the thiobarbituric acid reactive substance and glutathione peroxidase activity.

10.2.2.7 Analysis of Vitamins A, E, and C Using HPLC

Vitamins A, C, and E were analyzed by the method of Snezana et al. (2011).

10.2.2.8 Determination of Flavonoids

Flavonoids were determined from onion, ginger, garlic, tomato, watermelon seed, palm oil, and lemon by the method of Boham and Kocipal (1994).

10.2.2.9 Determination of Anti-oxidant Minerals (Cu, Zn, Mn, and Cr) by AAS Analysis

Cu, Zn, Mn, and Cr were determined by the method of Sibel et al. (2002).

10.2.2.10 Biochemical Analyses

The serum glucose was estimated by the glucose oxidase method using a Randox kit (Trinder 1969), serum total cholesterol (TC) was estimated by the enzymatic method using a Randox kit (Allain et al. 1974), serum high density lipoprotein-cholesterol (HDL-C) was estimated by the enzymatic method of Burstein et al. (1970) using a Randox kit, serum triglycerides were assayed by the method of Tietz (1990) using a Randox Kit, and serum low density lipoprotein-cholesterol (LDL-C) and very LDL-C (VLDL-C) were calculated using the Friedewald formula (Friedewald et al. 1972).

The atherogenic index (AI) was calculated as the ratio of LDL-cholesterol to HDL-cholesterol according to Abbott et al. (1988), the percentage protection against atherogenesis was calculated using the following equation:

$$\%\ \text{protection} = \frac{\text{AI of HC} - \text{AI of treated group}}{\text{AI of HC}} \times 100$$

The catalase activity was assayed using the Cayman's Catalase Assay Kit according to Johansson and Borg (1988), superoxide dismutase was assayed using the Cayman's Superoxide Dismutase Assay Kit according to Marklund (1980), glutathione peroxidase activity was assayed using the Cayman's Glutathione Peroxidase Assay Kit according to Paglia and Valentine (1967), lipid peroxidation as evidenced by the formation of thiobarbituric acid reactive substances was measured by the method of Niehans and Samuelson (1968), and insulin was estimated by the SPI-Bio rat insulin enzyme immunoassay kit according to Grassi and Pradelles (1991).

The insulin resistance index was calculated by the Homeostasis Model Assessment-Insulin Resistance (HOMA-IR) as described by Matthews et al. (1985).

The equation used is

$$\text{HOMA-IR} = \frac{\text{Fasting glucose (mmol/L)} \times \text{Fasting insulin } (\mu\text{U/mL})}{22.5}.$$

10.2.2.11 Statistical Analysis

Values are expressed as mean ± standard deviation of eight rats per group. The biochemical parameters were analyzed statistically using one way analysis of variance, followed by Dunnett's and Tukey multiple comparison test using GraphPad InStat software (version 3.10; San Diego, U.S.A). The differences were considered significant at $P < 0.05$.

10.3 RESULTS

Levels of anti-oxidant vitamins, minerals, and flavonoids from the constituted food supplements is present in Table 10.1.

The effect of an 8% w/w salt diet on the serum glucose, lipid profile, and atherogenic index is presented in Table 10.2. The results indicate a significant ($P < 0.001$) increase in the level of serum glucose, total cholesterol, triglyceride, LDL-C, and VLDL-C of the salt-treated groups as compared with the control group. The results also indicated a significant ($P < 0.001$) decrease in the level of HDL-C of the salt-treated groups as compared with the control group. There was a significant ($P < 0.001$) increase in the atherogenic index of the salt-treated groups as compared with the control group.

The effect of nutraceuticals on the systolic blood pressure (SBP), diastolic blood pressure (DBP), mean arterial blood pressure (MABP), and pulse rate (PR) are presented in Figures 10.1 through 10.4, respectively. The results indicated that salt-loading to the rats for 6 weeks increased the SBP, MABP, and PR, but no variation on the DBP between the salt-loading groups and the control group. Supplementation with nutraceuticals lowered the SBP, MABP, and PR. The salt-loading-untreated group showed a progressive increase of the SBP, MABP, and PR throughout the experimental period, while the control group showed no increase.

The results of the mean percentage weight gain of the rats in all the groups and mean percentage reduction in the SBP, MABP, and PR are presented in Figures 10.5 and 10.6, respectively. This present study observed that salt loaded untreated group and salt loaded treated group with nifedipine gained more weight recorded (42.4% ± 1.76%) and (37.02% ± 0.88%), respectively, than all other groups. The group that gained the least weight (18.3% ± 0.65%) was the group supplemented with 500 mg/kg of nutraceutical.

The results also showed that supplementation decreases the SBP, MABP, and PR, the rats supplemented with 500 mg/kg of

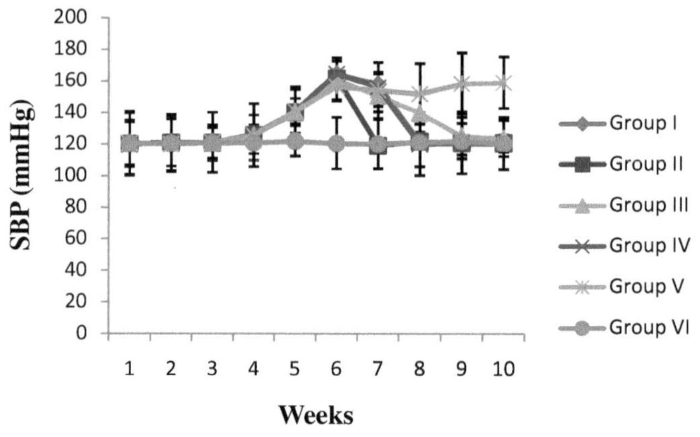

FIGURE 10.1 The effect of a salt diet and nutraceutical supplementation on the systolic blood pressure of salt-loaded rats week 1–6: salt diet only, week 6–10: salt diet plus supplements, group I: salt-loaded treated with 250 mg/kg of nutraceutical, group II: salt-loaded treated with 500 mg/kg of nutraceutical, group III: salt-loaded treated with 10 mg/kg of nifedipine, group IV: salt-loaded treated with 500 mg/kg of nutraceutical + 10 mg/kg of nifedipine, group V: salt-loaded untreated, and group VI: control group.

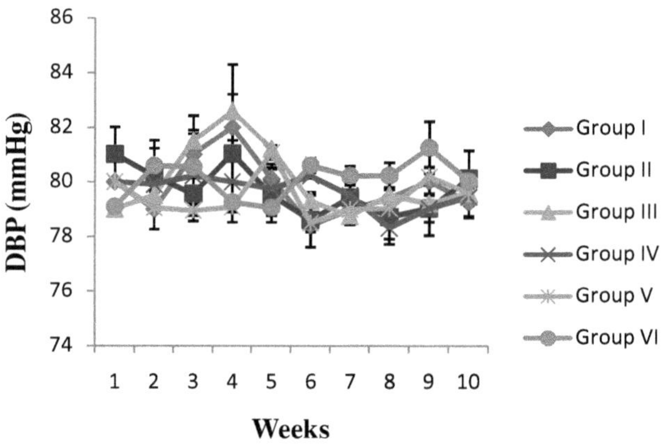

FIGURE 10.2 The effect of a salt diet and nutraceutical supplementation on the diastolic blood pressure of salt-loaded rats week 1–6: salt diet only, week 6–10: salt diet plus supplements, group I: salt-loaded treated with 250 mg/kg of nutraceutical, group II: salt-loaded treated with 500 mg/kg of nutraceutical, group III: salt-loaded treated with 10 mg/kg of nifedipine, group IV: salt-loaded treated with 500 mg/kg of nutraceutical + 10 mg/kg of nifedipine, group V: salt-loaded untreated, and group VI: control group.

TABLE 10.2
Biochemical Profile of Salt-Induced Metabolic Syndrome Rats Before Treatment

Group	Glc (mg/dL)	TC (mg/dL)	TAG (mg/dL)	HDL-C (mg/dL)	LDL-C (mg/dL)	VLDL-C (mg/dL)	AI
I–V	153.30 ± 3.40[a]	158.60 ± 9.70[a]	160.60 ± 11.50[a]	30.26 ± 0.40[a]	96.22 ± 7.70a	32.12 ± 0.30[a]	3.20 ± 0.40[a]
VI	86.40 ± 1.88	97.57 ± 1.90	98.83 ± 1.30	40.20 ± 0.56	37.60 ± 2.30	19.77 ± 0.20	0.94 ± 0.40

Note: Glc—glucose; groups I–V is salt-loaded and group VI is normal untreated. Values are expressed Mean ± SD; n = 3.

[a] $P < 0.001$ when comparing groups I–V with group VI-control by Dunnett's multiple comparison test.

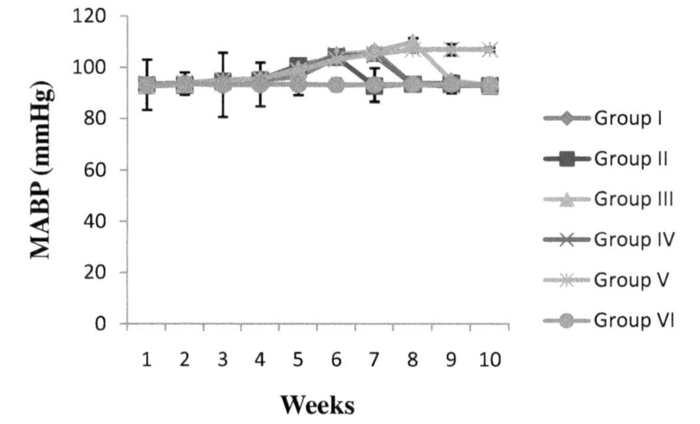

FIGURE 10.3 The effect of a salt diet and nutraceutical supplementation on the mean arterial blood pressure of salt-loaded rats week 1–6: salt diet only, week 6–10: salt diet plus supplements, group I: salt-loaded treated with 250 mg/kg of nutraceutical, group II: salt-loaded treated with 500 mg/kg of nutraceutical, group III: salt-loaded treated with 10 mg/kg of nifedipine, group IV: salt-loaded treated with 500 mg/kg of nutraceutical + 10 mg/kg of nifedipine, group V: salt-loaded untreated, and group VI: control group.

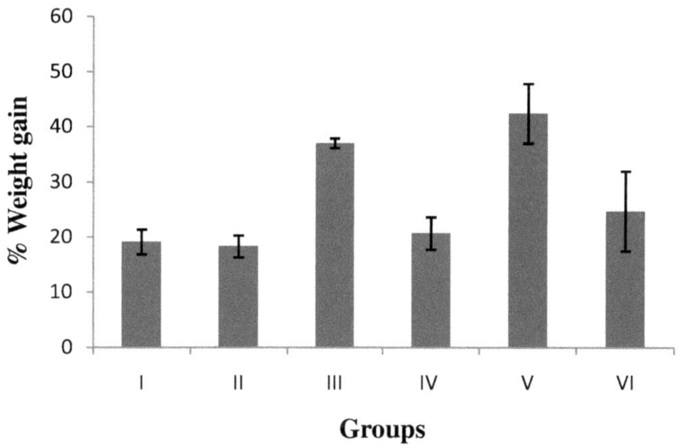

FIGURE 10.5 The mean percentage weight gain of salt-loaded rats treated with nutraceutical group I: salt-loaded treated with 250 mg/kg of nutraceutical, group II: salt-loaded treated with 500 mg/kg of nutraceutical, group III: salt-loaded treated with 10 mg/kg of nifedipine, group IV: salt-loaded treated with 500 mg/kg of nutraceutical + 10 mg/kg of nifedipine, group V: salt-loaded untreated, and group VI: control group.

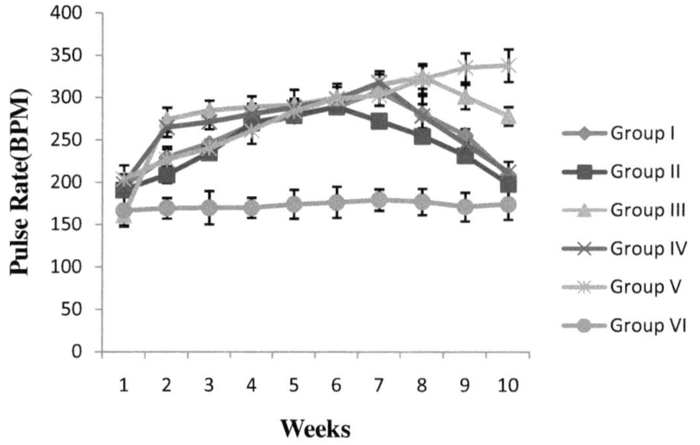

FIGURE 10.4 The effect of a salt diet and nutraceutical supplementation on the pulse rate of salt-loaded rats week 1–6: salt diet only, week 6–10: salt diet plus supplements, group I: salt-loaded treated with 250 mg/kg of nutraceutical, group II: salt-loaded treated with 500 mg/kg of nutraceutical, group III: salt-loaded treated with 10 mg/kg of nifedipine, group IV: salt-loaded treated with 500 mg/kg of nutraceutical + 10 mg/kg of nifedipine, group V: salt-loaded untreated, and group VI: control group.

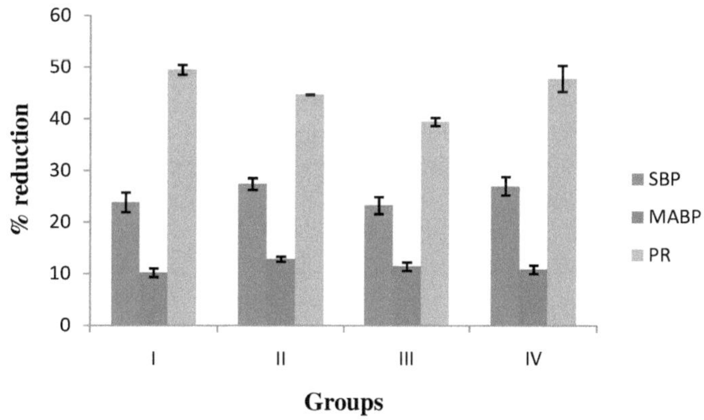

FIGURE 10.6 The mean percentage reduction in the SBP, MABP, and PR of salt-loaded rats supplemented with nutraceutical. The SBP-systolic blood pressure, MABP-mean arterial blood pressure, PR-pulse rate, group I: salt-loaded treated with 250 mg/kg of nutraceutical, group II: salt-loaded treated with 500 mg/kg of nutraceutical, group III: salt-loaded treated with 10 mg/kg of nifedipine, group IV: salt-loaded treated with 500 mg/kg of nutraceutical + 10 mg/kg of nifedipine, group V: salt-loaded untreated, and group VI: control group.

nutraceutical have the highest % reduction (27.42% ± 2.52%) and (12.85% ± 1.22%) in SBP and MABP, respectively, while the groups supplemented with 10 mg/kg of nifedipine have the lowest % reduction (23.22% ± 0.97%) and (10.23% ± 0.76%) in SBP and MABP, respectively.

It was observed that the rats supplemented with 500 mg/kg of nutraceutical + 10 mg/kg have the highest % reduction (47.78% ± 0.05%) in PR, while the group supplemented with 10 mg/kg of nifedipine have the lowest % reduction (39.46% ± 0.24%) in PR.

The results of the mean percentage protection against atherogenesis are presented in Figure 10.7. The results indicated (76.5% ± 2.13%) the mean average percentage protection in all the groups supplemented with nutraceuticals, while the non-supplemented group gave the lowest (13.02% ± 1.77%) protection against atherosclerosis.

The effect of nutraceuticals on the lipid profile and atherogenic index is presented in Table 10.3. The results indicated a significant ($P < 0.001$) decrease in the levels of total cholesterol, triglyceride, LDL-C, and VLDL-C of the groups supplemented with nutraceuticals as compared with the salt-loaded non-supplemented. There was no significant ($P > 0.05$) difference between the group supplemented with nutraceuticals and the control group.

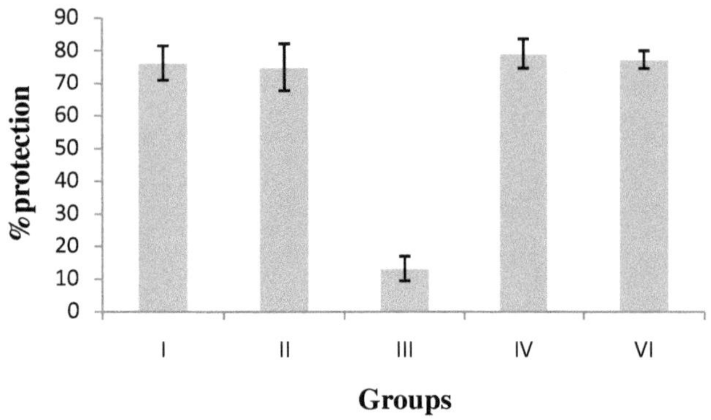

FIGURE 10.7 The mean percentage protection of nutraceuticals against atherosclerosis, group I: salt-loaded treated with 250 mg/kg of nutraceutical, group II: salt-loaded treated with 500 mg/kg of nutraceutical, group III: salt-loaded treated with 10 mg/kg of nifedipine, group IV: salt-loaded treated with 500 mg/kg of nutraceutical + 10 mg/kg of nifedipine, group V: salt-loaded untreated, and group VI: control group.

The results also show a significant ($P < 0.001$) increase in the levels of total cholesterol, triglyceride, LDL-C, and VLDL-C of the salt-loaded non-supplemented groups as compared with the control group. There was no significant ($P > 0.05$) difference between supplemented and non-supplemented groups.

The levels of HDL-C significantly ($P < 0.001$) increased in the groups supplemented with nutraceuticals as compared with the salt-loaded non-supplemented groups. A significant ($P < 0.05$) decrease was observed in the level of HDL-C of the salt-loaded non-supplemented groups as compared with the control group, but there was no significant ($P > 0.05$) variation between groups supplemented with nutraceuticals and between the salt-loaded non-supplemented groups.

The atherogenic index significantly ($P < 0.001$) decreased in the groups supplemented with nutraceuticals as compared with the salt-loaded non-supplemented groups. The atherogenic index of the salt-loaded non-supplemented groups as compared with the control group significantly ($P < 0.05$) increased. This present study observed no significant ($P > 0.05$) difference in atherogenic index between supplemented and non-supplemented groups.

The effect of supplementation with nutraceuticals on the markers of oxidative stress in the rats with metabolic syndrome is presented in Table 10.4. The results indicate a significant ($P < 0.001$) decrease in the MDA of the groups supplemented with nutraceuticals as compared with the salt-loaded non-supplemented groups. There was a significant ($P < 0.05$) variation between the groups supplemented with nutraceuticals and the control group. A significant ($P < 0.001$) increase in the MDA level of the salt-loaded non-supplemented groups as compared with the control group was observed.

The results also indicate a significant ($P < 0.001$) increase in the groups supplemented with nutraceuticals as compared with the salt-loaded non-supplemented groups. There was no significant ($P > 0.05$) difference in serum antioxidant enzymes, vitamin C and E, between supplemented and control groups. The results also show a significant ($P < 0.001$) decrease in the serum anti-oxidant enzymes and vitamins C and E of the salt-loaded non-supplemented groups as compared with the control group.

The effects of supplementation with nutraceuticals on the serum glucose, insulin, and insulin resistance in the rats with metabolic syndrome are presented in Table 10.5. The results indicate a significant ($P < 0.001$) decrease in the serum glucose of the supplemented groups as compared with the salt-loaded non-supplemented groups. There was no significant ($P > 0.05$) difference between the supplemented groups and the control group. A significant ($P < 0.001$) increase in the serum glucose levels occurred in the salt-loaded non-supplemented groups as compared with the control group.

A significant ($P < 0.001$) decrease in insulin levels and insulin resistance in the supplemented groups occurred as compared with the salt-loaded non-supplemented groups. There was no significant ($P > 0.05$) difference between the supplemented groups and the control group. The insulin levels and insulin

TABLE 10.3
Effect of Nutraceuticals on Lipid Profiles and Atherogenic Index in Rats with Metabolic Syndrome

Groups	TC (mg/dL)	TAG (mg/dL)	HDL-C (mg/dL)	LDL-C (mg/dL)	VLDL-C (mg/dL)	AI
I	106.0 ± 7.3[abc]	117.0 ± 9.6[abc]	42.0 ± 3.4[abc]	39.0 ± 8.7[abc]	24.0 ± 1.9[abc]	0.9 ± 0.2[abc]
II	107.9 ± 5.9[abc]	114.0 ± 4.8[abc]	43.0 ± 5.3[abc]	42.0 ± 1.4[abc]	23.0 ± 1.6[abc]	0.9 ± 0.2[abc]
III	178.4 ± 5.6[cd]	167.0 ± 7.7[cd]	33.0 ± 2.5[cd]	111.0 ± 6.1[cd]	34.5 ± 9.5[cd]	3.3 ± 0.6[cd]
IV	104.4 ± 5.5[abc]	118.0 ± 5.6[abc]	44.0 ± 3.3[abc]	36.0 ± 5.7[abc]	24.0 ± 1.6[abc]	0.8 ± 0.2[abc]
V	182.0 ± 15.8[ab]	169.0 ± 14.0[ab]	30.5 ± 4.4[ab]	117.0 ± 9.0[ab]	34.3 ± 2.9[ab]	3.8 ± 1.7[ab]
VI	102.4 ± 4.9[d]	110.5 ± 9.6[d]	42.8 ± 4.6[d]	37.7 ± 10.2[d]	22.8 ± 1.9[d]	0.9 ± 0.3[d]

Note: Group I: salt-loaded treated with 250 mg/kg of nutraceutical, group II: salt-loaded treated with 500 mg/kg of nutraceutical, group III: salt-loaded treated with 10 mg/kg of nifedipine, group IV: salt-loaded treated with 500 mg/kg of nutraceutical + 10 mg/kg of nifedipine, group V: salt-loaded untreated, and group VI: control group. Values are expressed as Mean ± standard deviation; n = 5.

[a] $P < 0.001$ when compared with salt-loaded untreated.
[b] $P > 0.05$ when compared with control group.
[c] $P < 0.001$ when compared with salt-loaded treated with 10 mg/kg of nifedipine.
[d] $P < 0.05$ when compared with control group by Tukey multiple comparison test.

TABLE 10.4
Effect of Supplementation with Nutraceuticals on Markers of Oxidative Stress in Rats with Metabolic Syndrome

	MDA nmol/mg	SOD u/mL	Catalase nmol/min/mL	GPX nmol/min/mL	Vit. E mg/dL	Vit. C mg/dL
I	0.30 ± 0.05[abc]	4.40 ± 1.00[abc]	7.10 ± 0.70[abc]	163.00 ± 16.06[abc]	0.70 ± 0.10[abc]	0.38 ± 0.02[abc]
II	0.26 ± 0.04[abc]	4.30 ± 1.30[abc]	14.10 ± 2.90[abc]	153.00 ± 9.10[abc]	0.90 ± 0.20[abc]	0.39 ± 0.05[abc]
III	1.94 ± 0.20[cd]	0.66 ± 0.03[cd]	1.90 ± 0.40[cd]	44.60 ± 5.70[cd]	0.10 ± 0.01[cd]	0.08 ± 0.01[cd]
IV	0.30 ± 0.07[abc]	4.80 ± 1.50[abc]	19.30 ± 2.90[abc]	149.00 ± 8.10[abc]	0.80 ± 0.10[abc]	0.40 ± 0.02[abc]
V	2.40 ± 0.40[ad]	0.55 ± 0.02[ad]	2.50 ± 0.02[ad]	40.10 ± 6.10[ad]	0.19 ± 0.01[ad]	0.07 ± 0.02[ad]
VI	0.22 ± 0.05[d]	4.10 ± 0.70[d]	12.20 ± 4.60[d]	119.90 ± 7.60[d]	0.82 ± 0.20[d]	0.37 ± 0.06[d]

Note: SOD: superoxide dismutase, GPx: glutathione peroxidase, group I: salt-loaded treated with 250 mg/kg of nutraceutical, group II: salt-loaded treated with 500 mg/kg of nutraceutical, group III: salt-loaded treated with 10 mg/kg of nifedipine, group IV: salt-loaded treated with 500 mg/kg of nutraceutical + 10 mg/kg of nifedipine, group V: salt-loaded untreated, and group VI: control group. Values are expressed as Mean ± standard deviation; n = 5.

[a] $P < 0.001$ when compared with salt-loaded untreated.

[b] $P > 0.05$ when compared with control group.

[c] $P < 0.001$ when compared with salt-loaded treated with 10 mg/kg of nifedipine.

[d] $P < 0.05$ when compared with control group by Tukey multiple comparison test.

TABLE 10.5
Effect of Supplementation with Nutraceuticals on Serum Glucose, Insulin, and Insulin Resistance in Rats with Metabolic Syndrome

Groups	Glucose (mmol/L)	Insulin (µU/mL)	HOMA-IR
I	5.43 ± 0.10[abc]	7.20 ± 2.48[abc]	1.74 ± 0.01[abc]
II	5.20 ± 0.09[abc]	6.50 ± 0.28[abc]	1.50 ± 0.02[abc]
III	7.80 ± 0.06[bc]	19.30 ± 0.5[bc]	6.69 ± 0.20[bc]
IV	5.18 ± 0.10[abc]	6.20 ± 0.74[abc]	1.43 ± 0.03[abc]
V	8.01 ± 0.09[ab]	22.30 ± 2.21[ab]	7.94 ± 0.01[ab]
VI	5.33 ± 0.08[b]	6.70 ± 1.24[b]	1.59 ± 0.04[b]

Note: Group I: salt-loaded treated with 250 mg/kg of nutraceutical, group II: salt-loaded treated with 500 mg/kg of nutraceutical, group III: salt-loaded treated with 10 mg/kg of nifedipine, group IV: salt-loaded treated with 500 mg/kg of nutraceutical + 10 mg/kg of nifedipine, group V; salt-loaded untreated, and group VI: control group. Values are expressed as Mean ± standard deviation; n = 5.

[a] $P < 0.001$ when compared with salt-loaded untreated.

[b] $P > 0.05$ when compared with control group.

[c] $P < 0.001$ when compared with salt-loaded treated with 10 mg/kg of nifedipine.

[d] $P < 0.05$ when compared with control group by Tukey multiple comparison test.

TABLE 10.6
Correlation Coefficient (r) of Blood Pressure and Weight Changes Against Serum Glucose, Lipid Profile, Insulin, Insulin Resistance Anti-oxidant Status, and Oxidative Stress Markers in Rats with Metabolic Syndrome

	SBP	PR	MABP	Weight
GLU	0.725***	0.636***	0.664***	0.995***
TC	0.692***	0.226***	0.628***	0.993***
TAG	0.842***	0.843***	0.795***	0.960***
HDL	−0.791***	−0.74***	−0.745***	−0.969***
LDL	0.964***	0.950*	0.615***	0.966***
VLDL	0.836**	0.833***	0.788**	0.961**
AI	0.824***	0.859***	0.774***	0.966***
Vit. E	−0.626***	−0.505***	−0.571***	−0.918***
Vit. C	−0.706**	−0.341**	−0.644**	−0.997***
SOD	−0.698***	−0.925***	−0.637**	−0.991***
CAT	−0.551**	−0.500**	−0.489**	−0.845***
GPX	−0.637*	−0.718**	−0.632**	−0.967***
Insulin	0.960***	0.936***	0.936***	0.829***
HOMA-IR	0.713***	0.685***	0.653**	0.849***
MDA	0.023***	0.799**	0.799***	0.975***

Note: Correlation coefficient (r) of blood pressure and weight changes against glucose, insulin, lipid profile, and oxidative stress markers. Glc: glucose, GPx: glutathione peroxidase, and SOD: superoxide dismutase.

***$P < 0.001$; *$P < 0.1$; **$P > 0.05$

resistance significantly ($P < 0.001$) increased in the salt-loaded non-supplemented groups as compared with the control group.

The results in Table 10.6 indicated a significant positive correlation between SBP, PR, MABP, weight and glucose, total cholesterol, triglyceride, LDL-C, VLDL-C, atherogenic index, MDA, insulin, and HOMA-IR, while HDL-C, vitamins C and E, SOD, GPx, and catalase indicated a significant negative correlation with SBP, PR, MABP, and body weight.

10.4 DISCUSSION

MS represents a condition involving obesity, dyslipidemia, hypertension, and diabetes mellitus as a consequence of the interaction between genetic and environmental factors (Haffner and Taegtmeyer 2003). MS is prevalent in most parts of the world, in which chronic state of the syndrome is associated with target organ damage, such as heart, kidney, and brain (Chen et al. 2004). Thus, the development of therapeutic/management strategies that may delay the onset and prevent complications associated with MS is critical to improve the lives of patients with MS. This study was designed to formulate an anti-oxidant-rich nutraceutical from locally available foodstuff and to evaluate whether supplementation with an anti-oxidant-rich nutraceutical could play a vital role in delaying the onset and progress in most of the traits of MS. In our model, a diet containing 8% salt was used to induce MS in Wistar albino rats for a period of 6 weeks. Our model fit the scenario that a high salt diet caused hypertension in the experimental rats and humans (Ogihara et al. 2002; Niu et al. 2005). The findings of this study indicate that salt-loading to the rats for 6 weeks increased the blood pressure and body weight. Salt-loading to the rats for 6 weeks significantly ($P < 0.001$) increased the hallmarks of the metabolic syndrome summarized as serum glucose, total cholesterol (TC), triglyceride (TAG), LDL-C, VLDL-C, AI, and HDL-C were significantly ($P < 0.001$) decreased. The mechanism of a high salt diet-induced metabolic syndrome could be attributed to an increased concentration of sodium in the circulatory system, which in turn activates a sympathetic nervous system and renin-angiotensin-aldosterone system, as well as increased signaling through the mineralocorticoid receptors. These may lead to increased production of reactive oxygen species and oxidative stress, and finally contribute to the etiopathology of insulin resistance, high blood pressure, impaired glucose homeostasis, and dyslipidemia. Another possible mechanism is that the salt diet is associated with the activation of adipokines (leptin, angiotensinogen, tumor necrosis factor α, transforming growth factor β, resistin, etc.) that may stimulate hepatic TAG synthesis, which in turn promotes the assembly and secretion of LDL, VLDL, and the reduction of HDL cholesterols. Obesity, insulin resistance, and diabetes may also be induced (Beltowski, 2003). The finding confirms the reports that salt loading to various strains of rats such as Sprague-Dawley rats (Niu et al. 2005), Wistar rats (Kagota et al. 2001), and Dahl salt-sensitive rats (Ogihara et al. 2002) results in increased mean arterial blood pressure, inhibition of insulin signaling, and induces insulin resistance.

Nutraceuticals may provide beneficial health effects beyond the traditional nutrients they contain (Wildman 2001). Anti-oxidant-rich nutraceuticals suppress the effects of the reactive oxygen species in order to delay the onset and progress of metabolic syndrome (Choi et al. 2010). The nutraceuticals that were locally available foodstuff are known to contain anti-oxidant vitamins, minerals, and flavonoids.

The elevation of the SBP indicates the risk of cardiovascular disease, which is more important than an increase in the DBP and is more difficult to control than the DBP (Hayman and Pavlik 2001). The lack of effect on the DBP observed in this study could be attributed to a salt diet which does not have much effect on the resting state of the heart during cardiac relaxation.

The blood pressure lowering effect of nutraceuticals could be attributed to the combined effects of the nutraceutical with a calcium channel blocking activity and diuretic effect. It may also be attributed to a high level of flavonoids in the nutraceutical, as flavonoids inhibit the activity of the angiotensin-converting enzyme which raises blood pressure. Other possible mechanisms could be due to the presence of anti-oxidant vitamins and minerals in the nutraceutical, thereby exhibiting anti-oxidant activity either by serving as the nucleophilic species or scavenge superoxide anions.

The highest percentage weight gain observed was in the non-supplemented rats with metabolic syndrome. The possible mechanism by which a salt diet increases body weight could be attributed to sodium levels activating the renin-angiotensin-aldosterone-system via an increase in sympathetic outflow, thereby stimulating adipokines that contribute to an increase in abdominal weight. The lowest percentage weight gain was observed in the rats supplemented with 500 mg/kg of nutraceutical. The weight lost could be attributed to the anti-obesity effects of the nutraceuticals through the modification of the lipid and carbohydrate metabolism, increased insulin sensitivity, and regulation of both appetite and adipocytokines. Other possible mechanisms could be attributed to the inhibitory effects of the nutraceutical in the proliferation, differentiation, angiogenesis in pre-adipocytes, and induction of apoptosis in mature adipocytes. It could also be associated with the up-regulation of the brown adipose tissue metabolism and down regulations of adiponectin and leptin gene expression in the adipocytes.

Studies have indicated a strong relationship between hypertension, dyslipidemia, insulin resistance, and hyperglycemia (Bergman et al. 2007; Saidu et al. 2012), all of which are consequences of the overproduction of reactive oxygen species. This present study found a significant ($P < 0.001$) decrease in serum Glu, TC, TAG, LDL-C, VLDL-C, and AI, and a significant ($P < 0.001$) increase in HDL-C of all the rats supplemented with nutraceuticals, compared with the non-supplemented rats. This could be attributed to the glycemic and hypolipidemic effects of nutraceuticals, in which the nutraceuticals reduce LDL cholesterol, particularly small dense LDL, which is associated with the increased risk of atherosclerosis. The lowering effect of nutraceuticals may be attributed to both the inhibition of hepatic fatty acid synthesis and increased catabolism of the TAG-rich lipoproteins. This increase in VLDL catabolism may result from the up-regulation of lipoprotein lipase expression and increased lipoprotein lipase activity because of the reduction in the serum apo C-III levels. The increase in HDL cholesterol may be associated with the increased expression of the apo A-I and A-II genes, and all these mechanisms can influence the insulin sensitivity in order to improve glucose homeostasis. Our finding is consistent with the work of Usman et al. (2011), Saidu et al. (2012), and Zahra et al. (2012). The highest average percentage protection ($76.5\% \pm 2.13\%$) against atherosclerosis was observed in all the rats supplemented with nutraceuticals. This could be attributed to the role of nutraceuticals in chain breaking, thereby donating the hydrogen atoms to free radicals in order to protect the cells from lipid peroxidation. The lowest percentage protection ($13.02\% \pm 1.77\%$) occurred in the non-supplemented rats. This could be attributed to a salt diet initiating the production of the reactive oxygen species that can oxidize the LDL cholesterol.

The MDA is a reliable marker of lipid peroxidation and peroxidative tissue injury (Janero 1990). It has been shown to be elevated in a salt-induced experimental rats model, suggesting that is a consequence rather than a cause of metabolic syndrome (Halliwell 1997). A significant decrease in the MDA levels was observed in all the rats supplemented with nutraceuticals. This could be attributed to the protective effects of nutraceuticals against lipid peroxidation. A significant ($P < 0.05$) difference in MDA levels between non-supplemented and normal control groups was observed. This could be attributed to active lipid peroxidation, which occurred in the rats with metabolic syndrome as a consequence of the overproduction of the reactive oxygen species.

The significant increase in the levels of anti-oxidant enzymes (superoxide dismutase, catalase, and glutathione peroxidase) and vitamins C and E was observed in all the rats supplemented with nutraceuticals as compared with the non-supplemented rats. This could be associated with the increase bioavailability of these micronutrients that are important components and co-factors of both anti-oxidant enzymes and vitamins.

Insulin resistance and hyperinsulinemia are key factors in the development of metabolic syndrome (Sironi et al. 2004). Indeed, insulin resistance and hyperinsulinemia are good predictors of metabolic syndrome (Sironi et al. 2004). Our findings indicate a significant ($P < 0.01$) decrease in insulin levels and insulin resistance in all the rats supplemented as compared with the non-supplemented rats. Nutraceuticals enhance the tyrosine kinase activity on the insulin receptors and inhibit phosphotyrosine phosphatase (an enzyme that inactivates insulin receptors) in order to improve the insulin sensitivity, as well as peripheral glucose uptake, and this indicates the combined action of several micronutrients in the nutraceuticals.

An attempt was also made to correlate the SBP, PR, MABP, and body weight with all the biochemical parameters assessed in this study. Our results show a strong positive correlation between SBP, PR, MABP, body weight and serum glucose, TC, TG, LDL-C, VLDL-C, AI, insulin, and MDA, and a strong negative correlation with HDL-C, catalase, superoxide dismutase, glutathione peroxidase, and vitamins C and E. This confirms the role of oxidative stress in the pathogenesis of metabolic syndrome and a promising therapeutic role of the nutraceuticals in reversing most of the markers of metabolic syndrome.

10.5 CONCLUSION

The supplementation of the rats with metabolic syndrome, with nutraceuticals formulated from locally available foodstuff (onion, tomato, garlic, ginger, watermelon seeds, palm oil, and lemon), which are sources of vitamins A, C, and E and Mn, Zn, Cr, Cu, and flavonoids, significantly improved the prognosis of the rats with salt-induced metabolic syndrome. Treatment with nutraceutical after induction significantly reverses some of the traits of MS and increases the levels of HDL-C and antioxidants.

Based on these findings, it can be concluded that supplementation with naturally occurring anti-oxidant vitamins, minerals, and flavonoids may reverse most of the traits of metabolic syndrome.

REFERENCES

Abbott, R. D., Wilson, P. W., Kannel, W. B., and Castelli, W. P. 1988. High density lipoprotein cholesterol, total cholesterol screening and myocardial infarction. The Framingham study. *Atherosclerosis* **8**: 207–211.

Allain, C. C., Poon, L. S., Chan, C. S. G., Richmond, W., and Fu, P. C. 1974. Enzymatic determination of total serum cholesterol. *Clinical Chemistry* **20**: P 470.

Andlauer, W., and Furst, P. 2002. Nutraceutical: A piece of history, present status and outlook. *Food Research International* **35**: 171–161.

Beltowski, J. 2003. Adiponectin and resistin—New hormones of white adipose tissue. *Medical Science Monitor* **9**: 55–61.

Bergman, R. N., Kim, S. P., and Hsu, I. R. 2007. Abdominal obesity: Role in the pathophysiology of metabolic disease and cardiovascular risk. *American Journal of Medicine* **120**: S3–S8.

Bilbis, L. S., Muhammad, S. A., Saidu, Y., and Adamu, Y. 2012. Effect of Vitamin A, C and E supplementation in the treatment of metabolic syndrome in albino rats. *Biochemistry Research International* **10**: 1–7.

Boham, A. B., and Koapal, A. C. 1994. Flavonoid and tannins from leaves of *Hawaiian vaccinium*. *Pacific Science* **48**: 458–460.

Burstein, M., Scholnick, H. R., and Morfin, R. 1970. Rapid method for the isolation of lipoproteins from human serum by precipitation with polyanions. *Journal of Lipid Research* **11**: 583–595.

Chen, J., Munter, P., and Hamm, L. L. 2004. The metabolic syndrome and chronic kidney disease in US adults. *Annals of Internal Medicine* **140**: 167–174.

Choi, U. K., Lee, O. H., Yim, J. H., Cho, C. W., Rhee, Y. K., Lim, S. I., and Kim, Y. C. 2010. Hypolipidemic and antioxidants effects of dandelion (*Taraxacum officinale*) root and leaf on cholesterol-fed rabbits. *International Journal of Molecular Sciences* **11**: 67–78.

Dillard, C. J., and German, J. B. 2000. Phytochemicals: Nutraceuticals and human health. *Journal of the Science of Food and Agriculture* **80**: 1744–1756.

Ford, E. S. 2004. The metabolic syndrome and mortality from cardiovascular disease and all-causes: Findings from the National Health and Nutrition Examination Survey II Mortality Study. *Atherosclerosis* **173**: 309–314.

Ford, E. S., Giles, W. H., and Dietz, W. H. 2002. Prevalence of the metabolic syndrome among US adults: Findings from the third National Health and Nutrition Examination Survey. *Journal of the American Medical Association* **287**: 356–359.

Franklin, S. S. (2005). Hypertension in the metabolic syndrome. In: Ishwarlal and Jailal, editors, *Metabolic Syndrome and Related Disorders*, Vol. 4, Irvine: Mary Ann Liebert, pp. 287–298.

Friedewald, W. T., Levy, R. I., and Fredrickson, D. S. 1972. Estimation of LDL-C in plasma without the use of the preparative ultracentrifuge. *Clinical Chemistry* **18** (6): 499–502.

Grassi, J., and Pradelles, P. 1991. Compounds labelled by the acetylcholinesterase of *Elecrophorus electricus*. Its preparation process and its use as a tracer or marquer in enzyme-immunological determinations. *United States patent*; 1: 047 (330).

Haffner, S., and Taegtmeyer, H. 2003. Epidemic obesity and the metabolic syndrome. *Circulation* **108**: 1541–1545.

Halliwell, B. 1997. Antioxidants and human disease: A general introduction. *Nutrition Reviews* **55**: S44–S49.

Hayman, D. J. and Pavlik, V. N. 2001. Patients with uncontrolled hypertension in the United State. *New England Journal*. doi:10.1056/NEJMoa010273.

Hayman, L. L. 2000. Abnormal blood lipids: Is it environment or is it genes? *Journal of Cardiovascular Nursing* **4**: 39–49.

Hopps, E., Noto, D., Caimi, G., and Averna, M. R. 2010. A novel component of the metabolic syndrome: The oxidative stress. *Nutrition, Metabolism, and Cardiovascular Diseases* **20**: 27–77.

Isezuo, S. A., Badung, S. L. H., and Omotoso, A. B. O. 2003. Comparative analysis of lipid profiles among patients with type 2 diabetes mellitus, hypertension and concurrent type 2 diabetes and hypertension: A view of metabolic syndrome. *Journal of the Nigeria Medical Association* **97**(6): 328–334.

Isezuo, S. A., and Ezunu, E. 2005. Demographic and clinical correlates of metabolic syndrome in native African type 2 diabetic patients. *Journal of the Nigeria Medical Association* **97**(6): 557–563.

Janero, D. R. 1990. Malondialdehyde and thiobarbituric acid-reactivity as diagnostic indices of lipid peroxidation and peroxidative tissue injury. *Free Radical Biology and Medicine* **9**: 515–540.

Johansson, L. H., and Borg, L. A. H. 1988. A spectrophotometric method for determination of catalase activity in small tissue samples. *Analytical Biochemistry* **174**: 331–336.

Kagota, S., Tamashiro, A., Yamaguchi, Y., Sugura, R., Kuno, T., Nakamura, K., and Kunitomo, M. 2001. Down regulation of vascular soluble guanylate cyclase induced by high salt intake in spontaneously hypertensive rats. *British Journal of Pharmacology* **134**: 737–744.

Kjeldsen, S. E., Naditch, B. L., Perlini, S., Zidek, W., Farsang, C., and Saba, E. 2008. Increased prevalence of metabolic syndrome in uncontrolled hypertension across Europe: The Global Cardiometabolic Risk Profile in Patients with Hypertension Disease survey. *Journal of Hypertension* **26**: 2064–2070.

Lakka, H. M., Laaksonen, D. E., Lakka, T. A., Niskanen, L. K., Kumpusalo, E., Tuomilehto, J., and Salonen, J. T. 2002. The metabolic syndrome and total and cardiovascular disease mortality in middle-aged men. *Journal of the American Medical Association* **288**: 2709–2716.

Marklund, S. 1980. Distribution of Cu/Zn superoxide dismutase and Mn superoxide dismutase in human tissues and extracellular fluids. *Acta Physiologica* **492**: 19–23.

Matthews, D. R., Hosker, J. P., Rudenski, A. S., Naylor, B. A., Treacher, D. F., and Turner R. C. 1985. Homeostasis model assessment: Insulin resistance and beta-cell function from fasting plasma glucose and insulin concentrations in man. *Diabetologia* **28**: 412–419.

Niehans, W. G., and Samuelsson, B. 1968. Formation of malondialdehyde from phospholipid arachidonate during microsomal lipid peroxidation. *European Journal of Biochemistry* **6**: 126–130.

Niu, T., Kristina, D. T., Paul, D. G., Michael, D. H., and Davis, R. M. 2005. Antioxidant treatment prevents renal damage and dysfunction and reduces arterial pressure in salt-sensitivity hypertension. *Hypertension* **45**: 934–939.

Norlin, J. B., Christoper, C., Curtis, D., and Eckhert. 1995. *Nutrient Requirement of Laboratory Animals*, 4th revised edition. *National Academic Press*, Washington, DC, 11–79.

Ogihara, T., Asano, T., and Ando, K. 2002. Angiotensin II-induced insulin resistance is associated with enhanced insulin signaling. *Hypertension* **40**: 872–879.

Ogihara, T., Asano, T., and Katagiri, H. 2004. Oxidative stress induces insulin resistance by activating the nuclear factor-kappa B pathway and disrupting normal subcellular distribution of phosphatidylinositol 3-kinase. *Diabetologia* **47**: 794–805.

Paglia, D. E., and Valentine, W. N. 1967. Studies on the quantitative and qualitative characterization of erythrocyte glutathione peroxidase. *Journal of Laboratory and Clinical Medicine* **70**: 158–169.

Papas, A. M. 1999. *Antioxidant Status, Diet, Nutrition, and Health.* Boca Raton, FL: CRC Press, pp. 89–105.

Park, Y. W., Zhu, S., Palaniappan, L., Heshka, S., Carnethon, M. R., and Heymsfield, S. B. 2003. The metabolic syndrome: Prevalence and associated risk factor findings in the US population from the Third National Health and Nutrition Examination Survey, 1988–1994. *Archives of Internal Medicine* **163**: 427–436.

Rahmouni, K., Mark, A. L., and Haynes, W. G. 2004. Adipose depot specific modulation of angiotensinogen gene expression in diet induced obesity. *American Journal of Physiology-Endocrinology and Metabolism* **286**: E891–E895.

Roberfroid, M. B. 2002. Global view on functional foods: European perspectives. *British Journal of Nutrition* **88**: 133–138.

Saidu, Y., Bilbis, L. S., Muhammad, S. A., and Nasiru, M. K. 2012. Serum lipid profile and antioxidants status of salt-induced hypertensive rats with an antioxidant rich nutraceutical. *Cameroon Journal of Experimental Biology* **8**: 47–54.

Sibel, S., Umit, D., Mustafa, S., and Latif, E. 2002. Determination of copper, iron, manganese, zinc and chromium by atomic absorption spectrophotometry in barking powder. *Journal of Food and Drug Analysis* **3**: 188–194.

Sironi, A. M., Gastaldelli, A., and Mari, A. 2004. Visceral fat in hypertension: Influence on insulin resistance and beta-cell function. *Hypertension* **44**: 127–133.

Snezana, S. M., Danijela, A. K., Danijela, C. N., and Milan, N. M. 2011. Rapid and reliable HPLC method of vitamin C in pharmaceutical samples. *Tropical Journal of Pharmaceutical Research* **10**: 105–111.

Suzuki, M., Kimura, Y., Tsushima, M., and Harano, Y. 2000. Association of insulin resistance with salt sensitivity and nocturnal fall of blood pressure. *Hypertension* **35**: 864–868.

Teoh, C. Y., and Davies, K. J. A. 2002. The broad spectrum of antioxidant defense and oxidant repair mechanisms. *Free Radical Research* **36**: 8–12.

Tietz, N. W. 1990. Serum triglyceride determination. In: *Clinical Guide to Laboratory Tests*, Second edition, Philadelphia, PA: W.B. Saunders Co, pp. 554–556.

Trinder, P. 1969. Determination of blood glucose in blood using glucose oxidase with an alternative oxygen acceptor. *Annals of Clinical Biochemistry* **6**: 24–25.

Umar, R. A., Hassan, S. W., and Ndukwe, O. F. 2010. Butrylcholinesterase activity in Nigerian type 2 diabetics with and without metabolic syndrome. *African Journal of Biotechnology* **9**(26): 4110–4113.

Usman, W., Saidu, Y., Ladan, M. J., and Ibrahim, N. D. 2011. Dissertation Report; Antioxidant Status of Alloxan-Induced Diabetic Rats Treated with a combination of a Hypoglycemic Drug and a Nutraceutical formulation.

Wildman, R. E. C. 2001. Classifying nutraceuticals. In: Wildman, R.E.C., editor, *Handbook of Nutraceuticals and Functional Foods*, Boca Raton, FL: CRC Press, pp. 13–30.

Zahra, B., Mahdieh, G., Parvin, M., Niloofar, S., and Fereidoun, A. 2012. Dietary total antioxidant capacity and the occurrence of metabolic syndrome and its components after a 3-year follow-up in adults: Tehran lipid and glucose study. *Nutrition and Metabolism* **9**: 70.

Zavaroni, I., Mazza, S., Dall'Aglio, E., Gasparini, P., Passeri, M., and Reaven, G. M. 1992. Revalence of hyperinsulinaemia in patients with high blood pressure. *Journal of Internal Medicine* **231**: 235–240.

Zheng, Y., Yamada, H., and Sakamoto. 2005. Roles of insulin receptor substrates in insulin-induced stimulation of renal proximal bicarbonate absorption. *Journal of the American Society of Nephrology* **16**: 2288–2295.

11 Pharmacognostic Studies on *Ormocarpum sennoides* (Willd.) DC.

A Bone Fracture Healing Plant from Fabaceae

M. M. Sudheer Mohammed, A. Narayanasamy, and S. Mahadevan

CONTENTS

11.1 INTRODUCTION

Historically, plants have yielded some of our most important drugs (Heinrich et al. 2012), but many of them are used without standardization, which in turn leads to adulteration. Substitute or counterfeit herbal materials often found in the market is a major problem in the herbal industry. Standardization of a crude drug is essential to avoid and identify the adulterants or substituents (Kumar et al. 2012). Standardization is performed by stepwise pharmacognostic and phytochemical studies.

Ormocarpum sennoides (Willd.) DC. (Family: Fabaceae) is a native of the Andaman Islands, India and Sri Lanka. It is a tall shrub, growing up to 3.0 meters in height, and it is known as "elumpotti" or "elumbotti" and "kattumurungai" in the Tamil language and also "setharikodi" among the Irula tribes in Tamil Nadu. Traditionally, the plant is used for many purposes, especially for bone setting. The leaves of the plant are ground with eggs into a paste, and then mixed with milk for drinking once in a day until the complete mending of a bone is a common practice (Rathakrishnan et al. 2009). The fresh leaves are consumed as such or prepared in a medicated candy to treat chest pain. The decoction of the root was considered as a good remedy for rheumatic fever (Kapoor, 2002). Also, the leaf powder is taken along with honey or milk to strengthen the bone, and its paste is tied onto the fracture site to heal the crack (Thamacin et al. 2014). However, pharmacognostic studies have not been reported so far with this plant. Thus, the present study is designed to evaluate the pharmacognostic profile of *Ormocarpum sennoides.*

11.2 MATERIALS AND METHODS

11.2.1 Procurement of Plant Material

"Elumpotti" twigs were collected from the Vengili Village, Vellore District, Tamil Nadu, India. The plant is identified as *Ormocarpum sennoides* (Willd.) DC. belonging to the family Fabaceae, and the same conformed with the voucher specimen in the Botanical Survey of India, Southern Regional Centre, Coimbatore, Tamil Nadu. The plant parts were collected from the same place for the present study. The plant materials were washed thoroughly until the impurities were removed and subjected to macroscopic, organoleptic, microscopic, and leaf constant studies. Shade-dried plant parts were milled to a coarse powder by mortar and pestle, and the powders were used for further studies such as fluorescence and secondary metabolite analysis.

11.2.2 Pharmacognostic Studies

Macroscopic features of the plant were recorded as per visual observation, such as habit, shape, size, texture, and other special features. For organoleptic studies, odor, color, and taste of the plant parts were recorded (Arya and Thakur 2012). Transverse sections of stem, root, and leaves were prepared with a razor blade, stained, and mounted following the standard microtechnique protocol as described by Brain and Turner (1975). Leaf constants such as stomatal number, stomatal index, vein islet numbers, vein termination numbers, and palisade ratio were enumerated by following the method described in The Ayurvedic Pharmacopoeia of India (2008). Fluorescence characteristics of the plant powders were recorded by treating them with various reagents and chemicals (Kokoski et al. 1958) under the visible and UV light (Chase Jr. and Pratt 1949).

11.2.3 Phytochemical Analysis

The successive extraction method was followed as described by Das et al. (2010) for the preparation of the extract, which in turn was subjected to various analyses. Twenty-five grams each of leaf, stem bark, and root bark powders were successively extracted for 48 hours with three organic solvents and an inorganic solvent (water). Each extract was centrifuged, and the supernatant was filtered and stored in sterile containers at room temperature for subsequent analysis. Preliminary phytochemical screening was carried out by using the standard procedures (Waldi 1965; Trease and Evans 2002; Mace 1963; Raaman 2006; Harborne 2009; Kokate 1999; Camporese et al. 2003), and the results are tabulated.

11.3 RESULTS

11.3.1 Pharmacognostic Studies

11.3.1.1 Macroscopic Features

Ormocarpum sennoides is distributed in the dry deciduous forests and pen forests of the Eastern Ghats. The forms include a shrub, scrambler, or small tree 1–5 m tall (Figure 11.1), mostly devoid of white hairs; Branchlets reddish-purple, glabrous or with viscid swollen-based hairs; Barks rough and fissured in outer surface at maturity; Wood is red streaked with black, whereas root is taproot with thin rootlets.

Leaves: Green, alternate-spiral, compound, imparipinnate, 3–10 cm long; Leaflets 6–17 in number and 0.6–5 × 0.4–2.2 cm in size, elliptic or oblong-elliptic, shortly mucronate, somewhat narrowed to the base, the margin entire, glabrous; Petiole 4–7 mm long; Rachis 2.5–6.5 cm long, glabrous or more or less glandular-setose; Stipules 2, up to 4 × 1.5 mm, ovate or lanceolate, acuminate with subulate tips. Midrib is prominent; lateral veins are very narrow, slightly darkened beneath. Inflorescence: Racemes glandular-setose, sticky, 2–13-flowered, 0.5–7 cm long; Peduncles 0.5–2 cm long; Pedicels 3–9 mm long; Bracts brown, scarious, 1.5–2 mm long, ovate, acuminate or 3-pointed, setose on the margins; Bracteoles 1–2 mm below the receptacle, 2–3 mm long, ovate-acuminate.

Flowers: Yellow, bisexual, regular, zygomorphic, up to 2 cm long; Calyx: Green, glabrous or with a few swollen-based hairs; receptacle cup (hypanthium) 3 mm long; calyx tube 2.5 mm long, lobes 5, upper lobes 4 mm long, lateral lobes 4.5–5 mm long, triangular or lanceolate, and lower lobe 6–7 mm long, very acute, lanceolate-boat-shaped. Corolla: Deciduous, 9–16 mm long; standard yellow, red-veined, or dark purple-red almost black, petals 5, ovate. Androecium: Stamens 10, diadelphous (5 + 5), uniform, exerted, anthers ending in stalked glands. Gynoecium: Style filiform; Stigma minute and terminal. Ovary: Ovary 5–8-ovulate, glabrous or with swollen-based hairs on only the sutures or all over the faces also. Pod: Dark blackish-green, oblong, opening from apex, narrow at both ends, flat, longitudinally ribbed; stipe 3–8 mm long; articles 9–20 × 5–6 mm, 3 mm thick, glabrous or covered with backwardly curved prickles 0.5–1.5 mm long. Seeds: 6 × 25 mm, compressed, brown, narrowly oblong. Flowering: December–January; Fruiting: January–April.

11.3.1.2 Organoleptic Studies

Organoleptic analysis exhibits the color, odor, and taste of various parts of the plant. Fresh leaves are green in color and are characteristic in odor with a bitter taste. Stems are dark red-purple in color, and they are odorless and tasteless. Roots are pinkish brown in color and are odorless with a sweet taste.

11.3.1.3 Anatomical Features of Leaf

The leaflet is concave on the adaxial side and slightly convex on the abaxial side. It is wide and thick along the median part and less thick along the margin. The upper epidermal cells are broad and wide. There are spindle-shaped darkly stained parenchyma cells beneath the upper epidermis. The lower epidermis is thick, and the cells are hemispherical in shape with papillate outer tangential walls. Differentiation of palisade and spongy parenchyma is not distinct. Both of the epidermises have anomocytic stomata and calcium oxalate crystals. Sessile secretory oil glands appear in the upper epidermis. The vascular bundle of the midrib is prominent. It includes six vertical rows of angular thick-walled xylem cells and a cluster of phloem cells. Calcium oxalate prismatic crystals are sparsely distributed in the mesophyll tissue (Figure 11.2).

FIGURE 11.1 Macro morphology of *Ormocarpum sennoides*.

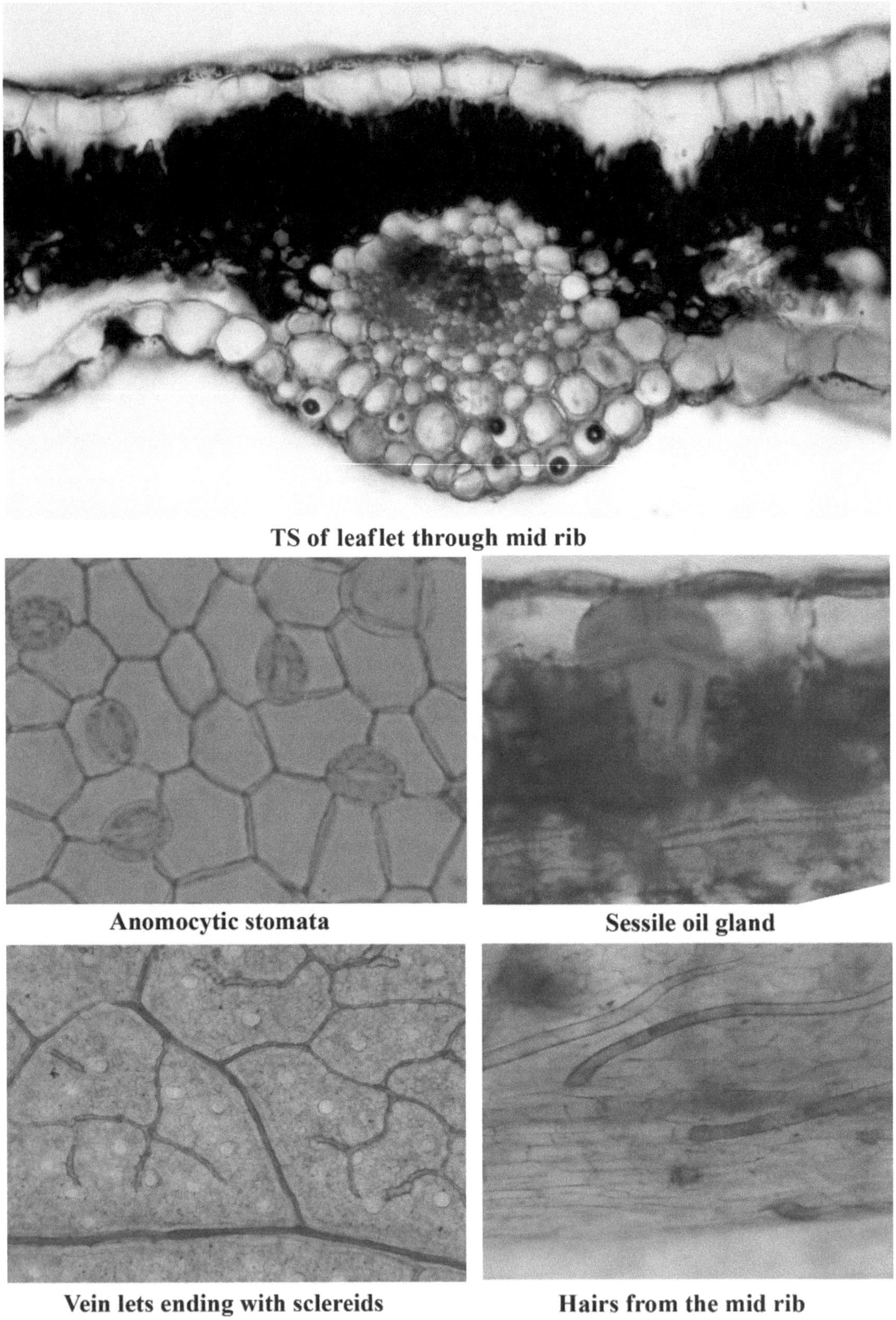

FIGURE 11.2 Anatomical features of a leaflet.

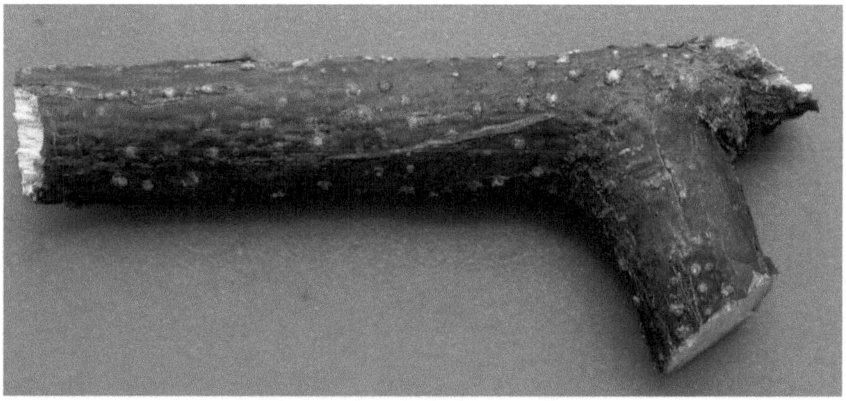

Ormocarpum sennoides **Aerial stem**

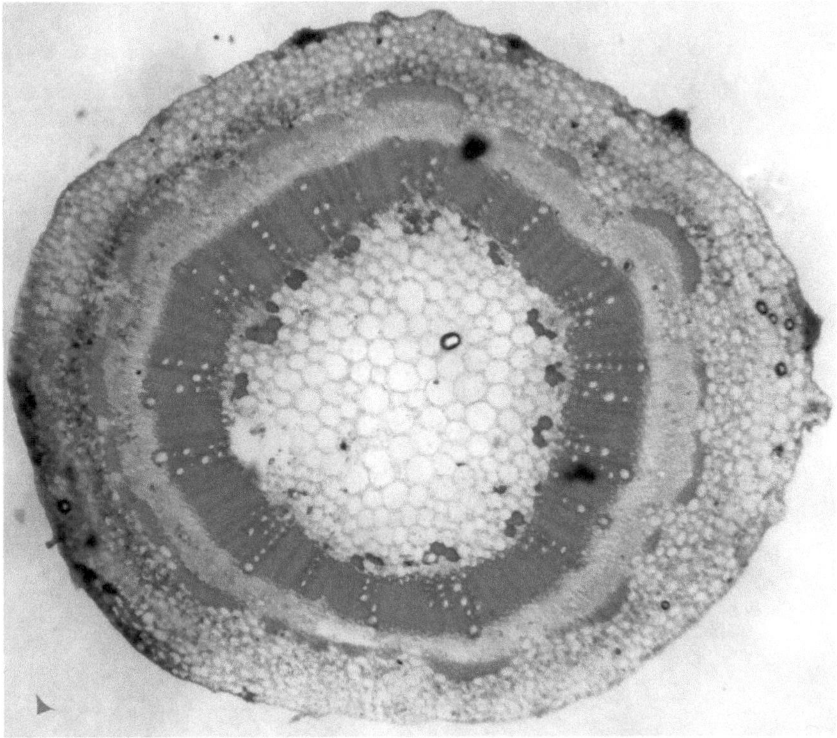

Cross section of the entire stem showing secondary growth

FIGURE 11.3 Anatomical features of a stem.

11.3.1.4 Anatomical Features of Stem

The stem is roughly circular. It consists of a thin, intact layer of epidermis and narrow cortex comprised of elliptical parenchyma cells. Tannin- and oil-filled cells and sclerenchyma cells are presented in the cortex. The vascular cylinder is thick and hollow and encloses wide pith. The phloem cells are compressed. The xylem includes vessels and narrow fibers. The vessels are wide, circular, and thin-walled. The pith is slightly angular in outline, and the cells are circular, thin-walled, and often filled with starch grains (Figures 11.3 and 11.4).

11.3.1.5 Anatomical Features of Root

The outer part of the root consists of periderm, which is moderately thick. The cortical cells are large and compact, and they are heavily loaded with starch grains. Sclereids often present in the cortex. The secondary phloem occurs inner to the cortex, which is narrow. The secondary xylem is undulate in outline due to the pressure of the wide circular vessels located in the peripheral part of the xylem cylinder. Xylem includes thick, wide medullary rays (Figures 11.5 and 11.6).

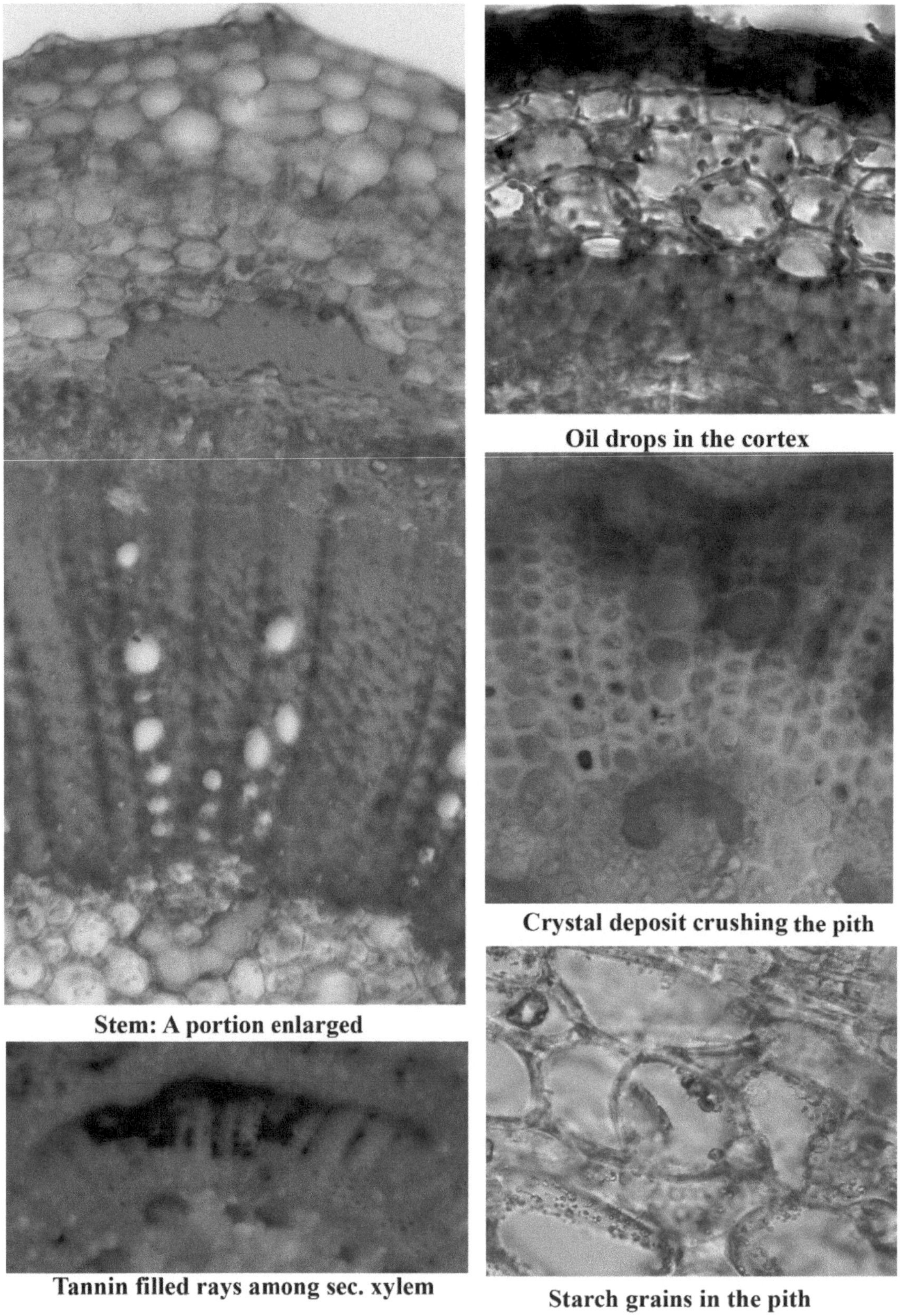

FIGURE 11.4 Cell inclusions in the stem.

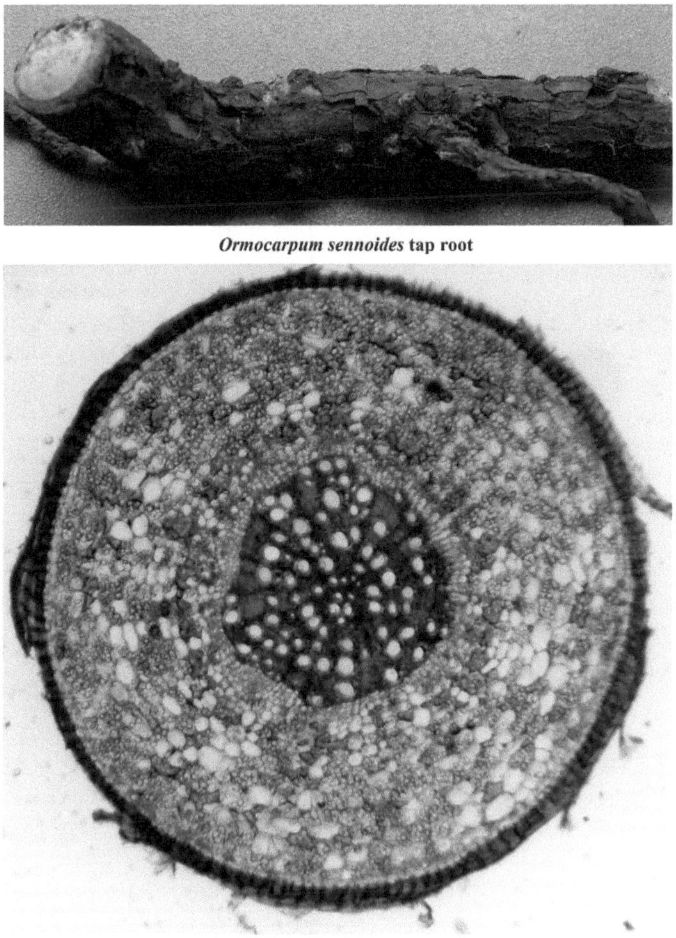

FIGURE 11.5 Anatomical features of a root.

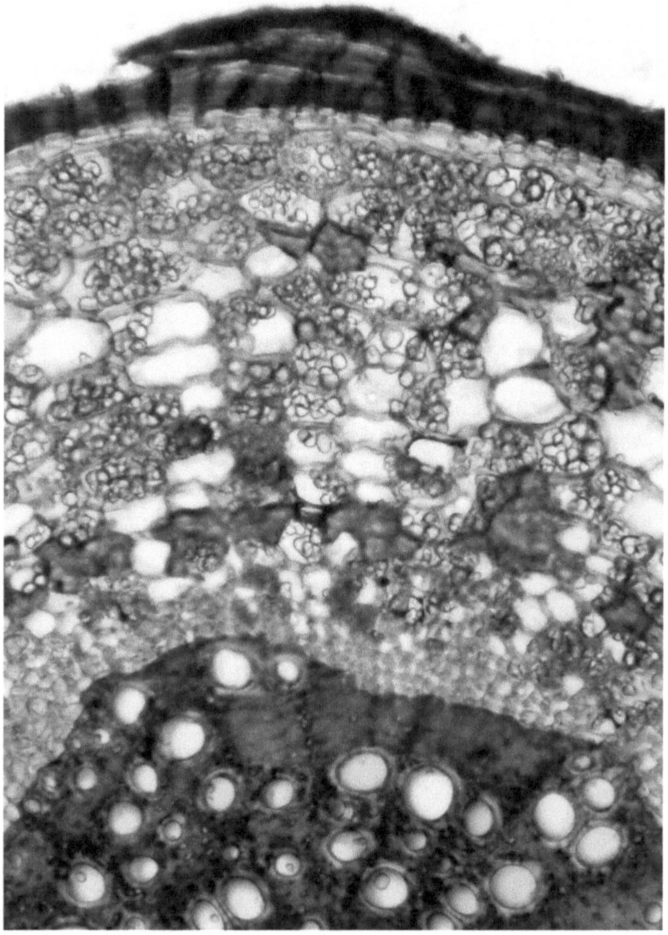

FIGURE 11.6 Internal features of the root tissue.

11.3.1.6 Leaf Constants

The quantitative measurements of the stomatal number, stomatal index, vein islet, vein termination number, and palisade ratio are presented in Table 11.1.

11.3.1.7 Fluorescence Analysis

The results of the fluorescence response of the powdered drugs and various extracts after being treated with different reagents and chemicals and emitting different colors under UV and visible lights were recorded. The color change for the crude powder and individual extracts are distinctive. The results are presented in Tables 11.2 and 11.3.

TABLE 11.1
Quantification of Leaf Constants

Leaf Constants	Determination/sq. mm
Stomatal number	Upper epidermis 38.6 ± 1.02
	Lower epidermis 44.3 ± 1.51
Stomatal index	Upper epidermis 17.87 ± 0.88
	Lower epidermis 18.97 ± 0.92
Vein termination number	12.66 ± 1.27
Vein islet number	9.33 ± 0.64
Palisade ratio	1.2 ± 0.17

11.3.2 Phytochemical Analysis

11.3.2.1 Qualitative Phytochemical Screening

Preliminary phytochemical screening revealed the presence of alkaloids, flavonoids, tannins, steroids, triterpenoids, and saponins in *O. sennoides*. But glycosides could not be detected in any parts, while steroids were absent in root extracts. A leaf shows different chemical constituents in different solvents than other parts of the plant. Ethanol extract yielded six compounds in leaves, whereas water extract yielded five compounds in the root. Ethanol and water extracts of the stem yielded four components. Petroleum ether and ethyl acetate extracts mostly showed the presence of steroids and triterpenoids. The results are presented in Table 11.4.

TABLE 11.2
Fluorescence Analysis of Leaf, Stem Bark, and Root Bark Powders

	Leaf		Stem		Root	
Reagents with Powder	**Visible Light**	**UV Light**	**Visible Light**	**UV Light**	**Visible Light**	**UV Light**
Leaf powder	Light green	Light brown	Light brown	Dark brown	Light brown	Light brown
Powder + conc. H_2SO_4	Dark brown	Light brown	Reddish black	Light green	Dark brown	Dark brown
Powder + conc. HNO_3	Reddish orange	Orange	Dark brown	Dark brown	Light green	Dark brown
Powder + 5% NaoH	Light brown	Red	Brownish red	Light green	Green	Dark brown
Powder + 5% $FeCl_3$	Dark green	Dark brown	Dark brown	Dark brown	Dark brown	Purple
Powder + 10% Hcl	Light yellow	Brown	Light brown	Light green	Light green	Creamy white
Powder + 30% NH_3	Light orange	Light brown	Brownish red	Light green	Green	Reddish orange
Powder + 70% Methanol	Green	Red	Yellowish green	Green	Green	Orange
Powder + 70% Ethanol	Dark green	Red	Light green	Green	Dark green	Orange
Powder + Water	Light orange	Light yellow	Light brown	Green	Creamy white	Creamy white

TABLE 11.3
Fluorescence Analysis of Leaf, Stem Bark, and Root Bark Extracts

	Leaf		Stem		Root	
Extracts	**Visible Light**	**UV Light**	**Visible Light**	**UV Light**	**Visible Light**	**UV Light**
Petroleum ether	Light yellow	Light green	Light yellow	Light green	Colorless	Colorless
Ethyl acetate	Dark green	Dark brown	Light green	Green	Pinkish red	Light green
Ethanol	Dark green	Reddish black	Light green	Green	Light orange	Green
Water	Brown	Light green	Brown	Green	Light orange	Green

TABLE 11.4
Phytochemical Screening of Leaf, Stem Bark, and Root Bark Extracts

		Leaf				Stem				Root			
Compounds	**Tests**	**PE**	**EA**	**ET**	**W**	**PE**	**EA**	**ET**	**W**	**PE**	**EA**	**ET**	**W**
Alkaloids	Dragendroff's test	–	–	+	+	–	–	+	–	–	–	+	+
Flavonoids	10% HCl and 5% NaOH test	–	–	+	+	–	+	+	+	–	–	–	+
Tannins	5% $FeCl_3$ test	–	–	+	+	–	–	+	+	–	–	+	+
Steroids	Liebermann–Burchard's test	+	–	+	–	–	+	–	–	–	–	–	–
Triterpenoids	Liebermann–Burchard's test	+	+	+	+	–	+	+	+	+	+	+	+
Saponins	Foam test	–	–	+	+	+	–	–	+	+	–	–	+
Glycosides	Borntrager's test	–	–	–	–	–	–	–	–	–	–	–	–

Abbreviations: PE—petroleum ether; EA—ethyl acetate; ET—ethanol; W—water
Notes: +, Denotes presence of compounds; –, Denotes absence of compounds

11.4 DISCUSSION

A growing demand of the medicinal plants together with confusion in the vernacular names and morphological similarity are considered as important reasons for adulteration (Prakash et al. 2013). *Ormocarpum sennoides* is morphologically more similar to its closely related species *Ormocarpum cochinchinense*. Further, *Ormocarpum cochinchinense* (Lour.) Merr as well as *Blepharis maderaspatensis* (L.) B. Heyne ex Roth (Acanthaceae) are also called "elumbotti" in the Tamil language. Moreover, *Moringa concanensis* Nimmo (Moringaceae) is also called "Kattu murungai" in the Tamil language. These reasons prompted to us to carry out this piece of work on *O. sennoides*.

Macroscopic studies revealed that *O. sennoides* possess yellow colored corolla with red veins. Luxative pods are green-dark blackish, opening from apex. These are the key features of this plant. The morphological characteristics of a particular plant will be very useful in identifying the plant at the species level (Moteriya et al. 2015). Organoleptic studies demonstrated that the roots are sweet and the stems are bitter in taste. Leaves, stems, and roots displayed green, dark red-purple, and pinkish brown colors, respectively, in normal daylight. The leaf only emitted some kind of odor. These characteristics could serve as important identification features of the plant.

The anatomical features established some important and peculiar characteristics of the plant. The stem possesses tannin- and oil-filled cells and sclerenchyma cells in the cortex and starch grains often present in pith. The root carries large cortical cells, which were loaded with starch grains. Sclereids often present in the cortex. Prominent medullary rays occurred in the xylem. In the leaf, anomocytic stomata were distributed in the epidermises. Sessile oil glands appeared in the upper epidermis. Lower epidermal cells have papillate outer tangential walls. The differentiation of palisade and spongy parenchyma is not distinct. Six vertical rows of xylem appeared in a vascular bundle. Microscopic studies are considered as reliable, simple, and the cheapest in establishing the identity of source materials (Shah et al. 2013). However, leaf constants of the plant provided accurate numerical data, which may be specific to the tested plant and is distinct from its closely related species, because it is a structural signature of each species.

In fluorescence analyses, different colors were observed, such as different shades of green, brown, yellow, orange, red, purple, and white in both the powders and extracts. The color change may be due to the presence and nature of the phytochemicals. Fluorescence analysis is a very important and useful tool for the identification of different constituents present in natural products. Still, many crude drugs are assessed qualitatively by using this parameter (Akbar et al. 2014).

The phytochemical screening of the crude extracts of leaf, stem bark, and root bark samples of *O. sennoides* demonstrated the presence of important secondary metabolites, such as alkaloids, flavonoids, tannins, steroids, triterpenoids, and saponins via standard tests. Glycosides could not be detected from this study by Borntrager's test. The detected compounds are known to have medicinal properties, which were confirmed already. For example, alkaloids derived from medicinal plants show biological activities like anti-inflammatory (Souto et al. 2011), anti-malarial (Dua et al. 2013) effects. Suhartono et al. (2012) reported that flavonoids possess antioxidant activity in some medicinal plants. Tannins are known to have anti-tumor and anti-viral activities (Kumari and Jain 2012). In addition, steroids are known to possess cardiotonic effects and also anti-bacterial and insecticidal properties (Bagrov et al. 2009).

Therefore, *O. sennoides* is expected to have many biological activities. Besides, phytochemical examination is also part of the quality control parameters. All the results of the present study will be useful in preventing the adulterants from the adulterants if any.

11.5 CONCLUSION

The findings of the present study will be helpful in authenticating and establishing quality control standards of the species. The current observations will also be useful in distinguishing the *Ormocarpum sennoides* from its closely related species, especially *Ormocarpum cochinchinense* and also from other genus or family members, as mentioned above. Future research is needed to evaluate the therapeutic actions of various phytochemicals of this species.

REFERENCES

Akbar, S., Hanif, U., Ali, J., and Ishtiaq, S. 2014. Pharmacognostic studies of stem, roots and leaves of *Malva parviflora* L. *Asian Pacific Journal of Tropical Biomedicine* 4(5):410–415.

Arya, V., and Thakur, R. 2012. Organoleptic and microscopic analysis of *Gentiana regeliana. Journal of Pharmacognosy and Phytochemistry* 1(2):32–35.

Bagrov, A. Y., Shapiro, J. I., and Fedorova, O. V. 2009. Endogenous cardio tonic steroids: Physiology, pharmacology, and novel therapeutic targets. *Pharmacological Reviews* 61(1):9–38.

Brain, K. R., and Turner, T. D. 1975. *The Practical Evaluation of Phytopharmaceuticals.* Bristol, UK: Wright-Scientechnica.

Camporese, A., Balick, M. J., Arvigo, R., Esposito, R. G., Morsellino, N., De Simone, F., and Tubaro, A. 2003. Screening of anti-bacterial activity of medicinal plants from Belize (Central America). *Journal of Ethnopharmacology* 87(1):103–107.

Chase Jr. C. R., and Pratt, R. 1949. Fluorescence of powdered vegetable drugs with particular reference to development of a system of identification. *Journal of the American Pharmaceutical Association* 38(6):324–331.

Das, K., Tiwari, R. K. S., and Shrivastava, D. K. 2010. Techniques for evaluation of medicinal plant products as antimicrobial agent: Current methods and future trends. *Journal of Medicinal Plants Research* 4(2):104–111.

Dua, V. K., Verma, G., Singh, B., Rajan, A.,Bagai, U., Agarwal, D. D., and Rastogi, A. 2013. Anti-malarial property of steroidal alkaloid conessine isolated from the bark of *Holarrhena antidysenterica. Malaria Journal* 12:194.

Harborne, J. B. 2009. *Phytochemical Methods: A Guide to Modern Techniques of Plant Analysis.* New Delhi, India: Springer (India) Pvt. Ltd.

Kapoor, S. 2002. *The Indian Encyclopedia*, Volume 1. Genesis Publishing Pvt. Ltd., Central Delhi, India.

Kokate, C. K. 1999. *Practical Pharmacognosy.* New Delhi, India: Vallabh Prakashan Publication.

Kokoski, C. J., Kokoski, R. J., and Slama, F. J. 1958. Fluorescence of powdered vegetable drugs under ultraviolet radiation. *Journal of the American Pharmaceutical Association (Scientific ed.)* 47(10):715–717.

Kumar, D., Kumar, K., Kumar, S., Kumar, T., Kumar, A., and Prakash, O. 2012. Pharmacognostic evaluation of leaf and root bark of *Holoptelea integrifolia* Roxb. *Asian Pacific Journal of Tropical Biomedicine* 2(3):169–175.

Kumari, M., and Jain, S. 2012. Tannins: An antinutrient with positive effect to manage diabetes. *Research Journal of Recent Sciences* 1(12):1–8.

Heinrich, M., Barnes, J., Gibbons, S. and Williamson, E. M. 2012. *Fundamentals of Pharmacognosy and Phytotherapy.* 2nd ed. Edinburgh, UK: Churchill Livingston, Elsevier Ltd.

Mace, M. E. 1963. Histochemical localization of phenols in healthy and diseased banana roots. *Physiologia Plantarum* 16(4):915–925.

Moteriya, P., Padalia, H., Rathod, T., Desai, D., and Chanda, S. 2015. Pharmacognostic standardization of *Madhuca indica* leaf and stem, an important medicinal plant. *International Journal of Pharmaceutical Sciences and Research* 6(2):705–711.

Prakash, O., Jyoti, Kumar, A., Kumar, P., and Manna, N. K. 2013. Adulteration and substitution in Indian medicinal plants: An overview. *Journal of Medicinal Plants Studies* 1(4):127–132.

Raaman, N. 2006. *Phytochemical Techniques.* New Delhi, India: New India Publishing Agency.

Shah, G., Chawla, A., Baghel, U. S., Rahar, S., Singh, P. S., and Dhawan, R. K. 2013. Pharmacognostic standardization of leaves of *Melaleuca leucadendron. Pharmacognosy Journal* 5(4):143–148.

Souto, A. L., Tavares, J. F., Da Silva, M. S., Diniz, M. D. F. F. M., de Athayde-Filho, P. F., and Barbosa Filho, J. M. 2011. Anti-inflammatory activity of alkaloids: An update from 2000 to 2010. *Molecules* 16(10):8515–8534.

Suhartono, E., Viani, E., Rahmadhan, M. A., Gultom, I. S., Rakhman, M. F., and Indrawardhana, D. 2012. Total flavonoid and anti-oxidant activity of some selected medicinal plants in South Kalimantan of Indonesian. *APCBEE Procedia* 4:235–239.

Kapoor, S. 2002. *The Indian Encyclopedia*, Volume 1. Genesis Publishing Pvt. Ltd., Central Delhi, India.

Rathakrishnan, T., Anandaraja, N., Ramasubramanian, M., Nirmala, L., and Thomas, M. I. 2009. *Traditional Agricultural Practices: Applications and Technical Implementations.* New Delhi, India: New India Publishing Agency.

Thamacin, A. M., Britto, J. S., Sinjumol, T., Mariat, G., Immanuel, S. M., Krishnakumar, J., and Raphael, R. M. 2014. Phytochemical evaluation of the ethanolic and aqueous extracts of *Ormocarpum sennoides* (Willd) DC. Prodr. (Leaf), *Zehneria scabra* (L.F.) Sond. (Tuber) and *Bauhinia Tomentosa* Linn. (Leaf). *International Journal of Pharmaceutical Sciences Review and Research* 28(1):22–25.

The Ayurvedic Pharmacopoeia of India. 2008. Part I, Volume-VI, 1st ed. New Delhi, India: Department of AYUSH, Ministry of Health and Family Welfare, Government of India.

Trease, G. E., and Evans, W. C. 2002. *Text Book of Pharmacognosy.* 15th ed. London, UK: W.B. Saunders Company Limited.

Waldi, D. 1965. Spray reagents for thin layer chromatography. In *Thin layer chromatography–A Laboratory Handbook*, ed. E. Stahl, pp. 483–502. New York: Academic Press.

12 Flowering Plants of Govind Pashu Vihar Wildlife Sanctuary, Western Himalaya

With Emphasis on High Altitude Threatened Medicinal Plants

Ramasamy Manikandan, Chithambaram Murugan, Nithya Sivagami Ponnurangam, and Mehala Devi Ramasamy

CONTENTS

12.1 INTRODUCTION

The Govind Pashu Vihar Wildlife Sanctuary is located between 31° 17' to 35° 55' N latitude and 77° 47' to 78° 37' E longitude in Purola Tehsil of the Uttarakashi District of Uttarakhand, which is situated in the Western Himalayas. The Western Himalayas and the Eastern Himalayas together is one of the largest centers of endemism in India, is one of the 34 biodiversity hotspots of the world (Synge 2005). According to Myers et al. (2000), biodiversity "hotspots" are the regions containing a high proportion of global biodiversity in a small area. Also biodiversity hotspots can be defined as areas having an exceptional concentration of endemic taxa. The sanctuary is situated in the Uttarakashi District of Uttarakhand, a newly formed state from Uttar Pradesh (Figure 12.1: Map 1, 2). As the protected areas are the prime centers of conservation and diversity, scientific information on all taxa needs to be collected and documented. Therefore, it is very important to make an assessment of the existing floristic diversity. Considering these reasons, Govind Pashu Vihar Wildlife Sanctuary was selected for an assessment of floristic diversity.

The Govind Pashu Vihar came into existence as a wildlife sanctuary in the year 1955 and is a part of the Upper Tons Valley. The core zone of the sanctuary was declared as a national park covering an area of 472.08 sq km in 1991. The sanctuary is named after the Bharat Ratna Govind Ballab Pant and is made up of the Rupin and Subin Ranges erstwhile of Tons Forest division. The sanctuary is divided into three ranges, namely, Rupin, Supin, and Sankri. Each range is in turn divided into two sections, viz., Himri, Parvat/Satta, Naitwar, Jakhol, Sankri, and Taluka, respectively (Figure 12.1: Map 2a), and each section further divided into five beats to control illicit activities including felling, poaching, etc. and to complete a developmental target. The inhabitants of the sanctuary are Rawain, Jaunsar, and Gujar. The present explored area shows diversity of vegetation, and it has been classified broadly as follows: 1. *sub-tropical pine forests*, 2. *Himalayan moist-temperate forests*, 3. *Himalayan dry-temperate forests*, 4. *sub-alpine forests*, and 5. *moist alpine scrubs* (Champion and Seth 1968).

12.1.1 PAST WORKS

Hooker and Thomson (1855) in their *Flora Indica* reported quite comprehensive information on the flora of the Garhwal Himalaya, which includes the present study area. Hooker and

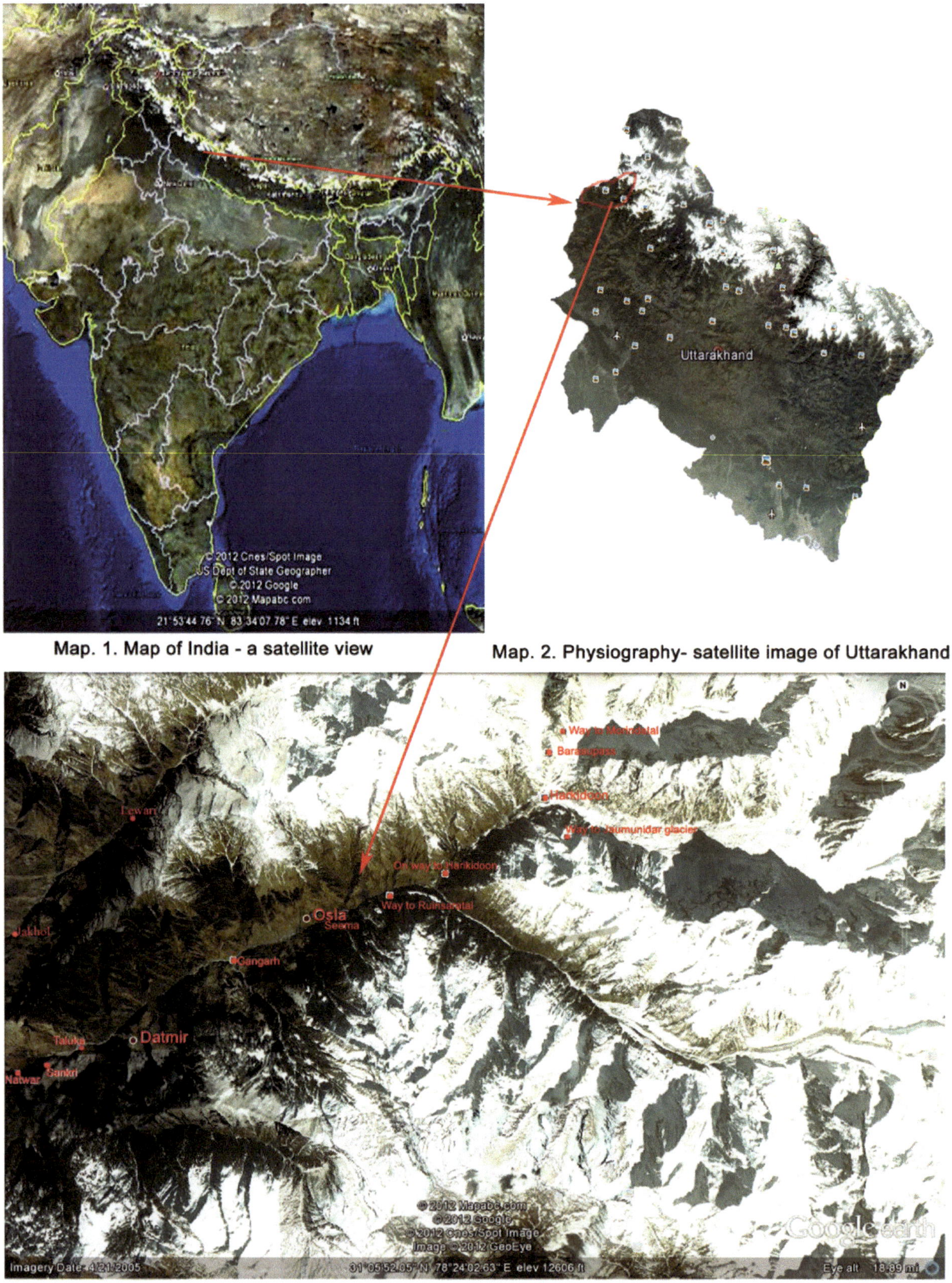

Map. 1. Map of India - a satellite view

Map. 2. Physiography- satellite image of Uttarakhand

Map. 2a. Location map of Govind Pashu Vihar Wildlife Sanctuary, Uttarkashi District, Uttarakhand (Google earth)

FIGURE 12.1 Map view of India.

his collaborators have dealt with many species in *The Flora of British India* (1872–1897). A few botanists have made botanical explorations into the region, such as Keshwan in 1897; Machinnon in 1889, 1897–1898, 1903; Gollen in 1889; and Gamble in 1893 (Burkill 1965). Atkinson's (1882) work on the flora of the Himalayas with special reference to Kumaon, Garhwal, Nepal, and Tibet and Duthie's (1903–1929) *Flora of Garhwal Himalaya* can be conveniently considered as one of the eminent contributions in the field. Some scholars like Deva and Naithani (1986), Garg (1987), Aswal et al. (1988), Dangwal (1993), and Nautiyal (1996) have made efforts to study the flowering plants of the Garhwal Himalaya in terms of certain families or groups of a given geographical entity. Important information can be derived from *Flowers of the Himalaya* (Polunin and Stainton 1984), *Blossoming Garhwal Himalaya* (Rawat et al. 1985), *Flowers of Himalaya* (Stainton 1988) and *Flowers of West Himalaya* (Dang 1993), Naithani's *Flora of Chamoli* (1984, 1985), *Flora of Tons Valley* by Rana et al. (2003), *Flowering Plants of Uttarakhand-* A checklist by Uniyal et al. (2007), and *Flora of Gangotri National Park, Western Himalaya* by Pusalkar and Singh (2012). Other important unpublished works include Badoni's (1989) *Herbaceous Flora of Uttarkashi District*, Bamola's (1993) *Lignosae Flora of Uttarkashi District*, and Kandwal's (2009) *Grass Flora of Uttarakhand*. Also botanists like Gibson (1954), Gupta (1955, 1957, 1960), Raizada (1958), Issar and Uniyal (1967), Naithani (1969), Rau (1974, 1975, 1981), Singh and Singh (1987, 1992), Gaur (1987, 1999), Rawat and Sharma (1992), Dangwal et al. (1993, 1995), Gaur and Painuli (1994), Uniyal et al. (1994, 1997), Rawat et al. (1994, 2001), Paramanand (2009), Manikandan et al. (2015), and Manikandan and Srivastava (2015a, 2015b) have contributed for the *Flora of Western Himalayas.*

12.2 MATERIALS AND METHODS

The present study was undertaken to explore this underexplored area periodically along different topographic and climatic gradients within the potential area of the Govind Pashu Vihar Wildlife Sanctuary, which is endowed with a rich floristic diversity. The duration of each study tour ranged from 10 to 20 days. The field observations have been noted in the field itself and any details which cannot be deduced from a herbarium specimen. Plant specimens have been processed through the conventional methods of drying, poisoning, mounting, stitching, and labeling (Fosberg and Sachet 1965; Bridson and Forman 1998). The collected specimens were identified by comparing with authentic specimens available in different herbaria such as BSD (Herbarium of Botanical Survey of India, Northern Regional Centre, Dehradun), CAL (Central National Herbarium, Howrah), LWG (Herbarium of National Botanical Research Institute, Lucknow), RRL/IIIM (Indian Institute of Integrative Medicine/Regional Research Laboratory, Jammu), and DD (Forest Research Institute, Dehradun) protologues, journals, and other relevant sources, such as *The Flora of British India* (Hooker, l.c.), *Flora of Tons Valley* (Rana et al. 2003), *Flowering Plants of Uttarakhand* (Uniyal et al. 2007), besides many other recent monographs and revisions. These specimens were deposited at the Herbarium of Botanical Survey of India, Northern Regional Centre, Dehradun (BSD).

12.3 RESULTS AND DISCUSSION

The floristic composition of the Govind Pashu Vihar Wildlife Sanctuary comprises 120 families, 464 genera, 804 species, 8 subspecies, and 14 varieties of angiosperms. Based on the number of genera within the family, the Asteraceae has the highest genera (52 genera), followed by Poaceae (45 genera), Fabaceae (22 genera), and according to the number of species against the genus, the same sequence, that is, the highest in Asteraceae (89 species), followed by Poaceae (56 species), and Fabaceae (41 species) have been recorded in the present work.

12.3.1 High Altitude Threatened Medicinal Plants

According to Walter and Gillett (1998), over 60,000 species have been evaluated for conservation status as per the internationally accepted criteria, of which, 34,000 are classified as globally threatened with extinction. The Botanical Survey of India has published Red data books (Nayar and Sastry 1987, 1988, 1990), which contain information on 622 threatened plants. Recently, on February 19, 2000, the 51st meeting of the International Union for Conservation of Nature and Natural Resources (IUCN) Council Species Survival Commission was held in Gland, Switzerland, which has come out with an improved version of the Red list categories and criteria version 3.1 (IUCN 2001). Based on IUCN criteria for the critically endangered, the endangered and vulnerable taxa are classified mainly based on the reduction in population size, extent of occurrence, and area of occupancy. The following threatened medicinal plants are recorded from the present study area (Table 12.1).

12.3.2 Factors Causing Loss of Biodiversity

The most important threats to the indigenous flora of the sanctuary is anthropogenic include deforestation, cattle grazing, timber and fuel wood cutting, soil erosion, construction schemes, plantation, plant collection, fire, tourist pressure, and the invasive species, result in the degradation of forests. Further, environmental factors such as seismic activity or violent stress strokes and floods chase landslides or wash away top soil and then alter or destroy the vegetation. The prospective conservation of such unmatched scenic beauty of the floristic composition as well as herbal wealth of the present study area is essential, and also it is a home for a lot of endemic plants and endangered animals which helps in maintaining genetic diversity.

12.3.3 Conservation Measures

Several measures are underway for the protection of the species reaching extinction and rare, endangered, and threatened species. These measures include survey and inventorization of plant resources in the wild, and to develop an accurate database for their planning and

TABLE 12.1
List of Threatened Medicinal Plants from Govind Pashu Vihar Wildlife Sanctuary, Western Himalaya

Name of the Species	Family	Local Names	Habit	Habitat	Altitude (m)	Phenology	Parts Used	Uses	Status
Acer caesium Wall. ex Brandis	Aceraceae	Kainjal, Marik	Tree	Moist temperate deciduous forest	2130–3050	April–July	Whole plant	Tradition medicine practice anti-rheumatic, emetic, purgative, anti-inflammatory.	VU
Aconitum falconeri Holmes ex Stapf.	Ranunculaceae	Meetha, Bis Bikh, Telia	Herb	Alpine meadows	3500–4800	July–September	Tuber	Sedative, carminative, anti-inflammatory, and root powder mixed with milk given to reduce cardio-toxic effect.	LC
Aconitum heterophyllum Wall. ex Royle	Ranunculaceae	Hatthis	Herb	Grassy slopes and riverbeds	3000–3500	August–September	Tuber	Crude extract of tuber mixed with honey taken to cure fever	CR
Aconitum violaceum Jacqeum. ex Stapf.	Ranunculaceae	Methu, Tilla Dudhia,	Herb	Open grassy slopes near alpine regions	3600–4500	July–October	Tuber	Paste of tuber applied on insect stings, contagious infections, and inflammation of the intestines.	VU
Acorus calamus L.	Araceae	Baji, Birch	Herb	Swampy, moist wet places	Up to 3500	March–May	Rhizome	Rhizome paste applied to treat headache, skin diseases, and wounds. Dried rhizomes/powders are used as insecticide	CR
Allium stracheyi Baker	Alliaceae	Seemori phul, Jambu	Herb	Rocks, shady open grassy, and dry slopes	2000–3800	July–October	Whole plant	Dried flowers used as a spice and curry powder locally known as "*Pharan*," which is much costlier than onion and helps in reducing blood cholesterol levels, acts as a tonic to the digestive system, and also tonify the circulatory system.	VU
Angelica glauca Edgew.	Apiaceae	Chora	Herb	Moist temperate forests	2500–4000	June–August	Flower root	Root powder mixed with milk given with empty stomach to cure severe stomach pain	EN
Berberis aristata DC.	Berberidaceae	Chothir, Kasmoi, Kingora	Shrub	Hillslopes near temperate forests	Up to 2800	March–June	Root stem	Exploited for berberine, a bitter alkaloid which is used opthalmia, febrifuge, and blood purifier	EN
Berberis pseudoumbellata Parker	Berberidaceae	–	Shrub	Open dry forests area	Up to 2700	April–June	Bark fruit	Powdered bark is used for the treatment of fever, backache, jaundice, and urinary tract infection while its fruit is valued as a tonic	LC
Carduus edelbergii Rech.*f.*	Asteraceae	Kandara	Herb	Forest margins, thickets in mountain hills	2000–4200	July–October	Whole plant	The plant is considered as an effective remedy for pleurisy. It is used as a febrifuge, diuretic and sudorific	VU
Cyananthus integer Wall. ex Benth.	Campanulaceae	Bhandai	Herb	Rocky boulders near alpine meadows	3200–4800	August–November	Whole plant	Endemic	LC
Cypripedium cordigerum D. Don	Orchidaceae	–	Herb	Open grassy field near temperate Himalayas	2200–3200	June–July	Flower	This species has a great ornamental value	EN

(Continued)

TABLE 12.1 (*Continued*)
List of Threatened Medicinal Plants from Govind Pashu Vihar Wildlife Sanctuary, Western Himalaya

Name of the Species	Family	Local Names	Habit	Habitat	Altitude (m)	Phenology	Parts Used	Uses	Status
Dactylorhiza hatagirea (D. Don) Soo	Orchidaceae	Punja Hattahaddi	Herb	Moist open grassy places in alpine meadows	2800–4200	June–October	Tuber	Paste is used to cure for tubers are used as nervine tonic and aphrodisiac. It is also used to treat cuts, wounds, cough, boil, stomach ache, diabetes, diarrhea, dysentery, paralysis, convalescence, impotence, and malnutrition and anemia.	CR
Delphinium denudatum Wall. ex Hook. *f.* & Thomson	Ranunculaceae	–	Herb	Open grassy fields	2400–3800	July–October	Tender shoot	Paste of tender shoot is given with tea to cure cold	CR
Dioscorea deltoidea Wall. ex Kunth	Dioscoreaceae	Gethi, Oakanchaw	Climber	Bushes, thickets in mountain hills	Up to 2800	May–June	Tuber	Used as a vermifuge and anthelmintic for purging out intestinal worms. Diosgenin is said to be a basic material for steroidal drugs preparation	CR
Fritillariaroylei Hook. *f.*	Liliaceae	Kakoli, Sheethkar	Herb	Alpine slopes	2600–4000	April–June	Bulb	The bulb is supposedly anti-asthmatic, anti-rheumatic, febrifuge, galactagogue, hemostatic, ophthalmic, and oxytocic. It is used in the treatment of TB and asthma.	EN
Goodyera fusca (Lindl.) Hook. *f.*	Orchidaceae	–	Herb	Boulders or moist rocky places	2200–3600	July–October	Flower	This species has a great ornamental value	EN
Habenaria pectinata D. Don	Orchidaceae	–	Herb	Moist shady places near temperate forests	Up to 2000	August–October	Leaf tuber	The leaves are crushed and applied in snakebites. Tubers mixed with condiments are used in arthritis	LC
Hedychium spicatum Buch.-Ham. ex Sm.	Zingiberaceae	Kapurachri	Herb	Moist shady places	2000–3400	July–October	Rhizome	Rhizome extract has anti-microbial activity which is used for making various types of drugs including anti-cancerous drug and essential oil	LC
Holboellia latifolia Wall.	Lardizabalaceae	Gomphal	Climbing shrub	Bushes near mixed temperate forests	1800–2800	March–June	Fruit	Edible, it is considered to be very palatable in the Himalayas	VU
Houttuynia cordata Thunb.	Saururaceae	–	Herb	Marshy areas	1500–2800	May–June	Whole plant	Prepared herbal medicine for its anti-viral, anti-bacterial, and anti-leukemic activities, used for diuresis and detoxification	EN
Jurinella macrocephala (Royle) Aswal & Goel	Asteraceae	Dhoop, Guggal	Herb	Alpine meadows	3200–4300	July–September	Root twig	Root extract is used as an incense, the whole plant is used in ceremonial and ritual purpose by tribal	EN

(*Continued*)

TABLE 12.1 (*Continued*)
List of Threatened Medicinal Plants from Govind Pashu Vihar Wildlife Sanctuary, Western Himalaya

Name of the Species	Family	Local Names	Habit	Habitat	Altitude (m)	Phenology	Parts Used	Uses	Status
Malaxis muscifera (Lindl.) Kuntze	Orchidaceae	Rsabhak	Herb	Moist shady slopes	2800–3600	July–September	Pseudo-bulb	It is used in sterility, hemorrhages, dysentery, fever, emaciation, burning sensation, rheumatism, and debility. Paste applied externally for insect bites	EN
Meconopsis aculeata Royle	Papaveraceae	Kaliharu, Thesu	Herb	Amidst boulders alpine meadows	3500–4800	June–October	Flower	Petals extract is given to cure fever	CR
Microstegium falconeri (Hook. *f.*) Clayton	Poaceae	–	Herb	Rocky places	Up to 2500	July–March	Whole plant	Grazing	LC
Morus serrata Roxb.	Moraceae	–	Tree	Bushes, thickets in mixed forests	1200–2700	March–May	Fruit root	Fruits edible, the juice of the root is used as an anthelmintic	EN
Nardostachys jatamansi (D. Don) DC.	Valerianaceae	Balchhar, Jatamansi	Herb	Alpine thickets or meadows	2500–4500	June–September	Whole plant	Plant extract given in heart-pain, fits, and rubbed in muscle fracture to relieve pain also regulate urination, menstruation, and digestion	CR
Paeonia emodi Wall. ex Royle	Paeoniaceae	Chandra	Herb	Ravine and hillslopes	1800–2600	May–October	Leaf	The leaves are used in urine disorder	CR
Paris polyphylla Sm.	Trilliaceae	Satwa	Herb	Shady moist places	1800–2800	May–June	Root	Roots are analgesic, anti-phlogistic, anti-pyretic, anti-spasmodic, anti-tussive, depurative, febrifuge, and narcotic. A decoction of the roots is used in the treatment of poisonous snakebites, boils and ulcers, diphtheria and, epidemic Japanese B encephalitis	CR
Parnassia nubicola Wall. ex Royle	Parnassiaceae	Phutkya	Herb	Amidst boulders near alpine meadows	3200–4800	June–October	Whole plant	Plant extract is used for the treatment of food poisoning. Rootstocks are used externally in snakebites	VU
Picrorhiza kurrooa Royle ex Benth.	Scrophulariaceae	Kutahi, Kowdai, Kutki	Herb	Hillslopes & rocky areas	3200–4000	June–September	Rhizome	Rhizomes are widely used an antibiotic and to treat liver ailments	EN
Podophyllum hexandrum Royle	Podophyllaceae	Fungli kakkidi	Herb	Boulders in alpine regions	3200–4800	April–July	Whole plant	Whole plant, but especially root, is cholagogue, cytostatic, and purgative. It contains podophyllin, which has an anti-miotic effect, so it is used especially in the treatment of ovarian cancer	CR
Polygonatum verticillatum (L.) All.	Convallariaceae	Sakkakul	Herb	Shady slopes	Up to 2700	May–June	Rhizome	It is valued as a salep, a strength giving food. Plant is diuretic and contains glucoside of digitalis group	EN
Rhododendron hypenanthum Balf. *f.*	Ericaeae	Altus, Althas Kodya	Shrub	Rocky places near alpine	3300–4500	May–October	Leaf	Dried leaves used as tea powder	VU

(*Continued*)

TABLE 12.1 (*Continued*)
List of Threatened Medicinal Plants from Govind Pashu Vihar Wildlife Sanctuary, Western Himalaya

Name of the Species	Family	Local Names	Habit	Habitat	Altitude (m)	Phenology	Parts Used	Uses	Status
Sauromatum venosum (Ait.) Schott	Araceae	Bhasma-kand	Herb	Shady places or undergrowth in deodar forests	Up to 2300	April–May	Whole plant	Ornamental	VU
Saussurea obvallata (DC.) Edgew.	Asteraceae	Brahma-kamal	Herb	Rocky slopes and amidst boulders	3500–4800	July–October	Flower	Cultural and ritual uses	EN
Schisandra grandiflora (Wall.) Hook. *f.* & Thomson	Schisandraceae	–	Straggling shrub	Bushes, thickets in temperate forests	2000–2900	May–August	Fruit	Edible	LC
Skimmia anquetilia Taylor & Air Shaw	Rutaceae	Kathurchar, Kedar pathri	Shrub	Open areas in *Quercus-Rhodendron* forests	2300–3500	April–November	Leaf	The leaves are used in religious ceremonies and also as incense "dhup"	EN
Thalictrum foliolosum DC.	Ranunculaceae	Pilijari, Pinjari	Herb	Shady places in mixed forests	2800–3500	July–September	Whole plant	Root is anti-periodic, diuretic, febrifuge, ophthalmic, purgative, salve, stomachic, and tonic. Used for atonic dyspepsia, peptic ulcers, indigestion, fevers, toothache, hemorrhoids, and ophthalmia. Leaf juice applied to boils and pimples	VU

Abbreviations: CR, critically endangered; VU, vulnerable; EN, endangered; LC, least concern.

monitoring purposes, the Protected Area Network has been created for in situ conservation, which has been designated as species-, habitats-, and ecosystems-oriented. Ex situ conservation of rare, endangered, threatened, and endemic species has been undertaken through botanic gardens, zoos, gene banks, etc., and as per the Biological Diversity Act, 2002, a state-wise list of species of plants which are on the verge of extinction along with guidelines to prohibit and regulate their collection, rehabilitate, and preserve these species has been proposed.

I. DICOTYLEDONES

1. RANUNCULACEAE

1. ***Aconitum balfourii*** Stapf., Rare on way to Jamnidhar Glacier, Barasu Pass, Deokyar, and Kedarakantha bugyal (3500–4000 m), August–October.
2. ***Aconitum falconeri*** Holmes ex Stapf., Rare in alpine meadows (3500–4000 m). July–October. "*Meetha*."
3. ***Aconitum heterophyllum*** Wall. ex Royle, Common in alpine regions, waste places, and along cultivated fields (3500–4000 m). Jakhol, Jamnidhar Glacier, Harkidoon, and Kedarakantha bugyal. August–September.
4. ***Aconitum violaceum*** Jacqeum. ex Stapf, Occasional in alpine zones, open places, and sandy soils (3500–4000 m). On way to Barasu Pass/Hattagad. July–October.
5. ***Actaea acuminata*** Wall. ex Royle, Common in mixed temperate forests (2300–3000 m). Kalajungle and Kedarkantha. June–October.
6. ***Anemone demissa*** Hook. *f.* & Thomson, Common almost throughout the alpine and subalpine regions (3300–4000 m). On way to Jamnidhar Glacier. May–September.
7. ***Anemone obtusiloba*** D. Don, Common in alpine and subalpine meadows (3200–4000 m). Kalhatta, Deokyar, Dhatmir-Harkidoon, and Jamnidhar. May–July. "*Ageli*," "*Thermayyis.*"
8. ***Anemone rivularis*** Buch.-Ham. ex DC., Common in alpine and subalpine regions (3500–4000 m). Harkidoon-Jamnidha, Barasu Pass-Ruinsara Tal, and Hatta Valley. On way to Kedarkantha. On way to Jakhol. May–July.
9. ***Anemone vitifolia*** Buch.-Ham. ex DC., Common in shady places and rocky crevices near mixed temperate (2200–2800 m). On way to Taluka and Sankri-Taluka. May–July.
10. ***Aquilegia pubiflora*** Wall. ex Royle, Scattered in open sandy soils, shady places, and in rocky crevices near alpine to subalpine regions (2700–3500 m). About 1 km below Seema Forest rest house. June–September.
11. ***Caltha palustris*** L., Common in open places, along the melting snows and streams near subalpine to alpine regions (2800–4500 m). Ruinsara Tal, Obragad, and Harkidoon. May–June.
12. ***Clematis connata*** DC., Common along the bridal path near mixed temperate (1700–2700 m). Harkidoon and Seema. July–October.
13. ***Clematis grata*** Wall., Common along the roadsides and on *Berberis* bushes near mixed temperate (1500–2500 m). Sankri and Jakhol. July–October.
14. ***Delphinium denudatum*** Wall. ex Hook. *f.* & Thomson, Common in open places near temperate regions (1500–2500 m). Near Sankri. On way to Jakhol. May–July.
15. ***Delphinium vestitum*** Wall. ex Royle, Common in temperate to alpine regions (2400–3500 m). Harkidoon and Seema. July–October.
16. ***Oxygraphis endlicheri*** (Walp.) Bennet & Chandra, Occasional in moist shady places and around sandy soils (3500–4000 m). Barasu Pass-Deokyar, on way to Jamnidhar Glacier. June–July.
17. ***Ranunculus diffusus*** DC., Common throughout the area (1500–2000 m). Taluka and Khimdar. May–October.
18. ***Ranunculus distans*** Royle, Abundant in moist places near temperate to subalpine regions (2000–3500 m). On way to Jamnidhar Glacier, Seema, and Naitwar near Sankri. July–August.
19. ***Ranunculus hirtellus*** Royle ex D. Don, Common throughout the area mostly near wet places (2000–3000 m). On way to Seema and Deokyar. May–August.
20. ***Ranunculus sceleratus*** L., Fairly common in damp places near temperate regions (1000–1500 m). Naitwar. May–September.
21. ***Thalictrum chelidonii*** DC., Occasional in moist shady hillslopes. On way to Harkidoon. June–September.
22. ***Thalictrum cultratum*** Wall., Common on higher altitude (3000–4000 m). Harkidoon-Barasu Pass and Jamnidhar. July–August.
23. ***Thalictrum elegans*** Wall. ex Royle, Common on sandy gravel soils near the alpine regions (3300–4500 m). On way to Barasu Pass. July–September.
24. ***Thalictrum foliolosum*** DC., Common in shady places near mixed temperate (2800–3500 m). Sankri-Taluka. July–September.
25. ***Thalictrum javanicum*** Blume, Occasional on hillslopes (2500–3000 m). Sankri. June–October.

26. ***Thalictrum platycarpum*** Hook. *f.* & Thomson, Rare on alpine meadows. Harkidoon. August–October.
27. ***Thalictrum reniforme*** Wall., Common on hillslopes (2500–3500 m). On way to Harkidoon and Seema. July–October.
28. ***Thalictrum saniculiforme*** DC., Common on mossy boulders and on mossy tree trunks near damp mixed temperate (1600–2700 m). Taluka and Seema. July–October.
29. ***Trollius acaulis*** Lindl. Common in alpine meadows (3500–4500 m). Harkidoon-Barasu Pass, Deokyar, and Jamnidhar. June–August.
30. ***Paeonia emodi*** Wall. ex Royle, Rare in ravine and on hillslopes (1800–2600 m). Sankri, Taluka, and Sankri-Jakhol. May–October.
31. ***Schisandra grandiflora*** (Wall.) Hook. *f.* & Thomson, Occasional in broad-leaved mixed temperate and coniferous forests (2000–2900 m). Sankri-Seema, Ruinsara Tal, and Jakhol. May–August.

2. MENISPERMACEAE

32. ***Cissampelos pareira*** L. var. **hirsuta** (Buch.-Ham. ex DC.) Forman, Common along the roadsides and in wet places (1500–2000 m). Naitwar-Taluka. June–October.
33. ***Stephania glabra*** (Roxb.) Miers, Common in ravines (1500–2000 m). Naitwar, near Mautad. May–August.

3. BERBERIDACEAE

34. ***Berberis apiculata*** (Ahrendt) Ahrendt, Occasional on the banks of streams and open dry places on hillslopes (2200–2500 m). On way to Harkidoon. May–August.
35. ***Berberis aristata*** DC., Common in mixed forests (1500–3000 m). Sankri. On way to Seema. March–October.
36. ***Berberis chitria*** Lindl., Common in exposed localities, often growing under dense shade in oak forests (1700–2500 m). Sankri. March–October.
37. ***Berberis coriaria*** Royle ex Lindl., Occasional in mixed forests (1700–2400 m). Sankri. May–June.
38. ***Berberis glaucocarpa*** Stapf, Common on exposed hillslopes (1700–2200 m). Sankri. May–June.
39. ***Berberis jaeschkeana*** C.K. Schneid., Occasional on open hillslopes in subalpine to alpine regions (3000–4200 m). Kalajungle. July–September.
40. ***Berberis lycium*** Royle, Common in exposed places and along roadsides (1500–2600 m). Sankri. March–July.
41. ***Berberis pachyacantha*** subsp. **zabeliana** (C.K. Schneid.) Jafri, Rare on hillslopes. April–June.
42. ***Berberis pseudoumbellata*** Parker, Very sparse on hillslopes, forming scattered patches in mixed temperate (1800–3000 m). Sankri-Taluka. On way to Seema. April–August.
43. ***Berberis umbellata*** Wall. ex G. Don, Rare on hillslopes, forming scattered patches (2700–3500 m). Sankri. On way to Kedarkantha. May–October.

4. PODOPHYLLACEAE

44. ***Sinopodophyllum hexandrum*** (Royle) T.S. Ying, Common in damp places near mixed temperate (2800–3500 m). Sankri-Kedarkanth, Jakhol-Deokyar, Jamindhar Glacier, Obragad, and Seema. April–July.

5. LARDIZABALACEAE

45. ***Holboellia latifolia*** Wall., Occasional in mixed temperate (1800–2800 m). Sankri-Seema. March–June.

6. PAPAVERACEAE

46. ***Meconopsis aculeata*** Royle, Occasional on amidst boulders and alpine meadows (3500–4500 m). Barasu Pass, Jamnidhar, Deokyar, and Kedarkantha. June–October.
47. ***Papaver dubium*** L., Rare in open places (1700–2000 m). Himri. April–May.

7. FUMARIACEAE

48. ***Corydalis cornuta*** Royle, Common in shady places near mixed temperate (2000–3600 m). Seema and Osla Village. May–September.
49. ***Corydalis govaniana*** Wall., Rare in shady places near temperate to alpine regions (3000–4000 m). Kalhattidhar. May–August.
50. ***Corydalis stracheyi*** Duthie ex Prain, Occasional in open grassy fields and amidst boulders (3600–4500 m). On way to Barasu Pass. July–October.
51. ***Corydalis vaginans*** Royle, Occasional on hillslopes and along roadsides in alpine regions (3500–4800 m). Barasu Pass. July–September.
52. ***Dactylicapnos macrocapnos*** (Prain) Hutch., Occasional on hillslopes (1700–2000 m). Himri-Naitwar and Seema. July–September.

8. **BRASSICACEAE**

53. ***Arabidopsis thaliana*** (L.) Heynh., Occasional on hillslopes and amidst grasses (3000–3800 m). Taluka-Seema. June–September.
54. ***Arabis amplexicaulis*** Edgew. Common in amidst grasses near subalpine-temperate regions (2800–3000 m). Kalkati-Dhar and Obragad. June–September.
55. ***Arabis pterosperma*** Edgew., Occasional in amidst grasses (3000–4000 m). Kalhattidhar and Deokyar. April–September.
56. ***Barbarea vulgaris*** W.T. Aiton, Occasional on hillslopes and open places (2700–3800 m). Seema-Harkidoon. May–September.
57. ***Capsella bursa-pastoris*** (L.) Medik., Common throughout the area on hillslopes and in waste places (1800–4000 m). Sankri, Harkidoon, and Istragad. March–September.
58. ***Cardamine flexuosa*** With., Occasional on moist shady places, along ditches and stream sides (up to 3100 m). On way to Ruinsara Tal. April–August.
59. ***Cardamine impatiens*** L., Occasional on hillslopes and in moist shady places (2000–2600 m). Obragad. April–July.
60. ***Crucihimalaya himalaica*** (Edgew.) Al-Shehbaz, Rare on hillslopes and amidst grasses (2700–3800 m). Sankri. June–September.
61. ***Draba lasiophylla*** Royle, Rare on amidst grasses in hillslopes near alpine and subalpine regions (3800–4500 m). On way to Harkidoon. June–September.
62. ***Eruca sativa*** Mill., Occasional in dry and open places (1000–1500 m). Naitwar near Sankri. January–April.
63. ***Erysimum hieraciifolium*** L., Common amidst grasses, on hillslopes and in open places (2000–3500 m). On way to Taluka and Sankri. May–September.
64. ***Lepidium capitatum*** Hook. *f.* & Thomson, Common on hillslopes and in open places (2000–2600 m). Naitwar. July–September.
65. ***Lepidium virgincum*** L., Common in open places and along roadsides (2000–2600 m). Naitwar-Sankri. On way to Jakhol. August–October.
66. ***Nasturtium officinale*** R. Br., Occasional along river side and moist sandy places (1000–1500 m). Naitwar. April–May.
67. ***Rorippa indica*** (L.) Hiern., Common on hillslopes (2000–2500 m). Sankri. On way to Taluka. May–July.
68. ***Turritis glabra*** L., Common on hillslopes (up to 2800 m). Obragad. May–August.

9. **VIOLACEAE**

69. ***Viola betonicifolia*** Sm., Common in moist shady places (2600–3000 m). Kedar River and Taluka-Seema. July–September.
70. ***Viola biflora*** L., Occasional on hillslopes (2800–3300 m). Kalhattidhar-Harkidoon and Deokyar. June–September.
71. ***Viola canescens*** Wall. in Roxb., Occasional on hillslopes and in shady moist places (1800–3000 m). Naitwar. April–October.
72. ***Viola pilosa*** Blume, Sanctuary, April–August.

10. **POLYGALACEAE**

73. ***Polygala persicariifolia*** DC., Common in forests, grasslands, and roadsides on slopes of hills (1200–2200 m). On way to Sankri and Naitwar. July–October.
74. ***Polygala tatarinowii*** Regel, Common in open grasslands, thickets, and along roadsides (up to 2300 m). On way to Taluka. August–November.

11. **CARYOPHYLLACEAE**

75. ***Arenaria festucoides*** Royle, Common on open hillslopes and sandy soils near alpine regions (3400–4000 m). Harkidoon, Barasu Pass, and Hattagad. June–October.
76. ***Arenaria neelgherrensis*** Wight & Arn., Common amidst grasses in subalpine and alpine slopes (3000–4000 m). Isragad. May–October.
77. ***Cerastium davuricum*** Fisch. ex Spreng., Scattered in subalpine meadows (3600–4500 m). Harkidoon rest house. July–October.
78. ***Cerastium fontanum*** Baumg., Occasional on open grassy hillslopes and rocky crevices (up to 3800 m). On way to Taluka and Seema. May–August.
79. ***Cerastium glomeratum*** Thuill., Common weed throughout the area (1700–3300 m). Sankri and Taluka-Seema. March–July.
80. ***Drymaria cordata*** (L.) Schult., Common on hillslopes and in shady places (1700–2200 m). Naitwar-Sankri. May–September.
81. ***Gypsophila cerastioides*** D. Don, Common on hillslopes, in shady places near subalpine, and alpine regions (2700–3500 m). Osla Village, Harkidoon, and Kalhattidhar. On way to Jamnidhar Glacier. May–October.
82. ***Sagina saginoides*** (L.) Karsten, Common in open places (1800–3500 m). Taluka-Seema. June–October.
83. ***Silene indica*** Roxb. ex Otth, Common along the streams and amidst grasses (2900–4000 m). Kalhattidhar-Harkidoon.

On way to Jamnidhar Glacier, Kedarkantha bugyal, and Barasu Pass. July–October.

84. ***Silene indica*** Roxb. ex Otth var. **edgeworthii** (Bocq.) Y. Nasir, Common on hillslopes in temperate and subalpine regions (2400–3000 m). Jakhol, Seema, and Harkidoon. June–October.
85. ***Silene setisperma*** Majumdar, Rare on open grassy slopes (3500–4200 m). Hatta Valley. July–August.
86. ***Silene viscosa*** (L.) Pers., Common in amidst grasses (2700–3600 m). On way to Kedartal. July–October.
87. ***Stellaria decumbens*** Edgew., Common on hillslopes in shady places (1500–3000 m). Taluka-Seema. July–October.
88. ***Stellaria media*** (L.) Villars, Common on hillslopes and along roadsides (1700–2500 m). Sankri. January-March.
89. ***Stellaria monosperma*** Buch.-Ham. ex D. Don, Common on hillslopes and in shady places (1800–2700 m). Taluka. On way to Seema. July–October.
90. ***Stellaria patens*** D. Don, Common on hillslopes in moist temperate to alpine regions (2600–3500 m). On way to Seema. On way to Harkidoon. May–October.

12. PORTULACACEAE

91. ***Portulaca oleracea*** L., Common along the roadside and in waste places (up to 2000 m). Naitwar. January–December.

13. TAMARICACEAE

92. ***Myricaria elegans*** Royle, Occasional on open hillslopes and near boulders (3500–4500 m). On way to Hatta Valley. On way to Jamnidhar Glacier. July–August.
93. ***Myricaria germanica*** (L.) Desv. subsp. **alopecuroides** (Schrenk) Kitam., Occasional on open hillslopes and in sandy soils near alpine ravines (3500–4000 m). On way to Jamnidhar Glacier and Harkidoon. May–September.

14. HYPERICACEAE

94. ***Hypericum dyeri*** Rehder, Common to grow on rocks in shady places (2000–2500 m). Sankri-Taluka and Seema. April–September.
95. ***Hypericum elodeoides*** Choisy, Common on open hillslopes and shady places near mixed temperate (2000–3000 m). On way to Taluka-Seema. July–October.
96. ***Hypericum himalaicum*** N. Robson, Occasional in rocky or grassy slopes and along roadsides (2200–2600 m). On way to Seema. August–October.
97. ***Hypericum hookerianum*** Wight & Arn., Common on rocky places and in shady places (1700–2300 m). Naitwar-Taluka, on way to Kalhattidhar. May–September.
98. ***Hypericum oblongifolium*** Choisy, Common on rocky places (1700–2300 m). Naitwar-Taluka, Salta, and Dhola area. May–September.
99. ***Hypericum perforatum*** L., Fairly common in open and disturbed areas. On way to Sankri, About 4 km below Seema. May–September.
100. ***Hypericum uralum*** Buch.-Ham. ex D. Don, Common on rocky shady places (1700–2200 m). Naitwar-Sankri, Taluka, and Dhola. May–September.

15. ACTINIDIACEAE

101. ***Saurauia napaulensis*** DC., Occasional in mixed temperate. Naitwar. March–October.

16. MALVACEAE

102. ***Abutilon indicum*** (L.) Sweet, Fairly common along roadsides and waste places (1200–1500 m). Naitwar. April–August.
103. ***Malva sylvestris*** L., Fairly common along roadsides and waste places (1200–1500 m). Sankri.
104. ***Malva verticillata*** L., Rare on hillslopes (1700–3000 m). Naitwar-Seema and Harkidoon. April–October.
105. ***Sida cordata*** (Burm. *f.*) Borss. Common on hillslopes and in waste places (1400–1800 m). Naitwar-Sankri. May–October.
106. ***Sida ovata*** Forssk., Occasional along roadsides near mixed temperate (up to 1800 m). Naitwar-Sankri. June–October.
107. ***Urena lobata*** L., Common in subtropical-temperate regions (800–1500). On way to Sankri. June–October.

17. TILIACEAE

108. ***Corchorus aestuans*** L., Common in waste places (ascending up to 1800 m). Naitwar near Sankri. August–February.
109. ***Triumfetta rhomboidea*** Jacq., Common throughout the area (ascending up to 1700 m). Naitwar-Sankri. August–January.

18. LINACEAE

110. ***Reinwardtia indica*** Dumort., Occasional in fringes of forests (1500–2500 m). Naitwar and Seema. December–June.

19. GERANIACEAE

111. ***Geranium collinum*** Steph. ex Willd., Common along irrigation ditches or damp places (2800–4800 m). On way to Barasu Pass, Hatta Valley, and Harkidoon. July–September.
112. ***Geranium nepalense*** Sweet, Common in shady places (1600–2800 m). Sankri, Taluka-Seema, and Naitwar. July–September.
113. ***Geranium ocellatum*** Cambess., Fairly common in moist shady places (1000–1600 m). Naitwar. On way to Obragad. March–May.
114. ***Geranium polyanthus*** Edgew. & Hook. *f.*, Common on open hillslopes (3600–4600 m). On way to Jamnidahr Glacier. August–October.
115. ***Geranium robertianum*** L., Common in moist shady places (1800–2800 m). On way to Seema. May–June.
116. ***Geranium wallichianum*** D. Don ex Sweet, Common throughout the temperate forests and in alpine regions (2300–3300 m). Seema. July–September.

20. OXALIDACEAE

117. ***Oxalis corniculata*** L., Common throughout the area (up to 3000 m). Naitwar-Sankri. March–November.

21. BALSAMINACEAE

118. ***Impatiens badrinathii*** P.K. Pusalkar & D.K. Singh, Common on moist shady places near mixed temperate (2500–3200 m). On way to Seema. July–September.
119. ***Impatiens glandulifera*** Royle, Common on hillslopes in shady places (2000–3000 m). On way to Sankri. July–September.
120. ***Impatiens leggei*** P.K. Pusalkar & D.K. Singh, Common on moist shady places in mixed temperate (2800–3600 m). On way to Seema. July–September.
121. ***Impatiens racemosa*** DC., Common on hillslopes in shady places (2200–3300 m). Sankri-Taluka and Naitwar. July–September.
122. ***Impatiens scabrida*** DC., Common on hillslopes and in shady places (2000–2800 m). Dodhital and Naitwar-Sankri. July–September.
123. ***Impatiens himalayensis*** R. Manik., Lakshmin. & S.K. Srivast., Rare on moist shady places (2500–3200 m). On way to Seema. July–September.
124. ***Impatiens sulcata*** Wall. Common on hillslopes (2000–3200 m). On way to Seema. July–September.
125. ***Impatiens thomsonii*** Hook. *f.*, Common along the roadsides and on hillslopes (1600–3000 m). On way to Sankri. July–September.

22. RUTACEAE

126. ***Aegle marmelos*** (L.) Correa, Occasional in dry places and planted near Hindu temples. May–June.
127. ***Boenninghausenia albiflora*** (Hook.) Reichnb. ex Meisn., Common in shady places (up to 2500 m). Naitwar-Sankri and Taluka. June–October.
128. ***Skimmia anquetilia*** Taylor & Air Shaw, Common in open areas and often found in *Quercus-Rhodendron* forests. Sankri and Taluka-Seema. April–November.
129. ***Zanthoxylum armatum*** DC., Common in evergreen thickets. Sankri and Naitwar-Sankri. March–September.

23. SIMAROUBACEAE

130. ***Ailanthus excelsa*** Roxb., Fairly common in wastelands near habitation and roadsides. Naitwar. April–June.

24. MELIACEAE

131. ***Melia azedarach*** L., Common in valleys (1000–1600 m). Naitwar, along Tons River. April–May.
132. ***Toona sinensis*** (A. Juss.) M. Roem, Common in the valleys (up to 850 m). Naitwar, near Mautad Village and Jakhol. April–September.

25. AQUIFOLIACEAE

133. ***Ilex dipyrena*** Wall., Common in mixed forests (2500–2800 m). Sankri and Jakhol. May–November.

26. CELASTRACEAE

134. ***Cassine albens*** (Retz.) Kosterm., Occasional in dry places near mixed temperate. Naitwar. September–March.
135. ***Euonymus echinatus*** Wall., Scattered in shady places, on tree trunks, and on rocky boulders near mixed temperate forests. On way to Seema. March–September.
136. ***Euonymus fimbriatus*** Wall., Grows in temperate forests. April–June.
137. ***Euonymus hamiltonianus*** Wall., Occasional in moist shady places. Sankri. March–November.
138. ***Euonymus tingens*** Wall. in Roxb., Common in moist shady areas. Sankri. April–September.

27. RHAMNACEAE

139. ***Hovenia acerba*** Lindl., Frequent in mixed forests (1600–1900 m). Sankri. June–August.
140. ***Rhamnus virgatus*** Roxb., Common throughout the area in mixed and temperate forests (1600–2400 m). Sankri. March–October.
141. ***Sageretia parviflora*** (Roem. ex Schult.) G. Don, Common throughout the area in mixed temperate (1600–2000 m). Sankri. April–October.

28. VITACEAE

142. ***Amplelocissus divaricata*** (Wall. ex M.A. Lawson) Planch., Common in mixed forests (1800–2800 m). Taluka-Seema. June–October.
143. ***Ampelocissus latifolia*** (Roxb.) Planch., Common in moist deciduous forests. On way to Sankri. May–October.
144. ***Ampelocissus rugosa*** (Wall.) Planch., Common on the bushes and trees in mixed forests (1600–2500 m). Naitwar. May–October.
145. ***Parthenocissus semicordata*** (Wall.) Planch., Common in mixed temperate (up to 1800 m). On way to Sankri. April–August.

29. LEEACEAE

146. ***Leea asiatica*** (L.) Ridsdale, Common in mixed temperate (up to 1200 m). Naitwar near Sankri. June–November.

30. SAPINDACEAE

147. ***Sapindus saponaria*** L., Occasional in mixed forests (1700–2200 m). Naitwar-Sankri. March–June.

31. HIPPOCASTANACEAE

148. ***Aesculus indica*** (Colebr. ex Cambess.) Hook., Common in mixed forests (2000–2800 m). Sankri, Taluka, and Seema. May–October.

32. ACERACEAE

149. ***Acer acuminatum*** Wall. ex D. Don, Common in mixed temperate and subalpine forests (2500–3500 m). Taluka-Seema, Jakhol, and Kedarkantha. April–November.
150. ***Acer caesium*** Wall. ex Brandis, Common in mixed temperate forests (2000–3300 m). Taluka-Seema and Sankri. March–October.
151. ***Acer oblongum*** Wall. ex DC., Rare in moist forests, especially along streams (2200–3000 m). February–May.
152. ***Acer sterculiaceum*** Wall., Occasional in mixed forests (2000–3300 m). Seema-Harkidoon. On way to Kedarkantha.

33. SABIACEAE

153. ***Meliosma simplicifolia*** (Roxb.) Walp. subsp. **yunnanensis** (Franchet) Beus., Occasional in moist deciduous forests (up to 1600 m). April–September.

34. ANACARDIACEAE

154. ***Cotinus coggygria*** Scop., Common in subtropical temperate regions (1200–2300 m). Sankri, Himri, and on way to Taluka. April–July.
155. ***Rhus chinensis*** Miller, Common near villages in mixed forests (1800–2800 m). Taluka-Seema. June–October.
156. ***Spondias pinnata*** (L.*f.*) Kurz, Occasional in moist mixed temperate. March–September.

35. CORIARIACEAE

157. ***Coriaria nepalensis*** Wall., Occasional throughout the valley (2000–3000 m). Naitwar. March–June.

36. MORINGACEAE

158. ***Moringa oleifera*** Lam., Occasional near human habitations. April–June.

37. FABACEAE

159. ***Aeschynomene indica*** L., Common in open moist places (up to 1400 m). Naitwar. August–October.
160. ***Astragalus chlorostachys*** Lindl., Rare on hillslopes near alpine regions (2800–4300 m). Kalhattidhar and Harkidoon. July–October.
161. ***Astragalus himalayanus*** Klotz., Common on hillslopes, along river banks and amidst grasses (2400–3300 m). Seema and Barasu Pass. On way to Jamnidhar Glacier. July–October.
162. ***Campylotropis speciosa*** (Royle ex Schindl.) Schindl., Occasional on hillslopes near mixed forests (1600–2000 m). Naitwar-Sankri and Himri. July–October.
163. ***Crotalaria albida*** Heyne ex Roth, Common in open places and along river banks (up to 2200 m). Naitwar. May–October.

164. ***Crotalaria calycina*** Schrank, Rare in open places near dry deciduous forests. On way to Naitwar. September–December.
165. ***Crotalaria humifusa*** Grah. ex Benth., Occasional on hills slopes amidst grasses (up to 1800 m). Naitwar. July–October.
166. ***Crotalaria prostrata*** Rottler ex Willd., Common on hillslopes near coniferous forests (1400–2200 m). Naitwar and Taluka. July–October.
167. ***Crotalaria sessiliflora*** L., Common along roadsides and near open grassy fields (up to 1600 m). On way to Sankri. May–October.
168. ***Crotalaria stipularia*** Desv., Rare on open grassy fields near mixed temperate forests (up to 1200 m). Naitwar. June–November.
169. ***Dalbergia sissoo*** Roxb., Common in exposed and along river banks. Naitwar. May–June.
170. ***Desmodium concinnum*** DC., Common in mixed forests (1400–1800 m). Naitwar-Sankri and Taluka. April–October.
171. ***Desmodium elegans*** DC., Common in mixed forests (1700–3000 m). Sankri and Taluka-Seema. April–October.
172. ***Desmodium heterocarpon*** (L.) DC., Common throughout the area (1400–2400 m). Jakhol, Dhola, and Naitwar-Taluka. July–October.
173. ***Desmodium microphyllum*** (Thunb.) DC., Common throughout the area (up to 2100 m). Naitwar-Dhola and Naitwar-Taluka. June–October.
174. ***Desmodium multiflorum*** DC. Common in mixed forests (1800–2500 m). Sankri and Taluka-Seema. June–October.
175. ***Erythrina suberosa*** Roxb., Occasional in dry places. March–June.
176. ***Flemingia macrophylla*** (Willd.) Prain, Common in moist places near mixed temperate (up to 1200 m). Naitwar-Sankri. July–October.
177. ***Flemingia procumbens*** Roxb., Common on hillslopes and amidst grasses (1600–2600 m). Naitwar-Sankri. July–October.
178. ***Flemingia strobilifera*** (L.) W.T. Aiton, Common on hillslopes near pine forests (up to 2000 m). Naitwar. July–October.
179. ***Hedysarum microcalyx*** Baker, Rare on open hillslopes and along roadsides (3000–4000 m). Seema-Harkidoon. June–July.
180. ***Indigofera cassioides*** Rottlerex DC., Sanctuary. March–May.
181. ***Indigofera hebepetala*** Benth. ex Baker, Common in mixed temperate forests (2000–2600 m). Taluka-Seema and Jakhol. May–September.
182. ***Indigofera heterantha*** Wall. ex Brandis, Common throughout the sanctuary except alpine regions (up to 3000 m). Naitwar-Sankri, Taluka-Seema, and Dhola-Changsil. May–October.
183. ***Lathyrus emodii*** (Wall. ex Fritsch) Ali, Rare on open grassy fields (2300–3800 m) Kalhattidhar and Harkidoon. June–September.
184. ***Lathyrus humilis*** (Ser.) Fisch. ex Spreng., Rare on hillslopes and amidst grasses (up to 3000 m). Kalhattidhar. June–September.
185. ***Lathyrus sphaericus*** Retz., Observ., Occasional on hillslopes and amidst grasses (up to 1600 m). Sankri. April–June.
186. ***Lespedeza gerardiana*** Grah. ex Maxim., Common on hillslopes, amidst grasses and open places near pin forests (up to 2000 m). Naitwar-Sankri. July–September.
187. ***Lespedeza juncea*** var. **sericea** (Thunb.) Forbes & Hemsl., Common throughout the area amidst grasses (up to 2300 m). Naitwar-Sankri, on way to Taluka. July–September.
188. ***Lotus corniculatus*** L., Common on hillslopes and amidst small grasses near subalpine to alpine regions (3000–3800 m). Seema, on way to Harkidoon. March–September.
189. ***Macrotyloma uniflorum*** (Lam.) Verdc., Occasional in fallow fields (1000–1800 m). Naitwar. September–December.
190. ***Millettia extensa*** (Benth.) Baker, Fairly common in mixed dry deciduous forests (up to 1500 m). Naitwar. April–November.
191. ***Parochetus communis*** Buch.-Ham. ex D. Don, Common in shady places, open hillslopes, and amidst grasses (2400–3500 m). Sankri-Taluka-Seema. May–September.
192. ***Thermopsis barbata*** Royle, Common on open hillslopes, amidst grass, and along the bridal path in sandy soil (3200–4000 m). Kalhattidhar and Harkidoon. May–June.
193. ***Trifolium repens*** L., Common on hillslopes and amidst grasses (1500–3500 m). Sankri and Seema-Harkidoon. April–September.
194. ***Trigonella emodi*** Benth., Common on hillslopes and amidst grasses (2000–3600 m). Harkidoon, Deokyar, Kalhattidhar, and Devasu. July–September.
195. ***Trigonella pubescens*** Edgew. ex Baker, Common on open moist places and grassy

fields (3200–4600 m). On way to Barasu Pass. April–August.

196. ***Vicia bakeri*** Ali, Common on hillslopes and climbing over bushes (1700–3500 m). On way to Sankri, Kalhattidhar, and on way to Seema. July–September.
197. ***Vicia sativa*** L., Occasional in deodar forests (1700–2500 m). Sankri. March–July.
198. ***Vigna radiata*** (L.) R. Wilczek, Occasional along the roadsides. Naitwar and Sankri. March–July.
199. ***Vigna vexillata*** (L.) A. Rich., Common in amidst grasses (up to 2000 m). Taluka. July–October.
200. ***Zornia gibbosa*** Span., Common on open places and amidst grasses (up to 2000 m). Naitwar-Himri. July–September.

38. CAESALPINIACEAE

201. ***Caesalpinia bonduc*** (L.) Roxb., Fairly common in mixed temperate. Sankri. September–February.
202. ***Caesalpinia decapetata*** (Roth.) Alston, Occasional in mixed temperate (up to 2000 m). Naitwar-Sankri. March–April.
203. ***Chamaecrista pumila*** (Lam.) V. Singh, Infrequent in open situations. Naitwar. July–October.

39. MIMOSACEAE

204. ***Acacia sinuata*** (Lour.) Merr., Fairly common in scrub forests (up to 950 m). Naitwar. April–December.
205. ***Albizia mollis*** (Wall.) Boivin, Occasional along roadsides near mixed temperate (1500–2300 m). Sankri. May–December.

40. ROSACEAE

206. ***Acomastylis elata*** (Wall. ex Royle) Bolle, Common on hillslopes near subalpine and alpine regions (3000–4000 m). Hatta valley, Kedarkantha, On way to Jamnidhar Glacier, Deokyar. June–October.
207. ***Agrimonia pilosa*** Ledeb., Common on hillslopes and in mixed forests (1600–2500 m). Naitwar-Sankri. May–September.
208. ***Cotoneaster acuminatus*** Lindl., Common in mixed forests to alpine regions (2500–3600 m). Seema. May–October.
209. ***Cotoneaster microphyllus*** Wall. ex Lindl., Common on hillslopes and on boulders near mixed forests to alpine regions (2300–4500 m). Kalhattidhar Glacier and Harkidoon. On way to Hatta Valley. May–October.
210. ***Cotoneaster obtusus*** Wall. ex Lindl., Fairly common in open and associated with *Rosa* and *Deutzia* spp. (2300–4500 m). Kalhattidhar Glacier. May–October.
211. ***Cotoneaster roseus*** Edgew., Occasional along the bridal path (2700–3500 m). On way to Harkidoon. May–October.
212. ***Duchesnea indica*** (Andr.) Focke, Common on hillslopes (1400–1800 m). Naitwar and Sankri-Himri. March–June.
213. ***Filipendula vestita*** (Wall. ex G. Don) Maxim., Occasional in open hillslopes near alpine regions (2700–3800 m). Seema-Harkidoon. May–October.
214. ***Fragaria nubicola*** Lindl. ex Lacaita Common on hillslopes and in shady places (up to 3000 m). Sankri, on way to Seema. April–June.
215. ***Geum roylei*** Bolle, Common on hillslopes and in shady places near mixed forests (up to 2800 m). Istragad, Seema, and Kedarkantha. April–October.
216. ***Potentilla arbuscula*** var. **pumila** (Hook. *f.*) Hand.-Mazz., Common on boulders, in sandy soil, and open places near alpine regions (4000–5000 m). Harkidoon, on way to Jamnidhar Glacier. July–September.
217. ***Potentilla argyrophylla*** Wall. ex Lehm., Common on hillslopes in subalpine to alpine regions (3000–4000 m). Deokyar, Barasu Pass, and Kalhattidhar. June–October.
218. ***Potentilla atrosanguinea*** Lodd., Common on hillslopes in subalpine to alpine regions (3000–4000 m). Barasu Pass, on way to Harkidoon. On way to Jamnidhar Glacier. June–October.
219. ***Potentilla cuneifolia*** Bertol., Common in shady places, on hillslopes, and open rocky places (3500–4500 m). On way to Barasu Pass. On way to Jamnidhar Glacier. June–October.
220. ***Potentilla eriocarpa*** Wall. ex Lehm., Common on hillslopes near glacial marians (4000–5000 m). On way to Barasu Pass. June–September.
221. ***Potentilla rigida*** Wall. ex Lehm., Common on rocky places and on boulders in alpine to subalpine regions (3500–4500 m). Harkidoon. June–September.
222. ***Potentilla supina*** L., Common in fallow fields, ditches, and along roadside depressions (up to 1800 m). Naitwar. June–September.
223. ***Prinsepia utilis*** Royle, Common in open places throughout the valley (1000–2500 m). Sankri and Naitwar-Taluka-Seema. December–July.

224. ***Prunus cerasoides*** D. Don, Common throughout the area (1400–2300 m). Naitwar (T.S. Rana & et al., *l.c.*), Himri-Taluka. April–December.
225. ***Prunus cornuta*** (Wall. ex Royle) Steud., Common throughout the area in cold temperate forests and associated with *Quercus semicarpifolia, Betula utilis*, etc. (2600–3500 m). Seema-Harkidoon, Khatakcha, and Khimdar. May–October.
226. ***Prunus persica*** (L.) Batsch, Common throughout the area and also widely cultivated (up to 3000 m). Sankri, Naitwar-Dhola, and Seema. January–October.
227. ***Pyracantha crenulata*** (D. Don) M. Roem., Common on open places (up to 2600 m). Naitwar. On way to Dhola. March–June.
228. ***Pyrus pashia*** Buch.-Ham. ex D. Don, Common throughout in mixed temperate especially on open hillslopes (up to 2300 m). Sankri. April–June.
229. ***Rosa brunonii*** Lindl., Common along bushes (1450–2800 m). Naitwar-Sankri and Taluka-Seema. April–February.
230. ***Rosa macrophylla*** Lindl., Common throughout the area in mixed forests to alpine regions (2300–3600 m). Sankri, Seema, and Harkidoon. May–October.
231. ***Rosa sericea*** Lindl., Common throughout the area in subalpine and alpine regions (3000–4500 m). Harkidoon. April–September.
232. ***Rosa webbiana*** Wall. ex Royle, Common throughout the area in subalpine and alpine regions (3000–4500 m). On way to Ruinsara Tal. April–October.
233. ***Rubus biflorus*** Buch.-Ham. ex Sm. in Rees, Occasional in mixed temperate forests (2200–3200 m). Sankri. April–July.
234. ***Rubus ellipticus*** Sm. Common throughout the area (up to 2300 m). Sankri-Taluka. March–June.
235. ***Rubus foliolosus*** D. Don, Fairly common in mixed forests (2200–3200 m). Sankri. May–October.
236. ***Rubus niveus*** Thunb., Common throughout the area (up to 2600 m). Naitwar, Jakhol, Sankri, and Seema. March–July.
237. ***Rubus paniculatus*** Sm. Common on climbing over bushes and trees (up to 2200 m). Naitwar and Sankri-Jakhol. May–July.
238. ***Sibbaldia parviflora*** Willd., Common on hillslopes, amidst grasses, and in moist shady places (3200–3800 m). Seema. On way to Harkidoon. May–September.
239. ***Sorbaria tomentosa*** (Lindl.) Rehder, Common throughout the area in mixed forests (2000–3000 m). Sankri, Naitwar, Dodital, and Taluka-Seema. June–October.
240. ***Sorbus aucuparia*** L., Occasional on open hillslopes near alpine regions (3400–4000 m). Harkidoon. On way to Maninda Tal. May–September.
241. ***Spiraea bella*** Sims., Common throughout the area in cold temperate forests to alpine regions (2800–3600 m). On way to Harkidoon. May–September.
242. ***Spiraea canescens*** D. Don, Very common among the hedges and fringes of the forests (2700–3500 m) Sankri and Taluka-Seema. On way to Khatakcha. May–September.
243. ***Spiraea vaccinifolia*** D. Don, Scattered along the bridal path and in open places near mixed temperate (1800–2800 m). Naitwar-Sankri. May–October.

41. SAXIFRAGACEAE

244. ***Astilbe rivularis*** Buch.-Ham. ex D. Don, Occasional in shady places near mixed forests (1600–2500 mJ). Sankri and akhol. June–October.
245. ***Bergenia ciliata*** (Haw.) Sternb., Common on rocky places and in moist shady places (2300–3200 m). Taluka-Seema. April–July.
246. ***Bergenia stracheyi*** (Hook. *f.* & Thomson) Engl., Common on rocky places and in gravel soil (3000–4500 m). On way to Jamnidhar Glacier. May–September.
247. ***Saxifraga asarifolia*** Sternb., Revis., Occasionally found growing on hillslopes and in moist shady places near alpine regions (3500–4500 m). Harkidoon. On way to Jamnidhar Glacier. May–September.
248. ***Saxifraga brunonis*** Wall. ex Sering, Common on hillslopes and marshy boulders near alpine regions (3500–4000 m). Jamnidhar Glacier, Huttagad, and Barasu Pass. On way to Maninda Tal. July–September.
249. ***Saxifraga diversifolia*** Wall. ex Sering, Common on hillslopes and in moist shady places near subalpine to alpine regions (3200–4000 m). Jamnidhar Glacier and Harkidoon. June–October.
250. ***Saxifraga filicaulis*** Wall. ex Sering, Common on hillslopes and in moist shady places near subalpine to alpine regions (3200–3800 m). Kalhattidhar and Harkidoon. July–September.

251. ***Saxifraga flagellaris*** subsp. **mucronulata** (Royle) Engl. & Irmsch., Occasional on hillslopes and amidst grasses near alpine regions (4000–4500 m). Harkidoon and Barasu Pass. May–September.
252. ***Saxifraga parnassifolia*** D. Don, Common on hillslopes and amidst grasses near alpine regions (3000–4000 m). On way to Harkidoon. August–October.
253. ***Saxifraga sibirica*** L., Common in amidst boulders and moist shady places (4000–4500 m) Barsu pass. May–September.

42. PARNASSIACEAE

254. ***Parnassia laxmannii*** Pallas ex Schult., Common in wet places (2300–3200 m). Seema. July–August.
255. ***Parnassia nubicola*** Wall. ex Royle, Common throughout the area in temperate to alpine regions (2000–4000 m). Hutta Valley and Kedarkantha. June–October.

43. HYDRANGEACEAE

256. ***Hydrangea heteromalla*** D. Don, Rare on moist shady places (2600–3200 m). On way to Seema. May–October.

44. PHILADELPHACEAE

257. ***Deutzia staminea*** R. Br. ex Wall., Common throughout the area in mixed temperate (2200–3500 m). Near Taluka. On way to Seema. May–July.
258. ***Philadelphus tomentosus*** Wall. ex G. Don, Common throughout the area in mixed temperate (2000–2700 m). Seema. May–July.

45. GROSSULARIACEAE

259. ***Ribes glaciale*** Wall. in Roxb., Common in alpine regions (3500–4000 m). Harkidoon, Barasu Pass, and Deokyar. May–September.
260. ***Ribes orientale*** Desf., Common on hillslopes (3200–4100 m). Harkidoon. April–June.

46. CRASSULACEAE

261. ***Hylotelephium ewersii*** (Ledeb.) H. Ohba, Common throughout the subalpine and alpine Himalaya on mossy boulders (3200–5000 m). On way to Harkidoon. On way to Jamnidhar Glacier. July–October.
262. ***Rhodiola crenulata*** (Hook. *f.* & Thomson) H. Ohba, Common on hill-slopes, rocky crevices and in shady places (3200–4000 m). On way to Harkidoon. May–September.
263. ***Rhodiola fastigiata*** (Hook. *f.* & Thomson) Fu, Common on hillslopes, rocky crevices, and in shady places (3800–4600 m). Harkidoon. May–September.
264. ***Rhodiola heterodonta*** (Hook. *f.* & Thomson) Boriss., Common in gravel soils and amidst boulders near the alpine regions (3500–4500 m). Deokyar. May–September.
265. ***Rhodiola imbricata*** Edgew., Common on moist boulders and along the streams, near alpine regions (3800–4500 m). Harkidoon. July–September.
266. ***Rhodiola quadrifida*** (Pallas) Fischer & Mey., Common on hillslopes and in rocky crevices (3600–4500 m). Harkidoon, On way to Maninda Tal. May–September.
267. ***Rhodiola sinuata*** (Royle ex Edgew.) Fu, Common on moist hillslopes (3200–4000 m). On way to Seema, Kalhattidhar, and Harkidoon. June–October.
268. ***Rhodiola wallichiana*** (Hook. *f.*) Fu, Common along streams in subalpine to alpine regions (3200–4200 m). Hatta Valley, Jamnidhar Glacier. June–October.
269. ***Rosularia rosulata*** (Edgew.) H. Ohba, Fairly common on stony and moist walls (1700–2500 m). Sankri. On way to Taluka. May–June.
270. ***Sedum multicaule*** Wall. ex Lindl., Common on hillslopes and on rocky crevices (3300–4500 m). Sankri-Taluka. June–October.
271. ***Sedum oreades*** (Decne.) Hamet., Rare on rocky crevices (3300–4500 m). Harkidoon. June–October.
272. ***Sedum trullipetalum*** Hook. *f.* & Thomson, Rare on mossy boulders in subalpine to alpine regions (3200–4500 m). On way to Harkidoon. On way to Maninda Tal and Barasu Pass. June–October.
273. ***Sinocrassula indica*** (Decne.) A. Berger, Common on rocks, dry stony and gravelly slopes (1200–3300 m). On way to Taluka. On way to Seema. July–October.
274. ***Tillaea schimperi*** (C.A. Mey.) M.G. Gilbert, Common on hillslopes and rocky places (1500–2000 m). Naitwar-Sankri. On way to Taluka. July–October.

47. DROSERACEAE

275. ***Drosera peltata*** J.E. Sm., Common in amidst grasses (2600–300 m). Seema-Kalhattidhar and Harkidoon. June–October.

48. MELASTOMACEAE

276. ***Osbeckia chinensis*** L., Occasional in damp localities (1200–1500 m). Naitwar. September–October.
277. ***Osbeckia stellata*** Buch.-Ham. ex D. Don, Common throughout the area (1500–2000 m). Sankri-Taluka. July–October.

49. LYTHRACEAE

278. ***Woodfordia fruticosa*** (L.) Kurz, Occasional on hillslopes and along the roadsides. April–June.

50. PUNICACEAE

279. ***Punica granatum*** L., Occasional in village areas and outskirts of forests. Naitwar. August–November.

51. ONAGRACEAE

280. ***Circaea alpina*** L. subsp. **imaicola** (Ascher. & Magn.) Kitam., Common on hillslopes amidst grasses (2500–3500 m). On way to Taluka. July–September.
281. ***Epilobium cylindricum*** D. Don, Common on hillslopes, in moist shady and watery-places (2500–4000 m). Harkidoon. May–September.
282. ***Epilobium latifolium*** L., Common on open hillslopes and in sandy soils (3500–4500 m). Ruinsara Tal, Harkidoon, and Barasu Pass. July–October.
283. ***Epilobium palustre*** L., Rare on hillslopes and in moist shady places (2000–3000 m). On way to Jamnidhar Glacier. July–September.
284. ***Epilobium royleanum*** Haussk., Common in moist shady places, along streams, and roadsides (2300– 3800 m). Seema-Harkidoon and Ruinsara Tal. July–October.
285. ***Ludwigia perennis*** L., Common on hillslopes, in moist shady and watery places (1500–2000 m). On way to Dhola. April–June.
286. ***Oenothera rosea*** L' Her. ex Aiton, Common throughout the area (up to 2500 m). Kaldiyan and Taluka. April–September.

52. PASSIFLORACEAE

287. ***Passiflora edulis*** Sims., Rare on hedges and along the Tons River (up to 1500 m). Naitwar. May–June.

53. CUCURBITACEAE

288. ***Herpetospermum penduncolosum*** (Ser.) Baill., Common along the roadsides (2000–2600 m). On way to Seema. August–September.
289. ***Mukia maderaspatana*** (L.) M. Roem., Occasional in disturbed areas near mixed temperate (up to 900 m). Naitwar. July–October.
290. ***Solena amplexicaulis*** (Lam.) Gandhi, Common along the bridal path near villages (1600–2500 m). On way to Dhola. On way to Sankri and Taluka. June–October.
291. ***Trichosanthes cucumerina*** L., Common along the bridal path (1500–2000 m). On way to Jakhol. August–September.
292. ***Trichosanthes tricuspidata*** Lour., Occasional in mixed temperate (up to 1250 m). Naitwar. On way to Jakhol. August–October.

54. BEGONIACEAE

293. ***Begonia picta*** Sm., Common on slopes, amidst grasses, and in moist shady places (up to 2000 m). On way to Sankri. June–October.

55. APIACEAE (= UMBELLIFERAE)

294. ***Acronema tenerum*** (DC.) Edgew., Common throughout the area from temperate to alpine regions (2600–4500 m). Seema-Harkidoon. July–October.
295. ***Angelica glauca*** Edgew., Common throughout the area in cold temperate to subalpine regions (2600–3300 m). Seema-Harkidoon. July–October.
296. ***Bupleurum candollei*** Wall. ex DC., Common throughout on open hillslopes and amidst grasses (2500–4000 m). On way to Seema. June–October.
297. ***Bupleurum hamiltonii*** N.P. Balakr., Common in moist shady places and amidst grasses (2000–3000 m). Naitwar-Seema. June–October.
298. ***Bupleurum lanceolatum*** Wall. ex DC., Common on hillslopes, along the bridal path, and amidst grasses (1500–3000 m). Sankri-Taluka. June–September.
299. ***Bupleurum marginatum*** Wall. ex DC., Common along the bridal path and amidst grasses (1500–2000 m). Kedarkantha. July–October.
300. ***Bupleurum rupestre*** Edgew., Common on hillslopes, along the bridal path, and amidst grasses (3000–4100 m). Harkidoon and Hatta Valley. August–November.
301. ***Centella asiatica*** (L.) Urb., Common on hillslopes, in moist places, and amidst grasses (up to 1700 m). Naitwar. April–October.
302. ***Chaerophyllum acuminatum*** Lindl., Occasional on open hillslopes (2500–4000 m). Harkidoon. June–September.

303. ***Chaerophyllum reflexum*** Lindl., Common on hillslopes, in shady places, and amidst grasses (1800–3000 m). Naitwar. July–September.
304. ***Heracleum lanatum*** Michx., Common on hillslopes in association with *Geranium* sp., Seneciosp., *Polygonum* sp., etc. (2000–3000 m). Hatta Valley and Harkidoon. On way to Jamnidhar Glacier. June–September.
305. ***Meeboldia achilleifolia*** (DC.) P.K. Mukh. & Constance, Common in open grassy slopes (3200–3800 m). Hatta Valley. July–October.
306. ***Pimpinella acuminata*** (Edgew.) C.B. Clarke, Common in moist shady places (1800–2500 m). Seema. August–October.
307. ***Pimpinella diversifolia*** DC., Common along the bridal path near mixed temperate (1500–2000 m). Naitwar, along Tons River. Seema. May–September.
308. ***Pleurospermum candollii*** (DC.) C.B. Clarke, Common on hillslopes and amidst grasses (3800–4500 m). Harkidoon-Maninda Tal. On way to Barasu Pass. August–October.
309. ***Pleurospermum densiflorum*** (Lindl.) C.B. Clarke, Common on hillslopes, amidst grasses, and along the bridal path (3000–4000 m). Seema-Harkidoon. July–October.
310. ***Pleurospermum stellatum*** (D. Don) C.B. Clarke, Common on open hillslopes and in sandy soils (3000–4000 m). Harkidoon. July–September.
311. ***Sanicula elata*** Buch.-Ham. ex D. Don, Common on open hillslopes and moist shady places (2000–3500 m). Taluka-Seema. July–October.
312. ***Schulzia garhwalica*** (H. Wolff) P.K. Mukh. & Constance, Rare in shady places and on hillslopes (3000–4000 m). On way to Barasu Pass. July–October.
313. ***Selinum vaginatum*** (Edgew.) C.B. Clarke, Common on hillslopes and amidst grasses (2000–4000 m). On way to Harkidoon and Ruinsara Tal. August–October.
314. ***Selinum wallichianum*** (DC.) Raizada & H.O. Saxena, Common on hillslopes and amidst grasses in association with *Geranium* sp., *Polygonum* spp., etc. (2000–4000 m). On way to Harkidoon. On way to Jamnidhar Glacier and Kedarkantha. July–October.
315. ***Tordyliopsis brunonis*** DC., Occasional in meadows, among shrubs and boulders near subalpine regions, and on hillslopes (3000–4400 m). On way to Barasu Pass. July–October.
316. ***Torilis japonica*** (Houtt.) DC., Occasional on hillslopes and amidst grasses (2500–3000 m). On way to Sankri and Taluka-Seema. June–October.
317. ***Tricholaser cachemiricum*** (C.B. Clarke) Alava, Occasional in outskirts of pine forests. Sankri. May–August.
318. ***Vicatia coniifolia*** DC., Occasional in amidst grasses (2000–3500 m). Taluka-Seema. May–September.

56. ARALIACEAE

319. ***Aralia cachemirica*** Decne., Occasional along the bridal path in mixed temperate (1800–2400 m). Sankri-Taluka. July–September.
320. ***Hedera nepalensis*** K. Koch, Common on climbing over trees (1600–2700 m). Sankri-Taluka and Jakhol. September–December.

57. CORNACEAE

321. ***Swida macrophylla*** (Wall.) Sojak, Common throughout the area near mixed temperate (1600–2500 m). Naitwar, Jakhol, Sankri, and Taluka. May–October.

58. ALANGIACEAE

322. ***Alangium chinensis*** (Lour.) Harms., Common throughout the area near mixed temperate (1600–2500 m). Naitwar and Sankri. May–October.

59. CAPRIFOLIACEAE

323. ***Leycesteria formosa*** Wall., Rare in mixed temperate and also found near the villages (1800–3000 m). Seema and Taluka. May–August.
324. ***Lonicera angustifolia*** Wall. ex DC., Fairly common in temperate forests (2500–3000 m). Naitwar. May–August.
325. ***Lonicera heterophylla*** Decne. Very common on dry rocky slopes in alpine zone (3500–4600 m). Ruinsara Tal. June–July.
326. ***Lonicera myrtillus*** Hook. *f.* & Thomson, Common on open hillslopes, in sandy graval soils, and on rocky crevices (3500–4500 m). Ruinsara below. On way to Jamnidhar Glacier. May–October.
327. ***Lonicera obovata*** Royle ex Hook. *f.* & Thomson, Common on hillslopes and in rocky places (3500–4500 m). Harkidoon. On way to Jamnidhar Glacier. May–October.
328. ***Lonicera purpurascens*** Walp., Occasional on open hillslopes and associated with in *Quercus semicarpifolia, Abies* forests

(3200–4000 m). On way to Harkidoon, and Ruinsara Tal. May–October.

329. ***Lonicera quinquelocularis*** Hardw., Fairly common in scrub forests (900–1600 m). Sankri. April–September.
330. ***Lonicera webbiana*** Wall. ex DC., Occasional in open hillslopes and in rocky places (2700–4000 m). Seema-Harkidoon. May–September.
331. ***Viburnum cotinifolium*** D. Don, Common throughout the area in mixed temperate (1800–3000 m). Naitwar and Harkidoon. April–October.
332. ***Viburnum erubescens*** Wall. ex DC., Occasional along the bridal path and also forming isolated patches near mixed temperate (1800–3000 m). Sankri. May–September.
333. ***Viburnum grandiflorum*** Wall. ex DC., Common on hillslopes (up to 3500 m). Kedarkantha. November–June.
334. ***Viburnum mullaha*** Buch.-Ham. ex D. Don, Common throughout the area in mixed temperate (2000–3500 m). On way to Taluka, Seema, and Naitwar. May–September.

60. RUBIACEAE

335. ***Argostemma sarmentosum*** Wall., Rare in moist shady places (1400–1800 m). Naitwar-Sankri. July–September.
336. ***Galium acutum*** Edgew., Common in amidst grasses near mixed temperate (2000–3700 m). On way to Jamnidhar Glacier. July–October.
337. ***Galium aparine*** L., Common in waste places and in moist shady places (1500–3400 m). On way to Dhola and Ruinsara Tal. February–August.
338. ***Galium asperifolium*** Wall. in Roxb., Common in moist shady places and found on *Lysimachia lobeloides* (1400–3500 m). On way to Salta and Naitwar. August–November.
339. ***Galium asperuloides*** Edgew., Common in moist shady places (1400–2700 m). Harkidoon. May–July.
340. ***Galium elegans*** Wall. ex Roxb., Common in moist shady places (1400–2500 m). Salta and Sankri. June–August.
341. ***Hedyotis attenuata*** Willd., Occasional on hillslopes (up to 1200 m). Naitwar. July–October.
342. ***Hedyotis corymbosa*** (L.) Lam., Common in moist shady areas, waste places, and along the cultivated fields (1400–2000 m). Naitwar. July–October.
343. ***Leptodermis lanceolata*** Wall., Common on hillslopes and near villages (1400–2500 m). Sankri and Naitwar. June–October.
344. ***Randia tetrasperma*** (Roxb.) Benth. & Hook. *f.* ex Brandis, Common in open and dry places (1200–1800 m). Naitwar. March–September.
345. ***Rubia manjith*** Roxb. ex Flem., Common in moist shady places (1400–3000 m). Naitwar-Jakhol-Seema. June–October.
346. ***Spermacoce pusilla*** Wall., Common in moist places (1400–1700 m). On way to Naitwar. March–November.

61. VALERIANACEAE

347. ***Valeriana hardwickii*** Wall., Common on hillslopes and amidst grasses (2000–3000 m). On way to Sankri. June–October.

62. DIPSACACEAE

348. ***Dipsacus inermis*** Wall., Common throughout the area in cold temperate mixed temperate (1800–3000 m). Sankri-Taluka. On way to Seema. July–October.
349. ***Morina longifolia*** Wall. ex DC., Very common on open hillslopes in cold temperate alpine regions (3000–4000 m). Harkidoon. June–October.

63. ASTERACEAE (=COMPOSITAE)

350. ***Achillea millefolium*** L., Rare on hillslopes (3000 m). Runsiyara Tal. July–October.
351. ***Artemisia capillaris*** Thunb., Common on hillslopes, open places, and mostly on sandy soils (2000–3500 m). Naitwar. July–September.
352. ***Artemisia dubia*** Wall. ex Bess., Occasional on hillslopes and along the bridal path (1500–2500 m). Sankri-Taluka. July–September.
353. ***Artemisia indica*** Willd. var. **elegantissima** (Pamp.) Y.R. Ling & Humpheries, Very common on hillslopes and amidst grasses (2700–3200 m). On way to Harkidoon. July–September.
354. ***Artemisia indica*** Willd. var. **indica**. Common on hillslopes and along the bridal path (1500–3000 m). Naitwar. March–December.
355. ***Artemisia japonica*** Thunb., Common on hillslopes (1700–3000 m). Sankri, Naitwar. July–October.
356. ***Artemisia myriantha*** Wall. ex Bess, Occasional on hillslopes and amidst grasses (1700–3000 m). Naitwar. October–December.

357. ***Artemisia roxburghiana*** Wall. ex Bess., Common on hillslopes (2500–3000 m). Kedarkantha bugyal. On way to Harkidoon. July–September.
358. ***Artemisia wallichiana*** Bess., Occasional on open hillslopes and mainly in sandy soils (2700–3500 m). On way Jamnidhar Glacier. July–September.
359. ***Tanacetum dolichophyllum*** (Kitam.) Kitam., Common on grassy hillslopes and along river banks (3000–4000 m). Hatta Valley, On way to Harkidoon, on way to Barasu Pass and Kedarkanth bugyal. July–September.
360. ***Aster diplostephioides*** (DC.) C.B. Clarke, Common on open hillslopes (3000–4500 m). Hatta Valley. On way to Barasu Pass. July–October.
361. ***Aster flaccidus*** Bunge, Occasional on open hillslopes (3200–4200). Seema and Hatta Valley. June–September.
362. ***Aster molliusculus*** (Lindl. ex DC.) C.B. Clarke, Occasional on open rocky hillslopes, shady places, and amidst grasses (2000–3000 m). On way to Jakhol. July–September.
363. ***Aster peduncularis*** Wall. ex Nees, Common on slopes in mixed temperate (1800–3000 m). On way to Taluka and Jakhol. On way to Seema. July–October.
364. ***Aster stracheyi*** Hook. *f.*, Rare on open hillslopes (3500–4500 m). On way to Barasu Pass. July–September.
365. ***Aster thomsonii*** C.B. Clarke, Occasional on hillslopes and in shady places (2000–3000 m). On way to Seema. July–October.
366. ***Brachyactis pubescens*** (DC.) Aitch. & C.B. Clarke, Occasional in crevices of walls near hillslopes. On way to Harkidoon. August–September.
367. ***Conyza aegyptiaca*** (L.) Aiton, Common in grassy fields, roadsides, and waste places (up to 2600 m). Sankri, Taluka, and Naitwar. September–January.
368. ***Conyza japonica*** (Thunb.) Less ex DC., Common moist and waste places. Naitwar and Sankri. May–June.
369. ***Erigeron acris*** L., Common on open grassy hillslopes (3400–4600 m). On way to Jamnidhar Glacier. June–July.
370. ***Erigeron annuus*** (L.) Pers., Common on hillslopes (3400–4600 m). Naitwar-Sankri. July–July.
371. ***Erigeron bellidioides*** (Buch.-Ham. ex D. Don) Benth. ex C.B. Clarke, Occasional in stony places (1500–2200 m). Sankri. May–June.
372. ***Erigeron multiradiatus*** (Lindl. ex DC.) C.B. Clarke, Common on open hillslopes (2500–3500 m). Taluka-Seema-Harkidoon. August–September.
373. ***Erigeron sublyratus*** DC., Common in open places (1000–1800 m). Sankri. April–July.
374. ***Myriactis javanica*** (Blume) DC., Occasional on hillslopes and on waste places (1500–2000 m). On way to Sankri and Jamnidhar Glacier. July–September.
375. ***Myriactis nepalensis*** Less., Fairly common weed in fallow fields (1500–3000 m). Sankri. On way to Kedarkantha. July–October.
376. ***Solidago virgaurea*** L., Common in mixed temperate (1600–3000 m). Hatta Valley. On way to Taluka. May–October.
377. ***Arctium lappa*** L., Few on open places and in sandy soils (2500 m). On way to Seema and Ruinsara Tal. July–September.
378. ***Carduus edelbergii*** Rech.*f.*, Common in open hillslopes (3200–4000 m). Harkidoon. August–October.
379. ***Circium wallichii*** DC., Common in waste and moist places (1600–3000 m). Sankri. May–October.
380. ***Jurinella macrocephala*** (Royle) Aswal & Goel, Common on hillslopes (3200–4300 m). On way to Jamnidhar Glacier. July–September.
381. ***Saussurea auriculata*** (DC.) Sch.-Bip., Common in open slopes and amidst shrubs near alpine regions (3000–4000 m). On way to Harkidoon and Kedarkantha bugyal. August–October.
382. ***Saussurea heteromalla*** (D. Don) Hand.-Mazz., Rare in dry, stony hillslopes (1000–3500 m). Sankri. May–August.
383. ***Saussurea obvallata*** (DC.) Edgew., Occasional in rocky slopes and amidst boulders near alpine regions (3500–4500 m). Hatta Valley. July–October.
384. ***Saussurea roylei*** (DC.) Sch.-Bip., Common in hillslopes near subalpine to alpine regions (3000–4500 m). On way to Barasu Pass. July–September.
385. ***Saussurea simpsoniana*** (Fielding & Gardner) Lipsch., Rare amidst boulders near alpine regions (4000–5000 m). Maninda Tal and Barasu Pass. July–October.
386. ***Saussurea stracheyana*** (Kuntze) Lipsch., Few in open places near alpine regions (3000–3500 m). Harkidoon. July–August.
387. ***Saussurea taraxacifolia*** (Lindl. ex Royle) Wall. ex DC., Occasional in hillslopes

(3200–4200 m). Barasu Pass and Ruinsara Tal, On way to Hatta Valley. July–October.

388. ***Serrulata pallida*** DC., Occasional in pine forests (1400–2000 m). Sankri. On way to Jakhol. May–June.
389. ***Tricholepis elongata*** DC., Common in shady places near mixed temperate (1500–2500 m). Sankri. May–September.
390. ***Cephalorrhynchus macrorhizus*** (Royle) Tuisl, Common on open hillslopes (2800–3800 m). On way to Harkidoon. August–October.
391. ***Chaetoseris cynaea*** (D. Don) C. Shih, Common on exposed shady slopes near mixed temperate forests (2500–3500 m). On way to Maninda Tal. On way to Ruinsara Tal. September–November.
392. ***Cicerbita violaefolia*** (Decne.) Beauv., Common on hillslopes and near cultivated fields (2900–3600 m). On way to Harkidoon and Kalhattidhar. July–September.
393. ***Crepis sancta*** (L.) Babco., Occasional on exposed hillslopes near mixed temperate (up to 3000 m). Sankri. March–June.
394. ***Dubyaea hispida*** DC., Common on exposed hillslopes and in moist shady places near alpine regions (3000–4000 m). On way to Jamnidhar Glacier and Harkidoon, On way to Maninda Tal. September–November.
395. ***Lactuca dolichophylla*** Kitam., Rare along roadsides in mixed temperate (up to 1600 m). On way to Jakhol. August–October.
396. ***Launaea aspleniifoila*** Hook. *f.*, Occasional in moist and waste places (up to 2200 m). Sankri. May–June.
397. ***Paramicrorhynchus procumbens*** (Roxb.) Kirp. Common along roadsides, sandy localities, riverbeds, and fields (up to 2600 m). Sankri. May–June.
398. ***Picris hieracioides*** L., Common on hillslopes (2000–2500 m). On way to Harkidoon. July–September.
399. ***Picrosia longifolia*** D. Don, Occasional in mixed temperate (1000–2000 m). On way to Taluka. May–June.
400. ***Prenanthes brunoniana*** Wall. ex DC., Common throughout the sanctuary (up to 3500 m). Taluka and Seema. On way to Harkidoon. July–September.
401. ***Sonchus wightianus*** DC., Common in moist places (1500–2500 m). Sankri. March–June.
402. ***Stenoseris graciliflora*** (Wall. ex DC.) C. Shih, Occasional along roadsides near mixed temperate (up to 1800 m). Kedarkantha. June–October.
403. ***Taraxacum eriopodum*** (D. Don) DC., Scattered on hillslopes (3300–4500 m). On way to Jamnidhar Glacier. July–September.
404. ***Taraxacum officinale*** Weber, Common in rocky and moist in deodar forests (1500–4000 m). Sankri. On way to Harkidoon. On way to Jamnidhar Glacier. Almost throughout the year.
405. ***Tragopogon gracilis*** D. Don, Few on open hillslopes and amidst grasses in coniferous forests (1600–3000 m). On way to Jakhol. April–June.
406. ***Youngia japonica*** (L.) DC., Occasional on shady hillslopes, moist sandy soils, and near cultivated fields (1600–2000 m). Sankri and Taluka-Seema. March–June.
407. ***Acanthospermum hispidum*** DC., Common along the road-sides and in waste places (1400–1700 m). Naitwar-Sankri. January–July.
408. ***Adenocaulon himalaicum*** Edgew., Common on hillslopes (2000–2500 m). Taluka-Seema. August–September.
409. ***Bidens bipinnata*** L., Common in moist places near mixed temperate (up to 950 m). Naitwar. September–April.
410. ***Galinsoga parviflora*** Cav., Icon., Common in wet places and fallow fields. Naitwar, Sankri, and Himri. May–December.
411. ***Sigesbeckia orientalis*** L., Common in waste places (up to 2500 m). On way to Sankri and Jakhol. June–October.
412. ***Xanthium strumarium*** L., Common as a weed in waste places (up to 1400 m). On way to Jakhol. June–October.
413. ***Anaphalis adnata*** Wall. ex DC., Common in open grassy hillslopes near mixed temperate (1500–3000 m). Naitwar. On way to Seema. July–October.
414. ***Anaphalis busua*** (Buch.-Ham. ex D. Don) DC., Common on open hillslopes (1500–3000 m). Ruinsara Tal. On way to Seema and Naitwar. Almost throughout the year.
415. ***Anaphalis contorta*** (D. Don) Hook. *f.*, Common throughout the area (1800–3500 m). On way to Seema and Deokyar. July–March.
416. ***Anaphalis margaritacea*** (L.) Benth. Common in open hillslopes near mixed temperate (2800–3500 m). On way to Seema. July–October.
417. ***Anaphalis nepalensis*** (Spreng.) Hand.-Mazz. var. **monocephala** (DC.) Hand.-Mazz., Common among rocks, stony screes, and on hillslopes (2000–4500 m). On way to Barasu Pass. July–October.

418. ***Anaphalis nepalensis*** (Spreng.) Hand.-Mazz. var. **nepalensis**., Common on hill-slopes in mixed temperate (1800–3500 m). On way to Barasu Pass and Harkidoon. July–October.
419. ***Anaphalis royleana*** DC., Common on hill-slopes and in open places (3000–4500 m). On way to Jamnidhar Glacier. August–October.
420. ***Anaphalis triplinervis*** (Sims.) C.B. Clarke, Common on hillslopes in mixed temperate (1800–3500 m). On way to Seema and Kedarkantha. July–October.
421. ***Blumea obliqua*** (L.) Druce, Common in moist shady places. Sankri. May–June.
422. ***Carpesium abrotanoides*** L., Common along roadsides and open waste places (1100–2500 m). On way to Sankri. September–November.
423. ***Carpesium nepalense*** Less., Common on open waste places (1600–2500 m). On way to Jamnidhar Glacier and Seema. On way to Naitwar. September–October.
424. ***Duhaldea cuspidata*** (DC.) Anderb., Occasional in mixed temperate (1500–2100 m). On way to Naitwar. On way to Sankri. June–September.
425. ***Inula orientalis*** Lam., Common in moist shady places (up to 3200 m). On way to Seema. June–September.
426. ***Leonopodium himalayanum*** DC., Scattered in open moist alpine slopes and near snow lines (3000–4500 m). On way to Jamnidhar. July–September.
427. ***Pseudognaphalium luteoalbum*** (L.) Hilliard & B.L. Burtt, Common in exposed areas. Naitwar. December–January.
428. ***Ainsliaea aptera*** DC., Common in shady places near mixed temperate (1700–2500 m). On way to Seema. May–September.
429. ***Ainsliaea latifolia*** (D. Don) Sch.-Bip., Common in shady places near mixed temperate (1700–2500 m). Kalajungle. May–September.
430. ***Gerbera maxima*** (D. Don) Beauv., Common in open grassy slopes (1800–3400). On way to Harkidoon. July–October.
431. ***Cremanthodium arnicoides*** (DC. ex Royle) R.D. Good, Common on hill-slopes and associated with *Senecio* sp., *Pedicularis* sp., *Thermopsis* sp., *Geranium* sp., etc. (2800–3800 m). Kalajungle and Harkidoon. July–October.
432. ***Ligularia amplexicaulis*** DC., Common on hillslopes and amidst boulders (3000–4000 m). On way to Harkidoon. July–October.
433. ***Ligularia fischeri*** (Ledeb.) Turcz., Fairly common in moist shady places (up to 2800 m). Seema. July–October.
434. ***Senecio graciliflorus*** DC., Common on hillslopes (3000–4000 m). On way to Ruinsara Tal. July–October.
435. ***Senecio laetus*** Edgew., Common almost throughout the sanctuary (1800–4000 m). Taluka. On way to Seema. On way to Kedarkantha. On way to Harkidoon. July–October.
436. ***Senecio nudicaulis*** Buch.-Ham. ex D. Don, Fairly common in exposed places (1600–2600 m). Sankri. On way to Jakhol. July–October.
437. ***Synotis alata*** (Wall. ex DC.) Jeffery & Chen, Fairly common in shady places near mixed temperate (2000–3500 m). On way to Kedarkantha and Harkidoon. July–September.
438. ***Synotis kunthiana*** (Wall. ex DC.) Jeffery & Chen, Common on hillslopes and in shady places near mixed temperate (2300–3500 m). Near Jamnidhar Glacier. July–September.
439. ***Synotis rufinervis*** (DC.) Jeffery & Chen, Common on hillslopes in mixed temperate (2500–3500 m). On way to Taluka. On way to Seema. July–October.

64. CAMPANULACEAE

440. ***Campanula colorata*** Wall., Common along the bridal path in mixed temperate (2400–3200 m). Naitwar, Sankri, and Seema. May–September.
441. ***Campanula latifolia*** L., Occasional on hillslopes and amidst grasses near alpine regions (3300–4000 m). Kalhattidhar and Harkidoon. July–October.
442. ***Codonopsis rotundifolia*** Benth., Occasional on hillslopes, along the bridal path and twining over bushes (2800–4000 m). On way to Harkidoon. July–September.
443. ***Cyananthus integer*** Wall. ex Benth., Common in hillslopes (2200–3400 m). Kalhattidhar. On way to Harkidoon. August–October.
444. ***Cyananthus lobatus*** Wall. ex Benth., Common in alpine meadows (3400–4000 m). Hatta Valley and Kedarkantha bugyal. July–October.

65. ERICACEAE

445. ***Cassiope fastigiata*** (Wall.) D. Don, Common on hillslopes near alpine regions (3500–4500 m). On way to Deokyar and Harkidoon. Above Ruinsara Tal. July–October.

446. ***Gaultheria trichophylla*** Royle, Common in grassy hillslopes and in moist shady places near subalpine-alpine regions (2800–4000 m). Deokyar. On way to Jamnidhar Glacier. July–October.
447. ***Lyonia ovalifolia*** (Wall.) Drude, Common along the river sides in mixed forests (1400–2800 m). Naitwar, Harkidoon, Sankri, and Seema. May–October.
448. ***Rhododendron arboreum*** Sm., Common throughout the area in mixed temperate forests (1600–2800 m). Seen from Sankri to Taluka. April–October.
449. ***Rhododendron campanulatum*** D. Don, Common on hillslopes and forming gregarious patches in alpine regions (3300–4000 m). On way to Jamnidhar Glacier and Harkidoon. On way to Maninda Tal. May–October.
450. ***Rhododendron hypenanthum*** Balf.*f.*, Common on hillslopes and in rocky places with *Betula* sp., *Juniperus* sp., etc. (3300–4500 m). Kedarkantha and Deokyar. On way to Jamnidhar Glacier, Kalajungle, and Ruinsara Tal. May–October.
451. ***Rhododendron lepidotum*** Wall. ex G. Don, Common on rocky places in association with *Rhododendron hypenanthum*, etc. near alpine regions (3300–4500 m). Kedarkantha. On way to Jamnidhar Glacier and Harkidoon. May–October.

66. PRIMULACEAE

452. ***Androsace geraniifolia*** Watt., Common on open grassy hillslopes near cold temperate to alpine regions (2500–4000 m). Sankri. May–October.
453. ***Androsace globifera*** Duby, Occasional on hillslopes, in open places and amidst grasses near alpine regions (2500–3200 m). On way to Harkidoon. May–October.
454. ***Androsace lanuginosa*** Wall., Common on open grassy hillslopes in temperate to subalpine regions (1800–3000 m). On way to Kalhattidhar. May–October.
455. ***Androsace poissonii*** R. Knuth, Common on open hillslopes (3200–4500 m). Barasu Pass. July–October.
456. ***Androsace rotundifolia*** Hardw., Common on open dry hillslopes and in sandy places near temperate to alpine regions (1800–3800 m). Seema. Near Bingad. May–September.
457. ***Androsace sarmentosa*** Wall., Common on open hillslopes and rocky crevices in cold temperate to alpine regions (2800–4000 m). On way to Taluka. May–September.
458. ***Lysimachia lobelioides*** Wall., Common in waste places and along the roadsides near mixed temperate (1500–2000 m). Naitwar. On way to Taluka. June–October.
459. ***Primula denticulata*** Smith, Common on hillslopes in cold temperate to alpine regions (2800–4000 m). Sankri, Hatta Valley, and Deokyar. May–September.
460. ***Primula involucrata*** Wall. ex Duby, Occasional on moist rocky places near alpine regions (3300–4500 m). On way to Barasu Pass. May–September.
461. ***Primula minutissima*** Jacq. ex Duby, Few on hillslopes in alpine regions (3500–4000 m). Deokyar. May–September.
462. ***Primula sessilis*** Royle ex Craib, Rare in humus rich soil and partially shaded places (up to 3500 m). Harkidoon and Deokyar. May–September.

67. MYRSINACEAE

463. ***Ardisia solanacea*** Roxb., Fairly common in damper parts, especially along shady ravines. Naiwar. February–November.

68. SYMPLOCOCACEAE

464. ***Symplocos paniculata*** (Thunb.) Miq., Common along the bridle path in mixed temperate (1600–2500 m). Naitwar, Taluka, Kedarkantha, and Sankri. April–June.
465. ***Symplocos ramosissima*** Wall. ex G. Don, Occasional in mixed temperate (1800–2600 m). On way to Sankri and Taluka. April–June.
466. ***Jasminum humile*** L., Common along the bridal path in mixed temperate (1600–2800 m). Sankri. Near Gangar Bridge. March–October.
467. ***Jasminum officinale*** L., Common along the bridal path, both open and shady places in mixed temperate (1800–2800 m). Sankri and Taluka. On way to Jakhol and Seema. April–August.
468. ***Olea ferruginea*** Royle, Occasional in the lower hills, often planted. April–October.
469. ***Syringa emodi*** Wall. ex Royle, Common in *Quercus semicarpifolia*, *Abies* sp., forests near cold temperate to alpine regions (3000–3600 m). On way to Harkidoon. May–October.

69. APOCYNACEAE

470. ***Ichnocarpus frutescens*** (L.) R. Br., Occasional hanging over wet rocks. Naitwar, near Mautad Village. September–April.

70. ASCLEPIADACEAE

471. ***Marsdenia roylei*** Wight & Arn., Common on open places and along the bridal path (1600–2500 m). Naitwar. May–July.
472. ***Vincetoxicum auriculatum*** (Royle ex Wight) A.P. Jagtap & N.P. Singh, Few along the bridal path and climbing over bushes in mixed temperate (up to 2600 m). On way to Seema. June–October.
473. ***Vincetoxicum hirudinaria*** Medic., Common on hillslopes and amidst grasses in cold temperate to alpine regions (2800–3600 m). Kalhattidhar. June–October.

71. BUDDLEJACEAE

474. ***Buddleja paniculata*** Wall. in Roxb., Frequent in exposed and rocky areas (1500–2200 m). Sankri and Naitwar. April–November.

72. GENTIANACEAE

475. ***Comastoma pedunculatum*** (Royle ex D. Don) Holub, Common on hillslopes and amidst grasses (3500–4500 m). On way Jamnidhar Glacier. June–October.
476. ***Gentiana argentea*** (D. Don) C.B. Clarke, Fairly common on hillslopes and amidst grasses (3800–4000 m). Hatta Valley. April–July.
477. ***Gentiana coronata*** (D. Don ex Royle) Griseb., Common on hillslopes and alpine meadows (3800–4800 m). Barasu Pass, Deokyar, Harkidoon, and Hatta Valley. July–November.
478. ***Gentiana stipitata*** Edgew., Occasional on hillslopes, in moist shady places, and amidst grasses (2800–3500 m). On way to Harkidoon. June–October.
479. ***Gentiana tubiflora*** (Wall. ex G. Don) Griseb., Common in open grassy fields and alpine meadows (4200–5200 m). Barasu Pass. August–October.
480. ***Gentianodes eumarginata*** Omer, Common in open moist grassy fields and alpine meadows (4000–5000 m). Barasu Pass, Deokyar, and Hatta Valley. May–September.
481. ***Gentianopsis paludosa*** (Munro ex Hook. *f.*) Ma, Scattered on hillslopes, in shady places, and amidst grasses (2300–3400 m). Seema-Harkidoon. July–September.
482. ***Halenia elliptica*** D. Don, Common in shady places and in open grassy hillslopes (2500–4000 m). Seema-Harkidoon and Kedarkantha. June–October.
483. ***Lomatogonium carinthiacum*** (Wulfen) Recb., Occasional in open grassy hillslopes (2500–4000 m). On way to Harkidoon. September–October.
484. ***Swertia alternifolia*** Royle ex D. Don, Common in open grassy fields (3500–4200 m). Harkidoon. July–September.
485. ***Swertia angustifolia*** Buch.-Ham. ex D. Don, Common on shady hillslopes (2000–3000 m). On way to Kedarkantha. September–October.
486. ***Swertia chirayita*** (Roxb. ex Fleming) Karsten, Common on hillslopes and in shady places (2400–3000 m). On way to Kedarkantha. July–October.
487. ***Swertia ciliata*** (D. Don ex G. Don) B.L. Burtt., Common on grassy hillslopes (2400–3000 m). Kalhatti, Kedarkantha, and Seema. On way to Harkidoon. On way to Jamnidhar Glacier and Sankri. June–October.
488. ***Swertia cordata*** (G. Don) C.B. Clarke, Common on hillslopes, in shady places, and amidst grasses. On way to Harkidoon and Seema. June–October.
489. ***Swertia cuneata*** Wall. ex D. Don, Common on grassy hillslopes and in shady places (3200–4000 m). Hattagad. July–October.
490. ***Swertia petiolata*** D. Don, Occasional on open hillslopes and among grassy fields (3500–4500 m). On way to Barasu Pass and Harkidoon. July–October.
491. ***Swertia tetragona*** (Edgew.) C.B. Clarke, Occasional in open grassy fields (3200–4200 m). On way to Harkidoon. August–October.
492. ***Tripterospermum volubile*** (D. Don) H. Hara, Occasional in mixed temperate (2000–2700 m). On way to Taluka. August–October.

73. BORAGINACEAE

493. ***Arnebia benthamii*** (Wall. ex G. Don) Johnston, Common on grassy slopes (3200–4500 m). Harkidoon. On way to Jamnidhar Glacier, Barasu Pass, and Deokyer. July–October.
494. ***Cynoglossum lanceolatum*** Forssk., Occasional on hillslopes and along the bridal path in mixed temperate (1600–2400 m). Naitwar. On way to Taluka. June–August.
495. ***Cynoglossum zeylanicum*** (Vahl ex Hornem.) Thunb. ex Lehm. Common along the roadsides and in waste places (1800–3000 m). Naitwar, near Mautad Village. Seema. July–October.

496. ***Hackelia uncinata*** (Royle ex Benth.) C.E.C. Fisch., Common on hillslopes and amidst grasses (3000–4000 m). On way to Jamnidhar Glacier. June–September.
497. ***Lindelofia longiflora*** (Benth.) Baill., Common on hillslopes and in shady places (3000–4000 m). Seema-Harkidoon. July–October.
498. ***Myosotis caespitosa*** C.F. Schultz, Occasional in open moist grassy places (2100–3400 m). Hatta Valley, Kedarkantha, and Barasu Pass. June–July.
499. ***Myosotis silvatica*** Ehrh. ex Hoffm., Common on hillslopes and in open places (3000–4000 m). On way to Harkidoon. June–October.

74. CONVOLVULACEAE

500. ***Cuscuta europaea*** L., Rare parasitic on bushes and along the roadsides (2800–3300 m). On way to Harkidoon. July–October.
501. ***Cuscuta gigantea*** Griff., Common parasitic on bushes and along the roadside (2700–3200 m). On way to Harkidoon. July–October.
502. ***Cuscuta reflexa*** Roxb., Occasional parasitic on bushes in waste places, along the roadside, and cultivated fields (1500–2700 m). Seema. July–October.
503. ***Dinetus racemosus*** (Roxb.) Buch.-Ham. ex Sweet, Occasional in thickets and hillslopes (up to 2600 m). On way to Sankri. May–October.
504. ***Evolvulus alsinoides*** (L.) L., Common along the roadsides and in waste places (1500–1800 m). Naitwar. Almost throughout the year.
505. ***Ipomoea nil*** (L.) Roth, Occasional climbing over bushes and along the roadsides (1500–2000 m). Naitwar. July–October.
506. ***Ipomoea purpurea*** (L.) Roth, Common in waste places and along the roadsides (1500–2000 m). Naitwar and Sankri-Taluka. July–October.

75. SOLANACEAE

507. ***Datura stramonium*** L., Common in waste places and along the roadsides (1600–2400 m). Sankri-Taluka. May–September.
508. ***Nicandra physalodes*** (L.) Gaertn., Common in waste places, cultivated fields, and along the roadsides (1500–2000 m). Naitwar, Taluka, and Dhatmir. July–October.
509. ***Physalis minima*** L., Occasional in waste places (up to 1500 m). Naitwar. July–August.
510. ***Solanum nigrum*** L., Common in moist shady places (1500–2000 m). Naitwar. May–October.
511. ***Solanum viarum*** Dunal, Few in waste places (1500–1700 m). Naitwar and Sankri. Throughout the year.

76. SCROPHULARIACEAE

512. ***Euphrasia himalayica*** Wettst., Common on hillslopes and amidst grasses (3200–3500 m). Kalhattidhar. August–October.
513. ***Euphrasia kurramensis*** Pennell, Common on hillslopes (3200–4600 m). On way to Barasu Pass. June–September.
514. ***Euphrasia pectinata*** Wall., Common in grassy slopes and thickets (up to 4000 m). Harkidoon and Hatta Valley. June–September.
515. ***Euphrasia platyphylla*** Pennell, Occasional on hillslopes and amidst grasses (3000–4000 m). Seema-Harkidoon. July–October.
516. ***Limnophila connata*** (Buch.-Ham. ex D. Don) Hand.-Mazz., Fairly common along the bridal path, in waste places, and marshy areas (1500–1700 m). Naitwar. Throughout the year.
517. ***Lindenbergia polyantha*** Royle ex Benth., Common on mossy rocky crevices and on mossy boulders (1500–2000 m). Naitwar. Throughout the year.
518. ***Lindernia ciliata*** (Colsm.) Pennell, Occasional in wet sandy places (up to 1200 m). Naitwar. August–December.
519. ***Lindernia crustacea*** (L.) F. Muell., Occasional in wet places. Naitwar. August–December.
520. ***Mazus dentatus*** Wall. ex Benth., Common in mixed temperate (up to 2500 m). On way to Taluka. March–October.
521. ***Mazus pumilus*** (Burm. *f.*) Steenis, Common in waste places and on grassy slopes (1500–2000 m). On way to Istragad. March–October.
522. ***Mazus surculosus*** D. Don, Common in waste places and on grassy slopes (1500–2500 m) Naitwar, Seema, and Jakhol. March–October.
523. ***Pedicularis bifida*** (Buch.-Ham. ex D. Don) Pennell, Common on hillslopes and amidst grasses (2000–3000 m). On way to Taluka. June–October.
524. ***Pedicularis cheilanthifolia*** Schrenkvar. **purpurea** (Pennel) Tsoong ex T. Husain & Agnihotri, Occasional in moist slopes in gullies and on moist plains near alpine regions (3500–4500 m). Harkidoon. On way to Maninda Tal. July–August.

525. ***Pedicularis gracilis*** Wall. ex Benth., Common on hillslopes and amidst grasses (2500–4000 m). Taluka and Seema. On way to Barasu Pass. July–October.
526. ***Pedicularis hoffmeisteri*** Klotzsch & Garcke, Common in *Quercussemicarpifolia-Abies* sp. forests on hillslopes and in open places near cold alpine regions (2700–3500 m). Seema-Harkidoon. On way to Hatta Valley and Ruinsara Tal. July–October.
527. ***Pedicularis mollis*** Wall. ex Benth., Rare on moist sandy areas (3800–4800 m). Harkidoon. July–September.
528. ***Pedicularis pectinata*** Wall. ex Benth., Common on hillslopes and amidst grasses (3000–4500 m). On way to Barasu Pass. On way to Jamnidhar Glacier, Hatta Valley, and Harkidoon. June–October.
529. ***Pedicularis punctata*** Decne., Common on open hillslopes and in moist places (3000–4000 m). On way to Barasu Pass and Harkidoon. June–October.
530. ***Picrorhiza kurrooa*** Royle ex Benth., Common on open hillslopes, sandy gravel soils, rocks, and amidst grasses (3200–4000 m). Barasu Pass, Harkidoon, and Deokyar. July–September.
531. ***Scrophularia calycina*** Benth., Common on hillslopes, amidst grasses, and in shady places (2800–3700 m). On way to Harkidoon. May–October.
532. ***Scrophularia edgeworthii*** Benth., Occasional in moist shady places (1600–3000 m). On way to Seema. August–October.
533. ***Scrophularia polyantha*** Royle ex Benth., Common in moist shady places (1600–2700 m). Seema and Sankri. July–October.
534. ***Torenia cordifolia*** Roxb., Occasional on moist places (up to 1000 m). Naitwar. July–October.
535. ***Verbascum thapsus*** L., Common on exposed hillslopes and along the roadsides (1500–3200 m). Naitwar and Seema. April–November.
536. ***Veronica anagallis-aquatica*** L., Common in stream banks and swampy places. Naitwar. February–July.

77. GESNERIACEAE

537. ***Chirita pumila*** D. Don, Common on rocks, cliffs, and among grassy clumps (up to 2200 m). Sankri. July–October.

78. ACANTHACEAE

538. ***Barleria cristata*** L., Common along the roadsides (up to 1000 m). On way to Naitwar and Jakhol. May–October.
539. ***Dicliptera bupleuroides*** Nees, Common at Naitwar. May–December.
540. ***Justicia diffusa*** Willd., Common in moist areas near mixed temperate. On way to Naitwar. August–October.
541. ***Rungia pectinata*** (L.) Nees, Common in mixed temperate (up to 900 m). Naitwar. July–August.
542. ***Strobilanthes pentstemonoides*** (Nees) T. Anderson var. **dalhousieanus** (Nees) J.R.I. Wood, Common in moist shady places and as forests undergrowth in mixed temperate (up to 2450 m). On way to Jakhol. July–October.
543. ***Strobilanthes wallichii*** Nees, Common in moist shady places and gregarious as the forest undergrowth near mixed temperate (950–3200 m). Sankri and Taluka. On way Jakhol. July–October.

79. VERBENACEAE

544. ***Verbena officinalis*** L., Common in open places and along roadsides (up 1500 m). Naitwar. May–December.

80. PHRYMACEAE

545. ***Phryma leptostachya*** L. var. **oblongifolia** (Koidz.) Honda, Common on way to Jakhol. July–September.

81. LAMIACEAE

546. ***Ajuga bracteosa*** Wall. ex Benth., Common on rocky slopes and along the roadsides. Naitwar. On way to Seema. April–October.
547. ***Ajuga brachystemon*** Maxim., Common on hillslopes (2500–3300 m). On way to Kalhattidhar. June–October.
548. ***Ajuga parviflora*** Benth., Common in the foothills, along roadsides, and nearby cultivated fields. Naitwar. March–October.
549. ***Anisomeles indica*** (L.) Kuntze, Common along banks of canals, irrigation channels, and roadsides. Naitwar. April–October.
550. ***Clinopodium umbrosum*** (M. Bieb.) C. Koch. Common in forest shade and damp places. Naitwar, on way to Taluka. On way to Seema. March–September.
551. ***Craniotome furcata*** (Link) Kuntze, Common in moist shady places (up to 3200 m). Taluka-Seema. July–October.
552. ***Dracocephalum wallichii*** Sealy, Common in alpine meadows (4200–4700 m). Maninda Tal, Barasu Pass. July–September.

553. ***Elsholtzia ciliata*** (Thunb.) Hyl., Common along the roadsides, disturbed areas. On way to Naitwar. August–December.
554. ***Elsholtzia eriostachya*** (Benth.) Benth. Common in moist shady places. Taluka, Seema, Kalhattidhar, and Harkidoon. July–October.
555. ***Elsholtzia flava*** (Benth.) Benth., Common along the roadsides. On way to Sankri. September–October.
556. ***Elsholtzia fruticosa*** (D. Don) Rehder, Occasional in forest openings and among scrubs. Sankri and Seema. July–November.
557. ***Elsholtzia strobilifera*** (Benth.) Benth., Common on hillslopes near alpine regions. Harkidoon. June–October.
558. ***Isodon coetsa*** (Buch.-Ham. ex D. Don) Kudo, Common along the roadsides, moist shady places. Naitwar, Taluka, Seema, and Kedarkantha. August–December.
559. ***Isodon lophanthoides*** (Buch.-Ham. ex D. Don) H. Hara, Common in moist shady places near mixed temperate (1000–2500 m). Dodital, Sankri, and Seema. August–October.
560. ***Isodon rugosus*** (Wall. ex Benth.) Codd., Common on scrubby and savanna slopes (up to 2700 m). Seema. February–September.
561. ***Lamium album*** L., Common among scrubs on humus rich soils (up to 3200 m). Taluka-Seema. August–October.
562. ***Leucas lanata*** Benth., Locally common in moist sunny and dry places (1500–2300 m). On way to Taluka and Istragad. January–June.
563. ***Mentha longifolia*** (L.) Huds., Common along canal banks and irrigation channels (up to 950 m). On way to Naitwar. July–November.
564. ***Micromeria biflora*** (Buch.-Ham. ex D. Don) Benth., Occasional along pathways near dry slopes (900–2000 m). Naitwar and Taluka. March–November.
565. ***Mosla ocymoides*** Buch.-Ham. ex Benth., Occasional in mixed deciduous forests (750–1200 m). Naitwar. August–October.
566. ***Nepeta discolor*** Royle ex Benth., Very common almost throughout the areas (850–1650 m). Hatta Valley. July–October.
567. ***Nepeta govaniana*** (Wall. ex Benth.) Benth., Occasional along streamlines and on humus rich soils (up to 3500 m). Seema. On way to Harkidoon. July–September.
568. ***Nepeta laevigata*** (D. Don) Hand.-Mazz., Very common in open shady places and near scrub forests (2200–4500 m). On way to Jamnidhar Glacier, Harkidoon, Seema, and Kalhattidhar. August–January.
569. ***Origanum vulgare*** L., Fairly common on dry open, scrubby, and savanna slopes (up to 3600 m). Harkidoon and Naitwar. August–November.
570. ***Perilla frutescens*** (L.) Britt., Common along the roadsides (800–1200 m). On way to Naitwar. September–November.
571. ***Phlomis bracteosa*** Royle ex Benth., Very common on alpine meadows and on rocky slopes (3500–4900 m). Kedarkantha bugyal, On way to Barasu Pass. July–December.
572. ***Phlomis macrophylla*** Wall. ex Benth., Common on hillslopes (1300–3000 m). On way to Seema. May–September.
573. ***Plectranthus barbatus*** Andrews, Occasional on open rocky places (up to 1800 m). On way Jakhol. July–October.
574. ***Prunella vulgaris*** L., Common in shady places and along the edges of rice fields. Sankri. On way to Kedarkantha. July–December.
575. ***Salvia hians*** Royle ex Benth., Common along the lesser water courses, on open grassy and rocky slopes. Naitwar. July–December.
576. ***Salvia lanata*** Roxb., Common along the roadsides (up to 2300 m). Sankri. March–April.
577. ***Salvia nubicola*** Wall. ex Sweet, Common in savanna slopes and among scrubs. On way to Taluka and Kedarkantha. July–March.
578. ***Scutellaria scandens*** Buch.-Ham. ex D. Don, Common on hillslopes (2000–3000 m). On way to Dhatmir. April–May.
579. ***Stachysbikramii*** Aswal & Melhotra, Very common on open slopes, shady places, forest openings, and along forest pathways. Seema, Harkidoon, and Sankri. July–January.
580. ***Teucrium royleanum*** Wall. ex Benth., Common on hillslopes and amidst stones by the side of dry streams (2200–3200 m). Seema. April–August.
581. ***Thymus serpyllum*** L. subsp. **quinquecostatus** (Celsk.) Kitmura, Common on dry rocky slopes, less so in scrubby areas (3200–4500 m). On way to Jamnidhar Glacier, Barasu Pass, and Kalhattidhar. February–May.

82. PLANTAGINACEAE

582. ***Plantago depressa*** Willd., Common along roadsides in moist places. Sankri. On way to Taluka and Naitwar. April–October.
583. ***Plantago erosa*** Wall. in Roxb., Occasional along roadsides and open areas. On way to Taluka and Naitwar. May–August.

83. AMARANTHACEAE

584. ***Achyranthes aspera*** L., Fairly common in the fallow fields and along roadsides (800–2500 m). Naitwar, Sankri, and Taluka. March–November.
585. ***Achyranthes bidentata*** Blume, Common in open places (800–2500 m). Naitwar, Sankri, and Taluka. October–December.
586. ***Cyathula tomentosa*** (Roth) Moq. Common in open moist areas and waste places (up to 2300 m). Naitwar and Sankri. On way to Taluka. September–December.

84. CHENOPODIACEAE

587. ***Acroglochin persicarioides*** (Poir.) Moq., Fairly common along riverbeds and roadsides (up to 850 m). On way to Seema. July–September.
588. ***Chenopodium ambrosioides*** L., Fairly common along riverbeds and roadsides (up to 850 m). On way to Naitwar. September–March.
589. ***Chenopodium hybridum*** L., Common along roadsides and near moist places (up to 2000 m). Taluka. July–October.
590. ***Chenopodium opulifolium*** Schrad. ex Koch & Ziz., Common along roadsides and near moist places (up to 2000 m). Taluka. July–October.

85. PHYTOLACCACEAE

591. ***Phytolacca acinosa*** Roxb., Fairly common along riverbeds and roadsides (900–2200 m). On way to Jakhol. On way to Taluka. May–September.

86. POLYGONACEAE

592. ***Aconogonon rumicifolium*** (Royle ex Bab.) H. Hara, Rare on moist hillslopes, along streams and among crevices (3000–4500 m). On way to Barasu Pass. On way to Jamnidhar Glacier. June–July.
593. ***Bistorta affinis*** (D. Don) Greene, Common on open slopes, edges, and rocky places (3000–4800 m). Harkidoon and Hattagad. June–September.
594. ***Bistorta amplexicaulis*** (D. Don) Green, Abundant in moist shady places and fallow fields. On way to Taluka and Seema. On way to Kedarkantha. On way to Jamnidhar Glacier. June–October.
595. ***Bistorta macrophylla*** (D. Don) Sojak, Common along the cultivated fields. Sankri and Hatta Valley. June–September.
596. ***Bistorta vivipara*** (L.) S.F. Gray, Common in moist shady areas and steep slopes (2500–4500 m). On way to Jamnidhar Glacier and Hatta Valley. June–August.
597. ***Fagopyrum esculentum*** Moench, Commonly cultivated for its nuts and has become naturalized almost throughout the sanctuary. Sankri, Jakhol, and Taluka to Seema. On way to Harkidoon. August–October.
598. ***Fagopyrum tataricum*** (L.) Gaertn., Common on hillslopes (2300–3500 m). Taluka to Seema. On way to Harkidoon. August–October.
599. ***Fallopia convolvulus*** (L.) A. Löve, Common as a weed in cultivated fields and moist shady (1500–3500 m). Naitwar. May–September.
600. ***Koenigia nepalensis*** D. Don, Common in moist places and under rocks with mosses (3500–4200 m). On way to Hatta Valley and Harkidoon. July–September.
601. ***Oxyria digyna*** (L.) Hill, Very common in wet soil and shady areas (2700–3600 m). Harkidoon and Deokyar. May–August.
602. ***Persicaria barbata*** (L.) H. Hara, Occasional along streams and riverbanks (800–1200 m). Taluka, on way to Naitwar. Almost throughout the year.
603. ***Persicaria capitata*** (Buch.-Ham. ex D. Don) H. Gross, Occasional on moist rocky places (1200–2200 m). Sankri. On way to Naitwar. May–August.
604. ***Persicaria chinensis*** (L.) H. Gross, Occasional on hillslopes and amidst grasses (1000–1800 m). Sankri and Taluka. September–November.
605. ***Persicaria hydropiper*** (L.) Spach, Common in ditches, on banks, and moist shady places (up to 2000 m). Taluka and Naitwar. April–September.
606. ***Persicaria nepalensis*** (Meisn.) H. Gross, Common along water canals and in moist shady rocky places (1200–3500 m). Seema, Naitwar, and Taluka. On way to Hatta Valley. June–September.
607. ***Polygonum aviculare*** L., Common as a weed in cultivated fields, on moist shady areas (up to 3500 m). On way to Jamnidhar Glacier. March–September.

608. ***Polygonum microcephalum*** var.**sphaerocephalum** (Wall. ex Meisn.) H. Hara, Common in marshy places (up to 1800 m). Naitwar. June–August.
609. ***Polygonum plebeium*** R. Br., Fairly common in open moist places and along roadsides (up to 1800 m). Naitwar. August–October.
610. ***Polygonum recumbens*** Royle ex Bab., Fairly common along dry streams near moist deciduous forests (1500–4000 m). On way to Naitwar and Taluka-Seema. September–March.
611. ***Polygonum vacciniifolium*** Wall. ex Meisn., Common on mountain slopes and rocky crevices (3000–4200 m). Harkidoon. August–October.
612. ***Rubrivena polystachya*** (Wall. ex Meisn.) M. Král, Occasional on rocks, slopes, slightly drier places (1500–3000 m). Taluka and Harkidoon. June–September.
613. ***Rumex acetosa*** L., Common on slopes and moist rocks (2000–4500 m). On way to Jamnidhar Glacier, Kedarkantha bugyal. July–September.
614. ***Rumex dentatus*** L., Fairly common in moist areas (up to 3000 m). Naitwar. January–May.
615. ***Rumex hastatus*** D. Don, Abundance along roadsides, on dry slopes, rocks, and exposed areas near habitation (700–2660 m). Sankri. June–October.
616. ***Rumex nepalensis*** Spreng., Common on slopes, waste, and moist shady places (900–2000 m). Sankri and Taluka. June–September.

87. PIPERACEAE

617. ***Peperomia tetraphylla*** (Forst.*f.*) Hook.*f.* & Arn., Occasional in mixed temperate. On way to Naitwar. April–November.

88. SAURURACEAE

618. ***Houttuynia cordata*** Thunb., Rare associated with *Acorus calamus* and sedges in marshy places (1500–2000 m). Near forest nursery, Naitwar, and Sankri. May–June.

89. LAURACEAE

619. ***Cinnamomum tamala*** (Buch.-Ham.) T. Nees et Nees, Rare in damp ravines. Naitwar. April–October.
620. ***Neolitsea pallens*** (D. Don) Momiya. & H. Hara, Common in shady valleys near Deodar, *Rhododendron* and *Quercus* forests. Sankri. On way to Taluka. March–July.
621. ***Neolitsea umbrosa*** (Nees) Gamble, Occasional in mixed temperate. Sankri and Taluka. March–October.
622. ***Persea gamblei*** (King ex Hook. *f.*) Kosterm., Common in mixed temperate. On way to Seema. March–October.
623. ***Persea odoratissima*** (Nees) Kosterm., Common in moist forests. March–June.

90. THYMELAEACEAE

624. ***Wikstroemia chamaedaphne*** (Bunge) Meisn., Fairly common in moist shady places. On way to Sankri. April–September.

91. ELAEAGNACEAE

625. ***Elaeagnus umbellata*** Thunb., Common in ravine, but occasional in *Quercus* forests, open dry places, gravelly hills, and along roadsides (850–1600 m). Naitwar. April–May.
626. ***Hippophae rhamnoides*** L. subsp. **tukestanica** Rousi, Occasional on open mountain slopes and moist sandy soil (900–1200 m). Sankri. June–September.
627. ***Hippophae salicifolia*** D. Don, Common in moist gravel or stony areas, often beside rivers or streams. On way to Seema. June–September.

92. VISCACEAE

628. ***Viscum articulatum*** Burm.*f.*, Common in deciduous forests. July–September.

93. EUPHORBIACEAE

629. ***Acalypha superba*** Forssk., Occasional in moist places. Naitwar. July–September.
630. ***Euphorbia cornigera*** Boiss. var. **cognata** (Klotzsch & Garcke) Binojk. & N.P. Balakr., Occasionally found growing in alpine meadows, margins of open forests (3200–4000 m). Seen, on way to Harkidoon. June–August.
631. ***Euphorbia indica*** Lam., Fairly common in open places, weed along streams and riverbanks (800–2000 m). Naitwar. On way to Sankri. May–September.
632. ***Euphorbia prostrata*** Aiton, Common in lawns, wastelands, and fields (up to 2000 m). Naitwar. May–October.
633. ***Euphorbia royleana*** Boiss., Occasional on rocky and exposed slopes (up to 1600 m). May–October.
634. ***Euphorbia thymifolia*** L., Common in open and wastelands (up to 1800 m). Naitwar. December–August.

635. ***Glochidion heyneanum*** (Wight & Arn.) Wight, Occasional in moist deciduous forests (up to 900 m). September–December.
636. ***Leptopus cordifolius*** Wall. ex Decne., Fairly common on moist rocks and along roadsides (up to 2000 m). Naitwar and Sankri. March–June.
637. ***Phyllanthus emblica*** L., Occasional along the roadsides near mixed temperate. February–October.
638. ***Phyllanthus pseudoparvifolius*** R.L. Mitra & Sanjappa, On shady banks near water, and on limestone rocks (to 2440 m). Taluka-Seema. August–October.
639. ***Phyllanthus urinaria*** L., Common in shade places and along streams near mixed temperate (to 1200 m). Naitwar. June–October.
640. ***Ricinus communis*** L., *Uses*: Oil obtained from the seeds is used in muscular pain and body ache.

94. BUXACEAE

641. ***Buxus wallichiana*** Baill., Common in mixed temperate (1800–3000 m). On way to Seema, Jakhol. May–September.
642. ***Sarcococca saligna*** (D. Don) Muell.-Arg. Common in moist shady places near deodar forests (2000–3500 m). Sankri, Taluka-Seema, and Jakhol. May–October.

95. URTICACEAE

643. ***Debregeasia longifolia*** (Burm.*f.*) Wedd., Fairly common in scrub forests (1500 m). Naitwar. July–January.
644. ***Debregeasia salicifolia*** (D. Don) Rendle, Fairly common along streams and shady depressions (up to 2000 m). Naitwar. March–August.
645. ***Elatostema monandrum*** (Buch.-Ham. ex D. Don) H. Hara, Common in mixed temperate (1200–2800 m). On way to Taluka. On way to Seema. June–September.
646. ***Elatostema sessile*** J.R. Forst. & G. Forst., Fairly common in moist forests (2400 m). Sankri. On way to Seema. June–November.
647. ***Girardinia diversifolia*** (Link) Friis, Occasional in moist deciduous forests (850–2200 m). Naitwar and Taluka. September–December.
648. ***Gonostegia hirta*** (Blume ex Hassk.) Miq. Common along roadsides (850–2200 m). Naitwar and Sankri. On way to Taluka. July–October.
649. ***Lecanthus peduncularis*** (Royle) Wedd., Common on hillslopes and moist shady places (800–2300 m). Naitwar and Taluka. On way to Seema. August–November.
650. ***Parietaria micrantha*** Ledeb., Common on rocky and moist shady places (800–2300 m). On way to Seema. August–October.
651. ***Pilea scripta*** (Buch.-Ham. ex D. Don) Wedd., Occasional in wet places, along old walls (1500–2000 m). On way to Taluka. August–November.
652. ***Pilea umbrosa*** Blume, Fairly common in moist shady places (1000–1800 m). On way to Sankri and Taluka. June–October.
653. ***Pouzolzia zeylanica*** (L.) Benn. & R. Br., Rare in moist shady places (up to 1200 m). Naitwar. On way to Sankri. September–December.
654. ***Urtica ardens*** Link, Common along the roadsides near mixed temperate (1000–2800 m). On way to Taluka. March–August.
655. ***Urtica dioica*** L., Common in moist shady places (1500 m). Naitwar. September–January.
656. ***Villebrunea frutescens*** (Thunb.) Blume, Common in moist shady places (1830 m). Sankri. May–October.

96. ULMACEAE

657. ***Celtis australis*** L., Occasional in moist mixed temperate (up to 2200 m). On way to Taluka. July–October.
658. ***Ulmus wallichiana*** Planch., Occasional in moist mixed temperate (up to 2200 m). On way to Taluka. July–October.

97. CANNABACEAE

659. ***Cannabis sativa*** L., Common throughout the sanctuary except glacier (800–3830 m). Sankri, Taluka, and Seema. June–September.

98. MORACEAE

660. ***Ficus benghalensis*** L., Occasionally planted along the roadsides in village areas. May–June.
661. ***Ficus hederacea*** Roxb., Fairly common, found creeping on moist rocks and other trees (up to 1800 m). Naitwar. May–June.
662. ***Ficus hispida*** L.*f.*, Infrequent in moist deciduous to semi-evergreen forests (up to 1600 m). Naitwar. November–May.
663. ***Ficus palmata*** Forssk., Common on hot dry slopes in clay-loam soils (up to 2500 m). Near Sankri and Naitwar. May–November.

664. ***Ficus sarmentosa*** Buch.-Ham. ex Sm., Occasional found creeping on rocks and in crevices or climbing on other trees with the help of adventitious roots (up to 2300 m). Taluka and Jakhol. May–September.
665. ***Morus serrata*** Roxb., Common in mixed temperate and along roadsides (1200–2700 m). Naitwar. Taluka. On way to Seema. March–May.

99. JUGLANDACEAE

666. ***Engelhardtia spicata*** Leschen. ex Blume, Common and often gregarious in mixed temperate (1000–1600 m). Naitwar. Near Matuad Village. March–June.
667. ***Juglans regia*** L., Common along roadsides and cultivated fields (1000–2300 m). Sankri and Taluka. April–October.

100. MYRICACEAE

668. ***Myrica esculenta*** Buch.-Ham. ex D. Don, Grows in open, mixed temperate on mountain slopes. August–May.

101. BETULACEAE

669. ***Alnus nepalensis*** D. Don, Common in mixed temperate (1000–1600 m). Naitwar and Sankri. October–December.
670. ***Alnus nitida*** (Spach) Endl., Common on roadside and along streams near mixed temperate (1000–1600 m). Sankri-Jakhol. September–October.
671. ***Betula alnoides*** Buch.-Ham. ex D. Don, Common in hillslopes near mixed temperate (2000–3000 m). Taluka-Seema. August–November.
672. ***Betula utilis*** D. Don, Common on hillslopes and found associated with *Rhododendron* spp., *Juniperus* sp., etc. (2300–4200 m). On way to Jamnidhar Glacier, Ruinsara Tal. August–November.
673. ***Carpinus viminea*** Lindley, Grows in mixed temperate. April–September.
674. ***Corylus colurna*** L., Common in hillslopes, often gregarious near mixed temperate (1600–3300 m). Seema. On way to Taluka. March–April.

102. FAGACEAE

675. ***Castanea sativa*** Mill., Occasional in mixed deciduous forests (1000–1830 m). Naitwar. April–August.
676. ***Quercus dilatata*** Lindl. ex Royle, Common in mixed temperate and associated with *Cedrus deodara, Quercus* sp., etc. Sankri. April–August.
677. ***Quercus leucotrichophora*** A. Camus, Common and often gregarious on hillslopes (1000–2700 m). April–September.
678. ***Quercus semecarpifolia*** Sm., Very common often associated with *Abies* sp., *Picea* sp., etc. at higher zones (2500–3000 m). Sankri, Seema, Taluka, and Jakhol. May–June.

103. SALICACEAE

679. ***Populus ciliata*** Wall. ex Royle, Common in exposed places and often cultivated. Naitwar. March–June.
680. ***Salix acmophylla*** Boiss., February–April
681. ***Salix flabellaris*** Andersson, Common in open hillslopes (up to 4000 m). On way to Maninda Tal. On way to Barasu Pass and Hattagad. June–September.
682. ***Salix denticulata*** Andersson, Common in open hillslopes (3500–4500 m). Harkidoon. On way to Jamnidhar Glacier. April–May.

II. MONOCOTYLEDONES

104. ORCHIDACEAE

1. ***Calanthe tricarinata*** Lindl., Occasional in mixed temperate (1500–2500 m). Sankri, Seema, and Gangad. May–July.
2. ***Cypripedium cordigerum*** D. Don, Rare in open hillslopes near mixed temperate forests (2200–3500 m). On way to Osla. June–July.
3. ***Cypripedium macranthos*** Sw. Common in open hillslopes (up to 2500 m). Isragad. July–October.
4. ***Dactylorhiza hatagirea*** (D. Don) Soo, Common in open hillslopes (3800–4000 m). Harkidoon. On way to Maninda Tal. July–October.
5. ***Epipactis helleborine*** (L.) Crantz, Occasional in open hillslopes (2800–4000 m). Harkidoon and Ruinsara Tal. June–August.
6. ***Goodyera fusca*** (Lindl.) Hook. *f.*, Common in moist rocky places (2200–3600 m). On way to Harkidoon, On way to Jamnidhar Glacier. August–October.
7. ***Goodyera repens*** (L.) R. Br., Occasional on moist rocky places (2000–3500 m). Seema. On way to Ruinsara Tal and Kedarkantha. July–September.
8. ***Habenaria pectinata*** D. Don, Rare in moist shady places near mixed temperate (up to 1800 m). Taluka. August–October.
9. ***Herminium lanceum*** (Thunb. ex Sw.) J. Vuijk., Occasional in hillslopes (2500–3500 m). Sankri-Taluka, Kalhattidhar, and Jakhol. July–August.

10. ***Herminium monorchis*** (L.) R. Br., Occasional in wet meadows (3200–4300 m). On way to Barasu Pass. On way to Jamnidhar Glacier. On way to Maninda Tal. July–August.
11. ***Malaxis muscifera*** (Lindl.) Kuntze, Rare in moist shady places (3000–4000 m). Seema-Harkidoon. July–September.
12. ***Peristylus elisabethae*** (Duthie) R.K. Gupta, Common in mixed temperate, open grasslands near alpine meadows (3000–4000 m). Seema-Harkidoon. August–October.
13. ***Platanthera clavigera*** Lindl., Common in open grassy fields (1500–3000 m). Barasu Pass and Harkidoon. August–October.
14. ***Platanthera latilabris*** Lindl., Common in open grassy fields (1500–3000 m). Taluka. July–August.
15. ***Ponerorchis chusua*** (D. Don) Soó, rare in moist shady places near alpine grasslands (1000–2200 m). Hatta Valley. July–October.
16. ***Satyrium nepalense*** D. Don, Common in open grassy fields (2000–4000 m). Kalhattidhar, Taluka, and Harkidoon. July–November.
17. ***Spiranthes sinensis*** (Pers.) Ames, Occasional on open grassy fields (2500–3700 m). Seema. May–September.

105. ZINGIBERACEAE

18. ***Hedychium coronarium*** J. König, Common along roadsides near mixed temperate (up to 2300 m). Taluka and Seema. August–November.
19. ***Hedychium spicatum*** Buch.-Ham. ex Sm., Common along roadsides near mixed temperate (up to 3200 m). Taluka. On way to Seema. July–November.
20. ***Roscoea alpina*** Royle, Common in mixed temperate (up to 1800 m). Sankri. August–October.
21. ***Roscoea purpurea*** J.E. Sm., Common in mixed temperate (up to 1550 m). Jakhol. July–August.

106. IRIDACEAE

22. ***Iris kemaonensis*** Wall. ex Royle, Fairly common in open grassy fields (up to 4500 m). Hatta Valley. July–October.
23. ***Iris milesii*** Foster, Very common in mixed temperate (up to 2300 m). Jakhol. May–July.

107. AGAVACEAE

24. ***Agave cantula*** Roxb., Occasional in dry and open places. August–September.

108. AMARYLLIDACEAE

25. ***Zephyranthes citrina*** Baker, Rare in moist places (up to 1500 m). Naitwar. September–December.

109. DIOSCOREACEAE

26. ***Dioscorea deltoidea*** Wall. ex Kunth, Occasional in bushes (1500–2150 m). Sankri, toward Kedarkantha. On way to Harkidoon, Jakhol, and Seema. May–June.
27. ***Dioscorea melanophyma*** Burkill & Prain, Fairly common in mixed temperate (1200–2400 m). On way to Sankri. August–September.

110. HAEMODORACEAE

28. ***Aletris pauciflora*** (Kl.) Hand.-Mazz., Rare in hillslopes (2700–4000 m). Barasu Pass and Hatta Valley. July–August.
29. ***Ophiopogon intermedius*** D. Don, Common in mixed temperate (up to 2000 m). Taluka. July–August.
30. ***Ophiopogon planiscapus*** Nakai, Occasional along roadsides near mixed temperate (850–2000 m). Taluka. July–August.
31. ***Paris polyphylla*** Sm., Rare in moist shady places near mixed temperate (up to 2400 m). Sankri. May–June.
32. ***Trillium govanianum*** Wall. ex Royle, Sanctuary. April–August

111. CONVALLARIACEAE

33. ***Polygonatum cirrhifolium*** (Wall.) Royle, Occasional in shady slopes near mixed temperate (1500–2200 m). Taluka. May–June.
34. ***Polygonatum verticillatum*** (L.) All., Occasional in shady slopes (1800–2660 m). Sankri. On way to Harkidoon, Jakhol, Dodhital, and Taluka. May–June.

112. ASPARAGACEAE

35. ***Asparagus filicinus*** Buch.-Ham. ex D. Don, Occasional in moist shady places (up to 2500 m). Sankri and Seema. May–June.
36. ***Maianthemum purpureum*** (Wall.) La Frankie, Common in open places and in thickets (up to 2400 m). On way to Kedarkantha. July–October.

113. ALLIACEAE

37. ***Allium humile*** Kunth, Common in open grassy fields (up to 2200 m). Deokyar. May–July.
38. ***Allium stracheyi*** Baker, Occasional in open grassy places (3200–4300 m). On way to Harkidoon. July–October.

39. ***Allium victorialis*** L., Common along roadsides (1200–2300 m). On way to Jakhol. June–August.
40. ***Allium wallichii*** Kunth, Occasional in open grassy places and stream banks (3200–4300 m). On way to Harkidoon and Kedarkatha bugyal. August–November.

114. LILIACEAE

41. ***Cardiocrinum giganteum*** (Wall.) Makino, Occasional in moist forests (2500–3200 m). On way to Seema. June–August.
42. ***Nomocharis oxypetala*** (Royle) Balf. *f.*, Occasional in open alpine meadow (3200–4800 m). Harkidoon, On way to Jamnidhar Glacier, Barasu Pass. June–August.

115. SMILACACEAE

43. ***Smilax aspera*** L., Occasional in mixed temperate (800–1800 m). Naitwar and Jakhol. March–August.
44. ***Smilax elegans*** Wall. ex Kunth, Occasional in mixed temperate (1000–1800 m). Sankri. On way to Seema. April–September.
45. ***Smilax ovalifolia*** Roxb., Occasional in shady and rocky places (up to 1500 m). Naitwar. February–May.

116. COMMELINACEAE

46. ***Commelina benghalensis*** L., Occasional in moist places and cultivated fields (up to 1200 m). Naitwar. Almost throughout the year.
47. ***Commelina caroliniana*** Walter, Occasional in marshy places and along roadsides (up to 1500 m). On way to Taluka. July–February.
48. ***Commelina maculata*** Edgew., Common in marshy places (up to 2800 m). Sankri and Taluka-Seema. June–October.
49. ***Cyanotis cristata*** (L.) D. Don, Occasional in moist places and along streams (up to 1000 m). On way to Naitwar and Taluka. July–December.
50. ***Cyanotis vaga*** (Lour.) Roem. & Schult., Common along streams and moist places (800–1800 m). Jakhol. July–September.
51. ***Murdannia divergens*** (C.B. Clarke) G. Brückn., Occasional in moist shady places and grassy fields (up 1200 m). Naitwar. August–November.
52. ***Murdannia nudiflora*** (L.) Brenan, Common in moist and marshy situations (up to 900 m). Naitwar and Jakhol. July–December.

117. JUNCACEAE

53. ***Juncus bufonius*** L., Occasional in moist grassy fields near mixed temperate (1000–1600 m). Sankri. May–June.
54. ***Juncus elegans*** Royle ex Buchenau, Occasional along roadsides near mixed forests (up to 2200 m). Taluka. July–September.
55. ***Juncus inflexus*** L., Common in moist places and on hillslopes (up to 3500 m). On way to Seema and Harkidoon. August–September.
56. ***Juncus leschenaultii*** J. Gay ex Laharpe, Common in marshy places (1800–3500 m). Seema-Harkidoon. June–October.
57. ***Juncus membranaceus*** Royle ex D. Don, Common in wet places in the hilly slopes (1800–3500 m). Seema-Harkidoon. May–October.
58. ***Juncus sphacelatus*** Decne.var. **himalensis** (Klotzsch) Jafri, Common in moist grassy fields (3400–4500 m). Seema-Harkidoon. May–August.
59. ***Juncus sphacelatus*** Decne., Common in moist grassy fields (3400–4500 m). On way to Harkidoon. August–October.

118. ARACEAE

60. ***Acorus calamus*** L., Occasional in marshy places and along river banks (up to 3400 m). Naitwar. May–July.
61. ***Arisaema jacquemontii*** Blume, Fairly common in forest openings (2300–3400 m). On way to Barasu Pass. On way to Seema. July–September.
62. ***Arisaema propinquum*** Schott, Common in moist shady places (2200–3500 m). On way to Seema. June–August.
63. ***Arisaema tortuosum*** (Wall.) Schott, Occasional in moist deciduous forests (800–1800 m). Sankri and Jakhol. July–November.
64. ***Sauromatum venosum*** (Ait.) Schott, Rare in open places and as an undergrowth in deodar forests (900–2060 m). Sankri. April–May.

119. CYPERACEAE

65. ***Carex cruciata*** Wahlenb., Occasional on rocky places (up to 1200 m). Jakhol. July–September.
66. ***Carex filicina*** Nees, Occasional in moist hill sides (up to 1200 m). On way to Naitwar and Sankri-Taluka. August–September.

67. ***Carex haematostoma*** Nees, Occasional in alpine meadows (3500–4800 m). On way to Maninda Tal. August–September.
68. ***Carex nivalis*** Boott, Occasional in meadows and rocky areas (4000–4800 m). On way to Maninda Tal. August–September.
69. ***Carex nubigena*** D. Don, Occasional in open grassy hillslopes (up to 2300 m). Sankri. On way to Jamnidhar Glacier. May–August.
70. ***Carex royleana*** Nees, Occasional on hillslopes (up to 3000 m). Taluka. August–September.
71. ***Carex setigera*** D. Don, Occasional in depressions (2100–3400 m). Sankri (T.S. Rana et al., *l.c*). May–June.
72. ***Carex wallichiana*** Spreng., Common in marshes and moist places (up to 2300 m). Sankri. May–June.
73. ***Cyperus alulatus*** J. Kern, Infrequent along the roadsides (up 900 m). On way to Naitwar. September–November.
74. ***Cyperus cyperoides*** (L.) Kuntze, Rare in moist open grasslands (up to 1000 m). On way to Naitwar. July–September.
75. ***Cyperus iria*** L., Occasional in moist places near streams (up to 1200 m). Naitwar. July–October.
76. ***Cyperus niveus*** Retz., Occasional in rock crevices and along roadsides (up to 1300 m). Naitwar. April–June.
77. ***Cyperus squarrosus*** L., Occasional in marshy places (up to 2700 m). Taluka-Seema. November.
78. ***Eriophorum comosum*** (Wall.) Wall. ex Nees, Occasional along the hillslopes and exposed rocky places (up to 2600 m). Sankri. On way to Naitwar. September–November.
79. ***Fimbristylis miliacea*** (L.) Vahl, Common in moist places (up to 1600 m). Naitwar-Sankri. May–October.
80. ***Isolepis setacea*** (L.) R. Br., Occasional along water courses, in marshy places with moving water (up to 3200 m). Sankri. On way to Naitwar. June–September.
81. ***Kobresia duthiei*** C.B. Clarke, Rare in open grassy slopes (up to 3450 m). On way to Harkidoon. August–October.
82. ***Kyllinga squamulata*** Thonn. *ex* Vahl, Rare in moist sandy areas (up to 850 m). Naitwar. August–October.
83. ***Pycreus sanguinolentus*** (Vahl) Nees *ex* C.B. Clarke, Fairly common in marshy places (2500–4000 m). Seema-Harkidoon. May–October.

120. POACEAE (= GRAMINEAE)

84. ***Agrostis pilosula*** Trin., Common along alpine meadows (2400–3700 m). Seema-Harkidoon. August–October.
85. ***Apluda mutica*** L., Common along roadsides and open grassy fields (up to 2000 m). Sankri. On way to Naitwar. On way to Taluka. May–June.
86. ***Arthraxon lancifolius*** (Trin.) Hochst., Fairly common in open grassy fields (up to 2800 m). Taluka-Seema. September–January.
87. ***Arundinella nepalensis*** Trin., Common along the roadsides and on open grassy places (up to 2300 m). On way to Naitwar and Sankri-Taluka. October–November.
88. ***Arundinella setosa*** Trin., Common on rocky slopes and in open forests, forest margins, grassy hillsides (up to 2300 m). On way to Naitwar and Sankri-Taluka. August–November.
89. ***Arundo donax*** L., Fairly common along streams and in sandy ravines (800–1500 m). Naitwar. On way to Naitwar. June–December.
90. ***Avena sativa*** L., Occasional in cultivated fields (up to 1600 m). Naitwar. May–June.
91. ***Brachypodium sylvaticum*** (Huds.) P. Beauv., Common in moist places (up to 1200 m). Naitwar. June–September.
92. ***Bromus japonicus*** Thunb. ex Murr., Common in open grassy hillslopes. June–August.
93. ***Bromus ramosus*** Huds., Common in open grassy hillslopes. October–February.
94. ***Calamagrostis pseudophragmites*** (Hall. *f.*) Koel., Common in open grassy hillslopes (1700–3700 m). Seema-Harkidoon. June–August.
95. ***Chrysopogon fulvus*** (Spreng.) Chiov., Common along riverbank (up to 1500 m). Naitwar, along Tons River. On way to Naitwar. May–June.
96. ***Chrysopogon gryllus*** (L.) Trin., Common in the hills and in dry localities (900–3300 m). Naitwar-Sankri and Taluka-Seema. June–November.
97. ***Cymbopogon martinii*** (Roxb.) Wats., Common in the hills and in dry localities. June–November.
98. ***Cynodon dactylon*** (L.) Pers., Common throughout the valley (up to 1600 m). Sankri. May–June.
99. ***Dactylis glomerata*** L., Common in the hills and in dry localities (1700–4000 m). Taluka-Seema. On way to Harkidoon. July–September.

100. ***Danthonia cachemyriana*** Jaub. & Spach, Common on dry rocky slopes and open grassy fields (3200–4200 m). Seema-Harkidoon. July–September.
101. ***Danthonia schneideri*** Pilg., Occasional on grassy slopes and open fields (2000–4000 m). On way to Seema, Harkidoon. July–October.
102. ***Dendrocalamus strictus*** (Roxb.) Nees, Occasional in moist shady ravines (up to 1500 m). Naitwar. March–November.
103. ***Digitaria horizontalis*** Willd., Occasional in marshy places (up to 1500 m). Naitwar. On way to Taluka and Jakhol. September.
104. ***Digitaria stricta*** Roth ex Roem. & Schult., Common in moist shady places (up to 1600 m). Naitwar. August–October.
105. ***Echinochloa colona*** (L.) Link, Frequent as weed in open situations and fallow fields (800–1600 m). Naitwar, Sankr, and Taluka. August–March.
106. ***Echinochloa frumentacea*** (Roxb.) Link, Cultivated near human habitation and often found as an escape (up to1800 m). Naitwar, Sankri, and Taluka. August–March.
107. ***Eleusine coracana*** (L.) Gaertn., Cultivated for its food grain, almost throughout the sanctuary. Naitwar-Sankri, Taluka-Seema, and Jakhol. August–October.
108. ***Elymus semicostatus*** (Nees ex Steud.) Meld., Common in open grassy fields. July–August.
109. ***Eragrostis nigra*** Nees ex Steud., Fairly common on hillslopes in exposed localities (up to 2800 m). Naitwar. On way to Seema. August–November.
110. ***Eulaliopsis binata*** (Retz.) C.E. Hubb., Common in dry places. April–July.
111. ***Heteropogon contortus*** (L.) P. Beauv. *ex* Roem. & Schult., Occasional in open grasslands and along roadsides (up to 1000 m). August–November.
112. ***Hierochloe laxa*** R. Br. ex Hook. *f.*, Common in alpine meadows (4800–5800 m). On way Barasu Pass. July–October.
113. ***Imperata cylindrica*** (L.) Raeuschel., Common in open grassy fields (up to 1600 m). Naitwar. July–March.
114. ***Melica scaberrima*** (Nees ex Steud.) Hook. *f.*, Common in open grassy fields (3600–4600 m). Harkidoon. On way to Ruinsara Tal. July–August.
115. ***Microstegium falconeri*** (Hook. *f.*) Clayton, Common on rocky places (up to 2500 m). On way to Taluka. July–March.
116. ***Milium effusum*** L., July–August.
117. ***Miscanthus nepalensis*** (Trin.) Hack., Common in open grassy fields (1600–3000 m). Taluka-Seema. August–October.
118. ***Muhlenbergia duthieana*** Hack., Common in open grassy fields (1600–3000 m). Taluka-Seema. August–October.
119. ***Neyraudia arundinacea*** (L.) Henrard, Common on hillslopes and along the roadsides (up to 1800 m). Jakhol. May–June.
120. ***Oplismenus undulatifolius*** (Ard.) Roem. & Schult., Occasional along the roadsides and open grassy slopes (up to 2500 m). Taluka-Seema. July–October.
121. ***Pennisetum flaccidum*** Griseb., Common in the higher alpine slopes (up to 3200 m). Seema. April–October.
122. ***Pennisetum orientale*** L.C. Rich. Common in open places (up to 1500 m). Naitwar. April–October.
123. ***Phalaris minor*** Retz., Occasional in moist and shady places (up to 1500 m). Naitwar. May–June.
124. ***Phleum alpinum*** L., Very common in open places (2000–4300 m). On way to Jamnidhar Glacier, Seema-Harkidoon. July–September.
125. ***Phleum paniculatum*** Huds., Common in moist and shady places (600–2000 m). Sankri. May–June.
126. ***Phragmites karka*** (Retz.) Trin. ex Steud., Occasional in marshy places (up to 2500 m). Sankri and Taluka-Seema. May–June.
127. ***Poa annua*** L., Occasional in marshy places (1400–2800 m). Naitwar-Sankri. May–September.
128. ***Poa nemoralis*** L., Fairly common in exposed places (1300–2300 m). Sankri. May–June.
129. ***Poa supina*** Schrad., Common in open grassy fields (1700–4300 m). Deokyar. May–July.
130. ***Pogonatherum paniceum*** (Lam.) Hack., Common on open grassy hillslopes, along roadsides, and streams (up to 2200 m). On way to Taluka. March–September.
131. ***Saccharum rufipilum*** Steud., Common on open grassy hillslopes (up to 2200 m). Taluka. July–September.
132. ***Saccharum spontaneum*** L., Common along riverbeds and streams (up to 1600 m). Naitwar. July–September.
133. ***Setaria megaphylla*** (Steud.) T. Durand & Schinz, Occasional in open grasslands (up to 1100 m). Naitwar. June–October.
134. ***Setaria pumila*** (Poir.) Roem. & Schult., Occasional in open grasslands

(up to 2300 m). Naitwar and Sankri. August–November.

135. ***Sinarundinaria falcata*** (Nees) C.S. Chao & Renvoize, Common in moist and shady places (up to 2400 m). Sankri. On way to Taluka. May–June.
136. ***Spodiopogon dubius*** Hack., Common in mountain slopes and forest margins (up to 2400 m). Jakhol. May–September.
137. ***Thamnocalamus spathiflorus*** (Trin.) Munro, Grows in coniferous and mixed subalpine forests.
138. ***Themeda anathera*** (Nees ex Steud.) Hack., Occasional on open grassy slopes and along roadsides (up to 1800 m). Taluka and Jakhol. June–October.
139. ***Tripogon filiformis*** Nees ex Steud., Common on open grassy slopes and on rocky places (2800–4400 m). Seema-Harkidoon. July–November.
140. ***Urochloa villosa*** (Lam.) T.Q. Nguyen, Common in open grassy hillslopes and moist places (up to 1200 m). On way to Naitwar. April–October.

12.4 CONCLUSION

The Govind Pashu Vihar Wildlife Sanctuary encompasses a rich biodiversity of both flora and fauna. It is a part of the Western Himalayas, which is one of India's largest centers for endemism. The sanctuary comprises 822 species of angiosperms, under 464 genera, and 120 families, including 7 subspecies and 12 varieties, and a complete inventory of all the recorded species including information on the high altitude threatened medicinal plants of the sanctuary, comprised of 39 species of angiosperms, and of these, 10 species are critically endangered, 13 endangered, 9 vulnerable, and 7 least concern is provided here.

ACKNOWLEDGMENTS

The authors are thankful to the director of the Botanical Survey of India, Kolkata, for providing the facilities to carry out the said project, and also, the first author is thankful to the scientist-in-charge of the Botanical Survey of India, Northern Regional Centre, Dehradun.

REFERENCES

AUTHORED BOOKS

Atkinson, E. T. 1882. *The Himalayan Gazetter.* Vol. 2 (Reprint 1973), Cosmo Publication, New Delhi.

Bridson, D., and L. Forman. 1998. *The Herbarium Handbook.* 3rd ed. Royal Botanic Gardens, Kew (Repr. 1999).

Burkill, I. H. 1965. *Chapters on the History of Botany in India.* Botanical Survey of India, Kolkata, India.

Champion, H. G., and S. K. Seth. 1968. *A Revised Survey of the Forest Types of India.* Government of India Press, New Delhi.

Dang, R. 1993. *Flowers of Western Himalaya.* Indus Publication, New Delhi.

Deva, S., and H. B. Naithani. 1986. *The Orchid Flora of North West Himalaya.* Scientific Publishers, New Delhi.

Duthie, J. F. 1903–1929. *Flora of the Upper Gangetic Plain and of the Adjacent Siwalik and Sub-Himalayan Tracts.* Superintendent of Government Printing, Kolkata, India.

Fosberg, F. R., and M.-H. Sachet. 1965. *Manual for Tropical Herbaria.* International Bureau for Plant Taxonomy and Nomenclature, Regnum Vegetabile vol. 39. IAAT, Utrecht, the Netherlands.

Garg, S. 1987. *Gentianaceae of North West Himalayas* (a revision). Today and Tomorrow's Print & Publishers, New Delhi.

Gaur, R. D. 1999. *Flora of the District Garhwal: North West Himalaya* (With Ethnobotanical Notes), Transmedia, Srinagar (Garhwal).

Hooker, J. D. 1872–1897. *The Flora of British India.* 7 vols. L. Reeve & Co., London, UK.

Hooker, J. D., and T. Thomson. 1855. *Flora Indica.* W. Pamplin, London, UK.

Naithani, H. B. 1984, 1985. *Flora of Chamoli.* vols. 1 & 2. Botanical Survey of India, Kolkata, West Bengal.

Nayar, M. P., and A. R. K. Sastry (eds.). 1987, 1988, 1990. *Red Data Book on Indian Plants.* Vols. 1–3. Botanical Survey of India, Kolkata, West Bengal (Repr. 2000).

Polunin, O., and A. Stainton. 1984. *Flowers of the Himalaya.* Oxford Press, New Delhi.

Pusalkar, P. K., and D. K. Singh. 2012. *Flora of Gangotri National Park, Western Himalaya, India.* Botanical Survey of India, Kolkata, India.

Rana, T. S., D. Bhaskar, and R. R. Rao. 2003. *Flora of Tons valley Garhwal Himalaya, Uttaranchal.* Bishen Singh Mahendra Pal Singh, Dehradun, India.

Rau, M. A. 1975. *High Altitude Flowering Plants of Western Himalaya.* Botanical Survey of India, Kolkata, India.

Rawat, J. K., J. K. Semwal, and A. N. Prohit. 1985. *Blossoming Garhwal Himalaya.* High Altitude Plant Physiology Research Centre, H.N.B. Garhwal University, Srinagar Garhwal.

Singh, J. S., and S. P. Singh. 1992. *Forests of Himalaya.* Gyanodaya Prakashan, Nainital, India.

Stainton, A. 1988. *Flowers of Himalaya–A supplement.* Oxford University Press, New Delhi.

Uniyal, B. P., B. Balodi, and B. Nath. 1994. *The Grasses of Uttar Pradesh – A checklist,* Bishen Singh Mahendra Pal Singh, Dehradun, India.

Uniyal, S. K., A. Swami, and B. P. Uniyal. 1997. *Cyperaceae of Uttar Pradesh – A Checklist.* Bishen Singh Mahendra Pal Singh, Dehradun, India.

Uniyal, B.P., J. R. Sharma, U. Choudhery, and D. K. Singh. 2007. *Flowering Plants of Uttarakhand – A checklist.* Bishen Singh Mahendra Pal Singh, Dehradun, India.

Walter, K. S., and H. J. Gillett. 1998. *The 1997 IUCN Red List of Threatened Plants.* IUCN, Gland, Switzerland.

CHAPTERS IN MULTI-AUTHORED BOOKS

Rau, M. A. 1974. Vegetation and phytogeography of Himalaya. In M.S. Mani (Ed.), *Ecology and Biogeography in India,* pp. 247–280. Dr. W. Junk B.V. Publishers, The Hague, the Netherlands.

Rau, M. A. 1981. Western Himalayan flora. In J.S. Lall (Ed.), *The Himalaya-Aspects of Changes,* pp. 50–63. Oxford University Press, New Delhi.

Rawat, D.S., B. S. Bhandari, and R. D. Gaur. 2001. Vegetational wealth: In O.P. Kandari and O.P. Gusain (Eds.), *Garhwal Himalaya-Nature, Culture & Society*. Transmedia, Srinagar, India.

Journals

Aswal, B. S., A. K. Goel, and B. N. Mehrotra. 1988. An inventory of family Asteraceae from Garhwal and Kumaon Himalaya. *J. Econ. Taxon. Bot.* 12(1): 1–37.

Dangwal, L. R., D. S. Rawat, and R. D. Gaur. 1993. Some interesting and rare plants of fabaceae from Garhwal Himalaya. *J. Indian Bot. Soc.* 72: 317–318.

Dangwal, L. R., D. S. Rawat, and R. D. Gaur. 1995. Some rare and less known legumes from Garhwal Himalaya. *Ind. J. Forest.* 18: 255–257.

Gaur, R. D. 1987. A contribution of flora of Srinagar Garhwal. *J. Econ. Taxon. Bot.* 9: 31–63.

Gaur, R. D., and P. Painuli. 1994. Some little known pteridophytes from Garhwal Himalaya. *Indian J. Forest.* 16(1): 88–90.

Gibson, J. T. M. 1954. The Harki Doon. *Himala. J.* 18: 93–102.

Gupta, R. K. 1955. Botanical exploration in the Bhillanga Valley of erstwhile Tehri-Garhwal State-1. *J. Bombay Nat. Hist. Soc.* 53: 581–594.

Gupta, R. K. 1957. Botanical exploration in the Bhillanga Valley of erstwhile Tehri-Garhwal State-2. *J. Bombay Nat. Hist. Soc.* 54: 878–894.

Gupta, R. K. 1960. Botanical trip to the source of river Ganga in Tehri-Garhwal Himalayas. *Indian J. Forest.* 86: 547–552.

Issar, S. K., and M. R. Uniyal. 1967. Orchids of Uttarakhand Himalayas. *Indian Forester* 67: 713–716.

Manikandan, R., and S. K. Srivastava. 2015a. Diversity, medicinal and threatened plants in Govind Pashu Vihar Wildlife Sanctuary, Western Himalaya. *Indian Forester* 141(9): 966–973.

Manikandan, R., and S.K. Srivastava. 2015b. Threatened medicinal plants protected *in situ* in Govind Pashu Vihar Wildlife Sanctuary, Western Himalaya. *Phyotaxonomy* 15: 90–95.

Manikandan, R., S. K. Srivastava, and P. K. Deroliya. 2015. Economically important plants from Govind Pashu Vihar Wildlife Sanctuary, Western Himalaya. *Ann. For.* 22(1): 57–75.

Myers, N., R. A. Mittermeier, C. G. Mittermeier, G. A. B. da Fonseca, and J. Kent. 2000. Biodiversity hotspots for conservation priorities. *Nature* 403: 853–858.

Naithani, B. D. 1969. Plant with the Kedarnath Parbat expedition, 1969. *Bull. Bot. Surv. India* 11: 224–233.

Raizada, M. B. 1958. New or noteworthy plants from the upper Gangetic Plain. *Proc. Nati Inst Sci.* 24: 198–204.

Rawat, D. S., L. R. Dangwal, and R. D. Gaur. 1994. Some interesting plant records from Garhwal Himalaya. *J. Bombay Nat. Hist. Soc.* 91:168–169.

Rawat, J. K., and S. K. Sharma. 1992. Conservation of biological diversity in the Uttaranchal. *Ind. Forest.* 118: 352–360.

Singh, J. S., and S. P. Singh. 1987. Forest vegetation of Himalaya. *Bot. Rev.* 53(1): 80–191.

Synge, H. 2005. Biodiversity hotspots revisited. *Plant Talk* 40: 33–36.

Online References

IUCN, 2001. *IUCN Red List Categories and Criteria* Version 3.1, Prepared by the IUCN Survival Commission. IUCN, Gland, Switzerland. https://www.iucnredlist.org/ (Assessed August 15, 2019).

Unpublished Documents

Badoni, A. K. 1989. *Herbaceous Flora of Uttarakashi District*, Unpublished D.Phil. thesis, HNB Garhwal University, Srinagar (Garhwal).

Bamola, B. K. 1993. *Lignosae Flora of Uttarakashi District*, Unpublished D.Phil. thesis, HNB Garhwal University, Srinagar (Garhwal).

Dangwal, L. R. 1993. *A Taxonomic Survey of Leguminous Plants of Garhwal Himalaya*. Unpublished D.Phil. thesis, HNB Garhwal University, Srinagar (Garhwal).

Kandwal, M. 2009. Grass Flora of Uttarakhand. Unpublished Ph. D. thesis, HNB Garhwal University, Srinagar (Garhwal).

Nautiyal, D. C. 1996. *A Taxonomic Survey of Poaceae of Garhwal Himalaya*. Unpublished D.Phil. thesis, HNB Garhwal University, Srinagar (Garhwal).

Paramanand, 2009. *Management Plan for Govind Wildlife Sanctuary and National Park*. Wildlife Preservation Organisation, Forest Department, Uttar Pradesh.

13 Anti-oxidant and Anti-microbial Activities of Flavonoids from the Fruit Extract of *Dacryodes edulis*

R. O. Omosimua, B. D. Macham, A. Onanuga, B. T. Thomas, C. O. Ugbomor, G. Iyappan, S. Ramalingam, A. S. Afolabi, and Sunday Asuquo Thomas

CONTENTS

13.1 INTRODUCTION: FLAVONOIDS AND *DACRYODES EDULIS*

Flavonoids are one of the largest classes of naturally occurring polyphenolic compounds (Miller 1996; Falcone Ferreyra et al. 2012). The presence of flavonoids in virtually all land-based plants lends support to the view that they are important in the development of plant life (Andersen and Markham 2006). They are found in almost all plant parts in the fruit, skin or peel, bark, flowers, and seeds of plants giving it color (Miller 1996; Panche et al. 2016). Some of the best food sources that contain flavonoids are red wine, apples, blueberries, bilberries, onions, soy products, and tea (Panche et al. 2016). Also, numerous medicinal plants contain a therapeutic amount of flavonoids and have played an important role in the treatment of diseases from ancient times to today (Wang et al. 2018). Among the many functions of plant flavonoids are those of UV protection, free radical protection, modulation of enzymatic activity, allelopathy, insect attraction or repulsion, nectar guides, probing stimulants, viral, fungal, and bacterial protection, nodulation in leguminous plants, pollen germination, etc., with even more potentials (Andersen and Markham 2006; Amalesh et al. 2011).

Dacryodes edulis commonly known as African pear belongs to the family Burseraceae, it is a dioecious, shade-loving plant which grows in non-flooded forests in the humid region (Ajibesin 2011; Omonhinmin 2012). They grow mainly as shrubs and trees and exude a resinous gum when the bark is cut (Ajibesin 2011). *D. edulis* is found in abundance especially in southeast Nigeria during the months of May to October. It is known as Ube and Elemi, respectively, by the Ibo and Yoruba ethnic groups in Nigeria (Omonhinmin 2012). They are commonly eaten in boiled or roasted form in combination with boiled corn when in season. However, due to the limited use of the fruit of *D. edulis*, the maximum benefit is not derived from it in its due season (Hart et al. 2006; Ibanga and Okon 2009). Various parts of *D. edulis* have been employed in the treatment of several infectious diseases (Ajibesin 2011), and a great deal of focus and attention have been given to other parts of the plant than the fruit for the treatment of diseases. Therefore, there is a need for us to examine more closely what foods we eat, and how diseases can be treated more efficiently with them, as food from plant sources contain compounds of potential medicinal use. Plant-derived medicines are also preferred as they have less side effects than synthetic drugs

(Varshochi and Asadollahi 2017; Yuan et al. 2016; Mutalib 2015). It is also imperative to constantly search for newer anti-microbial agents of plant origin with the aim of discovering potentially useful active ingredients that can serve as a source and a template for the synthesis of new anti-microbial agents, this is due to an increase in antibiotic-resistant strains and an increase in the incidence of re-emerging infectious diseases (Sarma 2017). There is also a need to search for newer anti-oxidant sources, especially from the common foods we eat, as they play vital roles in controlling those diseases related to oxidative stress (Poljsak 2011; Nourmohammadi et al. 2010; He et al. 2017). Exudates from the fruit of *D. edulis* are said to contain a wide variety of phytoconstituents including flavonoids which are implicated among many other functions as possible anti-oxidants and microbicides (Okwu and Nnamdi 2008; Ajibesin 2011). This study aimed at isolating crude flavonoid compounds from the underutilized fruit of *D. edulis*, commonly eaten as snacks in some parts of Africa, and validating the possibility of it being used as a possible anti-oxidant and microbicide agent.

13.2 MATERIALS AND METHODS

13.2.1 Collection and Processing of Plant Material

Fresh fruits of *Dacryodes edulis* were purchased in the month of January 2016 from Gwagwalada Market in the Gwagwalada Area Council of the Federal Capital Territory, Abuja in North Central Nigeria. The plant was identified with voucher number 1248 in the herbarium unit of the Department of Biological Sciences, Ahmadu Bello University, Zaria, Kaduna State, Nigeria. The fruits were washed and oven dried at 60°C for 4 days, peeled, and then oven dried for another day, the dried sample was subsequently powdered. The powdered plant sample (500 g) was weighed and defatted in 1 L petroleum ether (35°C–60°C) using a Soxhlet extractor. The defatted plant sample was extracted in 2.5 L absolute ethanol with continuous shaking for a week on a magnetic stirrer. The extract was then filtered out using a Whatman no. 1 filter paper and concentrated with a rotary evaporator at 70°C.

13.2.2 Phytochemical Screening

The preliminary phytochemical screening test was carried out on both the dried powdered sample and the ethanolic extract of the fruit of *D. edulis* to identify the various phytochemicals present using standard procedures (Harborne 1973; Trease and Evans 1989; Sofowora 1993).

13.2.3 Isolation of Flavonoids

The concentrated ethanolic extract was poured in excess of the distilled water with continuous stirring, and then filtered using Whatman no. 1 filter paper. The filtrate which comprises the water-soluble portion of the ethanolic extract was extracted with a liquid-liquid extractor using petroleum ether, benzene, and ethyl acetate as solvents. The ethyl acetate extract was concentrated to a small volume and refrigerated for 2 days in order to yield yellow crystals. The extract was treated with a few drops of sodium hydroxide solution to confirm for the presence of flavonoids (Tiwari et al. 2011).

13.2.4 DPPH Radical Scavenging Activity

The free radical scavenging activity was measured in terms of the hydrogen donating or free radical scavenging ability using the stable radical 1,1-Diphenyl-2-picrylhydrazyl (DPPH) (Blois 1958). About 0.1 mM DPPH in methanol was prepared, and 0.1 mL of this solution was added to 3.0 mL of the extract solution in methanol at different concentrations (0.03125, 0.0625, 0.125, 0.25, 0.5 mg/mL). Twenty minutes later, the absorbance was measured at 517 nm against the blank (methanol and DPPH). Radical scavenging activity was measured as the percentage of DPPH elimination after 20 minutes. Butylatedhydroxyl anisole, quercetin, and vitamin C were used as the reference compounds. The lower absorbance of the reaction mixture indicates a higher free radical scavenging activity, this was expressed as a percentage inhibition of the free radical by the sample using the formula below.

$$\%\ \text{Inhibition} = \frac{A_O - A_E}{A_O} \times 100 \quad (13.1)$$

where

$$A_O = \text{Absorbance of the control (blank)} \quad (13.2)$$

$$A_E = \text{Absorbance in the presence of the extract} \quad (13.3)$$

13.2.5 Anti-microbial Activity

The isolated flavonoids were tested for their possible anti-bacterial and anti-fungal activities against *Escherichia coli* American Type Culture Collection (ATCC) 2592, *Pseudomonas aeruginosa* American type culture collection (ATCC) 27853, *Proteus mirabilis* ATCC 29906, *Klebsiella pneumoniae* ATCC 13883, *Staphylococcus aureus* ATCC 29923, and *Candida albicans* ATCC 10231. The bacterial organisms were grown on nutrient broth at 37°C for 24 hours, while the fungus, *Candida albicans*, was grown in a Sabouroud dextrose liquid medium (Oxoid, England) at 25°C for 48 hours. A bacterial suspension with a turbidity equivalent to 0.5 McFarland standards was prepared (Bauer et al. 1966). The in vitro anti-microbial screening was carried out using Mueller-Hinton Agar (Oxoid, England) for the bacteria and Sabouroud dextrose agar for the fungi using the agar well diffusion method as described by Perez et al. (1990). Plates were incubated at 37°C for 24 hours for the bacteria, and the fungi were incubated at 25°C for 48 hours. The standard drugs used were Ciprofloxacin and Chloramphenicol against the bacterial organisms, while Nystatin was used against the fungus. At the end of the incubation period, the inhibition zones were measured in millimeters (mm). These studies were carried out in triplicate.

13.2.6 Minimum Inhibitory Concentration, Minimum Bactericidal and Fungicidal Concentration

Minimum inhibitory concentration (MIC) was done using double-fold dilutions by broth dilutions. The concentration of flavonoids and antibiotics were calculated as follows:

$$\text{Desired concentration} = \text{Starting concentration}/X$$

where X = dilution factor.

The lowest concentration of flavonoid-containing broth which inhibits bacteria growth was taken as the MIC of the extract (mg/mL) against each tested organism, this broth was subcultured onto agar plates. The minimum concentration of broth dilutions containing flavonoids which killed bacteria and fungi on agar plates was taken as the minimum bactericidal/fungicidal concentration (MBC) and (MFC).

13.2.7 Statistical Analysis

One-way analysis of variance (ANOVA) with Bonferroni multiple comparison tests were performed to test for significant differences between the treatment groups. A P value < 0.05 was taken as significant. All results were processed using GraphPad Prism version 6.05 for windows from GraphPad Software Inc. (http://www.graph-pad.com/).

13.3 RESULTS

13.3.1 Determination of Phytochemicals

Preliminary phytochemical screening revealed the presence of flavonoids in the dried powered sample and ethanol extract of *Dacryodes edulis* as seen in Table 13.1. Further screening for flavonoids of the isolated compound produced an intense yellow color on treating with two drops of sodium hydroxide, which became colorless on the addition of 1 mL dilute HCl, this indicates that the compound extracted was flavonoid.

13.3.2 Anti-oxidant Activity

The half maximum inhibitory concentration (IC_{50}) value of the crude flavonoid extract is lowest at 0.223 mg/mL, while the reference compounds [Butylated hydroxyanisole, quercitin, and vitamin C] had an IC_{50} value of 0.318, 0.334, and 0.299 mg/mL, respectively, the low IC_{50} value of the extracted flavonoid indicates a higher anti-oxidant activity than all the reference compounds as seen in Figure 13.1. The flavonoid extract has a very good anti-oxidant activity in comparison with the standard drug at concentrations as low as 5 mg/mL as shown in Figure 13.2. A P value < 0.05 indicates that there was a significant difference in treatment between the test flavonoids and reference compounds. Bonferroni multiple comparison tests also showed a significant difference between the flavonoids extract and all reference compounds used (butylated hydroxyl anisole, quercetin, and vitamin C).

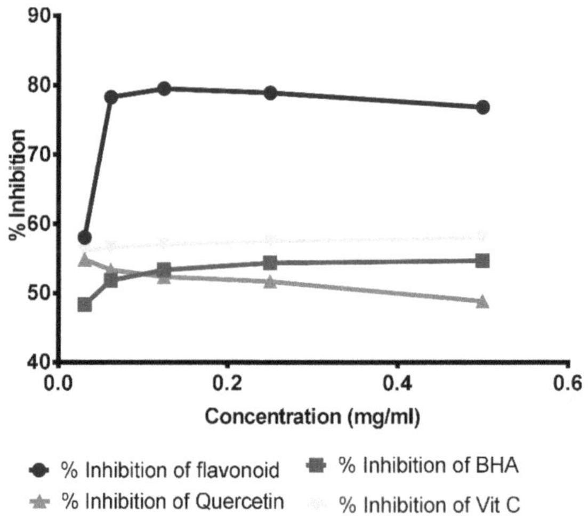

FIGURE 13.1 Relative DPPH free radical scavenging activity.

TABLE 13.1
Phytoconstituents Present in the Powdered Plant Material and Ethanolic Extract of *Dacryodes edulis*

Parameters	Phytoconstituents in Dried Sample	Phytoconstituents in Ethanolic Extract
Phenols	+	−
Saponins	+	+
Tannins	+	+
Steroids	+	−
Triterpenoids	−	+
Anthraquinones	−	−
Glycosides	−	−
Alkaloids	+	+
Terpenoids	+	−
Flavonoids	+	+

Note: + denotes phytochemical present and − denotes phytochemical absent.

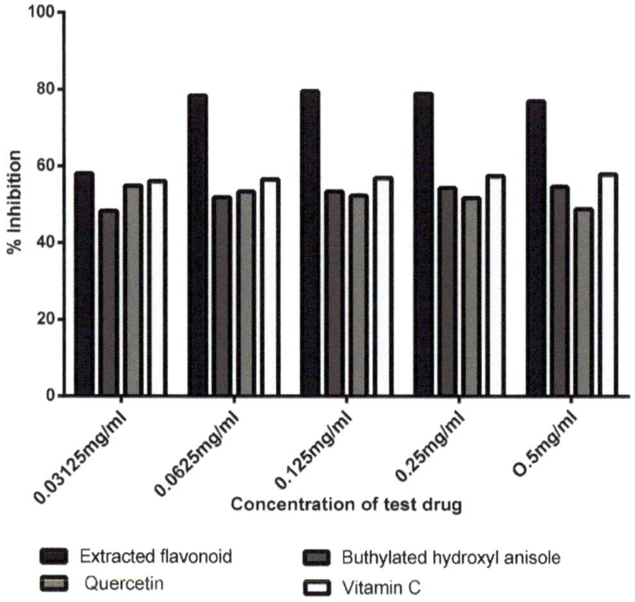

FIGURE 13.2 Comparison of DPPH free scavenging activity of extracted flavonoid against the standard compound.

13.3.3 Anti-microbial Activity

The result reveals that isolated flavonoids had anti-microbial activity at a low concentration of 5 mg for only the organism *P. mirabilis*. While the other microbes anti-microbial activity was observed at the 10 mg concentration and above, the zone of inhibition against the fungus *Candida albicans* showed slightly better activity at 20, 25, and 30 mg of the test flavonoid when compared to the standard drug Nystatin at 5 and 10 mg. This applies generally for the test drug (crude flavonoid), indicating a slight dose-dependent increase in anti-microbial activity, even though there was no difference in the zone of inhibition for the 20 and 25 mg treatment as seen on Figure 13.3. The anti-microbial activity results revealed that extracted flavonoids have more activity on the non-sporing Gram-positive organism tested (*S. aureus*) than the tested Gram-negative organisms, while the anti-fungal activity of the test drug appeared to be slightly above the standard compound at 5 and 10 mg concentrations.

The treatment revealed that there was a significant difference between treatment groups ($P < 0.05$; ANOVA test). However, when subjected to Bonferroni's multiple

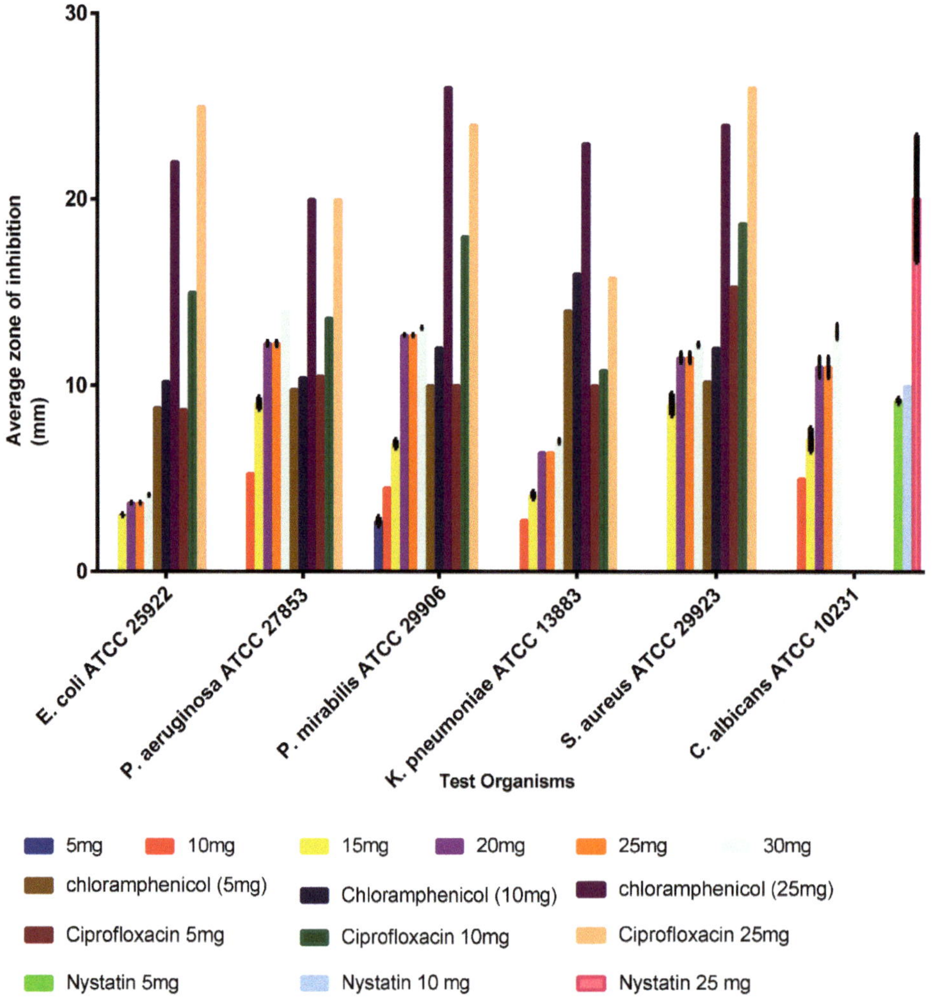

FIGURE 13.3 Anti-microbial activity of extracted flavonoid compound against standard drugs.

comparison post hoc tests, it was revealed that all treatments were not significantly different from each other.

Missing bars represent concentrations at which the test drug has no effect on the microorganism at that particular concentration.

13.3.4 MIC and MBC/MFC of Flavonoid Extract

A minimum concentration of 15 mg/mL inhibited the growth of *E. coli* ATCC 25922, *P. aeruginosa* ATCC 27853, *K. pneumoniae* ATCC 13883, *S. aureus* ATCC 29923, and *C. albicans* ATCC 10231, except in *P. mirabilis* ATCC 29906, where MIC was 7.5 mg/mL (Tables 13.2 and 13.3). In all concentrations of crude flavonoid tested for their ability to kill these pathogens, *E. coli* ATCC 25922 still showed visible growth at all concentrations when subcultured onto solid media. However, *P. aeruginosa* ATCC 27853, *K. pneumoniae* ATCC 13883, *S. aureus* ATCC 29923, *C. albicans* ATCC 10231, and *P. mirabilis* ATCC 29906 showed no visible growth on solid media at 30 mg/mL.

13.3.5 Discussion

The result indicates that flavonoids extracted from the fruit of *D. edulis* have the ability to inhibit free radicals, as they had a much better effect at scavenging free radicals between 55%–85% in comparison to the reference compounds: Butylated hydroxyl anisole, quercetin, and vitamin C at the same concentration. However, as seen in Figure 13.2, it is not dose-dependent. The anti-oxidant property of the leaves has initially been attributed to its high level of flavonoids (Omogbai and Eneh 2011; Ajibesin 2011). The high anti-oxidant activity observed in the flavonoid of this fruit even at low concentrations suggests that it is useful in preventing diseases resulting from oxidative stress (Kuete and Efferth 2010). It has also been shown that boiling increases the anti-oxidant potency of the seed extract (Ogunmoyole et al. 2012), and as such, heating may also not affect the anti-oxidant potency of the fruit. The anti-microbial activity of flavonoids from the fruit of *D. edulis* had a dose-dependent activity against *B. subtilis*, *S. aureus*, *E. coli*, *K pneumonia*, and *C. albicans*, but was most effective against the fungus *C. albicans* at concentrations as low as 5 mg/mL. Various parts of *D. edulis* have been tested for their activity against pathogenic bacteria with inhibitory activity: leaves (Onocha and Oloyede 2011), seeds (Omogbai and Eneh 2011), and oil (Obame et al. 2008). A great deal of the anti-microbial and anti-oxidant screenings have focused only on the leaves and oils from the resins. However, in a previous study, polyphenols isolated from the fruit were also observed to have favorable anti-oxidant activity (Atawodi et al. 2009). This study showed that the fruit of *D. edulis*, which is commonly eaten as a snack, is not only a good source of anti-oxidant, but also has the potential use as an anti-microbial agent.

TABLE 13.2
Minimum Inhibitory Concentration (mg/mL) of Flavonoid Extract

Organisms	30	15	7.5	3.8	1.9	0.9	0.46	0.23
E. coli ATCC 25922	–	–	+	+	+	+	+	+
P. aeruginosa ATCC 27853	–	–	+	+	+	+	+	+
P. mirabilis ATCC 29906	–	–	–	+	+	+	+	+
K. pneumoniae ATCC 13883	–	–	+	+	+	+	+	+
S. aureus ATCC 29923	–	–	+	+	+	+	+	+
C. albicans ATCC 10231	–	–	+	+	+	+	+	+

Note: Where – indicates no growth and + indicates growth of the organism in Mueller-Hinton broth for bacterial organisms and potato dextrose broth for the fungus.

TABLE 13.3
Minimum Bactericidal/Fungicidal Concentration (mg/mL) of Flavonoid Extract

Organisms	30	15	7.5	3.8	1.9	0.9	0.46	0.23
E. coli ATCC 25922	+	+	+	+	+	+	+	+
P. aeruginosa ATCC 27853	–	+	+	+	+	+	+	+
P. mirabilis ATCC 29906	–	+	+	+	+	+	+	+
K. pneumoniae ATCC 13883	–	+	+	+	+	+	+	+
S. aureus ATCC 29923	–	+	+	+	+	+	+	+
C. albicans ATCC 10231	–	+	+	+	+	+	+	+

Note: Where – indicates no growth and + indicates growth of the organism in Mueller-Hinton agar for bacterial organisms and potato dextrose agar for the fungus.

13.4 CONCLUSION

D. edulis is economically viable, as it can be used for a wide variety of purposes such as chewing sticks, carpentry, and local mortar and pestles. The fruit is also known to have immense nutritional value. However, in Africa, people mostly consume the fruit without true knowledge of its medicinal value, as other parts of the plants apart from the fruit are mostly focused on as medicines. In this study, the fruit of *D. edulis* was effective in scavenging free radicals and, as such, will be effective in combating some free radical-mediated diseases. However, it is important to carry out in vivo anti-oxidant activity in order to determine its specific mechanism of action.

It will be of interest to carry out further work in the isolation and characterization of the particular flavonoid compounds responsible for the anti-microbial activity, as well as understanding the mechanism by which it carries out its antimicrobial activity.

ACKNOWLEDGMENTS

The authors wish to acknowledge the Biotechnology and Chemistry Advanced Research Centre, Sheda Science and Technology Complex, and the Department of Microbiology, Olabisi Onabanjo, Ago Are, Nigeria, for providing the necessary materials and facilities to carry out the research work. Also, the Department of Pharmaceutical Microbiology and Biotechnology, Faculty of Pharmacy, Niger Delta University, Wilberforce Island, Bayelsa State and the Plant Genetic Engineering Laboratory, Department of Biotechnology, Bharathiar University, Coimbatore, India, for their valuable guidance during the research and also their contribution in manuscript preparation.

REFERENCES

Ajibesin, K. K. 2011. *Dacryodes edulis* (G Don) HJ Lam: A review on its medicinal, phytochemical and economical properties. *Research Journal of Medicinal Plant* 5(1):32–41.

Amalesh, S., Das, G., and Das, K. S. 2011. Roles of flavonoids in plants. *International Journal of Pharmaceutical Science and Technology* 6(1):12–35.

Andersen, O. M., and Markham, K. R. 2006. *Flavonoids. Chemistry, Biochemistry and Applications.* Boca Raton, FL: Taylor & Francis Group.

Atawodi, S. E., Atawodi, J. C., Idakwo, P., Pfundstein, B., Haubner, R., Wurtele, G., Spiegelhalder, B., Bartsch, H., and Owen, R. W. 2009. Evaluation of the polyphenol composition and antioxidant activity of African variety of *Dacryodes edulis* (G.Don) HJ Lam fruit (abstract). *Journal of Medicinal Food* 12(6):1321–1325.

Bauer, A. W., Kirby, W. M. M., Sherris, J. C., and Truck, M. 1966. Antibiotic susceptibility testing by a standardized single disk method. *American Journal of Clinical Pathology* 45(4):493–496.

Blois, M. S. 1958. Antioxidant determinations by the use of stable free radical. *Nature* 181:1199–1201.

Falcone Ferreyra, M. L., Rius, S. P., and Casati, P. 2012. Flavonoids: Biosynthesis, biological functions, and biotechnological applications. *Frontiers in Plant Science* 3(9):1–15.

Harborne, J. B. 1973. *Phytochemical Methods.* London, UK: Chapman and Hall Ltd.

Hart, A. D., Barimalaa, I. S., and KiinKabari, D. B. 2006. Utilization of the Nigerian pear (*Dacryodes edulis*) as a spread. *Nigerian Food Journal* 24(1):114–117.

He, L., He, T., Farrar, S., Ji, L., Liu, T., and Ma, X. 2017. Antioxidants maintain cellular redox homeostasis by elimination of reactive oxygen species. *Cellular Physiology and Biochemistry* 44(2):532–553.

Ibanga, O. I., and Okon, D. E. 2009. Mineral and antinutrients in two varieties of African pear (*Dacryodes edulis*). *Journal of Food Technology* 7(4):106–110.

Kuete, V., and Efferth, T. 2010. Cameroonian medicinal plants: Pharmacology and derived natural products. *Frontiers in Pharmacology*: 1–19. doi:10.3389/fphar.2010.00123.

Miller, A. L. 1996. Antioxidant flavonoids: Structure, function and clinical usage. *Alternative Medicine Review* 1(2):103–111.

Mutalib, Y. 2015. Comparison between phytotherapy and conventional drug therapy used in urolithiasis management in Hawler City, Kurdistan Region\Iraq. *Journal of Pharmacognosy and Phytochemistry* 83(41):83–86.

Nourmohammadi, I., Athari-Nikazm, S., Vafa, M. R., Bidari, A., Jazayeri, S., Hoshyarrad, A., Hoseni, F., and Faihi-Radmandi, M. 2010. Effect of antioxidant supplementation on oxidative stress in rheumatoid arthritis patients. *Journal of Biological Sciences* 10(1):63–66.

Obame, L. C., Koudou, J., Kumulungui, B. S., Bassolé, I. H. N., Edou, P., Ouattara, A. S., and Traore, A. S. 2008. Volatile components, antioxidant and antimicrobial properties of the essential oil of *Dacryodes edulis* G. Don from Gabon. *Journal of Applied Sciences* 8(19):3532–3535.

Ogunmoyole, T., Kade, I. J, Johnson, O. D., and Makun, O. J. 2012. Effect of boiling on the phytochemical constituents and antioxidant properties of African pear *Dacryodes edulis* seeds in vitro. *African Journal of Biochemistry Research* 6(8):105–114.

Omogbai, B., and Eneh, T. 2011. Antibacterial activity of *Dacryodes edulis* seed extracts on food-borne pathogens. *Bayero Journal of Pure and Applied Sciences* 4(1):17–21.

Omonhinmin, A. C. 2012. Ethnobotany of *Dacryodes edulis* (G. Don) HJ Lam in Southern Nigeria 1: Practices and applications among the Yoruba speaking people. *Ethnobotany Research and Applications* 10:175–184.

Okwu, D. E., and Nnamdi, F. U. 2008. Evaluation of the chemical composition of *Dacryodes edulis* and *Raphia hookeri* Mann and Wendl exudates used in herbal medicine in South Eastern Nigeria. *African Journal of Traditional, Complementary and Alternative Medicine* 5(2):194–200.

Onocha, P. A., and Oloyede, G. K. 2011. UV—Visible analysis of free radical scavenging activities of lear extracts of *Dacryodes edulis* G. Don Lam (Burseraceae): A comparison of three assay methods. *Advances in Environmental Biology* 5(9):2913–2918.

Panche, A. N., Diwan, A. D., and Chandra, S. R. 2016. Flavonoids: An overview. *Journal of Nutritional Science*, 5. https://doi.org/10.1017/jns.2016.41

Perez, C., Paul, M., and Bazerque, P. 1990. An antibiotic assay by the agar well diffusion method. *Acta biologiae et medicinae experimentalis* 15:113–115.

Poljsak, B. 2011. Strategies for reducing or preventing the generation of oxidative stress. *Oxidative Medicine and Cellular Longevity*. https://doi.org/10.1155/2011/194586.

Sarma, N. 2017. Emerging and re-emerging infectious diseases in South East Asia. *Indian Journal of Dermatology* 62(5):451–455.

Sofowora, A. 1993. *Medicinal Plants and Traditional Medicine in Africa*. Ibadan, Nigeria: Spectrum Books Ltd.

Tiwari, P., Kumar, B., Kaur, M., and Kaur, H. 2011. Phytochemical screening and extraction: A review. *Internationale Pharmaceutica Sciencia* 1(1):98–106.

Trease, G. E., and Evans, W. C. 1989. *Pharmacognosy*. Bailliere Tindall, London.

Varshochi, F., and Asadollahi, K. 2017. Alhagi; traditional and modern medicine effective against kidney stones. *Journal of Nephropharmacology* 6(1):15–16.

Wang, T. Y., Li, Q., and Bi, K. S. 2018. Bioactive flavonoids in medicinal plants: Structure, activity and biological fate. *Asian Journal of Pharmaceutical Sciences* 13(1):12–23.

Yuan, H., Ma, Q., Ye, L., and Piao, G. 2016. The traditional medicine and modern medicine from natural products. *Molecules* 21(5):559.

14 Anti-diabetic and Cytotoxicity Studies of Ethanol Extract of *Mundulea sericea*

A Threatened Medicinal Plant

S. Gangadevi, K. Kalimuthu, and P. Viswanathan

CONTENTS

14.1 INTRODUCTION

The leading global risks for mortality in the world are high blood pressure (13% of deaths), tobacco use (9% of deaths), high blood glucose (6% of deaths), physical inactivity (6% of deaths), and being overweight and obesity (5% of deaths). These risks are responsible for raising the risk of chronic diseases such as heart disease, diabetes, and cancers. They affect individuals across all income groups: high, middle, and low (WHO 2009). Diabetes mellitus is a major chronic metabolic disorder and an extremely serious condition from both clinical and public health standpoints. Nearly 5% of the world's population is affected by diabetes. According to the World Health Organization projections, the diabetic population is likely to increase to 300 million or more by the year 2025 (King et al. 1998). Ramachandran et al. (2010) also estimated that diabetes mellitus affected at least 285 million people worldwide in 2010, and the number is expected to reach 438 million by the year 2030. Another global risk for mortality is cancer. According to the GLOBOCAN 2018 survey, it is estimated that there are around 18.1 million new cases of cancer and 9.6 million deaths from cancer (Bray et al. 2018). Many chemo drugs are available for this, but they provoke more side effects, however, a plant-based remedy is different from chemo drugs with less or no side effects. Plants are capable of synthesizing secondary metabolites (Wittstock and Gershenzon 2002), which can be used for new drug discovery and development. Medicinal plant extracts are a potential source for the development of new agents that are effective against infections currently difficult to treat (Chung et al. 2004; Nair and Chanda 2004; De and Ifeoma 2002).

Mundulea sericea belong to the family Leguminosae, and this species is listed in the International Union for Conservation of Nature (IUCN) red list category (least concern). This species has presented a great deal of chemical constituents in plant parts, and the whole plant is used for various medicinal purposes. *Mundulin* is an isolated compound which has many biological activities including anti-bacterial, anti-fungal, cytotoxic, and cancer chemopreventative ones (Mazimba et al. 2011; Cao et al. 2004). There are many reports on the bioactivity of *M. sericea*, like insecticidal (Magaji et al. 2012; Langat et al. 2012; Anjili et al. 2014), anti-microbial (Mazimba et al. 2012), analgesic (Iyer et al. 2004), anti-oxidant potential, and chemical profiling (Khyade and Waman 2017). Hence, the present study aims to analyze the anti-diabetic and cytotoxicity studies of the bark ethanol extract of the *Mundulea sericea*.

14.2 MATERIALS AND CHEMICALS

Hexane, petroleum ether, chloroform, ethyl acetate, ethanol and methanol, sodium phosphate buffer, sodium chloride, porcine pancreatic α-amylase enzyme, starch solution, dinitrosalicylic acid, distilled water, Tris buffer, α-glucosidase enzyme, Eagles Minimum Essential Medium, fetal bovine serum, trypsin-ethylenediaminetetraacetic acid, and dimethylsulfoxide were used for this study.

14.2.1 Plant Collection, Identification, and Extract Preparation

Mundulea sericea was collected from Kuppapalayam, the Western Ghats region, Coimbatore, and Tamil Nadu. The species name (BSIS/RC/5/23/2017/Tech) was authenticated by Dr. C. Murugan, scientist & head of office, Botanical Survey of India, Southern Centre Region, Coimbatore, and India. The bark of *M. sericea* was dried under a shade house condition and ground with a mortar and pestle. One hundred grams of powder were subjected to Soxhlet apparatus by the successive extraction method (Elwekeel et al. 2013). Different polar solvents such as hexane, petroleum ether, chloroform, ethyl acetate, ethanol, and methanol were used for this extraction, and after extraction these solvents were dried through a rotary evaporator and stored. The crude ethanol extract was used for further studies.

14.2.2 Anti-diabetic Analysis

14.2.2.1 α-Amylase Inhibition Assay

A *Mundulea sericea* bark ethanol (MSBE) extract with different concentrations and a 0.02 M sodium phosphate buffer 500 µL (pH 6.9 with 0.006 M sodium chloride) with a porcine pancreatic α-amylase enzyme (0.5 mg/mL) were incubated at 25°C for 10 minutes (Worthington 1993). After incubation, 500 µL of a 1% starch solution in a 0.02 M sodium phosphate buffer (pH 6.9 with 0.006 M sodium chloride) was added to the reaction mixture. Subsequently, the reaction mixture was incubated at 25°C for 10 minutes and 1.0 mL of dinitrosalicylic acid was added. The reaction was stopped by incubating the mixture in boiling water for 5 minutes, and then subsequently cooled at room temperature. With 10 mL of distilled water, the reaction mixture was diluted, and the absorbance was measured at 540 nm in a spectrophotometer. The inhibitory activity data were expressed in a percentage. The IC_{50} value was calculated by the 50% inhibition concentration of the sample extracts.

14.2.2.2 α-Glucosidase Inhibition Assay

One milliliter of a starch substrate (2% w/v maltose or sucrose) with a 0.2 M Tris buffer pH 8.0 and various concentrations of an MSBE extract were incubated for 5 minutes at 37°C (Apostolidis et al. 2007). The reaction was initiated by adding 1 mL of an α-glucosidase enzyme (1 U/mL), followed by incubation for 10 minutes at 37°C. Then, the reaction mixture was heated for 2 minutes in a boiling water bath to stop the reaction. The amount of liberated glucose was measured by the glucose oxidase peroxidase method.

14.2.3 In Vitro Anti-cancer Activity

The anti-cancer activity of the MSBE extract was studied by various cancer cell lines through the standard protocol of Mosmann (1983) and Monks et al. (1991). The monolayer cells were detached with trypsin-ethylenediaminetetraacetic acid to make single cell suspensions, and viable cells were counted using a hemocytometer and diluted with a medium containing 5% fetal bovine serum to give a final density of 1 × 105 cells/mL. One hundred microliters per well of cell suspension were seeded into 96-well plates at a plating density of 10,000 cells/well and incubated to allow for cell attachment at 37°C, 5% CO_2, 95% air, and 100% relative humidity. After 24 hours, the cells were treated with serial concentrations of the test samples. They were initially dissolved in neat dimethylsulfoxide, and an aliquot of the sample solution was diluted twice the desired final maximum test concentration with serum-free medium. An additional four serial dilutions were made to provide a total of five sample concentrations. Aliquots of 100 µL of these different sample dilutions were added to the appropriate wells already containing 100 µL of the medium, resulting in the required final sample concentrations. Following a sample addition, the plates were incubated for an additional 48 hours at 37°C, 5% CO_2, 95% air, and 100% relative humidity. The medium without samples was served as a control, and a triplicate was maintained for all concentrations.

14.3 RESULTS

14.3.1 Anti-diabetic Activity

The in vitro anti-diabetic activity of the MSBE extract was examined in α-amylase and α-glucosidase enzymes, and the enzymes inhibition percentage was noticed as a dose-dependent manner and standard that is given in Figures 14.1 and 14.2. Among the enzyme inhibition assays, the α-glucosidase enzyme inhibition was a significantly different result than the α-amylase enzyme inhibition. The maximum percentages of inhibition, 59.22 and 65.27, were observed at 500 µg/mL concentrations in α-amylase and α-glucosidase, which are compared with a standard acarbose inhibition percentage that was noticed as eighty percentage. Both enzyme inhibition percentages were gradually decreased with sample concentration increases. The lowest percentage of enzymes inhibitions, 18.35 and 18.74, were observed in α-amylase and α-glucosidase, respectively (Figures 14.1 and 14.2).

14.3.2 In Vitro Cytotoxicity Study

Figure 14.3 shows the viability percentages of various cancer cell lines in the presence of MSBE crude extract. The result of the Vero cell viability had the maximum percentage of cell death observed at a 250 µg/mL concentration, and the cell viability was increased when the sample concentration decreased (Figure 14.4). The highest percentage of cell death, 43.55, and lowest cell death, 93.60, were observed in the 250 and 31.25 µg/mL concentrations, respectively. Likewise, all other tested cancer cell viabilities were a dose-dependent manner, and the significant cancer cell death percentage (18.79) was observed in the HeLa cell line at a 250 µg/mL concentration, followed by Hep-G2, A549, and MCF-7 cell lines that observed

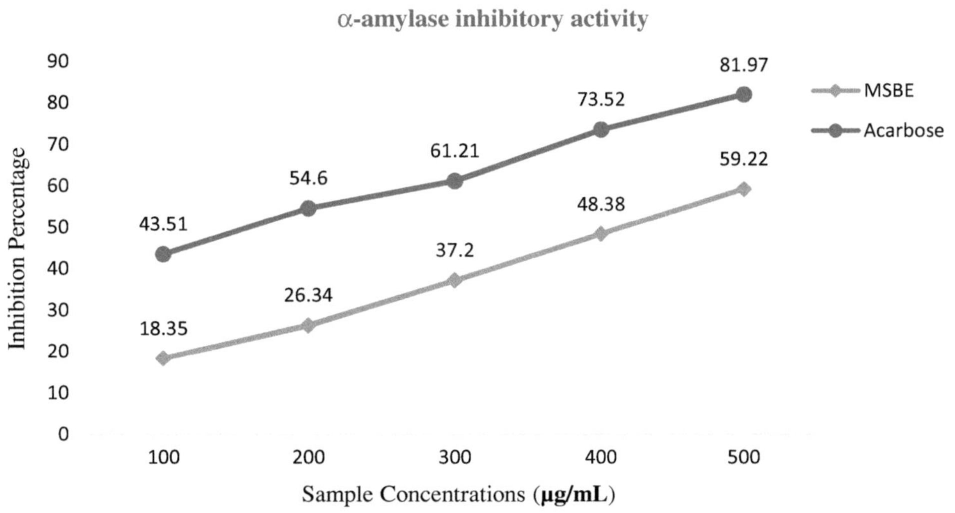

FIGURE 14.1 α-amylase enzyme inhibitory activity of the MSBE extract.

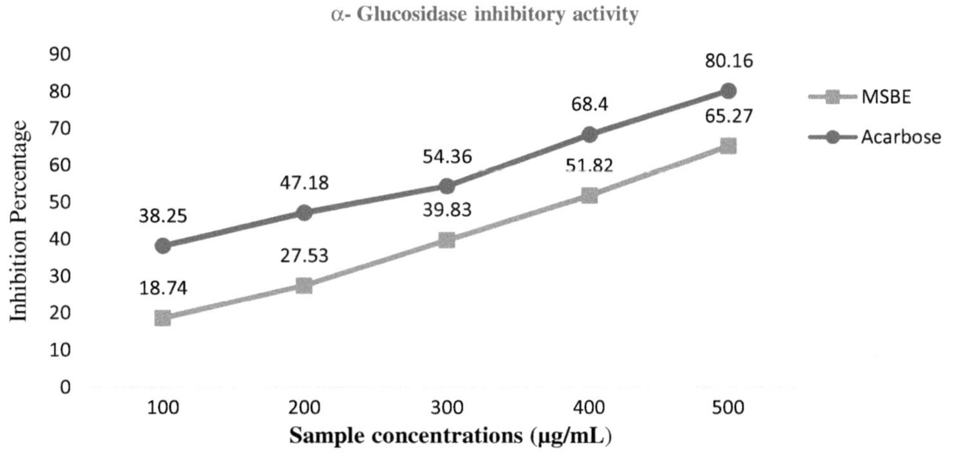

FIGURE 14.2 α-glucosidase enzyme inhibitory activity of the MSBE extract.

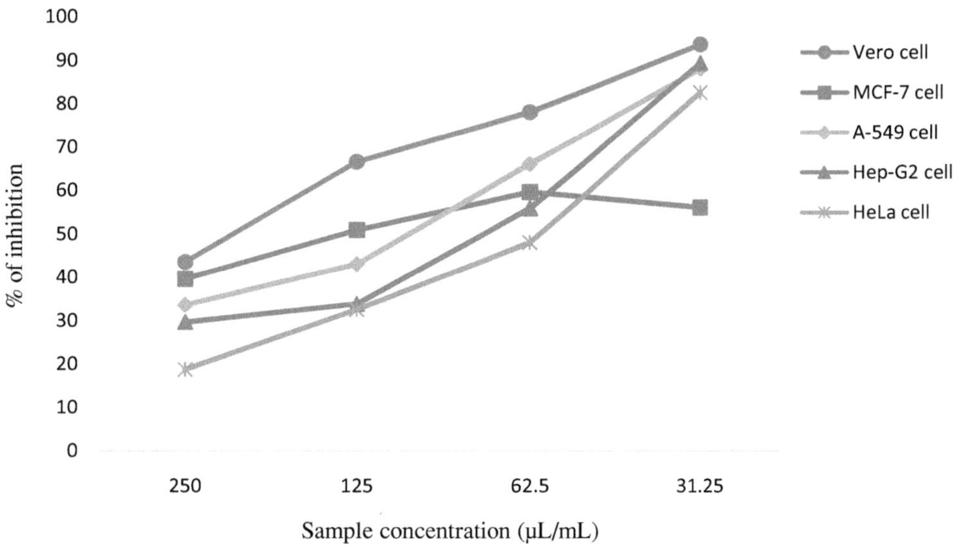

FIGURE 14.3 Cytotoxicity cells viability assay of the MSBE extract by various cancer cells.

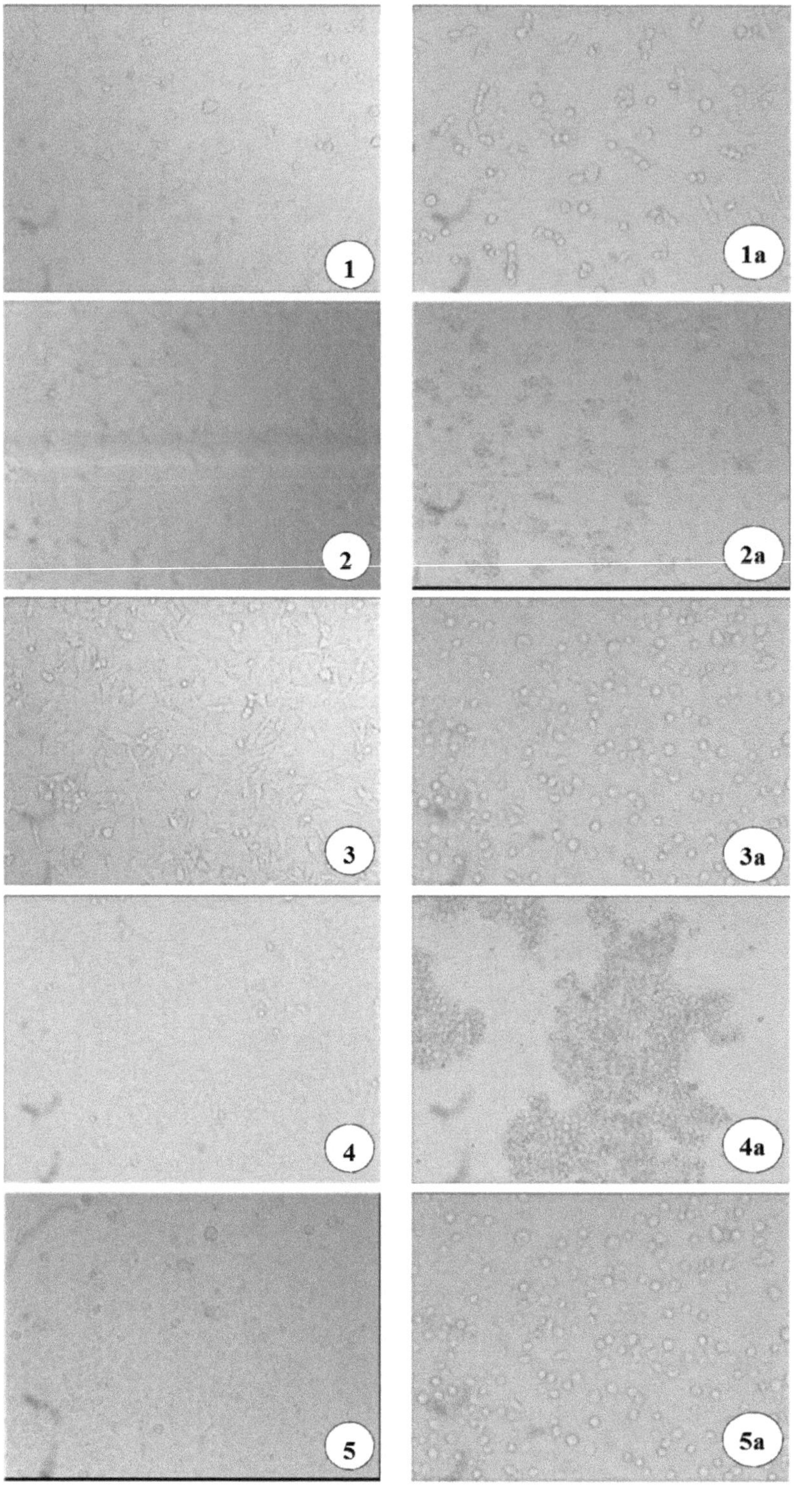

FIGURE 14.4 Photograph of various cancer cell lines. 1—Vero normal cell; 1a—Vero cytotoxicity 2—MCF-7 normal cell; 2a—MCF-7 cytotoxicity 3—A549 normal cell; 3a—A549 cytotoxicity 4—Hep-G2 normal cell; 4a—Hep-G2 cytotoxicity 5—HeLa normal cell; 5a—HeLa cytotoxicity.

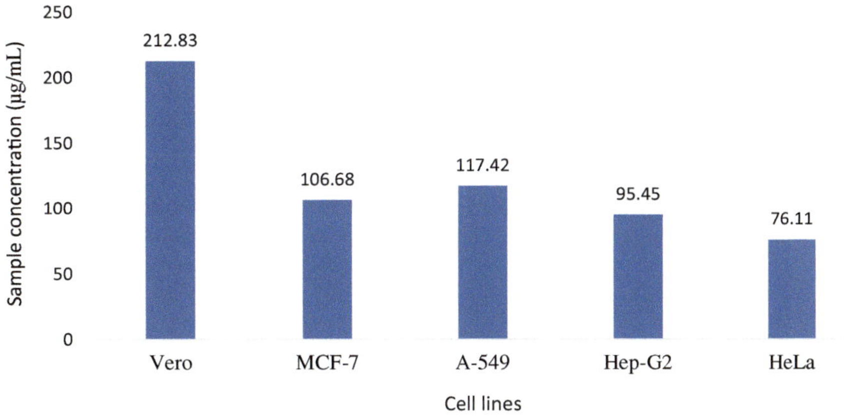

FIGURE 14.5 Cytotoxicity IC_{50} values of various cancer cell lines.

significant cancer cell death occur in ascending orders at 250 µg/mL concentrations and which cell death percentages were 29.3, 35.55, and 39.59, respectively (Figure 14.3). The maximum percentages of cell viability of all cancer cells were observed at 31.35 µg/mL concentrations and the viability percentages were 89.37 (Hep-G2), 88.05 (A549), 82.54 (HeLa), and 55.09 (MCF-7). The IC_{50} values of the Vero cell line along with each cancer cell line were illustrated in Figure 14.5, and the values were 212.83, 106.68, 117.42, 95.45, and 76.11 µg/mL concentrations of Vero, MCF-7, A-549, Hep-G2, and HeLa cancer cell lines, respectively (Figure 14.5). Among the various cytotoxicity activities, the most potential cytotoxicity was observed in the HeLa cell line (Figure 14.5), which also reported significant cell death.

14.4 DISCUSSION

In diabetes, two main carbohydrate hydrolyzing enzymes (α-amylase and α-glucosidase) are responsible for postprandial hyperglycemia. α-amylase begins the process of carbohydrate digestion by the hydrolysis of 1,4-glycosidic linkages of polysaccharides (starch, glycogen) to disaccharides, and α-glucosidase catalyzes the disaccharides to monosaccharides, which lead to postprandial hyperglycemia (Hara and Honda 1990; Matsui et al. 2007). Hence, inhibitors of α-amylase and α-glucosidase are useful in the control of hyperglycemia as they delay carbohydrate digestion, which consequently reduces the postprandial plasma glucose level. The herbal drugs are used as complementary approaches in existing medications for the treatment of diabetes, and its complications are growing worldwide, and many plants in different countries are known to have anti-diabetic effects (Hasani-Ranjbar et al. 2008). The ancient Indian literature reports that more than 800 plants have anti-diabetic properties, while ethnopharmacological surveys indicate that more than 1200 plants can be used for hypoglycemic activity (Mishra et al. 2010). The present work was aimed to analyze the α-amylase and α-glucosidase enzymes inhibition activities of the MSBE extract. Both assays showed a dose-dependent manner, and the maximum inhibition percentage was observed at a higher concentration of 500 µg/mL concentrations. The highest percentage of inhibition, 59.22 and 65.27, was observed in α-amylase and α-glucosidase assays, respectively, which was compared with the standard of acarbose (81.97 and 80.16). This plant has entire parts that have reported to contain many secondary metabolites, and these are used by various pharmacological activities (Khyade and Waman 2017). In particular, terpenes are used for the inhibition of the amylase activity (Bezerra et al. 2002; Costa et al. 2000).

Cancer is one of the most dangerous, fast, progressive diseases of the present century with a huge mortality rate, even in the developed countries. Roughly 60%–75% of currently used anti-cancer agents are derived from natural sources (Cragg et al. 2011; Newman et al. 2013). In this regard, the cytotoxic effects of the MSBE extracts were evaluated against four cancer cell lines and the normal line of the Vero cell line using the 3-(4,5-dimethylthiazaol-2-yl)2,5-diphenyl tetrazolium bromide (MTT) assay. The cell viability percentage of the Vero cell line was observed at 43.55 at a 250 µg/mL concentration and an IC_{50} value of 212.83 µg/mL concentration. Among the various cytotoxicity assays, the HeLa cell line reported significant cytotoxicity activity and the IC_{50} value of 76.11 µg/mL concentration. This is in agreement with the earlier studies on *Baccharoides anthelmintica* (L.) Moench and *Turnera ulmifolia* (Chinnadurai et al. 2016; Kalimuthu et al. 2016).

14.5 CONCLUSION

Our study showed that the ethanol extract from the bark of *Mundulea sericea* is a source of phytochemicals having notable anti-diabetic and anti-cancer activities. This indicates their effectiveness in the treatment of diabetes and cancer. The findings showed that the ethanol extract of *M. sericea* can be used as national anti-diabetic and anti-cancer agents.

REFERENCES

Anjili, C. O., Mugambi, R., Siele, D. K., Langat, B., Kamanyi, K., and Nyasende, S. 2014. The effects of *Mundulea sericea* (Fabales: Fabaceae) water extracts on *Phlebotomus duboscqi* (Diptera: Psychodidae) eggs and larvae. *African Journal of Pharmacology and Therapeutics* 3(2): 47–50.

Apostolidis, E., Kwon, Y. I., and Shetty, K. 2007. Inhibitory potential of herb, fruit, and fungal enriched cheese against key enzymes linked to type 2 diabetes and hypertension. *Innovative Food Science & Emerging Technologies* 8: 46–54.

Bezerra, J. C., Silva, I. A., Ferreira, H. D., Ferri, P. H., and Santos, S.C. 2002. Molluscicidal activity against *Biomphalaria glabrata* of Brazilian Cerrado medicinal plants. *Fitoterapia* 73: 428–430.

Bray, F., Ferlay, J., Soerjomataram, I., Siegel, R. L., Torre, L. A., and Jemal, A. 2018. Global cancer statistics 2018: GLOBOCAN estimates of incidence and mortality worldwide for 36 cancers in 185 countries. *CA: A Cancer Journal for Clinicians* 68(6): 394–424.

Cao, S., Schilling, J. K., Miller, J. S., Andriantsiferana, R., Rasamison, V. E., and Kingston, D. G. I. 2004. Cytotoxic compounds from *Mundulea chapelieri* from the Madagascar Rainforest. *Journal of Natural Products* 67(3): 454–456.

Chinnadurai, V., Kalimuthu, K., Prabakaran, R., and Sharmila Juliet, Y. 2016. Antiangiogenesis and anticancer activity of leaf and leaf callus extracts from *Baccharoides anthelmintica* (L.) Moench (Asteraceae). *British Journal of Pharmaceutical Research* 13(5): 1–9.

Chung, P. Y., Chung, L. Y., Ngeow, Y. F., Goh, S. H., and Imiyabir, Z. 2004. Antimicrobial activities of Malaysian plant species. *Pharmaceutical Biology* 42(4–5): 292–300.

Costa, T. R., Fernandes, O. F., Santos, S. C., et al. (2000). Antifungal activity of volatile constituents of *Eugenia dysenterica* leaf oil. *Journal of Ethnopharmacology* 72(1–2): 111–117.

Cragg, G. M., Kingston, D. G., and Newman, D. J. 2011. *Anticancer Agents from Natural Products*. New York: CRC Press.

De, N., and Ifeoma, E. 2002. Antimicrobial effects of components of the bark extracts of Neem (*Azadirachta indica* A. Juss). *Technology and Development* 8: 23–28.

Elwekeel, A., Elfishawy, A., and AbouZid, S. 2013. Silymarin content in *Silybum marianum* fruits at different maturity stages. *Journal of Medicinal Plants Research* 7: 1665–1669.

Hara, Y., and Honda, M. 1990. The inhibition of α-amylase by tea polyphenols. *Agricultural and Biological Chemistry* 54(8): 1939–1945.

Hasani-Ranjbar, S., Larijani, B., and Abdollah, M. 2008. A systematic review of Iranian medicinal plants useful in diabetes mellitus *Archives of Medical Science* 4: 285–292.

Iyer, M., Karode, R., Deshmukh, R., and Deshmukh, V. 2004. Antinociceptive activity of *Mundulea sericea* leaves. *Journal of Natural Remedies* 4(2): 127–130.

Kalimuthu, K., Chinnadurai, V., Prabakaran, R., Subramaniam, P., and Sharmila Juliet, Y. 2016. Antimicrobial, antioxidant and anticancer activities of *Turnera ulmifolia* (Yellow Alder) callus. *British Journal of Pharmaceutical Research* 9(5): 1–12.

King, H., Aubert, R. E., and Herman, W. H. 1998. Global burden of diabetes, 1995–2025: Prevalence, numerical estimates, and projections. *Diabetes Care* 21: 1414–1431.

Khyade, M. S., and Waman, M. B. 2017. Chemical profile and antioxidant properties of *Mundulea sericea. Pharmacognosy Journal* 9(2): 213–220.

Langat, B. K., Siele, D. K., Wainaina, C., Mwandawiro, C., Ondicho, J., and Tonui, W. K. 2012. Larvicidal effect of *Mundulea sericea* (Leguminosae) plant extract against *Aedes aegypti* L. (Diptera: Culicidae). *African Journal of Pharmacology and Therapeutics* 1(3): 106–109.

Magaji, B. T., Bamaiyi, L. J., Igheghe, L., and Mani, U. 2012. Bioassay of *Mundulea sericea* ethanol leaf extract and leaf powder against *Callosobruchus maculatus* (F.) on stored cowpea. *Global Journal of Science Frontier Research Agriculture and Veterinary Sciences* 12: 30–36.

Matsui, T., Tanaka, T., Tamura, S., et al. 2007. α-Glucosidase inhibitory profile of catechins and theaflavins. *Journal of Agricultural and Food Chemistry* 55(1): 99–105.

Mazimba, O., Masesane, I. B., and Majinda, R. R. T. 2011. A flavanone and antimicrobial activities of the constituents of extracts from *Mundulea sericea. Natural Product Research* 5. doi:10.1080/14786419.2011.616504.

Mazimba, O., Masesane, I. B., Majinda, R. R., and Muzila, A. 2012. GC-MS analysis and antimicrobial activities of the non-polar extracts of *Mundulea sericea. South African Journal of Chemistry* 65(1): 50–52.

Mishra, S. B., Rao, C. V., and Ojha, S. K. 2010. An analytical review of plants for anti-diabetic activity with their Phytoconstituents and mechanism of action. An analytical review of plants for anti-diabetic activity with their phytoconstituents & mechanism of action. *International Journal of Pharmaceutical Sciences and Research* 1(1): 29–46.

Monks, A., Scudiero, D., Skehan, P., et al. 1991. Feasibility of high flux anticancer drug screen using a diverse panel of cultured human tumour cell lines. *Journal of the National Cancer Institute* 83: 757–766.

Mosmann, T. 1983. Rapid colorimetric assay for cellular growth and survival: Application to proliferation and cytotoxicity assays. *Journal of Immunological Methods* 65: 55–63.

Nair, R., and Chanda, S. V. 2004. Antibacterial activity of some medicinal plants of Saurashtra region. *Journal of Cell and Tissue Research* 4: 117–120.

Newman, P., Witchalls, J., Waddington, G., and Adams, R. (2013). Risk factors associated with medial tibial stress syndrome in runners: a systematic review and meta-analysis. *Open Access Journal of Sports Medicine* 13(4): 229–241.

Ramachandran, A., Ma, R. C., and Snehalatha, C. 2010. Diabetes in Asia. *The Lancet*, 375: 408–418.

Wittstock, U., and Gershenzon, J. 2002. Constitutive plant toxins and their role in defense against herbivores and pathogens. *Current Opinion in Plant Biology* 5(4): 300–307.

World Health Organization (WHO). 2009. *Global Health Risks: Mortality and Burden of Disease Attributable to Selected Major Risks*. Global Health Risks, Geneva, Switzerland, 1–70.

Worthington, V. 1993. Alpha amylase. V. Worthington (Ed.), *Worthington Enzyme Manual Freehold*. Worthington Biochemical Corp, Lakewood, NJ, 36–41.

15 Evaluation of In Vitro and In Vivo Anti-oxidant and Anti-inflammatory Potential of *Aloe vera* Gel Extract

S. Govindarajan, Spoorthy N. Babu, and Ayesha Noor

CONTENTS

15.1 INTRODUCTION

Oxidative stress and inflammation are some of the major causes for rheumatoid arthritis, diabetes, cardiovascular disease, obesity, cancer, etc. (Hajjar and Gotto 2013). The increase in oxidative stress-related effects are involved in an inflammatory process which includes direct production of free radicals and other reactive oxygen species (ROS) by stimulated immune cells. Dysregulation in inflammation leads to the pathogenesis of arthritis. The progression of arthritis is characterized by immune cells such as macrophage and T cells, which infiltrate the tissue and synovial membrane of the joints affected with inflammation (Kalaiselvan and Rasool 2015), leading to the accumulation of pro-inflammatory cytokines such as tumor necrosis factor alpha (TNF-α), interleukin 6 (IL-6), interleukin 1 beta (IL-1β), etc. The ROS share an interrelationship with the progression of inflammation, and these can act as mediators for the arthritis pathogenesis. There have been numerous advancements made for treating arthritis, but the long-term usage of available drugs leads to gastrointestinal bleeding, renal disorders, and morbidity (Umar et al. 2012). Due to these side effects and also due to the low efficacy and high cost of available drugs, it has become prudent to look for alternative plant-based treatments for arthritis. Plant-based phenolic constituents are known to decrease the ROS level and regulate the inflammatory stimuli through various mechanisms, such as the decrease of the free radicals formation and other ROS pathways (Salimonu et al. 1982). They also act through the inhibition of ROS-generating pro-inflammatory enzymes (Shaikh, Pund, and Gacche 2016) and metal chelating activity.

Plants are ancient medicinal resources and possess rich medicinal value due to the active phytoconstituents (Altemimi et al. 2017). Plant phytoconstituents have numerous pharmacological activities, such as anti-oxidant, anti-inflammatory, anti-diabetic, anti-cancer, anti-microbial, etc., which are used for the treatment of various medical ailments (Mehta et al. 2017). The plant *Aloe vera* has been shown to possess multiple biological effects such as anti-oxidant, anti-inflammatory, anti-diabetic, anti-cancer, anti-microbial, wound healing, etc. (Hamman 2008; Choi and Chung 2003).

Aloe vera (*Aloe barbadensis* miller) is an immemorial plant which has been known for centuries for its benefits in health, beauty, medicine, and skin care (Radha and Laxmipriya 2015). It belongs to the Aloeaceae family and is a shrubby, perennial, xerophytic, succulent plant (Ahlawat and Khatkar 2011). Most of the in vitro and in vivo studies have been reported individually related to either the anti-oxidant activity (Moniruzzaman et al. 2012; Jairath, Chatli, and Biswas 2016) or the anti-inflammatory activity of the *Aloe vera* gel extract (Langmead, Makins, and Rampton 2004). Furthermore, our preliminary studies using the *Aloe vera* gel extract have shown anti-diabetic activity (Noor et al. 2008). To extend and correlate these findings, we have assessed the in vitro and in vivo anti-oxidant and anti-inflammatory effect of the *Aloe vera* gel extract in the inflamed paws of Freund's adjuvant arthritis-induced rats with respect to oxidative damage and inflammatory mediators.

15.2 MATERIALS AND METHODS

All standard chemicals and Enzyme-Linked Immunosorbent Assay (ELISA) kits for the estimation of TNF-α, IL-6 levels were procured from Sigma-Aldrich (USA). sPLA2 levels were estimated through a Cayman Assay kit (USA). The Complete Freund's Adjuvant was procured from Sigma-Aldrich. The diclofenac drug was purchased from Cipla, India. All other chemicals used were of analytical grade and purchased from Sisco Research Laboratories, Mumbai, India.

15.2.1 Plant Collection, Identification, and Extract Preparation

Aloe vera leaves were collected from the southern part of India, and they were authenticated by the Botanical Survey of India, Southern Regional Centre, Coimbatore, India (BSI/SRC/5/23/2018/Tech/728). The gel extract was prepared according to the protocol of Noor et al. (2008). The determination of both the anti-oxidant and anti-inflammatory was done by in vitro as well as in vivo methods. The extract was further dissolved and stirred overnight in 70% methanol, centrifuged at 5000 rpm at 4° for 15 minutes, and the supernatant was collected. This supernatant was used for the determination of in vitro anti-oxidant assays.

15.2.2 Determination of Total Polyphenol Content

The methanolic extract of the *Aloe vera* gel extract was analyzed for total phenolic content by the Folin-Ciocalteu colorimetric method (Saeed, Khan, and Shabbir 2012). We added 10 μL (1 mg) of the sample to 20 μL of the Folin-Ciocalteu reagent and 50 μL of 25% sodium carbonate solution, and made up to 1 mL with double distilled water. We incubated for 60 minutes at room temperature in a dark place. Gallic acid was used as a standard reference and the absorbance was measured at 725 nm by using a Beckman Coulter DU 730 Life Science UV/V spectrophotometer. The results were expressed as gallic acid equivalents (mg GAE/g of dry extract).

15.2.3 Identification of Phenolic Compounds Profile Using RP-HPLC

The identification of individual phenolic compounds was carried out according to Raja and Venkataraman (2016) by reversed phase high-performance liquid chromatography (RP-HPLC) C18 column (Waters, 150 × 3.9 mm, 5 μm). Sample volume 20 μL of *Aloe vera* gel extract was injected and analyzed at a linear gradient with the flow rate of 1 mL/min. The mobile phase used was solvent-A: Milli-Q distilled water and solvent-B: acetonitrile with the linear gradient program of 0–5 minutes, solvent A-B (88%–12%); 5–10 minutes (70%–30%), 10–20 minutes (45%–55%); 20–25 minutes (40%–60%), 25–30 minutes (30%–70%). The eluent was monitored at the wavelength of 290 nm. The identification of phenolic compounds was done by comparing retention time and area of peaks in the extracts against those of the standard compounds from Sigma-Aldrich. The data acquisition and processing were performed using the Breeze data (Waters) system.

15.2.4 Determination of Total Flavonoid Content

The total flavonoid content was analyzed by the aluminum tri chloride colorimetric method (Suman Chandra et al. 2014). One milligram of the sample was added to 25 μL of potassium acetate (1 M) and 25 μL of aluminum tri chloride (10%), and made up to 1.2 mL with distilled water. This was incubated for 30 minutes at room temperature. The quercetin was used as the standard reference and the absorbance was measured at 415 nm. The results were expressed as quercetin equivalents (mg QE/g of dry extract).

15.2.5 Determination of In Vitro Anti-oxidant and Anti-inflammatory Activities

15.2.5.1 Determination of Total Anti-oxidant Capacity

Total anti-oxidant capacity was estimated through the phosphomolybdenum method (Maswada 2013). One milligram of the sample was added to 1 mL of 1:1:1 reagent which contained H_2SO_4 (0.6 M), 28 mM sodium phosphate, 4 mM ammonium molybdate, and 100 μL of methanol. This was boiled for 90 minutes in a boiling water bath. Ascorbic acid was used as a standard reference and was measured at an absorbance of 695 nm. The results were expressed as ascorbic acid equivalents (mg AAE/g of dry extract).

15.2.5.2 ABTS Radical Scavenging Activity

An 2,2'-azino-bis(3-thylbenzothiazoline-6-sulfonic acid (ABTS) assay was performed according to Zheleva-Dimitrova, Nedialkov, and Kitanov (2010). Various concentrations of the methanolic extract of the *Aloe vera* gel (0.1, 0.25, 0.5, 1, 2 mg mL) were mixed with 1 mL of diluted ABTS solution and incubated in the dark for 15 minutes at 37°C, and the absorbance was noted at 734 nm against the methanol solvent blank. The ascorbic acid was used as a standard reference compound. The ABTS scavenging activity was calculated as follows:

$$\text{ABTS radical-scavenging activity (\%)} = \left[(\text{Ac} - \text{As})/\text{Ac}\right] \times 100$$

where Ac denotes the absorbance of the control reaction (containing all reagents except the sample), and As denotes the absorbance of the test sample.

15.2.5.3 DPPH Scavenging Activity

The free radical scavenging activity at various concentrations of the methanolic extract of the *Aloe vera* gel (0.1, 0.25, 0.5, 1, 2 mg/mL) was evaluated by a DPPH assay (Aksoy et al. 2013). The different concentrations of the extract were vortexed with 1.0 mL of 0.1 mM of the DPPH in methanol. The mixture was incubated in the dark for 30 minutes at room temperature. The degree of inhibition of the DPPH was monitored by a decrease in absorbance, which was measured at 517 nm against the methanol solvent blank. The DPPH solution without the extract served as the control. The scavenging activity was calculated as:

$$\text{DPPH radical-scavenging activity (\%)} = \left[(\text{Ac} - \text{As})/\text{Ac}\right] \times 100$$

where Ac denotes the absorbance of the control reaction (containing all reagents except sample), and As denotes the absorbance of the sample.

15.2.5.4 Metal Chelating Assay

Metal chelating activity was measured as described by Wong et al. (2014) by adding 0.2 mL of 0.1 mM $FeSO_4$ and 0.4 mL of 0.25 mM ferrozine with different concentrations of the *Aloe vera* gel extract (Wong et al. 2014). After incubating at room temperature for 10 minutes, the absorbance was recorded at 562 nm with the ascorbic acid as the standard reference compound. The metal chelating activity was calculated as follows:

$$\text{Metal chelating activity} = \left[(\text{Ac} - \text{As})/\text{Ac}\right] \times 100$$

where Ac denotes the absorbance of the control reaction (containing all reagents except the sample), and As denotes the absorbance of the sample.

15.2.5.5 Hydrogen Peroxide Scavenging Activity

H_2O_2 assay was carried out as described by Mukhopadhyay et al. (2016). The *Aloe vera* gel extract (0.1–2 mg) was added to 0.25 mL of the ferrous ammonium sulfate (1 mM), and 0.5 mL of hydrogen peroxide (5 mM) was incubated for 5 minutes at room temperature. After the addition of 1.5 mL of 1,10-phenanthroline (1 mM), it was incubated for 10 minutes, and the absorbance was measured at 510 nm against a blank solution without hydrogen peroxide solution. The ascorbic acid was used as a standard reference compound. The H_2O_2 scavenging activity was calculated as follows:

$$H_2O_2 \text{ radical-scavenging activity (\%)} = \left[(\text{Ac} - \text{As})/\text{Ac}\right] \times 100$$

where Ac denotes the absorbance of the control reaction (containing all reagents except the sample), and As denotes the absorbance of the sample.

15.2.6 Inhibition of Albumin Denaturation

The anti-inflammatory activity of the *Aloe vera* gel extract of different concentrations (0.1–2 mg/mL) was studied by using the inhibition of bovine serum albumin denaturation technique, which was studied according to the method of Reshma, Arun, and Brindha (2014). The reaction mixture contained 100 μL of the gel extract with 500 μL of 1% bovine serum albumin and was initially incubated at room temperature for 10 minutes and further heated up for 20 minutes at 51°C. The absorbance was measured at 660 nm after cooling the samples to room temperature. Aspirin was taken as the positive control. The percentage inhibition of protein denaturation was calculated using (%) inhibition = [(Ac − As)/Ac] × 100, where Ac denotes the absorbance of the control reaction (containing all reagents except the sample), and As denotes the absorbance of the sample.

15.2.7 Determination of In Vivo Anti-oxidant and Anti-inflammatory Activities

Adult male healthy Wistar rats with a body weight ranging from 180 to 200 g were maintained in an airconditioned room (25°C ± 1°C) with a 12-hour light/12-hour dark cycle. The animals were acclimatized to the environment prior to the experimental use by supplying water ad libitum and a standard laboratory diet of food provided. The animal experimental study was approved by Institutional Animal Ethics Committee (VIT/IAEC/12/July23/03). Following, are the groups used for the study:

Group-I: Normal control rats
Group-II: Inflammatory control rats
Group-III: Inflammatory-induced rats treated with *Aloe vera* extract (300 mg/kg bw)
Group-IV: Inflammatory-induced rats treated with diclofenac (10 mg/kg bw)

For animal experiments, the *Aloe vera* gel extract dose (300 mg/Kg body weight (bw)/animal) was chosen based upon our previous study and the standard drug taken was diclofenac (Oudghiri and Azza 2015). The extract was resuspended in water and administered orally.

15.2.7.1 Complete-Freund's Adjuvant-Induced Rat Paw Edema Assay

The rat paw edema assay protocol was a slightly modified method of Rasool, Latha, and Varalakshmi (2000). Group I served as the control. In group-II, arthritis was induced by Freund's Adjuvant (0.1 mL) into the right hind foot paw which served as the inflammatory control. Group-III rats were Freund's Adjuvant-induced arthritis rats treated with an *Aloe vera* extract (300 mg/kg bw), and group-IV were adjuvant-induced arthritis rats treated with the standard drug diclofenac (10 mg/kg bw), respectively.

The treatments of group-III and group-IV Wistar rats were started after 7 days post-injection of the administration of the Complete-Freund's Adjuvant and continued for 14 days. The arthritis was assessed by means of physical measurement, wherein the paw volume was assessed by the measurement of the hind foot paw by using the Vernier scale. The difference in paw volume on day 1 and day 14 was considered as the inflammatory edema. The volume changes in the standard and test groups were compared with the control and the percentage of inhibition was calculated according to Sarkhel (2016). The inhibition of the paw volume of the treated and control groups were calculated as follows:

$$\% \text{ inhibition of edema} = \left(\mathrm{T} - \mathrm{T_0}/\mathrm{T}\right) \times 100$$

Where T = thickness of the paw in the inflammatory control group and T_0 represents the thickness of paw volume in the treated groups, respectively.

After the experimental period, the rats were autopsied under mild ether anesthesia, and the blood was collected. The blood was collected by cardiac puncture in an ethylenediaminetetraacetic acid (EDTA)-coated vial and centrifuged at 1800 rpm for 20 minutes at 4°C, and the plasma was separated and used for the estimation of TNF-α, IL-6, secretory phospholipase A2 (sPLA2) levels through respective ELISA kits. The liver and pancreas were removed and washed with ice-cold phosphate-buffered saline. The tissue homogenate was prepared in 50 mM phosphate buffer pH 7.4 at 4°C by ultrasonication for 3 minutes. The tissue homogenate was centrifuged at 4°C for 15 minutes at 8000 rpm, and the supernatant was used for the estimation of the superoxide dismutase and catalase activity.

15.2.7.2 Assessment of Anti-oxidant Enzymes Activity in Pancreas and Liver

15.2.7.2.1 Estimation of Superoxide Dismutase

Superoxide dismutase (SOD) activity was measured by slightly a modified method of Murthy et al. (2005). The assay was performed using 20 µL of tissue homogenate that was suspended in 2.8 mL of the 50 mM Tris buffer pH 8.2. Pyrogallol was used as the substrate at a concentration of 15 mM and 100 µL solution was added for each replicate. The absorbance was measured at 420 nm. The control solution contained 100 µL of the 50 mM Tris buffer with 2.8 mL of the 15 mM of pyrogallol. The degree of inhibition of the auto-oxidation of the pyrogallol at alkaline pH by the SOD was used to measure the enzyme activity. The SOD activity was expressed as 50% inhibition of auto-oxidation of the pyrogallol. The specific activity was expressed as units/mg of protein.

15.2.7.2.2 Estimation of Catalase Activity

Catalase activity was carried out as described by Sepici-Dincel et al. (2007). A tissue homogenate of 100 µL was mixed with 900 µL of the 50 mM phosphate buffer and 100 µL of the 30 mM H_2O_2. The absorbance was measured at 240 nm against the blank without H_2O_2. Units of catalase activity are expressed as the amount of enzyme that decomposes 1 µM of H_2O_2 per minute. The specific activity was expressed in terms of U/mg of protein.

15.2.8 Statistical Analysis

Data were expressed as the means ± Standard Deviation (SD). All data were analyzed using GraphPad Prism 6.0 (GraphPad Software, Inc. La Jolla, CA, USA). The significance was analyzed using one-way analysis of variance (ANOVA) followed by the Bonferroni test. Data were considered statistically significant for $p < 0.05$.

15.3 RESULTS

15.3.1 Total Phenolic Content and Identification of Phenolic Constituents in *Aloe vera* Gel Extract by Using RP-HPLC

The total phenolic content was found to be 7.21 ± 0.5 mg GAE/gm in the *Aloe vera* gel extract. The total polyphenols

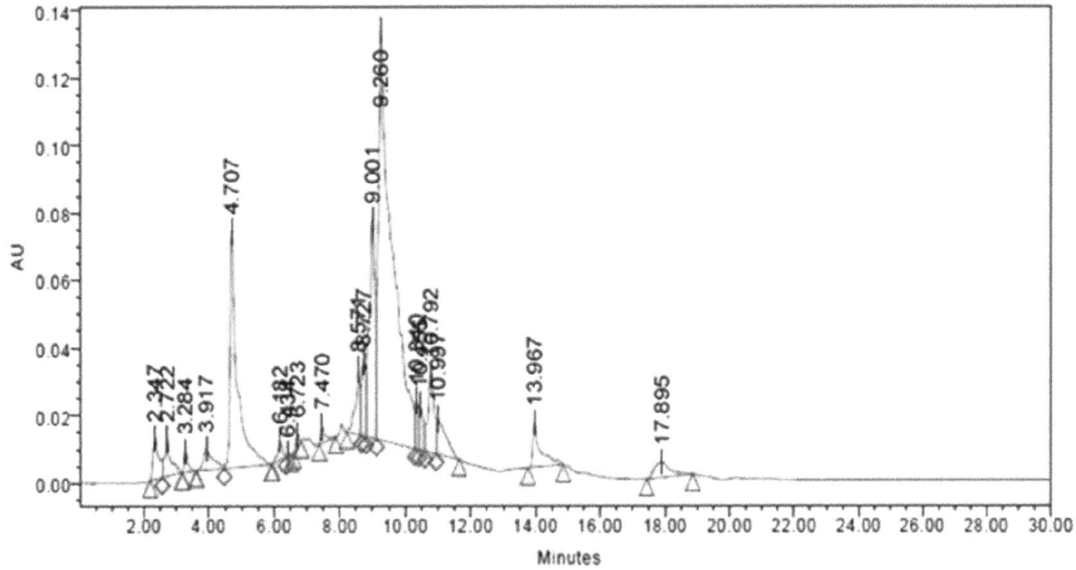

FIGURE 15.1 HPLC chromatogram of the *Aloe vera gel* extract at 290 nm. The HPLC profile of the *Aloe vera* gel extract: (1) ascorbic acid, (2) gallic acid, (3) catechin, (4) naringin, (5) taxifolin, (6) pelargonidin chloride, (7) myrcetin, (8) luteolin, (9) quercetin, (10) naringenin, and (11) kaempferol. Column: RP-HPLC C18 column (Waters, 150 × 3.9 mm, 5 μm); eluent: Milli Q distilled water and solvent-B: acetonitrile of 0–5 minutes, solvent A-B (88%–12%); 5–10 minutes (70%–30%), 10–20 minutes (45%–55%); 20–25 minutes (40%–60%); and 25–30 minutes (30%–70%); Flow rate: 1 mL/min; detection 290 nm.

were calculated using the standard gallic acid curve. The identification of the different phenolic compounds present in the *Aloe vera* gel extract has been performed by RP-HPLC. Through RP-HPLC analysis, the chromatogram revealed the presence of myrcetin, quercetin, ascorbic acid, gallic acid, catechin, naringin, taxifolin, pelargonidin chloride, luteolin, narinigenin, and kaempferol in the *Aloe vera* gel extract by comparison with the retention times of the standards. The chromatogram of the RP-HPLC of the extract is shown in Figure 15.1. Table 15.1 represents the various phenolic compounds present in the *Aloe vera* gel extract.

TABLE 15.1
Phenolic Constituents of the *Aloe vera* Gel Extract Identified Through RP-HPLC

S. No	Compounds	Gel Extract	Retention Time (RT)
1.	Ascorbic acid	+	2.34
2.	Gallic acid	+	2.72
3.	Catechin	+	4.70
4.	Naringin	+	6.72
5.	Taxifolin	+	7.4
6.	Pelargonidin chloride	+	7.4
7.	Myrcetin	+	8.72
8.	Luteolin	+	9.0
9.	Quercetin	+	9.26
10.	Naringenin	+	10.79
11.	Kaempferol	+	10.99

Note: All experiments performed in triplicates.

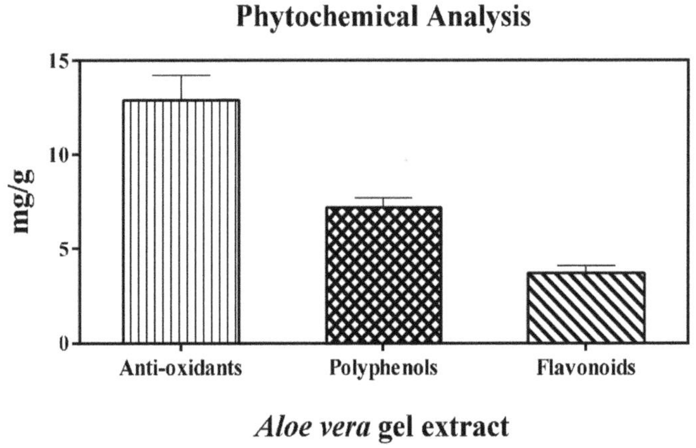

FIGURE 15.2 Phytochemical analysis of the *Aloe vera* gel. The total polyphenols, total flavonoids, and total anti-oxidant content in the *Aloe vera* gel extract. All experiments performed in triplicates and expressed as mean ± SD.

The total flavonoids were found to be 3.71 ± 0.4 mg of quercetin equivalents (QE) per gram of dry extract of the *Aloe vera* gel (Figure 15.2).

15.3.2 Determination of In Vitro Anti-oxidant Potential of *Aloe vera* Gel Extract

The total anti-oxidant capacity of the *Aloe vera* gel showed 12.9 ± 1.3 mg AAE/gm by the Phosphomolybdenum method. The total anti-oxidant content was calculated using the standard ascorbic acid curve with linear regression. In this study, the analysis of the in vitro anti-oxidant activities was evaluated by using various assay methods, such as ABTS, DPPH, H_2O_2 scavenging, and metal chelating property. It was observed that as the concentration increased, the ability to scavenge the free radicals such as ABTS, DPPH, H_2O_2, and the metal chelating properties was also increased. At the highest concentration (2 mg/mL), the extract showed better ABTS scavenging potential, followed by the metal chelating activity, H_2O_2, and DPPH assay, respectively, as shown in Figure 15.3.

15.3.3 Determination of In Vitro Anti-inflammatory Property of *Aloe vera* Gel Extract

In this study, it was observed that the *Aloe vera* gel extract showed better inhibition of heat-induced albumin denaturation in a concentration-dependent manner compared to the standard drug aspirin (Figure 15.4). The maximum inhibition of 90.6% was observed at 2 mg/mL. Aspirin, a standard anti-inflammatory drug, showed the maximum inhibition of 86.4% at the concentration of 2 mg/mL as compared to the control.

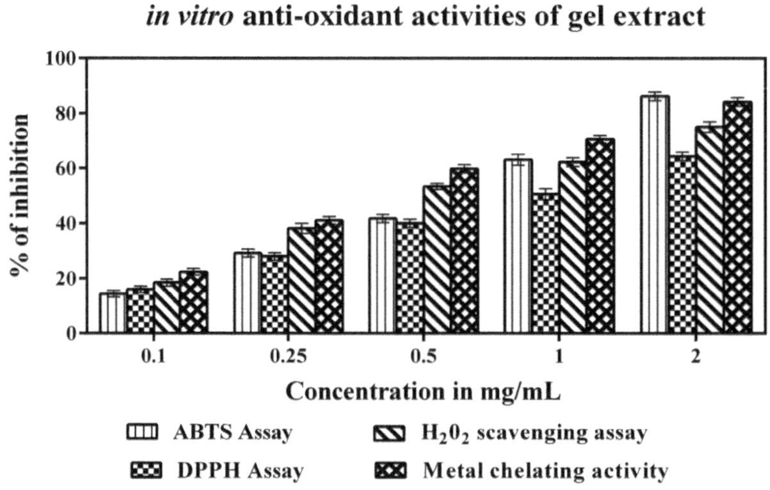

FIGURE 15.3 Assessment of the in vitro anti-oxidant property of the *Aloe vera* gel extract. Radical scavenging activities of ABTS, DPPH, H_2O_2, and metal chelating activity in the *Aloe vera* gel extract. All experiments were performed in triplicates and expressed as mean ± SD.

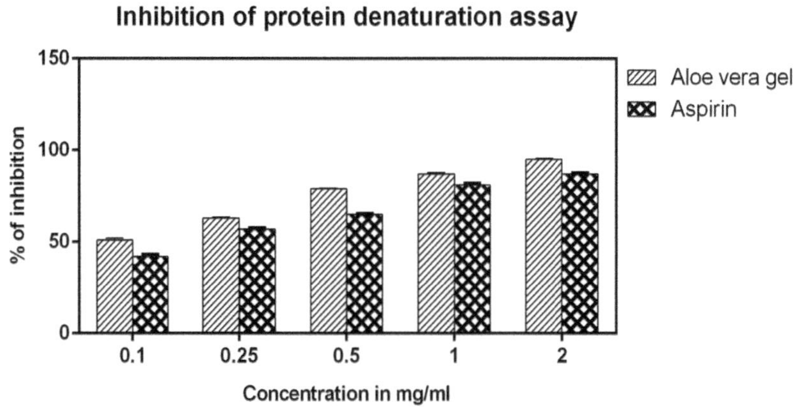

FIGURE 15.4 In vitro evaluation of anti-inflammatory potential of the *Aloe vera* gel extract. Inhibition of albumin denaturation of the *Aloe vera* gel extract was determined by various concentrations varying from 0.1 mg/mL to 2 mg/mL. The experiments were performed in triplicates and expressed as mean ± SE.

15.3.4 Effect of *Aloe vera* Gel Extract on In Vivo Anti-inflammatory Property

The *Aloe vera* gel extract's anti-inflammatory property was studied in Freund's Adjuvant-induced rat paw edema model and the assessment of inflammatory markers/mediators, such as TNF-α, IL-6, and s-PLA2 levels, respectively. The effect of the *Aloe vera* gel extract on the paw edema in the experimental rats is shown in Figure 15.5a. The paw thickness of the rats when measured revealed an increase in diameter from day 7, which increased further up to day 14. After the treatment with the *Aloe vera* extract (group-III), the paw edema reduced significantly by 69.5% and 63.4% in the rats treated with diclofenac (group-IV), respectively (Figure 15.5a). The *Aloe vera* gel extract reversed much of the swelling that developed prior to the treatment compared to the adjuvant-induced rats (group-II).

In the adjuvant-induced rats (group-II), it was observed that the inflammatory marker levels of TNF-α, IL-6, and sPLA2 significantly increased ($p < 0.0001$) when compared to the normal group (group-I). Upon the treatment with the *Aloe vera* gel extract in the adjuvant-induced rats (group-III), the levels of all the three inflammatory mediators decreased significantly (TNF-α ~ 64.2%, IL-6 ~ 67.7%, and sPLA2 levels ~ 68%) when compared to group-II and group-IV (TNF-α ~ 43.15%, IL-6 ~ 44.7%, and sPLA2 ~ 66.5%, respectively) ($p < 0.0001$) (Figure 15.5b–d).

15.3.5 Effect of *Aloe vera* Gel Extract on In Vivo Anti-oxidant Property

The results obtained in our study showed that the SOD and catalase levels in the pancreas were decreased in the rats injected with Freund's Adjuvant (group-II). Upon feeding of the *Aloe vera* gel extract to the arthritis-induced rats (group-III), both the SOD and catalase levels in the pancreas were significantly increased by 60.6% and 52.4%, respectively, when compared to the arthritis-induced rats (group-II) ($p < 0.0001$) (Figure 15.6a and b).

In the diclofenac treated group, the levels of the SOD and catalase increased by around 54% and 30.1%, respectively. Similar results were observed in the liver after the *Aloe vera* gel extract treatment (group-III), the SOD and catalase levels were increased by 64.1% and 46.1%, respectively, when compared to the diclofenac-treated rats (group-IV) (48.5% and 28.8%, respectively) and the arthritis-induced rats (group-II) ($p < 0.0001$) (Figure 15.6c and d). Overall, in this study, it was observed that the *Aloe vera* gel extract showed significantly better anti-oxidant potential when compared to the diclofenac ($p < 0.001$).

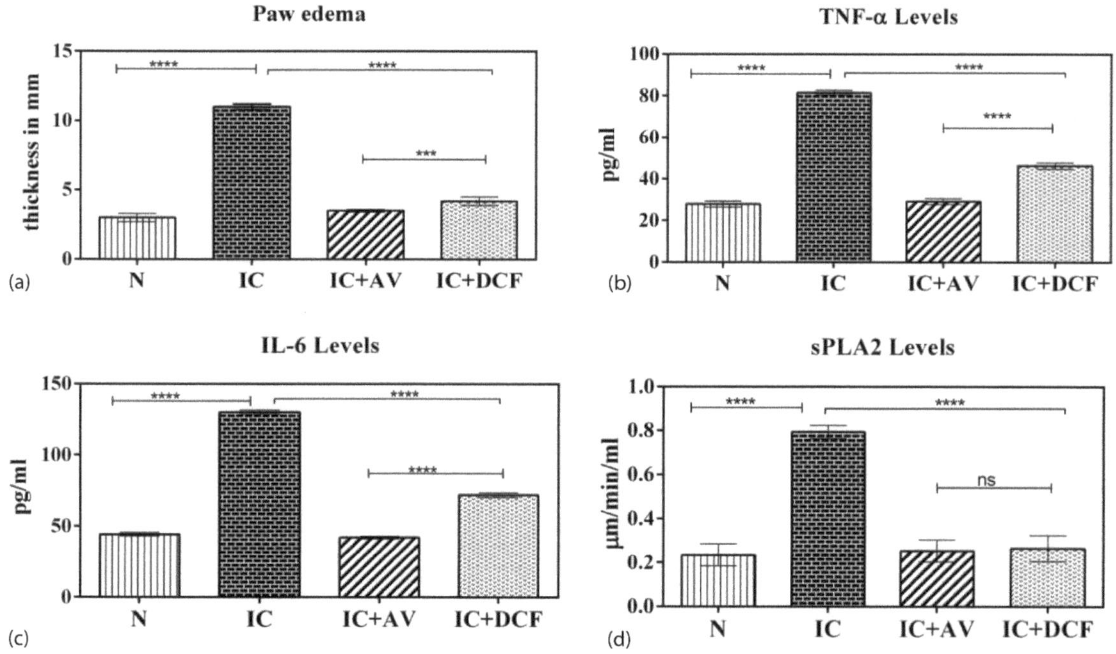

FIGURE 15.5 (a–d) Effect of the *Aloe vera* gel extract and diclofenac on inflammatory markers in the serum of the control and experimental rats. (a) paw edema, (b) TNF-α levels, (c) IL-6 levels, and (d) sPLA2 levels. Results expressed as mean ± SD for six animals. The statistical analysis was calculated by using ANOVA, followed by the Bonferroni test. Data were considered statistically significant for $p < 0.05$.

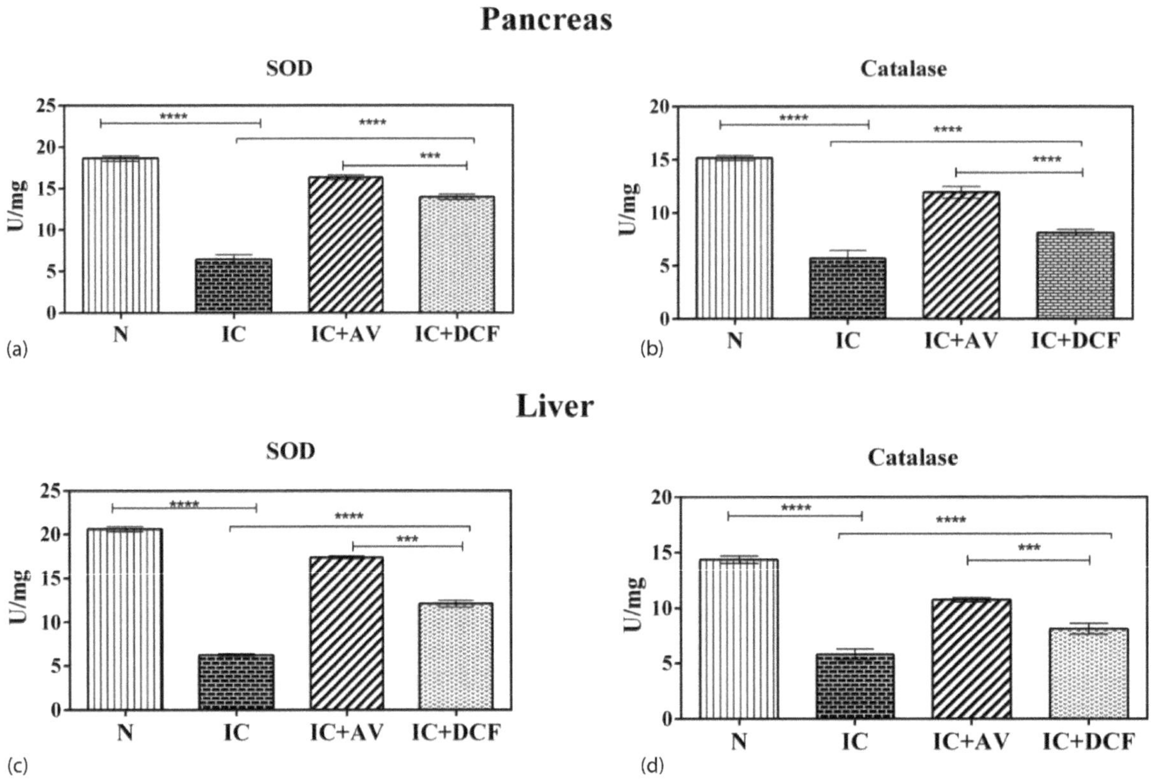

FIGURE 15.6 (a–d) Effect of the *Aloe vera* gel extract on the anti-oxidant status in the pancreas and liver of the control and experimental rats. (a) SOD levels in the pancreas, (b) catalase levels in the pancreas, (c) SOD levels in the liver, and (d) catalase levels in the liver. The results were expressed as mean ± SD. The statistical analysis was performed by using ANOVA, followed by the Bonferroni test. Data were considered statistically significant for $p < 0.05$.

15.4 DISCUSSION

Freund's Adjuvant-induced paw edema and the carrageenan-induced paw edema are the most widely used in vivo models to study the anti-inflammatory potential in the search for new drugs for the treatment of arthritis. The Freund's Adjuvant model involves the induction of arthritis by injecting the adjuvant (*Mycobacterium butyricum* suspended paraffin oil) either in the paw or tail of the rats. This leads to a cross reaction between the joint and the antigen of the *Mycobacterium* (Bendele et al. 1999), which damages the cartilage and bones in the joints due to pro-inflammatory mediators. In this study, we evaluated the in vitro anti-oxidant and anti-inflammatory potential of the *Aloe vera* gel extract. Further, the adjuvant-induced arthritis rats with respect to the oxidative damage caused due to the ROS production and inflammatory marks/mediators were evaluated.

The phytochemical and RP-HPLC analyses have shown the presence of various phenolic compounds present in the *Aloe vera* extract (Figures 15.1 and 15.2, Table 15.1). Plant polyphenols are known to possess anti-oxidant properties, and they are known to prevent cell damage caused by oxidative stress and scavenge free radicals (Mehta et al. 2017). Various in vitro assays performed in this study also indicate that the *Aloe vera* gel extract has shown the anti-oxidant property through the ability to scavenge free radicals, such as ABTS, DPPH, H_2O_2, and metal chelating property. Our results corroborate with Ghasemzadeh, Jaafar, and Rahmat (2010), that the link of the anti-oxidant activity of the medicinal plants to their phenolic constituents having scavenging properties play into the neutralizing and scavenging free radicals. This can be attributed to the presence of flavonoids and phenolic compounds present in the *Aloe vera* gel extract (Table 15.1). These scavenging methods (ABTS, DPPH, H_2O_2, and metal chelating activity) are well known by their mechanisms of action and would be complementary to the study of the anti-oxidant potential of various medicinal plants. It has also been observed that the anti-oxidant property of the gel extracts is dose-dependent, i.e., as the concentration increases, the anti-oxidant potential also increases as shown in Figure 15.3. Besides having the anti-oxidant potential, the phenolic constituents also play an effective role in the alleviation of inflammation (Mehta et al. 2017). Protein denaturation is one of the known causes of arthritic disease and other inflammatory disorders (Sangita Chandra et al. 2012). So we evaluated the anti-inflammatory activity of the *Aloe vera* gel extract using a protein denaturation assay. It was observed that as the concentration of the *Aloe vera* gel extract increases, the inhibition of the heat-induced albumin denaturation was also increased when compared to the standard drug aspirin (Figure 15.4). This anti-inflammatory effect may be due to the synergistic effect of the *Aloe vera* gel polyphenolic constituents. The results are in agreement with Bhat, Rizvi, and Kumar (2014), who reported the anti-inflammatory effect using the arthritis-induced paw edema by *Cuminum cyminum* seeds.

In this study, the anti-inflammatory potential of the *Aloe vera* extract with respect to its ability to reduce the damage caused through oxidative stress was evaluated. An important aspect of the inflammation is that there is a disturbance in the homeostasis of the anti-inflammatory and pro-inflammatory cytokines (Kalaiselvan and Rasool 2015). The increased levels of pro-inflammatory cytokines, such as TNF-α, IL-6, monocyte chemoattractant protein 1, etc. propagate the progression of the inflammation through the infiltration of neutrophils, leading to destruction and damage of the joints and bones (Zhang et al. 2013). In this study, we have observed the levels of TNF-α, IL-6, and sPla2 in inflammation and the effect of the *Aloe vera* gel extract on these inflammatory mediators. TNF-α is one of the pleiotropic cytokines that is seen mostly in cases of inflammatory disorders. Cells such as synoviocytes and T-cells produce pro-inflammatory cytokines upon activation, thus leading to synovial inflammation in the joints (Kalaiselvan and Rasool 2015). Another pleiotropic cytokine is IL-6, which plays a crucial role in the maturation of B cells. IL-6 is also involved in the production of the auto-antibodies and plays a role in arthritic progression (Srirangan and Choy 2010). IL-6 is involved in the progression of joint inflammation through elevated levels of Vascular Endothelial Growth Factor (VEGF). It was observed in this study that the levels of TNF-α and IL-6 were elevated in the plasma of the inflammatory control rats. Oral feeding of the *Aloe vera* extract has shown a significant reduction in the levels of these inflammatory cytokines (Figure 15.3b and c). There are reports which suggest that phenolics such as quercetin, catechin, rutin, kaempferol, myricetin, and luteolin have anti-inflammatory properties which may help in the reduction of the inflammation and alleviation of rheumatoid arthritis (Lee and Kim 2010). The presence of the phenolic constituents, listed in Table 15.1, in the *Aloe vera* gel extract may play a role in decreasing the levels of TNF-α and IL-6, thereby leading to the reduction in inflammation as seen in the reduction of paw edema (Figure 15.3a).

sPLA2 enzymes that are active during the inflammation and transmembrane signaling are one of the major targets for drug development against inflammation (Reid 2005). They help in the progression of inflammation through mast cells degranulation, release of pro-inflammatory cytokines. The levels of sPLA2 have been found to be high in the joints of patients affected with arthritis (Coulthard et al. 2011). In this study, it was observed that the levels of sPLA2 in the adjuvant-induced arthritis rats were elevated, and upon treatment with the *Aloe vera* gel extract, these levels were significantly reduced as shown in Figure 15.5d. Plant phenolic and flavonoids have been reported to modulate the levels of sPLA2, cyclooxygenase, and lipoxygenase which are known for their role in the inflammation and progression of rheumatoid arthritis (Kim et al. 2004). The reduction in levels of sPLA2 in this study might be due to the presence of phenolic constituents, such as catechin, naringin, taxifolin, pelargonidin chloride, myrcetin, luteolin, quercetin, naringenin, and kaempferol in the *Aloe vera* gel extract.

In rheumatoid arthritis patients, the synovial fluid contains high levels of free radicals and the presence of TNF-α, IL-6, and IL-1β which are also involved in the hydroxyl radicals and hydrogen peroxide production (Quinonez-Flores et al. 2016). To reduce the effects of toxicity of the free radicals, the body has a coordinated radical scavenging organization comprised of anti-oxidants, such as catalase, SOD, glutathione, etc. (Kalaiselvan and Rasool 2015). The SOD is the first line of defense and catalase is the second line of defense, wherein the SOD reacts with the free radicals forming hydrogen peroxide and the catalase reacts with the hydrogen peroxide leading to the liberation of water, thereby eliminating the free radicals. In our study, it was observed that there was a decrease of the SOD and catalase levels in the liver and pancreas of the adjuvant-induced arthritis rats (Figure 15.4). This can be due to more free radical formation due to the elevated levels of the pro-inflammatory cytokines and more phagocytosis, leading to the weakening of the anti-oxidant system in the body indicating a weak defense system against oxidative damage, which further leads to the progression of arthritis (Rasool, Latha, and Varalakshmi 2000; Kalaiselvan and Rasool 2015). Upon the administration of the *Aloe vera* gel extract, the levels of the SOD and catalase significantly increased as shown in Figure 15.6, compared to the standard drug diclofenac. The phenolic compounds and flavonoids present in the *Aloe vera* gel extract are known to possess anti-oxidant properties. The polyphenols may also help in inhibiting the neutrophil infiltrations, release of lysosomal enzymes by scavenging the free radicals formed due to inflammation (Zhang and Tsao 2016; Hussain et al. 2016). The polyphenolic constituents, such as kaempferol and quercetin help in activating the anti-oxidant enzyme levels and thereby leading to the decrease in pro-inflammatory cytokines and enzymes (Oliviero et al. 2018). The presence of polyphenolic constituents in the *Aloe vera* gel extract may be responsible for the reduction of free radicals and inflammatory markers/mediators as observed through both in vitro and in vivo studies.

15.5 CONCLUSION

The present study demonstrates that the *Aloe vera* gel extract possesses significant anti-oxidant activity against various anti-oxidant assays and potent anti-inflammatory effects. The presence of various phenolic constituents, such as gallic acid, quercetin, myrcetin, kaempferol, and luteolin may be responsible for anti-oxidant and anti-inflammatory activities. The anti-inflammatory and anti-oxidant properties of *Aloe vera* might be due to the synergistic effects of these phytoconstituents. On the basis of our results, it was concluded that the *Aloe vera* gel extract possesses significant anti-oxidant and anti-inflammatory effects as observed by in vitro and in vivo studies. These beneficial properties of *Aloe vera* depend on the climate, season, mineral content, geographical origin, etc., and therefore studies are needed to check that any toxic constituents, if any, are at permissible levels. Further studies are recommended to isolate the bioactive constituents of *Aloe vera* and the mechanisms of action through which these constituents of *Aloe vera* help in reducing inflammation through the decrease of the formation of free radicals.

ACKNOWLEDGMENTS

The authors thank the Department of Science and Technology (IR/SO/LU-03/2004), New Delhi, India, for the financial support. The authors are thankful to the management of the Vellore Institute of Technology (VIT) for providing the necessary facilities.

CONFLICT OF INTEREST

The authors declare no conflict of interest.

REFERENCES

Ahlawat, Kulveer Singh, and Bhupender Singh Khatkar. 2011. "Processing, food applications and safety of *Aloe vera* products: A review." *Journal of Food Science and Technology* 48 (5): 525–533. doi:10.1007/s13197-011-0229-z.

Aksoy, Laçine, Erdi Kolay, Yasin Ağilönü, Zeyneb Aslan, and Mustafa Kargioğlu. 2013. "Free radical scavenging activity, total phenolic content, total antioxidant status, and total oxidant status of endemic *Thermopsis turcica*." *Saudi Journal of Biological Sciences* 20 (3): 235–239. doi:10.1016/j.sjbs.2013.02.003.

Altemimi, Ammar, Naoufal Lakhssassi, Azam Baharlouei, Dennis Watson, and David Lightfoot. 2017. "Phytochemicals: Extraction, isolation, and identification of bioactive compounds from plant extracts." *Plants* 22 (6): 1–23. doi:10.5912/jcb722.

Bendele, Alison, Jennifer Mccomb, Ty Gould, Tracy Mcabee, Gina Sennello, Elizabeth Chlipala, and Michael Guy. 1999. "Animal models of arthritis: Relevance to human disease." *Toxicologic Pathology* 27 (1): 134–142. doi:10.1177/019262339902700125.

Bhat, Sangeeta P., Waseem Rizvi, and Anil Kumar. 2014. "Effect of *Cuminum cyminum* L. seed extracts on pain and inflammation." *Journal of Natural Remedies* 14 (2): 186–192. doi:10.18311/jnr/2014/81.

Chandra, Sangita, Priyanka Chatterjee, Protapaditya Dey, and Sanjib Bhattacharya. 2012. "Evaluation of in vitro anti-inflammatory activity of coffee against the denaturation of protein." *Asian Pacific Journal of Tropical Biomedicine* 2 (1): S178–S180. doi:10.1016/S2221-1691(12)60154-3.

Chandra, Suman, Shabana Khan, Bharathi Avula, Hemant Lata, Min Hye Yang, Mahmoud A. Elsohly, and Ikhlas A. Khan. 2014. "Assessment of total phenolic and flavonoid content, antioxidant properties, and yield of aeroponically and conventionally grown leafy vegetables and fruit crops: A comparative study." *Evidence-Based Complementary and Alternative Medicine*. doi:10.1155/2014/253875.

Choi, Seongwon, and Myung Hee Chung. 2003. "A review on the relationship between *Aloe vera* components and their biologic effects." *Seminars in Integrative Medicine* 1 (1): 53–62. doi:10.1016/S1543-1150(03)00005-X.

Coulthard, Liam G., Jaclyn Costello, Brent Robinson, Ian A. Shiels, Stephen M. Taylor, and Trent M. Woodruff. 2011. "Comparative efficacy of a secretory phospholipase A2 inhibitor with conventional anti-inflammatory agents in a rat model of antigen-induced arthritis." *Arthritis Research and Therapy* 13 (2): R42–R51. doi:10.1186/ar3278.

Ghasemzadeh, Ali, Hawa Z.E. Jaafar, and Asmah Rahmat. 2010. "Antioxidant activities, total phenolics and flavonoids content in two varieties of Malaysia young ginger (*Zingiber Officinale* Roscoe)." *Molecules* 15 (6): 4324–4333. doi:10.3390/molecules15064324.

Hajjar, David P., and Antonio M. Gotto. 2013. "Biological relevance of inflammation and oxidative stress in the pathogenesis of arterial diseases." *American Journal of Pathology* 182 (5): 1474–1481. doi:10.1016/j.ajpath.2013.01.010.

Hamman, Josias H. 2008. "Composition and applications of *Aloe vera* leaf gel." *Molecules* 13 (8): 1599–1616. doi:10.3390/molecules13081599.

Hussain, Tarique, Bie Tan, Yulong Yin, Francois Blachier, Myrlene C.B. Tossou, and Najma Rahu. 2016. "Oxidative stress and inflammation: What polyphenols can do for us?" *Oxidative Medicine and Cellular Longevity*. doi:10.1155/2016/7432797.

Jairath, Gauri, Manish Kumar Chatli, and Ashim Kumar Biswas. 2016. "Comparative study on in vitro and in vivo evaluation of antioxidant potential of apple peel extract and *Aloe vera* gel." *Journal of Food Processing and Preservation* 40 (4): 607–614. doi:10.1111/jfpp.12639.

Kalaiselvan, Sowmiya, and Mahaboob Khan Rasool. 2015. "The anti-inflammatory effect of triphala in arthritic-induced rats." *Pharmaceutical Biology* 53 (1): 51–60. doi:10.3109/13880209.2014.910237.

Kim, Hyun Pyo, Kun Ho Son, Hyeun Wook Chang, and Sam Sik Kang. 2004. "Anti-inflammatory plant flavonoids and cellular action mechanisms." *Journal of Pharmacological Sciences* 96 (3): 229–243. doi:10.1254/jphs. CRJ04003X.

Langmead, Louise, Richard J. Makins, and David S. Rampton. 2004. "Anti-inflammatory effects of *Aloe vera* gel in human colorectal mucosa in vitro." *Alimentary Pharmacology and Therapeutics* 19 (5): 521–527. doi:10.1111/j.1365-2036.2004.01874.x.

Lee, Je Hyuk, and Gun Hee Kim. 2010. "Evaluation of antioxidant and inhibitory activities for different subclasses flavonoids on enzymes for rheumatoid arthritis." *Journal of Food Science* 75 (7): H212–H217. doi:10.1111/j.1750-3841.2010.01755.x.

Maswada, Hanafey F. 2013. "Assessment of total antioxidant capacity and antiradical scavenging activity of three Egyptian wild plants." *Journal of Medical Sciences (Faisalabad)* 13 (7): 546–554. doi:10.3923/jms.2013.546.554.

Mehta, Jignasu P., Pravin H. Parmar, Sohil H. Vadia, Mahendra K. Patel, and Chandrabhanu B. Tripathi. 2017. "In-vitro antioxidant and in-vivo anti-inflammatory activities of aerial parts of cassia species." *Arabian Journal of Chemistry* 10: s1654–s1662. doi:10.1016/j.arabjc.2013.06.010.

Moniruzzaman, Mohammed, Begum Rokeya, Sohel Ahmed, Amrita Bhowmik, Md Ibrahim Khalil, and Siew Hua Gan. 2012. "In vitro antioxidant effects of *Aloe barbadensis* Miller extracts and the potential role of these extracts as antidiabetic and antilipidemic agents on streptozotocin-induced type 2 diabetic model rats." *Molecules* 17 (11): 12851–12867. doi:10.3390/molecules171112851.

Mukhopadhyay, Debanjan, Pinakpani Dasgupta, Debalina Sinha Roy, Shauroseni Palchoudhuri, Ipsita Chatterjee, Shahnaz Ali, and Sujata Ghosh Dastidar. 2016. "A sensitive in vitro spectrophotometric hydrogen peroxide scavenging assay using 1,10-phenanthroline." *Free Radicals and Antioxidants* 6 (1): 124–132. doi:10.5530/fra.2016.1.15.

Murthy, Kotamballi N. Chidambara, Vanitha A. Vinothkumar, Rajesha, Mahadeva Swamy, Rao P. Sowmya, and Gokare A. Ravishankar. 2005. "In vivo antioxidant activity of carotenoids from *Dunaliella salina*—A green microalga." *Life Sciences* 76 (12): 1381–1390. doi:10.1016/j.lfs.2004.10.015.

Noor, Ayesha, Soosaih Gunasekaran, Amirtham Soosai Manickam, and Mookambeshwaran Arunachalam Vijayalakshmi. 2008. "Antidiabetic activity of *Aloe vera* and histology of organs in streptozotocin-induced diabetic rats." *Current Science* 94 (8): 1070–1076.

Oliviero, Francesca, Anna Scanu, Yessica Zamudio-Cuevas, Leonardo Punzi, and Paolo Spinella. 2018. "Anti-inflammatory effects of polyphenols in arthritis." *Journal of the Science of Food and Agriculture* 98 (5): 1653–1659. doi:10.1002/jsfa.8664.

Oudghiri, Mounia, and Zora Azza. 2015. "In vivo anti-inflammatory and antiarthritic activities of aqueous extracts from *Thymelaea hirsuta*." *Pharmacognosy Research* 7 (2): 213. doi:10.4103/0974-8490.150510.

Quinonez-Flores, Celia Maria, Susana Aidee Gonzalez-Chavez, Danyella Del Rio Najera, and Cesar Pacheco-Tena. 2016. "Oxidative stress relevance in the pathogenesis of the rheumatoid arthritis: A systematic review." *BioMed Research International*. doi:10.1155/2016/6097417.

Radha, Maharjan H., and Nampoothiri P. Laxmipriya. 2015. "Evaluation of biological properties and clinical effectiveness of *Aloe vera*: A systematic review." *Journal of Traditional and Complementary Medicine* 5 (1): 21–26. doi:10.1016/j.jtcme.2014.10.006.

Raja, Cprasanna, and Krishnan Venkataraman. 2016. "Aloe vera phytochemicals inhibits dipeptidyl peptidase iv (Dpp-Iv), an anti-diabetic target." *International Journal of Pharma and Bio Sciences* 7 (3): 120–128.

Rasool, MahaboobKhan, Lenin Mary Latha, and Palaninathan Varalakshmi. 2000. "Effect of *Withania somnifera* on lysosomal acid hydrolases in adjuvant-induced arthritis in rats." *Pharmacy and Pharmacology Communications* 6 (4): 187–190. doi:10.1211/146080800128735863.

Reid, Robert. 2005. "Inhibitors of secretory phospholipase A2 group IIA." *Current Medicinal Chemistry* 12 (25): 3011–3026. doi:10.2174/092986705774462860.

Reshma, Kallur Purushothaman Arun, and Pemaiah Brindha. 2014. "In vitro anti-inflammatory, antioxidant and nephroprotective studies on leaves of *Aegle marmelos* and *Ocimum sanctum*." *Asian Journal of Pharmaceutical and Clinical Research* 7 (4): 121–129.

Saeed, Naima, Muhammad R. Khan, and Maria Shabbir. 2012. "Antioxidant activity, total phenolic and total flavonoid contents of whole plant extracts *Torilis leptophylla* L." *BMC Complementary and Alternative Medicine* 12: 221–233. doi:10.1186/1472-6882-12-221.

Salimonu, Lekan Samusa, Abiodun O. Johnson, Akintunde I. Williams, Gloria Iyabo Adeleye, and Babatunde O. Osunkoya. 1982. "Phagocyte function in protein-calorie malnutrition." *Nutrition Research* 2 (4): 445–454. doi:10.1016/S0271-5317(82)80052-3.

Sarkhel, Sumana. 2016. "Evaluation of the anti-inflammatory activities of *Quillaja saponaria* mol. saponin extract in mice." *Toxicology Reports* 1 (3): 1–3. doi:10.1016/j.toxrep.2015.11.006.

Sepici-Dincel, Aylin, Şereften Açikgöz, Cemal Çevik, Meltem Sengelen, and Erdem Yeşilada. 2007. "Effects of in vivo antioxidant enzyme activities of myrtle oil in normoglycaemic and alloxan diabetic rabbits." *Journal of Ethnopharmacology* 110 (3): 498–503. doi:10.1016/j.jep.2006.10.015.

Shaikh, Rafik U., Mahesh M. Pund, and Rajesh N. Gacche. 2016. "Evaluation of anti-inflammatory activity of selected medicinal plants used in Indian traditional medication system in vitro as well as in vivo." *Journal of Traditional and Complementary Medicine* 6 (4): 355–361. doi:10.1016/j.jtcme.2015.07.001.

Srirangan, Srinivasan, and Ernest H. Choy. 2010. "The role of interleukin 6 in the pathophysiology of rheumatoid arthritis." *Therapeutic Advances in Musculoskeletal Disease* 2 (5): 247–256. doi:10.1177/1759720X10378372.

Umar, Sadiq, Jamil Zargan, Khalid Umar, Sayeed Ahmad, Chandra Kant Katiyar, and Haider A. Khan. 2012. "Modulation of the oxidative stress and inflammatory cytokine response by thymoquinone in the collagen induced arthritis in Wistar rats." *Chemico-Biological Interactions* 197 (1): 40–46. doi:10.1016/j.cbi.2012.03.003.

Wong, Fai Chu, Ann Li Yong, Evon Peir Shan Ting, Sim Chyi Khoo, Hean Chooi Ong, and Tsun Thai Chai. 2014. "Antioxidant, metal chelating, anti-glucosidase activities and phytochemical analysis of selected tropical medicinal plants." *Iranian Journal of Pharmaceutical Research* 13 (4): 1409–1415.

Zhang, Hua, and Rong Tsao. 2016. "Dietary polyphenols, oxidative stress and antioxidant and anti-inflammatory effects." *Current Opinion in Food Science* 8: 33–42. doi:10.1016/j.cofs.2016.02.002.

Zhang, Yuqin, Wei Xu, Huang Li, Xun Zhang, Yufa Xia, Kedan Chu, and Lidian Chen. 2013. "Therapeutic effects of total alkaloids of *Tripterygium wilfordii* Hook f. On collagen-induced arthritis in rats." *Journal of Ethnopharmacology* 145 (3): 699–705. doi:10.1016/j.jep.2012.11.018.

Zheleva-Dimitrova, Dimitrina, Paraskev Nedialkov, and Gerassim Kitanov. 2010. "Radical scavenging and antioxidant activities of methanolic extracts from *Hypericum* species growing in Bulgaria." *Pharmacognosy Magazine* 6 (22): 74–78. doi:10.4103/0973-1296.62889.

16 Monoterpenes Modulating IL-10
A Systematic Review

Saravanan Shanmugam, Jullyana S. S. Quintans, Parimelazhagan Thangaraj, Luciana Scotti, Marcus T. Scotti, Adriano A. S. Araújo, and Lucindo J. Quintans-Júnior

CONTENTS

16.1 INTRODUCTION

Oxidative stress plays a significant role in the pathogenesis of inflammatory disorders, inflammation is the immune system's response to fight against the external pathogens or internal injury. The inflammatory response is triggered by innate immune receptors that recognize pathogens and damaged cells. The overexpression of the pro-inflammatory mediators, such as reactive oxygen mediator, neutrophil infiltration, and cytokines has been known to contribute to the inflammatory cascade in the pathological process of colitis (Neurath 2014). Moreover, the nuclear transcription factor kappa B (NF-κB) is identified as one of the pivotal regulatory components in inflammatory diseases and inhibition of its activity may alleviate the severity of the inflammation process (Liu et al. 2017).

Cytokines are a group of small soluble proteins that are secreted by a wide variety of cell types, and they perform several activities which play a main role in the stimulation of immune cell proliferation and differentiation, which leads to the termination of the invading pathogens or repairing the damaged tissue by activating the downstream signaling pathways through interaction with their receptors on the cell membrane (Pizarro and Cominelli 2007). Cytokines encompass a wide range of molecules that are essential for the communication between cells of the immune system and other non-immune cells. These cytokines are classified as pro and anti-inflammatory cytokines. Some cytokines that function primarily to promote the inflammation are known as pro-inflammatory cytokines, such as interleukin (IL)-1, IL-6, and tumor necrosis factor (TNF)-α, while others suppress the activity of the pro-inflammatory cytokines and are called anti-inflammatory cytokines, for example, IL-4, IL-10, and IL-13, and they are potent activators of B lymphocytes (Ye et al. 2016). Hence, a "balance" between the effects of pro and anti-inflammatory cytokines is thought to determine the outcome of a disease, whether it is in the state of an acute or chronic condition (Dinarello 2000).

Terpenes are the most widespread group of naturally occurring organic chemicals. The fundamental structure of the terpenes contains an isoprenyl unit (2-methyl-1,3-butadiene) linked in a head-to-tail fashion with the general structural formula $(C_5H_8)n$, where n is the number of linked isoprenyl units. Based on the number of carbon atoms, the terpenes are being classified as: hemiterpenes (5 carbons, 5C), monoterpenes (10C), sesquiterpenes (15C), diterpenes (20C), sesterterpenes (25C), triterpenes (30C), tetraterpenes (40C), and polyterpenes (nC), that comprise a larger number of isoprene units (>45C) (Bilia et al. 2014; Galata et al. 2014; Asbahani et al. 2015; Pereira et al. 2018). Monoterpenes are a major class of secondary metabolites in aromatic and medicinal plants. They are classified into two major groups: oxygenated monoterpenes and monoterpene hydrocarbons. They exert a wide range of biological activities with anti-inflammatory, anti-cancer, anti-diabetic, cardioprotective, hepatoprotective activities, and other human ailments (Brito et al. 2013; Oliveira et al. 2016; Srinivasan and Muruganathan 2016; Saad and Abdelgaleil 2018; Santos et al. 2018). Several studies have proved monoterpene as a potential resource for anti-oxidant and anti-inflammatory (Guimarães et al. 2015; Suntres et al. 2015) actions. The lipophilic characteristic features of these monoterpenes show a promising relationship to the modulation of cytokines with their speedy action and nature of absorption (Spelman et al. 2006). Monoterpenes have also been acknowledged to stimulate an increase in anti-inflammatory cytokines, such as IL-1R, IL-4, IL-6, IL-10, IL-11, and IL-13 (Held et al. 2007; Lima et al. 2013a). There has been an increasing interest in the beneficial effects of monoterpenes in inflammatory diseases. Especially, they decrease the pro-inflammatory cytokines and at the same time increase the anti-inflammatory cytokine levels. In this present review, we discussed the monoterpenes, mainly which are involved in the modulation of the IL-10 as an anti-inflammatory mechanism process.

16.2 CYTOKINES AND INTERLEUKINS

Cytokines are a subtype of growth factors that are produced by hematopoietic and immune cell types and include interferons and interleukins. Our body immune function depends on the biologic activities of numerous small glycoprotein messengers termed cytokines (originally called lymphokines and monokines). Cytokines are peptides that have a fundamental role in communication within the immune system and in allowing the immune system and host tissue cells to exchange information. They also play a vital role in mediating the cross-talk between the nervous and immune systems (Haroon et al. 2012; McInnes 2017). These cytokines are usually less than 80 kDa in size which regulate a wide range of biological functions including innate and acquired immunity, hematopoiesis, inflammation and repair, and proliferation through mostly extracellular signaling, also they are prominently involved in inflammatory responses and defense against viral infections (Chung 2009; Sherbet 2011). Based on the functions of the cytokines, they are classified as two main classes: pro- and anti-inflammatory cytokines, cytokines of neutrophil and eosinophil recruitment and activation, cytokines derived from T-helper (Th) and T-regulatory cells, and cytokines of T-cell recruitment and growth factors. Cytokine receptors are linked to multiple signaling pathways in the cytoplasm and nucleus, leading to transcriptional and post-transcriptional activation of many factors including cytokines (Chung 2009). Cytokines exist in broad families that are structurally related, but exhibit diverse functions [e.g., the TNF/TNF receptor superfamily, IL superfamily (McInnes 2017)]. Currently, cytokines are organized by their numerical order of interleukins (numbered from IL-1 to IL-38). Based on the functional profile of an immune response, cytokine production is broadly regulated by T-helper 1 cells (Th1), which generally mediate a pro-inflammatory cellular immune response and T-helper 2 cells (Th2), which enhance anti-inflammatory and humoral immune reactions (Hou et al. 2017; Quintans et al. 2019).

16.3 INTERLEUKIN-10

IL-10 is a pleiotropic cytokine, initially discovered as an activity produced by murine type 2 helper T cells. It was first called cytokine synthesis inhibitory factor because of its ability to inhibit the production of certain cytokines by Th1, also, it inhibits many effector cells and disease processes, and its levels are inversely correlated with disease incidence and severity (Ekmekcioglu et al. 2015; Kubo et al. 2017). IL-10 is synthesized by a wide range of cell types, including B cells, monocytes, DCs, nK cells, and T cells. Interleukin designates any soluble protein or glycoprotein product of leukocytes that regulates the responses of other leukocytes.

IL-10 is a potent anti-inflammatory cytokine which reduces inflammation in several disease models. It inhibits pro-inflammatory cytokine production and Th1 and Th2 cell activation, which is likely attributable to the effects of IL-10 on antigen presenting cells (APCs) and its direct effects on T cell function (Kubo et al. 2017). Several reports indicate that IL-10 promotes tissue protection. In addition, a single injection of systemically administered IL-10 improved locomotor recovery and reduced TNF-α expression (David et al. 2012). Also, this promotes hematopoietic cell growth, differentiation, and activity, also plays a vital role to the proper functioning of the central nervous system, cardiorespiratory system, and liver, as well as to bone remodeling, lipid metabolism, and embryogenesis and maintenance of pregnancy (Ekmekcioglu et al. 2015).

IL-10 is an important regulatory cytokine with anti-inflammatory properties and has a central role in infections by limiting the immune response to pathogens, and thereby preventing damage to the host, so producing both immunosuppressive and immunostimulatory properties (Saraiva and O'Garra 2010). Moreover, IL-10 can operate on both innate and adaptive immune cells and has a broad spectrum of immunomodulatory activities that suppress proliferation, cytokine secretion, and costimulatory molecule expression of pro-inflammatory immune cells (Shouval et al. 2014). This anti-inflammatory cytokine belongs to type-II cytokine family with nine members: IL-10, IL-19, IL-20, IL-22, IL-24, IL-26, and the more distantly related IL-28A, IL-28B, and IL-29 (Ouyang et al. 2011). Several cells produce IL-10, such as macrophages, dendritic cells, B cells, and various subsets of CD4 and CD8 T cells (Couper et al. 2008). IL-10 carries out its immunosuppressive role largely by inhibiting NF-κB-activated transcription of genes encoding the pro-inflammatory cytokines, particularly, TNF, IL-1, IL-6, IL-8, and IL-12. There is also evidence that IL-10 destabilizes cytokine mRNAs, including the IL-10 mRNA itself. In addition, IL-10 can inhibit the activation of the Ras (protein)/mitogen-activated protein kinase (Ras/MAPK) signaling pathway and the tyrosine phosphorylation of Vav1, again downregulating transcriptional activity. The heterodimeric IL-10 receptor is expressed mainly on hematopoietic cells. The IL-10R1 chain has an extracellular domain that resembles that in the interferon (IFN) receptors and a long cytoplasmic domain that associates primarily with Jak1. The IL-10R2 chain has a shorter cytoplasmic domain that associates primarily with Tyk2 (Mak et al. 2006; Fioranelli and Grazia 2014) (Figure 16.1).

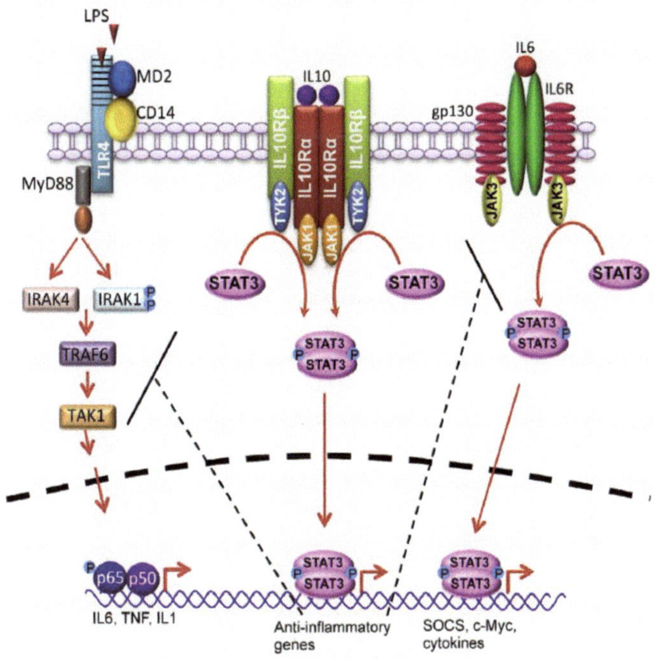

FIGURE 16.1 IL-10 Signaling pathways for the inflammation mechanism. (Adapted from Shouval, D.S. et al., Interleukin 10 receptor signaling: Master regulator of intestinal mucosal homeostasis in mice and humans, in: Chapter Five in *Advances in Immunology*, 177–210, 2014. With permission.)

16.4 IL-10 AND MONOTERPENES

Inflammation is a protective measure taken by the organism to eliminate the injurious stimuli; however, the use of anti-inflammatory substances can be an effective tool in the therapeutic treatment of the diseases. Research focusing on innovative therapeutic agents obtained from natural sources is increasingly growing (Carbone et al. 2018). Among them, essential oils, which mainly contain monoterpenes, are being widely used to prevent and treat human disease (Pereira et al. 2018; Severino et al. 2015). Also, as a potential source for production of cytokine inhibitors (de Lima et al. 2011; de Oliveira et al. 2011; Nonato et al. 2011). In this present review, we discussed about 19 research articles that reported the interaction between the monterpenes and IL-10. All the reports indicated that the monoterpene compounds from the plants and plant products significantly increase the production of IL-10 anti-inflammatory cytokine by various mechanisms (Table 16.1).

The IL-10/IL-10R signaling pathway. IL-10 signals through the heterotetrameric I-10R complex which is comprised of IL-10Ra (IL-10R1 in humans) and IL-10Rb (IL-10R2 in humans). Binding of IL-10 to its receptor leads to JAK1- and TYK2-mediated phosphorylation of STAT3. The phosphorylation of STAT3 forms drives the expression of anti-inflammatory mediators (Shouval et al. 2014).

Geniposide, an iridoid glycoside monoterpenoid compound, contains a cyclopentane ring, which is usually fused to a six-membered oxygen heterocycle (Quintans et al. 2019). It is abundantly found in the plants of *Gardenia jasminoides* Ellis fruit, which is commonly used as a traditional medicine in many Asian countries for its anti-phlogistic and anti-pyretic effects (Lee et al. 2005). Iridoids are found in a wide variety of plants, especially, Apocynaceae, Lamiaceae, Loganiaceae, Rubiaceae, Scrophulariaceae, and Verbenaceae families (Viljoen et al. 2012). In this present review, we found five geniposide/alcoholic compounds which reported various activities including adjuvant arthritis (Dai et al. 2014; Chen et al. 2015) and inflammation (Wang et al. 2012; Xiaofeng et al. 2012; Deng et al. 2013). Freund's Complete Adjuvant was used as a chemical agent for the induction of the rheumatoid arthritis, which is a chronic, inflammatory, and systemic auto-immune disease, which is characterized by synovial inflammation and joint destruction. There were 30, 60, and 120 mg/kg of geniposide that played a vital role in the increasing of the IL-10 and IL-4 and reduced the rheumatoid arthritis in the treated rats. The cytokine IL-10 has complex, broadly anti-inflammatory effects on a wide variety of cells. IL-10 expression is strongly regulated by p38 MAPK at both transcriptional and post-transcriptional levels which mainly regulates the inflammatory response in cells (Chen et al. 2015). Geniposide also inhibits airway inflammation and hyperresponsiveness. The production of anti-inflammatory cytokines can potentially counteract pro-inflammatory cytokines, affecting and modifying the intensity of the inflammatory process. Pre-treatment with geniposide (40 and 80 mg/kg) could significantly inhibit the production of TNF-α and IL-6, but also promote the release of IL-10 induced by lipopolysaccharide (LPS) and abridge the LPS-induced acute lung injury in mice (Xiaofeng et al. 2012).

Paeoniflorin (PF), a monoterpene glucoside, is one of the principal bioactive components of the Paeonia root. PF has been used for 2000 years as an immune modulator, and an anti-allergic agent (He and Dai 2011), and it displays effective anti-inflammatory properties in several animal models (Wu et al. 2010; Shi et al. 2014). Additionally, PF can also normalize the Bacillus Calmétte-Guérin plus LPS-induced liver injury, complete Freund's adjuvant (CFA)-induced arthritis, dextran sulfate sodium (DSS)-induced colitis, acute myocardial infarction, concanavalin A-induced hepatitis, ovalbumin (OVA)-induced asthma, and imiquimod-induced psoriasis by significant downregulation of pro-inflammatory cytokines, such as TNFα, IL-1β, IL-6, IL-12, IL-5, IL-13, IL-17, and INF-γ as reported in recent studies. The pre-treatment of mice with PF produced a significant inhibition of TNF-α and IL-1β levels and significantly upregulated IL-10 and IL-4 levels in the cortex and hippocampus. Further, the beneficial effects of the PF treatment were associated with a modulation of inflammatory responses in APP/PS1 mice (Quintans et al. 2019). In this present review, we found three research articles which studied the PF in different animal model experiments including hepatic fibrosis (Li et al. 2009), allergen-induced murine dendritic (Shi et al. 2014) cells, and rheumatoid

TABLE 16.1
Characteristics of Studies Selected in the Systematic Review

Authors, Year & Country	Substance/Chemical Group	Animals (Strain/Sex)	Dose (Route)	Model	Cytokines Involved
Chen et al. (2015), China	Geniposide/alcoholic	Rats (SD/M)	30, 60, and 120 mg/kg (i.g.)	Adjuvant arthritis	Increased IL-10
Dai et al. (2014), China	Geniposide/alcoholic	Rats (SD/M)	30, 60, 120 mg/kg (i.g.)	Adjuvant arthritis	Increase IL-4 and TGF-β1
Deng et al. (2013), China	Geniposide/alcoholic	Mice (B/F)	20, 40, 80 mg/kg	Allergic airway inflammation, (Asthma)	Reduced IL-4, IL-13
Wang et al. (2012), China	Geniposide/alcoholic	Rats (SD/M)	15, 30, or 60 mg/kg (i.v.)	Neuronal cell death induced by oxygen and glucose deprivation	Reduced IL-10
Xiaofeng et al. (2012), China	Geniposide/alcoholic	Mice (B/M)	20, 40, or 80 mg/kg) (i.p.)	LPS-induced acute injury	Increase IL-10
Lima et al. (2013a), Brazil	Carvacrol/alcoholic	Mice (S/M) or Mice (C/M)	50 and 100 mg/kg (i.p.)	Paw inflammation model	Increased IL-10
Choi et al. (2010), Korea	(S)-cis-verbenol/alcoholic	Rats (SD/M)	100 mg/kg (i.p.)	Focal cerebral ischemia	TGF-β was not changed
Trinh et al. (2011), Korea	Eucalyptol (1,8-cineole)/α-terpineol/ alcoholics	Mice (ICR/M and F)	10% v/v (i.v.)	Bacterial vaginosis and vulvovaginal candidiasis	Increased IL-10
Lima et al. (2013b), Brazil	1,8-cineole (Eucalyptol)/ alcoholic	Mice (S/M)	100, 200, and 400 mg/ kg (p.o.)	Acute pancreatitis	Increased IL-10
Dong and Chen (2013), China	Catalpol/alcoholic	Rats (SD/M)	30, 60, 120 mg/kg (p.o)	Diabetic nephropathy	Decrease TGF-β1
Fu et al. (2014), China	Catalpol/alcoholic	Mice (B/M)	2.5, 5, and 10 mg/kg (i.p.)	LPS-induced acute injury	Inhibited IL-4 and Increased IL-10
Bas et al. (2006), Spain	Scrovalentinoside	Mice (S/F) and Rats (W/F)	100, 50, and 25 μM, 5 and 10 mg/kg	Delayed-type hypersensitivity	Inhibited the production IL-4, IFN-γ, LTB4
Saravanan et al. (2014), India	Swertiamarin	Rats (SD/F)	2, 5, or 10 mg/kg	Delayed-type hypersensitivity	Increased IL-10 and IL-4
Imamura et al., (2014) Japan	Perillyl alcohol	Mice (B/M)	75 mg/kg (i.p.)	Allergic airway inflammation	Reduced IL-4, IL-10, IL-13, IL-17
Zhong et al. (2013), China	p-cymene	Mice (C/F)	53.58, 107.16, 214.32 mg/kg (i.p.)	LPS-induced acute injury	Increased IL-10
Li et al. (2009), China	Paeoniflorin	Mice (B/F)	30 mg/kg (p.o.)	Hepatic fibrosis in mice	Reduced IL-13
Shi et al. (2014), China	Paeoniflorin	Mice (B/F)	25, 50, 100, and 200 μg/mL	Allergic contact dermatitis	Increased IL-10 and TGF-β
Wu et al. (2007), China	Paeoniflorin	Rats (SD/M)	25, 50, and 100 mg/kg (i.g.)	Adjuvant arthritis	Increased IL-4 and TGF-β1
Bas et al. (2007b), Spain	Scropolioside A	Mice (S/F)	6.25, 12.5, 25, 50, 100 μM	Delayed-type hypersensitivity	Decreased IL-1β, IL-2, IL-4,

Note: S-Swiss; B-Balb/c; C-C57BL/6J; CN-C57BL/6 N; D-DBA/1J; FI-Fischer; 344W-Wistar; F-Female; M-Male; SD-Sprague–Dawley; SW-Swiss Webster i.v. - intravenous; i.p. - intraperitoneal; i.g.-intragastric.

arthritis (Wu et al. 2007). All of these studies reported that the compound PF at various concentrations increases the anti-inflammatory cytokines, such as IL-4 and IL-10, which reduces the inflammation in the cells.

1,8-cineole (eucalyptol) is a major monoterpenoid present in many plant volatile/essential oils, principally, the *Rosmarinus* and *Eucalyptus* species which are mainly used by the pharmaceutical industry in drug formulations, as a percutaneous penetration enhancer, and for its decongestant and antitussive effects, and in aromatherapy as a skin stimulant (Elaissi et al. 2012; Lima et al. 2013b). Santos et al. (2004) identified 1,8-cineole as a substance with analgesic, anti-inflammatory, and hepatoprotective properties, and to attenuate the colonic damage in rats with acute trinitrobenzene sulfonic acid-induced colitis, an animal model of the inflammatory bowel disease of humans. We found two previous studies of this 1,8-cineole compound in cerulein-induced acute pancreatitis (Lima et al. 2013b) and the amelioration of bacterial vaginosis and vulvovaginal candidiasis in mice by inhibiting the bacterial growth and NF-κB activation (Trinh et al. 2011). Lima et al. (2013b) reported that 1,8-cineole (100, 200, and 400 mg/kg, per oral (p.o.) significantly ($p < 0.05$) lowered the cerulein-induced elevation of the serum amylase and lipase at all doses, a significant decrease in pancreatic edema was observed only at 200 and 400 mg/kg. Meanwhile, cerulein significantly ($p > 0.05$) elevated the serum levels of TNF-α, IL-1β, and IL-6, whereas the level of serum IL-10 was decreased. Treatment with 1,8-cineole (100, 200, and 400 mg/kg, p.o.) and thalidomide (200 mg/kg, p.o.) effectively decreased the levels of TNF-α, IL-1β, and IL-6, but the levels of IL-10 were increased. Also, the authors concluded that these changes in cytokine levels were significantly mitigated by 1,8-cineole pre-treatment, suggesting that it may have an important role in the inhibition of the pro-inflammatory response in Acute Pancreatitis (AP). By creating a balance between pro-inflammatory and anti-inflammatory cytokines, 1,8-cineole might exercise a therapeutic benefit for the control of acute pancreatitis. Trinh et al. (2011) also studied this compound and recommended that 1,8-cineole inhibited the expressions of pro-inflammatory cytokines (IL-1β, IL-6, TNF-α), cyclooxygenase (COX)-2, Inducible Nitric Oxide Synthase (iNOS), and the activation of NF-κB and increased the expression of the anti-inflammatory cytokine IL-10.

Catalpol, an iridoid glycoside extracted from the roots of *Rehmannia glutinosa*, is a nutraceutical substance already used in food supplementation for neurodegenerative disorders and plays a vital role in the regulation of anti-inflammatory and pro-inflammatory cytokine levels and has been shown to have anti-oxidant, anti-apoptosis, and anti-inflammatory activities (Zhang et al. 2007, 2009; Fu et al. 2014). Catalpol was found to inhibit LPS and IFN-r-induced inflammatory responses in astrocytes (Bi et al. 2013). Catalpol can also downregulate the messenger RNA (mRNA) expressions of TGF-1 and CTGF in the renal cortex and lower blood glucose concentrations in diabetic rats with nephropathy. It stimulated the synthesis of Extracellular Matrix (ECM) in the glomeruli through the TGF-1/Smad protein cell signaling pathway (Lan 2011). TGF-1 mRNA expression and protein levels are increased in the glomeruli in various experimental animal diabetic models (Nakamura et al. 1993). Only two reports are available on the catalpol compound. The evidence that pre-treatment of catalpol has a protective effect on LPS-induced acute lung injury in mice, and that the potential mechanism of this action may be related to the inhibition of Toll-like Receptor 4 (TLR4) expression, which subsequently leads to the suppression of NF-κB and MAPK activation and pro-inflammatory and anti-inflammatory cytokines production (Fu et al. 2014). Another study investigated the effect of catalpol on diabetic nephropathy in rats (Dong and Chen 2013). The obtained results from this study reveal that catalpol 30, 60, 120 mg/kg could significantly reduce the Kidney Weight Index (KWI), improve the kidney function and pathological change, decrease the tissue level of Angiotensin II (Ang II), TGF-1, Connective Tissue Growth Factor (CTGF), fibronectin (FN), and collagen type IV (Col IV), catalpol ameliorated the early diabetic nephropathy. It improved the renal function and decreased the overproduction of the FN and Col IV expressions in the renal cortex of the diabetic rats. The mechanism may be related to depressing TGF-1 and CTGF mRNA expression and decreasing the excretion of Ang II, TGF-1, and CTGF.

Some other monoterpenes were reported earlier for the management of the IL-10, including carvacrol, (S)-cis-verbenol, scrovalentinoside, swertiamarin, perillyl alcohol, *p*-cymene, and scropolioside A. Carvacrol (5-isopropyl-2-methylphenol) is a phenolic monoterpene found in the essential oil of the family Lamiaceae, which includes the genera Origanum and Thymus (Baser 2008). Previous studies have demonstrated that carvacrol has anti-inflammatory properties (Wagner et al. 1986; Botelho et al. 2008; Landa et al. 2009; Hotta et al. 2010; Guimarães et al. 2012). Lima et al. (2013a) reported the anti-inflammatory effects of carvacrol: evidence for a key role of interleukin-10. In this study, the authors administrated 50 and 100 mg/kg of carvacrol into the mice, which attenuated the paw edema and reduced the IL-1 band prostaglandin E 2, but not TNF-a, local levels. Similarly, carvacrol (100 mg/kg) reduced the COX-2 and IL-1b mRNA expression. The levels of IL-10, an anti-inflammatory cytokine, and the IL-10 mRNA expression in the inflamed paw were enhanced by carvacrol. In fact, IL-10 acts as a potent anti-inflammatory cytokine by suppressing the activation and function of immune cells. Accordingly, a crucial effect of IL-10 is its ability to selectively block the expression of pro-inflammatory cytokines (Moore et al. 2001), including IL-1β (de Waal Malefyt et al. 1991; Karam et al. 2007). Thus, the authors investigated the possibility that IL-10 production contributes to the anti-inflammatory effect of carvacrol.

(S)-cis-verbenol is a natural metabolite which bark beetles transformed from (-)-alphapinene of a host pine tree. Essential oils containing (S)-cis-verbenol have been reported to have biological activities such as anti-microbial and anti-fungicidal, and also, they reduce ischemia/hypoxia-induced cell death in neuroblastoma SH-SY5Y cells (Tabanca et al. 2001; Alma et al. 2004; Chang et al. 2007). Our search results explored one study about the compound (S)-cis-verbenol which reduced cerebral ischemic injury that was caused by 1.5-hour middle cerebral artery occlusion followed by 24-hour reperfusion (Choi et al. 2010). Furthermore, (S)-cis-verbenol significantly prevented neuronal cell death caused by oxygen-glucose deprivation (1 hour) and subsequent re-oxygenation (5 hour). Furthermore, (S)-cis-verbenol reduced the expression levels of pro-inflammatory cytokines in ischemic brain and immunostimulated glial cells. Especially, (S)-cis-verbenol does not change the TGF-β level and increase the anti-inflammatory cytokines in ischemic brain and immunostimulated glial cells.

p-cymene (4-isopropyltoluene or 1-isopropyl-4-methylbenzene) is a biological compound naturally present in the essential oils of *Chenopodium ambrosioides* L., oregano, thyme, and other herbs (Alma et al. 2003; Cavalli et al. 2004; Zhong et al. 2013). Furthermore, *p*-cymene pre-treatment had marked effects on the LPS-induced cytokine secretion in vitro and in vivo. In addition, *p*-cymene inhibited the gene expression of TNF-α, IL-1β, and IL-6 by blocking MAPK and IκB pathways. Also, *p*-cymene markedly suppressed the production of TNF-α and IL-1β and increased IL-10 secretion. *Scrophularia auriculata* L. ssp. *pseudoauriculata* (Senn) Bolos et Vigo (S. valentina Rouy) is a Mediterranean species used in folk medicine in the treatment of inflammatory skin diseases. Giner et al. (1998) found this species to be rich in saponins and iridoids, reporting the isolation and identification of two saponins, verbascosaponin and verbascosaponin A, and two iridoids, scropolioside A and scrovalentinoside. Scrovalentinoside inhibited the pro-inflammatory mediators production and improved the anti-inflammatory effect by increasing the result of the production of the anti-inflammatory cytokine IL-10 (Bas et al. 2007a). In another study, it was revealed that Swertiamarin is a secoiridoid glycoside found in *Enicostema axillare* (Lam) A. Raynal, a medicinal plant used as a depurative in the Indian system of traditional medicine. Many previous studies evaluated swertiamarin for its anti-diabetic, nephroprotective, anti-dyslipidemic, anti-atherogenic, anti-oxidant, anti-microbial, neuritogenic, hepatoprotective, anti-nociceptive, anti-edematogenic, anti-cholinergic, and COX-2- and TNF-inhibiting properties (Yamahara et al. 1991; Chiba et al. 2011; Hudecová et al. 2012; Vaidya et al. 2012a, 2012b). The in vivo immunomodulatory activity of swertiamarin (2, 5, and 10 mg/kg b.w.) was evaluated in a model of sheep red blood cells by assessing its effect on organ weight, hemagglutinating antibody titer, plaque-forming cells, quantitative hemolysis of sheep red blood cells, and delayed type hypersensitivity. The in vitro immunomodulatory potential was studied on isolated splenocytes, neutrophils, and peritoneal macrophages. The in silico immunomodulatory effects were evaluated by the docking of swertiamarin on pro-inflammatory cytokines to confirm its potential. In in vitro studies, treatment with swertiamarin modulated the mRNA and protein expression of IFN-γ, IL-10, and IL-4 significantly ($P \leq 0.05$) and also favored a Th2-mediated response on concanavalin A-induced splenocytes (Saravanan et al. 2014). The compound perillyl alcohol suppresses antigen-induced immune responses in the lung. Perillyl alcohol (PA) is an isoprenoid which inhibits farnesyl transferase and geranylgeranyl transferase, key enzymes that induce conformational and functional changes in small G proteins to conduct signal production for cell proliferation. Thus, it has been tried for the treatment of cancers (Imamura et al. 2014). PA exists naturally in plants in a pure or mixed form. Further, its derivative, limonene, is included in natural foods such as lemons and oranges and has a similar biological effect to PA (Schulz et al. 1994; Del Toro-Arreola et al. 2005). PA suppressed antigen-induced systemic immune responses and attenuated allergic airway inflammation in a mouse model. Considering that it exists in nature, PA or its derivatives could be a novel preventive or therapeutic option for asthma with minimal side effects.

Our review described 11 monoterpenes that regulate cytokine release and the levels of others inflammatory mediators. Several different inflammatory markers were evaluated as a target of monoterpenes across the studies included within this review. Predominantly, all the monoterpenes have the anti-inflammatory property by increasing the IL-10 anti-inflammatory cytokines. Many of the monoterpenes have a potential anti-inflammatory action on cytokine production by blocking the NF-κB and MAPK signaling pathways. In these signaling pathways, the monoterpenes neither regulate the pro-inflammatory cytokines nor reduce the levels of the cytokines, which increase the development of the anti-inflammatory cytokines to ameliorate the anti-inflammation process.

16.5 MOLECULAR DOCKING STUDIES ON IL-10 AND MONOTERPENES

Molecular docking is one of the major bioinformatic tools used in the drug-discovery process. This is one of the virtual screening methods that mainly focuses on the structures of targets and ligands. It predicts the binding geometries as well as the binding energy of the ligand-target complex. The mechanism of action and molecular target of the IL-10 were studied in this present review. Molecular docking analysis can be conducted to study the interaction of monoterpene compounds with molecular targets of anti-inflammatory activity, especially IL-10. Further, this structure-activity relationship can be used to develop new derivative natural compounds with higher anti-inflammatory activity.

The structure of the receptors was downloaded from Protein Data Bank (http://www.rcsb.org/pdb/home/home.do) (IDPDB 1J7V-HUMAN IL-10/IL-10R1 COMPLEX). After analyzing the structure, we classify and separate the proteins. The 3D conformers of ligands, in the Spatial Data File (SDF) format, were obtained from PubChem (2019). The ligands were submitted to molecular docking using the Molegro Virtual Docker v. 6.0.1 (Thomsen and Christensen 2006). All the water compounds were deleted from the receptor, and the method was prepared using the default parameter settings on the same software: GRID (grid file or bucket grid is a point access method which splits a space) of 15 Å of radius. A MolDock score [GRID] algorithm was used as the score function, and the search algorithm was MolDock (Thomsen and Christensen 2006).

16.5.1 Molecular Docking for IL-10R1 and IL-10 with Cineole and Carvacrol

IL-10 is the foremost member of the type-II cytokine family, comprising IL-19, IL-20, IL-22, IL-24, IL-26, IL-28, and IL-29 (Walter 2014). IL-10 was first described by Fiorentino et al. as an inhibitor of cytokine synthesis and initially termed "cytokine synthesis inhibitory factor," as it was released by Th2 cells and inhibited IFNg production by Th1 cells (Fickenscher et al. 2002; Trivella et al. 2010). Human and mouse IL-10 have roughly 73% sequence homology and are secreted as 178-amino acid proteins (Windsor et al. 1993). While both human IL-10 and mouse IL-10 are comprised of non-covalently linked homodimers, the receptor for IL-10 is a heterotetramer complex comprising two IL-10R1 molecules (encoded by the IL-10ra gene) and two IL-10R2 molecules (encoded by the IL-10rb gene) (Kotenko et al. 1997; Moore et al. 2001).

In our study, we have observed that the ligands investigated for stable complexes with IL-10R1 and its receptor exhibit negative enthalpic energy. Both compounds develop interactions with residues of macromolecules in the steric nature (red lines between apolar atoms) and/or hydrogen bonds (blue). We highlight the interactions of the ligands with the IL-10R1 receptor, which presented hydrologic bonds in all complexes. The energies indicate a tendency in the coupling of molecules, especially in cineole (–3.42 kcal/mol) and carvacrol (–46.89 kcal/mol). When compared to cineole, carvacrol interacts with the IL-10R1 receptor with more negative enthalpic energy, which reveals that the compound carvacrol is able to stimulate IL-10 expression that contributed to control the inflammation process. Cineole has steric interactions (between non-polar atoms) with LEU 98, MET 88, LEU 65, and LEU 69. Meanwhile, carvacrol has steric interactions (between non-polar atoms) with LEU 98 and TYR 72 (Figure 16.2).

Further, our results show the binding energies obtained in cineole (−49.86 kcal/mol) and carvacrol (−65.31 kcal/mol) in the formation of the ligand–macromolecule complexes with IL-10 receptor protein, and Figure 16.3 reports the ligand maps calculated with the best conformations between the two molecules. When Carvacrol (CARV) is complexed with the IL-10, we find a negative energy value of −65.31 kcal/mol, indicating that binding with the target is more possible and favorable, also indicating that the CARV is a better ligand then the cineole compound (Figure 16.3).

Our previous reports also revealed that the attenuation of cancer pain by carvacrol aided through a docking study which exposed favorable interaction overlay of the CARV with IL-10 and $GABA_A$ (Guimarães et al. 2014). This monoterpene is also able to stimulate IL-10 expression (Lima et al. 2013a), an anti-inflammatory cytokine that contributed to control the of neuropathic pain (Milligan et al. 2012) management. From all the above results, we concluded that it is possible to propose that the anti-inflammatory properties of carvacrol and cineole are dependent on their capability to diminish the production of pro-inflammatory cytokines including IL-1β and prostanoids, also possibly increasing the production of IL-10 release.

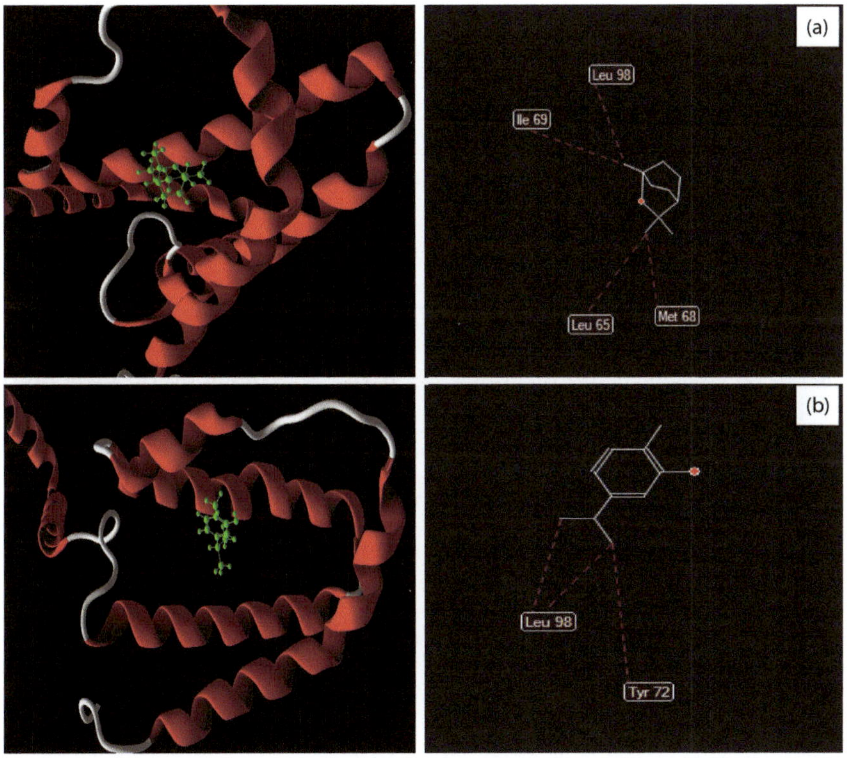

FIGURE 16.2 Molecular docking for IL-10R1 with (a) cineole and (b) carvacrol.

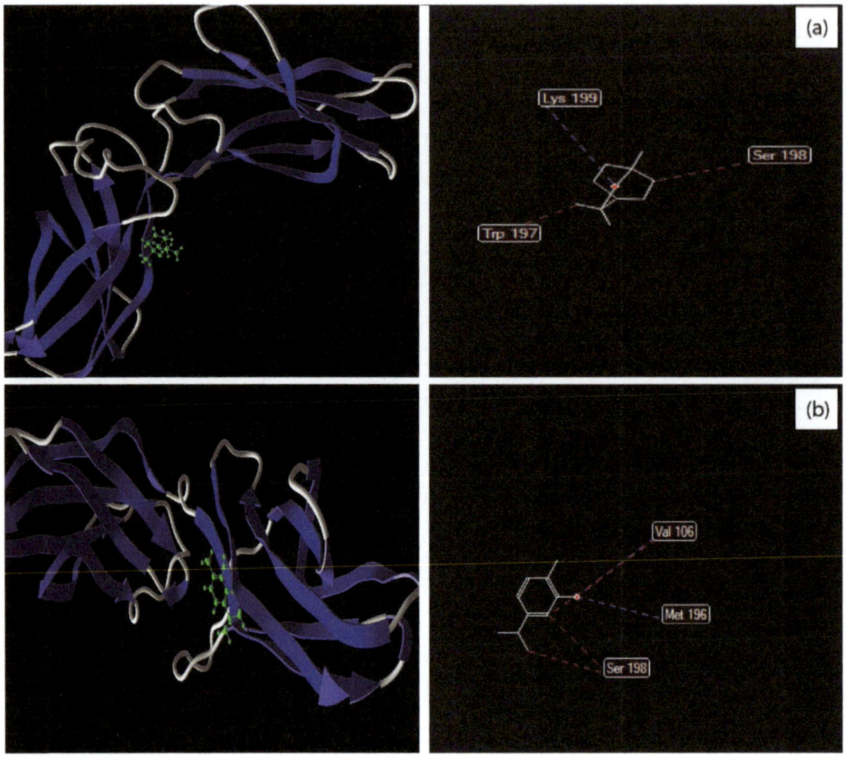

FIGURE 16.3 Molecular docking for IL-10 with (a) cineole and (b) carvacrol.

16.6 CONCLUSION

All of these 11 monoterpenes have many pharmacological properties, especially monoterpenes prove their ability to attenuate the inflammatory response by modulating the production of cytokines. In this review, we summarized the present knowledge on monoterpenes that possess anti-inflammatory effects and can modulate the release of anti-inflammatory cytokines, specifically IL-10. Particularly, the monocyclic monoterpenes, such as carvacrol and *p*-cymene increased the levels of the IL-10, which play a prominent function in the anti-inflammatory effect. Also, all of these monoterpenes control the NF-κB signaling pathway, which is one of the most important pathways for the anti-inflammatory action. This present compelling evidence proves the anti-inflammatory effect of monoterpenes, which also helps the future prospective research to fully explore their potential as compounds supporting anti-inflammatory and other human ailments.

ACKNOWLEDGMENTS

This work was supported by grants from Conselho Nacional de Desenvolvimento Científico e Tecnológico (CNPq) and Coordenação de Aperfeiçoamento de Pessoal de Nível Superior (CAPES) from Brazil.

ABBREVIATIONS

COX	Cyclooxygenase
DNCB	1-Chloro-2, 4-dinitrobenze
FCA	Freund's complete adjuvant
GE	Geniposide
IL	Interleukin
IFN	Interferons
LPS	Lipopolysaccharide
MHC	Major histocompatibility complex
NKT cell	Natural killer T cell
NF-κ	B factor nuclear kappa B
PA	Perillyl alcohol
PC	*p*-Cimene
PF	Paeoniflorin
PGE	Prostaglandin E
RA	Rheumatoid arthritis
PPAR	Peroxisome proliferator activated receptor
TGF-β	Transforming growth factor beta
TNF	Tumor necrosis factor
TPN	α-terpineol
VEGF	Vascular endothelial growth factor

REFERENCES

Alma, M. H., Mavi, A., Yildirim, A., Digrak, M., and Hirata, T., 2003. Screening chemical composition and in vitro antioxidant and antimicrobial activities of the essential oils from *Origanum syriacum* L. growing in Turkey. *Biological & Pharmaceutical Bulletin*, 26 (12), 1725–1729.

Alma, M. H., Nitz, S., Kollmannsberger, H., Digrak, M., Efe, F. T., and Yilmaz, N., 2004. Chemical composition and antimicrobial activity of the essential oils from the gum of Turkish pistachio (*Pistacia vera* L.). *Journal of Agricultural and Food Chemistry*, 52 (12), 3911–3914.

Bas, E., Recio, M. C., Abdallah, M., Máñez, S., Giner, R. M., Cerdá-Nicolás, M., and Ríos, J. L., 2007a. Inhibition of the pro-inflammatory mediators' production and anti-inflammatory effect of the iridoid scrovalentinoside. *Journal of Ethnopharmacology*, 110, 419–427.

Bas, E., Recio, M. C., Giner, R. M., Manez, S., Cerda-Nicolas, M., Rıos, J. L. 2006. Anti-inflammatory activity of 5-O-demethylnobiletin, a polymethoxyflavone isolated from *Sideritis tragoriganum*. *Planta Medica*, 72, 136–142.

Bas, E., Recio, M. C., Giner, R. M., et al. 2007. Demethylnobiletin inhibits delayed-type hypersensitivity reactions, human lymphocyte proliferation and cytokine production. *British Journal of Pharmacology.* 152(8): 1272–1282. doi:10.1038/sj.bjp.0707500.

Baser, K. H. C., 2008. Biological and pharmacological activities of carvacrol and carvacrol bearing essential oils. *Current Pharmaceutical Design*, 14 (29), 3106–3119.

Bi, J., Jiang, B., Zorn, A., Zhao, R.-G., Liu, P., and An, L.-J., 2013. Catalpol inhibits LPS plus IFN-γ-induced inflammatory response in astrocytes primary cultures. *Toxicology In Vitro: An International Journal Published in Association with BIBRA*, 27 (2), 543–550.

Bilia, A. R., Guccione, C., Isacchi, B., Righeschi, C., Firenzuoli, F., and Bergonzi, M. C., 2014. Essential oils loaded in nanosystems: A developing strategy for a successful therapeutic approach. *Evidence-Based Complementary and Alternative Medicine: eCAM*, 2014, 651593.

Botelho, M. A., Rao, V. S., Montenegro, D., Bandeira, M. A. M., Fonseca, S. G. C., Nogueira, N. A. P., Ribeiro, R. A., and Brito, G. A. C., 2008. Effects of a herbal gel containing carvacrol and chalcones on alveolar bone resorption in rats on experimental periodontitis. *Phytotherapy Research: PTR*, 22 (4), 442–449.

Brito, R. G., Santos, P. L., Prado, D. S., Santana, M. T., Araújo, A. A. S., Bonjardim, L. R., Santos, M. R. V., de Lucca Júnior, W., Oliveira, A. P., and Quintans-Júnior, L. J., 2013. Citronellol reduces orofacial nociceptive behaviour in mice—Evidence of involvement of retrosplenial cortex and periaqueductal grey areas. *Basic & Clinical Pharmacology & Toxicology*, 112 (4), 215–221.

Carbone, C., Martins-Gomes, C., Caddeo, C., Silva, A. M., Musumeci, T., Pignatello, R., Puglisi, G., and Souto, E. B., 2018. Mediterranean essential oils as precious matrix components and active ingredients of lipid nanoparticles. *International Journal of Pharmaceutics*, 548 (1), 217–226.

Cavalli, J.-F., Tomi, F., Bernardini, A.-F., and Casanova, J., 2004. Combined analysis of the essential oil of *Chenopodium ambrosioides* by GC, GC-MS and 13C-NMR spectroscopy: Quantitative determination of ascaridole, a heat-sensitive compound. *Phytochemical Analysis: PCA*, 15 (5), 275–279.

Chang, H.-J., Kim, H. J., and Chun, H. S., 2007. Quantitative structure-activity relationship (QSAR) for neuroprotective activity of terpenoids. *Life Sciences*, 80 (9), 835–841.

Chen, J. Y., Wu, H., Li, H., Hu, S. L., Dai, M. M., and Chen, J., 2015. Anti-inflammatory effects and pharmacokinetics study of geniposide on rats with adjuvant arthritis. *International Immunopharmacology*, 24, 102–109.

Chiba, K., Yamazaki, M., Kikuchi, M., Kakuda, R., and Kikuchi, M., 2011. New physiological function of secoiridoids: Neuritogenic activity in PC12h cells. *Journal of Natural Medicines*, 65 (1), 186–190.

Choi, I.-Y., Lim, J. H., Hwang, S., Lee, J.-C., Cho, G.-S., and Kim, W.-K., 2010. Anti-ischemic and anti-inflammatory activity of (S)-cis-verbenol. *Free Radical Research*, 44 (5), 541–551.

Chung, K. F., 2009. Cytokines. *Asthma and COPD*, 327–341.

Couper, K. N., Blount, D. G., and Riley, E. M., 2008. IL-10: The master regulator of immunity to infection. *Journal of Immunology (Baltimore, Md.: 1950)*, 180 (9), 5771–5777.

Dai, M. M., Wu, H., Li, H., Chen, J., Chen, J. Y., Hu, S. L., and Shen, C., 2014. Effects and mechanisms of Geniposide on rats with adjuvant arthritis. *International Immunopharmacology*, 20, 46–53.

David, S., López-Vales, R., and Wee Yong, V., 2012. Harmful and beneficial effects of inflammation after spinal cord injury: Potential therapeutic implications. *Handbook of Clinical Neurology*, 109, 485–502.

de Lima, F. O., Nonato, F. R., Couto, R. D., Barbosa Filho, J. M., Nunes, X. P., Ribeiro dos Santos, R., Soares, M. B. P., and Villarreal, C. F., 2011. Mechanisms involved in the antinociceptive effects of 7-hydroxycoumarin. *Journal of Natural Products*, 74 (4), 596–602.

de Oliveira, C. M., Nonato, F. R., de Lima, F. O., Couto, R. D., David, J. P., David, J. M., Soares, M. B. P., and Villarreal, C. F., 2011. Antinociceptive properties of bergenin. *Journal of Natural Products*, 74 (10), 2062–2068.

Del Toro-Arreola, S., Flores-Torales, E., Torres-Lozano, C., Del Toro-Arreola, A., Tostado-Pelayo, K., Guadalupe Ramirez-Dueñas, M., and Daneri-Navarro, A., 2005. Effect of D-limonene on immune response in BALB/c mice with lymphoma. *International Immunopharmacology*, 5 (5), 829–838.

de Waal Malefyt, R., Abrams, J., Bennett, B., Figdor, C. G., and de Vries, J. E., 1991. Interleukin 10(IL-10) inhibits cytokine synthesis by human monocytes: An autoregulatory role of IL-10 produced by monocytes. *The Journal of Experimental Medicine*, 174, 1209–1220.

Deng, Y., Guan, M., Xie, X., Yang, X., Xiang, H., Li, H., Zou, L., Wei, J., Wang, D., and Deng, X., 2013. Geniposide inhibits airway inflammation and hyperresponsiveness in a mouse model of asthma. *International Immunopharmacology*, 17 (3), 561–567.

Dinarello, C. A., 2000. Proinflammatory cytokines. *Chest*, 118 (2), 503–508.

Dong, Z. and Chen, C. X., 2013. Effect of catalpol on diabetic nephropathy in rats. *Phytomedicine*, 20, 1023–1029.

Ekmekcioglu, S., Kurzrock, R., and Grimm, E. A., 2015. Hematopoietic growth factors and cytokines. *The Molecular Basis of Cancer*, 789–808.e4.

El Asbahani, A., Miladi, K., Badri, W., Sala, M., Addi, E. H. A., Casabianca, H., Mousadik, A. El, Hartmann, D., Jilale, A., Renaud, F. N. R., and Elaissari, A., 2015. Essential oils: From extraction to encapsulation. *International Journal of Pharmaceutics*, 483 (1–2), 220–243.

Elaissi, A., Rouis, Z., Mabrouk, S., Salah, K. B. H., Aouni, M., Khouja, M. L., Farhat, F., Chemli, R., and Harzallah-Skhiri, F., 2012. Correlation between chemical composition and antibacterial activity of essential oils from fifteen *Eucalyptus* species growing in the Korbous and Jbel Abderrahman arboreta (North East Tunisia). *Molecules (Basel, Switzerland)*, 17 (3), 3044–3057.

Fickenscher, H., Hör, S., Küpers, H., Knappe, A., Wittmann, S., and Sticht, H., 2002. The interleukin-10 family of cytokines. *Trends in Immunology*, 23 (2), 89–96.

Fioranelli, M. and Grazia, R. M., 2014. Twenty-five years of studies and trials for the therapeutic application of IL-10 immunomodulating properties. From high doses administration to low dose medicine new paradigm. *Research Article Journal of Integrative Cardiology Journal of Integrative Cardiology*, 1,2–6.

Fu, K., Piao, T., Wang, M., Zhang, J., Jiang, J., Wang, X., and Liu, H., 2014. Protective effect of catalpol on lipopolysaccharide-induced acute lung injury in mice. *International Immunopharmacology*, 23 (2), 400–406.

Galata, M., Sarker, L. S., and Mahmoud, S. S., 2014. Transcriptome profiling, and cloning and characterization of the main monoterpene synthases of *Coriandrum sativum* L. *Phytochemistry*, 102, 64–73.

Giner, R. M., Villalba, M. L., Recio, M. C., Máñez, S., Gray, A. I., and Ríos, J. L., 1998. A new iridoid from *Scrophularia auriculata* ssp. pseudoauriculata. *Journal of Natural Products*, 61 (9), 1162–1163.

Guimarães, A. G., Oliveira, M. A., Alves, R. dos S., Menezes, P. dos P., Serafini, M. R., de Souza Araújo, A. A., Bezerra, D. P., and Quintans Júnior, L. J., 2015. Encapsulation of carvacrol, a monoterpene present in the essential oil of oregano, with β-cyclodextrin, improves the pharmacological response on cancer pain experimental protocols. *Chemico-Biological Interactions*, 227, 69–76.

Guimarães, A. G., Scotti, L., Scotti, M. T., Mendonça Júnior, F. J. B., Melo, N. S. R., Alves, R. S., De Lucca Júnior, W., Bezerra, D. P., Gelain, D. P., and Quintans Júnior, L. J., 2014. Evidence for the involvement of descending pain-inhibitory mechanisms in the attenuation of cancer pain by carvacrol aided through a docking study. *Life Sciences*, 116 (1), 8–15.

Guimarães, A. G., Xavier, M. A., de Santana, M. T., Camargo, E. A., Santos, C. A., Brito, F. A., Barreto, E. O., Cavalcanti, S. C. H., Antoniolli, A. R., Oliveira, R. C. M., and Quintans-Júnior, L. J., 2012. Carvacrol attenuates mechanical hypernociception and inflammatory response. *Naunyn-Schmiedeberg's Archives of Pharmacology*, 385 (3), 253–263.

Haroon, E., Raison, C. L., and Miller, A. H., 2012. Psychoneuroimmunology meets neuropsychopharmacology: Translational implications of the impact of inflammation on behavior. *Neuropsychopharmacology*, 37 (1), 137–162.

He, D.-Y. and Dai, S.-M., 2011. Anti-inflammatory and immunomodulatory effects of *Paeonia lactiflora* Pall., a traditional chinese herbal medicine. *Frontiers in Pharmacology*, 2, 10.

Held, S., Schieberle, P., and Somoza, V., 2007. Characterization of α-terpineol as an anti-inflammatory component of orange juice by in vitro studies using oral buccal cells. *Journal of Agricultural and Food Chemistry*, 55 (20), 8040–8046.

Hotta, M., Nakata, R., Katsukawa, M., Hori, K., Takahashi, S., and Inoue, H., 2010. Carvacrol, a component of thyme oil, activates PPARα and γ and suppresses COX-2 expression. *Journal of Lipid Research*, 51 (1), 132–139.

Hou, R., Garner, M., Holmes, C., Osmond, C., Teeling, J., Lau, L., and Baldwin, D. S., 2017. Peripheral inflammatory cytokines and immune balance in Generalised Anxiety Disorder: Case-controlled study. *Brain, Behavior, and Immunity*, 62, 212–218.

Hudecová, A., Kusznierewicz, B., Rundén-Pran, E., Magdolenová, Z., Hasplová, K., Rinna, A., Fjellsbø, L. M., et al. 2012. Silver nanoparticles induce premutagenic DNA oxidation that can be prevented by phytochemicals from *Gentiana asclepiadea*. *Mutagenesis*, 27 (6), 759–769.

Imamura, M., Sasaki, O., Okunishi, K., Nakagome, K., Harada, H., Kawahata, K., Tanaka, R., Yamamoto, K., and Dohi, M., 2014. Perillyl alcohol suppresses antigen-induced immune responses in the lung. *Biochemical and Biophysical Research Communications*, 443 (1), 266–271.

Karam, M. C., Hamdan, H. G., Abi Chedid, N. A., Bodman-Smith, K. B., and Baroody, G. M., 2007. Interleukin-10 reduces hyperalgesia and the level of Interleukin-1 beta in BALB/c mice infected with *Leishmania major* with no major effect on the level of Interleukin-6. *Journal of Neuroimmunology*, 183 (1–2), 43–49.

Kotenko, S. V, Krause, C. D., Izotova, L. S., Pollack, B. P., Wu, W., and Pestka, S., 1997. Identification and functional characterization of a second chain of the interleukin-10 receptor complex. *The EMBO Journal*, 16 (19), 5894–5903.

Kubo, T., Morita, H., Sugita, K., and Akdis, C. A., 2017. Introduction to mechanisms of allergic diseases. *Middleton's Allergy Essentials*, 1–27.

Lan, H. Y., 2011. TGF-β/Smad signaling in diabetic nephropathy. *Clinical and Experimental Pharmacology & Physiology*, 32, 236–243.

Landa, P., Kokoska, L., Pribylova, M., Vanek, T., and Marsik, P., 2009. In vitro anti-inflammatory activity of carvacrol: Inhibitory effect on COX-2 catalyzed prostaglandin E2 biosynthesisb. *Archives of Pharmacal Research*, 32 (1), 75–78.

Lee, I.-A., Lee, J. H., Baek, N.-I., and Kim, D.-H., 2005. Antihyperlipidemic effect of crocin isolated from the fructus of *Gardenia jasminoides* and its metabolite crocetin. *Biological & Pharmaceutical Bulletin*, 28 (11), 2106–2110.

Li, X., Shen, J., Zhong, Z., Wen, H., Luo, Q., and Wei, W., 2009. Paeoniflorin: A monomer from traditional Chinese medical herb ameliorates *Schistosoma japonicum* egg-induced hepatic fibrosis in mice. *The Journal of Parasitology*, 95 (6), 1520–1524.

Lima, M. da S., Quintans-Júnior, L. J., de Santana, W. A., Martins Kaneto, C., Pereira Soares, M. B., and Villarreal, C. F., 2013a. Anti-inflammatory effects of carvacrol: Evidence for a key role of interleukin-10. *European Journal of Pharmacology*, 699 (1–3), 112–117.

Lima, P. R., de Melo, T. S., Carvalho, K. M. M. B., de Oliveira, Í. B., Arruda, B. R., de Castro Brito, G. A., Rao, V. S., and Santos, F. A., 2013b. 1,8-cineole (eucalyptol) ameliorates cerulein-induced acute pancreatitis via modulation of cytokines, oxidative stress and NF-κB activity in mice. *Life Sciences*, 92 (24–26), 1195–1201.

Liu, D., Gao, L., Zhang, J., Huo, X., Ni, H., and Cao, L., 2017. Anti-inflammatory and anti-oxidant effects of licorice flavonoids on ulcerative colitis in mouse model. *Chinese Herbal Medicines*, 9 (4), 358–368.

Mak, T. W., Saunders, M. E., Mak, T. W., and Saunders, M. E., 2006. Cytokines and cytokine receptors. *The Immune Response*, 463–516.

McInnes, I. B., 2017. Cytokines. In: *Kelley and Firestein's Textbook of Rheumatology*. Elsevier, 396–407.

Milligan, E. D., Penzkover, K. R., Soderquist, R. G., and Mahoney, M. J., 2012. Spinal interleukin-10 therapy to treat peripheral neuropathic pain. *Neuromodulation: Technology at the Neural Interface*, 15 (6), 520–526.

Moore, K. W., Malefyt, R. D. W., Coffman, R. L., and Garra, A. O., 2001. Interleukin-10 and the interleukin-10 receptor. *Annual Review of Immunology*, 19, 683–765.

Nakamura, T., Fukui, M., Ebihara, I., Osada, S., Nagaoka, I., Tomino, Y., and Koide, H., 1993. mRNA expression of growth factors in glomeruli from diabetic rats. *Diabetes*, 42 (3), 450–456.

Neurath, M. F., 2014. Cytokines in inflammatory bowel disease. *Nature Reviews Immunology*, 14 (5), 329–342.

Nonato, F. R., Nogueira, T. M. O., de Almeida Barros, T. A., Lucchese, A. M., Oliveira, C. E. C., Santos, R. R. dos, Soares, M. B. P., and Villarreal, C. F., 2011. Antinociceptive and anti-inflammatory activities of *Adiantum latifolium* Lam.: Evidence for a role of IL-1β inhibition. *Journal of Ethnopharmacology*, 136 (3), 518–524.

Oliveira, M. G. B., Brito, R. G., Santos, P. L., Araújo-Filho, H. G., Quintans, J. S. S., Menezes, P. P., Serafini, M. R., et al. 2016. α-Terpineol, a monoterpene alcohol, complexed with β-cyclodextrin exerts antihyperalgesic effect in animal model for fibromyalgia aided with docking study. *Chemico-Biological Interactions*, 254, 54–62.

Ouyang, W., Rutz, S., Crellin, N. K., Valdez, P. A., and Hymowitz, S. G., 2011. Regulation and functions of the IL-10 family of cytokines in inflammation and disease. *Annual Review of Immunology*, 29 (1), 71–109.

Pereira, I., Severino, P., Santos, A. C., Silva, A. M., and Souto, E. B., 2018. Linalool bioactive properties and potential applicability in drug delivery systems. *Colloids and Surfaces B: Biointerfaces*, 171, 566–578.

Pizarro, T. T. and Cominelli, F., 2007. Cytokine therapy for crohn's disease: Advances in translational research. *Annual Review of Medicine*, 58 (1), 433–444.

PubChem [online], 2019. Available from: https://pubchem.ncbi.nlm.nih.gov/ [Accessed July 15, 2019].

Quintans, J. S. S., Shanmugam, S., Heimfarth, L., Araújo, A. A. S., Almeida, J. R. G. d. S., Picot, L., and Quintans-Júnior, L. J., 2019. Monoterpenes modulating cytokines—A review. *Food and Chemical Toxicology*, 123, 233–257.

Saad, M. M. G. and Abdelgaleil, S. A. M., 2018. Effectiveness of monoterpenes and phenylpropenes on *Sitophilus oryzae* L. (Coleoptera: Curculionidae) in stored wheat. *Journal of Asia-Pacific Entomology*, 21 (4), 1153–1158.

Santos, F. A., Silva, R. M., Campos, A. R., De Araújo, R. P., Lima Júnior, R. C. P., and Rao, V. S. N., 2004. 1,8-cineole (eucalyptol), a monoterpene oxide attenuates the colonic damage in rats on acute TNBS-colitis. *Food and Chemical Toxicology: An International Journal Published for the British Industrial Biological Research Association*, 42 (4), 579–584.

Santos, P. L., Matos, J. P. S. C. F., Picot, L., Almeida, J. R. G. S., Quintans, J. S. S., and Quintans-Júnior, L. J., 2018. Citronellol, a monoterpene alcohol with promising pharmacological activities—A systematic review. *Food and Chemical Toxicology*, 123, 459–469.

Saraiva, M. and O'Garra, A., 2010. The regulation of IL-10 production by immune cells. *Nature Reviews Immunology*, 10 (3), 170–181.

Saravanan, S., Pandikumar, P., Prakash Babu, N., Hairul Islam, V. I., Thirugnanasambantham, K., Gabriel Paulraj, M., Balakrishna, K., and Ignacimuthu, S., 2014. In vivo and in vitro immunomodulatory potential of swertiamarin isolated from *Enicostema axillare* (Lam.) A. Raynal that acts as an anti-inflammatory agent. *Inflammation*, 37 (5), 1374–1388.

Schulz, S., Bühling, F., and Ansorge, S., 1994. Prenylated proteins and lymphocyte proliferation: Inhibition by d-limonene related monoterpenes. *European Journal of Immunology*, 24 (2), 301–307.

Severino, P., Andreani, T., Chaud, M., Benites, C., Pinho, S., and Souto, E., 2015. Essential oils as active ingredients of lipid nanocarriers for chemotherapeutic use. *Current Pharmaceutical Biotechnology*, 16 (4), 365–370.

Sherbet, G. V., 2011. Cytokines. In: *Growth Factors and Their Receptors in Cell Differentiation, Cancer and Cancer Therapy*, 115–132.

Shi, D., Ma, A., Zheng, H., Huo, G., Yan, H., Fu, H., Qiu, Y., and Liu, W., 2014. Paeoniflorin inhibits the maturation and immunostimulatory function of allergen-induced murine dendritic cells. *International Immunopharmacology*, 19 (2), 221–232.

Shouval, D. S., Ouahed, J., Biswas, A., Goettel, J. A., Horwitz, B. H., Klein, C., Muise, A. M., and Snapper, S. B., 2014. Interleukin 10 receptor signaling. In: *Advances in Immunology*. 177–210.

Spelman, K., Burns, J., Nichols, D., Winters, N., Ottersberg, S., and Tenborg, M., 2006. Modulation of cytokine expression by traditional medicines: A review of herbal immunomodulators. *Alternative Medicine Review: A Journal of Clinical Therapeutic*, 11 (2), 128–150.

Srinivasan, S. and Muruganathan, U., 2016. Antidiabetic efficacy of citronellol, a citrus monoterpene by ameliorating the hepatic key enzymes of carbohydrate metabolism in streptozotocin-induced diabetic rats. *Chemico-Biological Interactions*, 250, 38–46.

Suntres, Z. E., Coccimiglio, J., and Alipour, M., 2015. The bioactivity and toxicological actions of carvacrol. *Critical Reviews in Food Science and Nutrition*, 55 (3), 304–318.

Tabanca, N., Kirimer, N., Demirci, B., Demirci, F., and Can Başer, K. H., 2001. Composition and antimicrobial activity of the essential oils of *Micromeria cristata* subsp. *phrygia* and the enantiomeric distribution of borneol. *Journal of Agricultural and Food Chemistry*, 49, 4300–4303.

Thomsen, R. and Christensen, M. H., 2006. MolDock: A new technique for high-accuracy molecular docking. *Journal of Medicinal Chemistry*, 49 (11), 3315–3321.

Trinh, H.-T., Lee, I.-A., Hyun, Y.-J., and Kim, D.-H., 2011. Artemisia princeps Pamp. Essential oil and its constituents eucalyptol and α-terpineol ameliorate bacterial vaginosis and vulvovaginal candidiasis in mice by inhibiting bacterial growth and NF-κB activation. *Planta Medica*, 77 (18), 1996–2002.

Trivella, D. B. B., Ferreira-Júnior, J. R., Dumoutier, L., Renauld, J.-C., and Polikarpov, I., 2010. Structure and function of interleukin-22 and other members of the interleukin-10 family. *Cellular and Molecular Life Sciences*, 67 (17), 2909–2935.

Vaidya, H., Giri, S., Jain, M., and Goyal, R., 2012a. Decrease in serum matrix metalloproteinase-9 and matrix metalloproteinase-3 levels in Zucker fa/fa obese rats after treatment with swertiamarin. *Experimental and Clinical Cardiology*, 17, 12.

Vaidya, H., Prajapati, A., Rajani, M., Sudarsanam, V., Padh, H., and Goyal, R. K., 2012b. Beneficial effects of swertiamarin on dyslipidaemia in streptozotocin-induced type 2 diabetic rats. *Phytotherapy Research: PTR*, 26 (8), 1259–1261.

Viljoen, A., Mncwangi, N., and Vermaak, I. 2012. Anti-inflammatory iridoids of botanical origin. *Current Medicinal Chemistry*, 19 (14), 2104–2127. https://doi.org/10.2174/092986712800229005.

Wagner, H., Wierer, M., and Bauer, R., 1986. In vitro inhibition of prostaglandin biosynthesis by essential oils and phenolic compounds. *Planta Medica*, (3), 184–187.

Walter, M. R., 2014. The molecular basis of IL-10 function: From receptor structure to the onset of signaling. *Current Topics in Microbiology and Immunology*, 380, 191–212.

Wang, J., Hou, J., Zhang, P., Li, D., Zhang, C., and Liu, J., 2012. Geniposide reduces inflammatory responses of oxygen-glucose deprived rat microglial cells via inhibition of the TLR4 signaling pathway. *Neurochemical Research*, 37 (10), 2235–2248.

Windsor, W. T., Syto, R., Tsarbopoulos, A., Zhang, R., Durkin, J., Baldwin, S., Paliwal, S., Mui, P. W., Pramanik, B., and Trotta, P. P., 1993. Disulfide bond assignments and secondary structure analysis of human and murine interleukin 10. *Biochemistry*, 32 (34), 8807–8815.

Wu, H., Wei, W., Song, L., Zhang, L., Chen, Y., and Hu, X., 2007. Paeoniflorin induced immune tolerance of mesenteric lymph node lymphocytes via enhancing beta 2-adrenergic receptor desensitization in rats with adjuvant arthritis. *International Immunopharmacology*, 7, 662–673.

Wu, S.-H., Wu, D.-G., and Chen, Y.-W., 2010. Chemical constituents and bioactivities of plants from the genus Paeonia. *Chemistry & Biodiversity*, 7 (1), 90–104.

Xiaofeng, Y., Qinren, C., Jingping, H., Xiao, C., Miaomiao, W., Xiangru, F., Xianxing, X., et al. 2012. Geniposide, an iridoid glucoside derived from *Gardenia jasminoides*, protects against lipopolysaccharide-induced acute lung injury in mice. *Planta Medica*, 78 (6), 557–564.

Yamahara, J., Kobayashi, M., Matsuda, H., and Aoki, S., 1991. Anticholinergic action of *Swertia japonica* and an active constituent. *Journal of Ethnopharmacology*, 33, 31–35.

Ye, H., Wang, Y., Yan, J., and Jenson, A. B., 2016. Characterization of the anti-inflammation mechanism for the AO herbal extract. *Experimental and Molecular Pathology*, 101 (3), 341–345.

Zhang, A., Hao, S., Bi, J., Bao, Y., Zhang, X., An, L., and Jiang, B., 2009. Effects of catalpol on mitochondrial function and working memory in mice after lipopolysaccharide-induced acute systemic inflammation. *Experimental and Toxicologic Pathology: Official Journal of the Gesellschaft fur Toxikologische Pathologie*, 61 (5), 461–469.

Zhang, X. L., Jiang, B., Li, Z. B., Hao, S., and An, L. J., 2007. Catalpol ameliorates cognition deficits and attenuates oxidative damage in the brain of senescent mice induced by D-galactose. *Pharmacology, Biochemistry, and Behavior*, 88 (1), 64–72.

Zhong, W., Chi, G., Jiang, L., Soromou, L. W., Chen, N., Huo, M., Guo, W., Deng, X., and Feng, H., 2013. p-Cymene modulates in vitro and in vivo cytokine production by inhibiting MAPK and NF-κB activation. *Inflammation*, 36 (3), 529–537.

17 UV/V is Spectrophotometric Characterization of the Leaf Polyphenolics Content in *Elaeocarpus tectorius* and its Therapeutic Potential against Selected Urinary Tract Infection Pathogens

M. Ashwini Lydia, Suman Thamburaj, Gayathri Jagadeesan, Gayathri Nataraj, Kasipandi Muniyandi, Saikumar Sathyanarayanan, and Parimelazhagan Thangaraj

CONTENTS

17.1 INTRODUCTION

Antibiotics are the major drugs that have been used to act against numerous infective contagious diseases for decades. The wide use of these has caused a certain resistance, which is a common phenomenon of the microorganisms that are actually the causative agents. This urged the health care organizations and pharmaceutical industries all over the world to develop plant-based herbal products and formulations, as there is a drift in strategy. Meanwhile, there are substantial alternative sources of natural anti-microbials from plants that possess different modes of actions. Of those, many of them have been used in traditional medicinal systems for centuries and were found to have competitive effects as compared to a few commercial synthetic antibiotics (Abdallah 2011). In spite of the advancement and immense success in the development of medicine, science, and technology, still the spread and control of infectious diseases undergo failure. The World Health Organization stated that infectious diseases are the second leading issue for the cause of deaths in the world (WHO 2002). The emergence and rapid spread of new infections have now caused a switchover to the use of natural products with plants as the raw source having antibiotic actions, as they are known to be the largest biochemical and pharmaceutical stores on our planet and also carry active biochemical compounds with anti-oxidant and anti-microbial properties (Blois 1958; Cowan 1999). The biological activity of the plants can be predicted with respect to their use by people in traditional indigenous systems of medicine.

Urinary tract infections are the most common microbial infection occurring in the urinary tract. Frequently, women suffer from this infection during menstruation and pregnancy. So it occurs more predominantly in women than men. It is approximated that about 35% of healthy women suffer with UTIs at various stages in their life. About 5% of women each year experience a UTI with the problems of painful urination (dysuria) and frequency (Nabbugodi et al. 2015). The major causative agent for a UTI is the uropathogenic bacteria *E. coli* (Abraham and Miao 2015). The causative agents for community-acquired urinary tract infections are mostly the uropathogenic *Escherichia coli* from the gut of about 80%–85%, then *Pseudomonas aeruginosa* with about 11%, *Staphylococcus aureus* with 5%–10%, followed by *Candida albicans* the yeast fungal pathogen as 9%, and they were the test pathogens used for the study (Nicolle 2008; Salvatore et al. 2011). They act as hosts for the frequent causes of urinary infections. The increasing multidrug resistance of UTI-causing microbial strains has made the treatment of UTIs difficult and has led to the greater use of costly broad-spectrum synthetic drugs (Tenney et al. 2018). This problem needs an improved effort, which has resulted in the search for valuable microbicide agents to be used against UTIs that are caused by pathogenic microorganisms resistant to current antibiotic drugs (Toner et al. 2016).

Though synthetic antibiotics tend to act against the pathogens, medicinal plants that are traditionally used by healers are valued.

Elaeocarpus is the largest genus belonging to the Elaeocarpaceae family, consisting of mostly tree species. There are about 350 species estimated to exist, and they are widely spread in Southeast Asian countries. In India, generally this species grows in the Himalayan region and temperate zones of the Western Ghats. *Elaeocarpus* species have many biological active molecules such as indolizilidine alkaloids, triterpenes, tannin such as geraniin and 3, 4, 5-trimethoxy geraniin, grandisines, rudrakine fatty acids, ellagic acid derivatives, cytotoxic compounds flavonoids, and quercitin (Muthuswamya 2014). *E. tectorius* is a species that is one among the traditional plants used by the tribal people of the Nilgiris District, which is well known for its unique vegetation and diversity. The indigenous traditional knowledge of the utilization of *E. tectorius* is used in the treatment of leprosy, pneumonia, rheumatism, ulcers, piles, and dropsy. In view of the above information, *E. tectorius* was taken as the experimental plant to seek out the variations of the total polyphenolic contents in the leaf extracts and their therapeutic potential. This is the first report of antimicrobial activity of this particular plant against selected UTI pathogens.

17.2 MATERIALS AND METHODS

17.2.1 Collection of Plant Leaves

The *E. tectorius* leaves were collected from Coonoor, The Nilgiris District, Tamil Nadu, India, and were then identified with their morphological characteristics and description by using the voucher specimen in the Botanical Survey of India Southern Regional Centre, Coimbatore, Tamil Nadu. The plant specimen was given the reference no. BSI/SRC/5/23/2018/Tech./2853. The leaves were first washed with distilled water to remove the adhering external dusts. Then they were subjected to shade drying under normal environmental conditions and were further powdered finely using a blender.

17.2.2 Extraction and Processing

Extraction of the leaf sample in different solvents is important for carrying out further analysis and comparative studies. Soxhlet apparatus was used for extraction using solvents such as petroleum ether, dichloromethane, ethyl acetate, methanol, and water. One hundred milligrams of the leaf samples were extracted using 300 mL of each solvent. Each extract was thereafter concentrated using a rotary vacuum evaporator (Equitron Ev11-ABS.051), and then air dried. The dried extracts were stored in small glass vials at 20°C for further studies. The amount of crude extracts recovered after successive extraction was weighed, and the yield percentage was calculated based on the following formula,

$$\text{Extraction yield}(\%) = (\text{Mass of extract} / \text{Mass of raw material}) \times 100$$

17.2.3 Preliminary Phytochemical Analysis

Plants are the biological factories containing various biochemical phytocompounds. The screening of those compounds is the important part of the featured study. The leaf powder of *E. tectorius* was screened for the presence of phytochemical constituents, such as alkaloids, phenolics compounds, tannins, flavonoids, terpenoids, saponins, glycosides, and steroids using standard methods (Thangaraj 2016).

17.2.4 Determination of Total Phenolic, Flavonoid, and Tannin Contents

Anti-oxidant compounds such as phenolics flavonoids and tannins were estimated quantitatively. The total phenolic content of the *E. tectorius* leaf extracts was determined according to the method described by Siddhuraju and Becker (2003). Initially, 0.5 mL of the Folin-Ciocalteu reagent was added to the plant sample, followed by 2.5 mL of 5% sodium carbonate, it was incubated in the dark at 27°C for 40 minutes, and the blue absorbance was measured at 725 nanometers. For the tannins, the extracts were treated with polyvinyl polypyrrolidone. The tannin content of the plant extract was calculated by the following formula (Siddhuraju and Manian 2007),

$$\text{Tannins}\% = \text{Total phenolics} - \text{Nontannin phenolics}$$

The results of the total phenolics and tannins were expressed in gallic acid equivalents. The flavonoid content of *E. tectorius* leaf extracts was quantified according to the method given by Zhishen et al. (1999). To the plant extracts, 150 µL of 5% sodium nitrite and 10% aluminum chloride was added. This was incubated at 27°C for 6–10 minutes, and then 2 mL of 4% sodium hydroxide was added, and the absorbance of pink color was measured at 510 nanometers. The results were expressed as rutin equivalents, and all the estimations were made in triplicates.

17.2.5 In Vitro Antioxidant Assays

17.2.5.1 DPPH Scavenging Assay

The anti-oxidant activities of the *E. tectorius* leaf extracts were determined using the standard protocols. The ability of the plant extracts to scavenge the stable free radical 2,2-diphenyl-1-picrylhydrazyl (DPPH) was evaluated by the method of Blois (1958) with minor modifications (Thangaraj 2016). Plant extracts were diluted to various concentrations, and the volume was adjusted to 100 µL using methanol. About 5 mL of a 0.1 mM methanolic solution of DPPH was added to the aliquots of different extracts of the plant sample and standard rutin. The complex was then incubated for 20 minutes at room temperature, which resulted in the delocalization of electrons and also gave rise to a violet color formation; the absorbance was measured at 517 nm. The DPPH percentage inhibition was estimated using the following formula:

$$\%\text{DPPH inhibition} = (\text{A0} - \text{A1}) / \text{A0} \times 100,$$

where A0 and A1 were the absorbances of the blank and extracts or the absorbance taken place before (A0) and after the reaction (A1), respectively.

17.2.5.2 Ferric Reducing/Antioxidant Power Assay

The ferric reducing power of the plant leaf extracts was estimated based on the standard protocol given by Pulido et al. (2000), the ability to reduce the TPTZ (2,4,6-tripyridyl-s-triazine) (TPTZ)-Fe (III) complex to TPTZ-Fe (II) complex. The ferric reducing anti-oxidant power (FRAP) reagent (900 µL), which is to be freshly used in contact with the sample (10 µL of aliquots of plant extracts, 1 mg/mL of respective organic solvents), is prepared by three constituents 10 mM TPTZ in 40 mM HCl, 20 mM $FeCl_3.6H_2O$, and 0.3 M acetate buffer (pH 3.6) in the necessary quantity. Then the mixed solution is subjected to incubation at 37°C for 30 minutes. The ferric reduction is then measured at 595 nm. The results were expressed as mM Fe (II) equivalent/g extract.

17.2.5.3 Phosphomolybdenum Assay

The assay to determine the total anti-oxidant capacity was analyzed by the standard method of Prieto et al. (1999). The standard used was ascorbic acid. About 1 mg per mL of the sample concentration of each solvent extract and the standard ascorbic acid were added with 3 mL of the reagent solution that consisted of 0.6 M sulfuric acid, 28 mM sodium phosphate, and 4 mM ammonium molybdate. The mixture was then incubated in a water bath at 95°C for about 90 minutes, the green phosphomolybdenum complex formation was noted after the incubation, and it was allowed to cool until it reached room temperature; the absorbance of the mixture was then monitored spectrophotometrically at 695 nm against the reagent, which was taken as the blank. The results were expressed as mg ascorbic acid equivalents (AAE)/g extract.

17.2.6 Anti-microbial Activity

17.2.6.1 Test Microbial Pathogens

Escherichia coli (MTCC-433), *Staphylococcus aureus* (MTCC-737), *Pseudomonas aeruginosa* (MTCC-741), and *Candida albicans* (MTCC-227) were the pure cultures maintained using a nutrient agar medium for bacteria and a Sabouraud dextrose agar medium for fungi and used for the anti-microbial testing. They were obtained from Microbial Type Culture Collection (MTCC), Institute of Microbial Technology, Chandigarh.

17.2.6.2 Disk-Diffusion Agar Method

The modified disk diffusion agar method (Olurinola 1996) was followed to verify the anti-microbial activity of *E. tectorius* leaf extracts against selected bacterial and fungal pathogens. The sterile swabs were used to swab the bacterial cultures from the broth onto the plates containing Mueller Hinton agar medium, similarly the fungal yeast pathogen '*Candida*' were swabbed in the plate containing Sabouraud dextrose agar medium. Wells were made on the seeded plates with the help of a sterilized cork borer (6 mm Hi-Media). The collected different samples were further dissolved in 4% Dimethyl sulfoxide (DMSO). The diluted samples (100 µL of 10 mg/mL concentration) were dispended into the wells, and the plates were incubated aerobically at 37°C for 18–24 hours for bacterial pathogens and 28°C for 48 hours for fungal pathogens. In the same way, positive and negative control wells were made with Ampicillin-15 µg (bacteria), Miconazole-30 µg (*Candida*), and 4% DMSO, respectively. The entire microbial assay was carried out under strict aseptic conditions. The zones of inhibition (mm) of the samples were measured after incubation, and the activity index was also calculated. Triplicates were maintained for each pathogen, and the readings were taken in three different fixed directions, and the average values were recorded.

17.2.6.3 Determination of Minimum Inhibitory Concentration and Minimum Bactericidal/Fungicidal Concentration of *E. tectorius* Leaf Extracts

The determination of the minimum inhibitory concentrations (MICs) were carried out in a *E. tectorius* leaf for serial extract in the broth dilution method, for those that showed a $\geq$ 8 mm diameter zone of inhibition as determined through the well diffusion method. The extract with 5 mg/mL concentration was gradually diluted in the 900 μL sterile Mueller-Hinton bacterial broth and Sabouraud dextrose broth for *Candida* sp. Then, 100 μL of the culture (1×10^5 CFU/mL) was placed in all the broths, and one was kept as the control for each test microorganism. The culture tubes were then incubated at 37°C. The OD values were taken using a spectrophotometer (Shimadzu UV-spectrophotometer, UV-1800) at 570 nm. From the respective values obtained, the percentage of growth inhibition and MICs were calculated against each test pathogen. By sub-culturing, the test dilution used in the MIC and the minimum bactericidal/fungicidal concentration (MBC/MFC) was determined, thereafter by centrifugation (Eppendorf centrifuge 5430 R), the cultured pellets were suspended in 100 μL of the sterile broth. The total suspension was swabbed onto the desired plates containing the media and allowed to incubate for a further 24–48 hours at 37°C, and then onto a fresh solid medium and incubated further for 24 hours. The concentration of the plant extract that completely killed the organisms was considered as the MBC/MFC (CLSI 2008, 2009).

17.2.7 Ultraviolet/Visible Spectral Measurement

Ultraviolet (UV)-visible spectroscopy is performed for the qualitative analysis of phyto-compounds in plant extracts and for the identification of certain classes of polyphenolic compounds. This technique is very simple compared to other techniques and not time-consuming. The UV-visible (Vis) spectra of the diluted *E. tectorius* ethyl acetate and methanol leaf extracts (100 μL from 1 mg/mL concentration of extracts) were measured in the range of 200–1000 nm using a UV-Vis double beam spectrophotometer 1800 PC (Shimadzu, Japan).

17.2.8 Thin Layer Chromatography Separation

The Silica gel 60 F254 (Merck, Germany) was used for the thin layer chromatography (TLC) analysis of *E. tectorius* ethyl acetate and the methanol leaf extract. The silica plates (8 × 3 cm size) were activated at 120°C for 2 minutes and then used. The extracts to be separated were spotted (5 μL from 1 mg/mL concentration of extracts) with a capillary tube, 1.0 cm above from the end of the silica gel plate, which was in contact with the solvent hexane: ethyl acetate: methanol (60:30:10%). After spotting, the TLC plates were kept in a chamber containing the solvent. A chromatogram was developed by the movement of the solute (mobile phase) across a thin layer of silica (stationary phase). After the development of a chromatogram, the resolved spots were revealed by detection under UV light at a longer wavelength (365 nm) and a shorter wavelength (254 nm), which helped to detect the compound (colored compounds and fluorescence compounds).

17.2.9 Statistical Analyses

The results were expressed as mean ± standard deviation (SD). The data were statistically analyzed using the SPSS version 17.0 by means of one way Analysis of Variance (ANOVA), followed by Duncan's test for total polyphenolics and anti-oxidant studies. Mean values were considered statistically significant when $p < 0.05$.

17.3 RESULTS AND DISCUSSION

17.3.1 Extract Yield Percentage

The leaf samples of *E. tectorius* were extracted using Soxhlet apparatus in a successive method. The yield of crude extract varies from each solvent extraction; it is based upon the nature and polarity of the solvents. Among the five solvent extractions of *E. tectorius* leaf powder, the maximum amount of yield was obtained in the methanolic extracts of the leaf (18.09%) (Table 17.1), whereas the dichloromethane extract of the leaf (1.2%) revealed the minimum yield. As the yield of higher polar solvents is considerably high as compared to the lower polar solvents, it may indicate that *E. tectorius* may have active components, such as polyphenolics, anti-oxidants, and anti-microbials.

Soxhlet extraction has been used for over a century and is a standard technique that is widely used for extracting valuable bioactive compounds from various natural sources, especially plants (De Castro and Priego-Capote 2010). The extraction of the bioactive components from the plant materials is a most important standard method and the first step used in different commercial sectors, such as pharmaceutical, food, and chemical industries (Azmir et al. 2013).

TABLE 17.1
Yield of the *E. tectorius* Leaf Extracts

Plant Materials Used	Solvents	Yield Extracts (%)
Leaf powder	Petroleum ether	1.70
	Dichloromethane	1.20
	Ethyl acetate	6.72
	Methanol	18.09
	Aqueous	3.22

17.3.2 Preliminary Phytochemical Screening

Preliminary phytochemical screening has revealed the presence of phytochemicals considered as active chemical constituents. Important phytochemicals such as alkaloids, phenolics, tannins, flavonoids, terpenoids, glycosides, and steroids were present in the *E. tectorius* leaf (Table 17.2). The result of the preliminary phytochemical analysis shows that the plant is rich in secondary metabolites. Medicinal plants having alkaloids are used in medicines for reducing headaches and fevers (Okwu and Okwu 2004). The terpenoids that are present are reported to have anti-inflammatory, anti-malarial, inhibition of cholesterol synthesis, anti-oxidant, and anti-microbial properties (Mahato and Sen 1997). Steroids have been reported for their wound healing properties, these are anti-inflammatory, analgesic, and anti-oxidant (Ayinde et al. 2007). The presence of glycosides in whole leaves might play a role in the cardioprotective potential of *E. tectorius* (Brian et al. 1985). Further, the presence of different phytoconstituents in the plant materials may be responsible for anti-microbial, anti-oxidant, and other biological properties. The phytochemical screening may be useful for the identification of the bioactive principles and may lead to drug discoveries. Further, these tests facilitate their quantitative estimation and qualitative separation for pharmacologically active chemical compounds (Parivuguna 2008). The present study that was undertaken revealed the presence of a number of bioactive compounds which can be used as a lead compound for synthesizing drugs for various ailments.

TABLE 17.2
Phytochemical Screening of *E. tectorius*

Phytochemical Constituents	Leaf Powder
Alkaloids	+++
Phenolic compounds	+++
Tannins	++
Flavonoids	+
Saponins	–
Glycosides	++
Terpenoids	++
Steroids	+++

Note: (+): presence of chemical compound; (–): absence of chemical compound; (+) < (++) < (+++): based on the intensity of characteristic color.

17.3.3 Total Polyphenolic Contents of *E. tectorius* Leaf and Fruit Extracts

The total polyphenolic contents (phenolics, tannins, and flavonoids) of different extracts of the *E. tectorius* leaf were analyzed and presented in Table 17.3. In this study, the successive ethyl acetate leaf extract of *E. tectorius* (658.66 ± 26.06 mg gallic acid equivalents (GAE)/g of extract) shows a higher total phenolic content, while the methanol leaf extract (48 ± 4.45 mg GAE/g of extract) has a higher tannin content. However, the flavonoid content was found to be higher in the water leaf extract (226.25 ± 5.12 mg RE/g of extract). The petroleum ether leaf extract showed the lower presence of the total polyphenolic contents (phenolics, tannins, and flavonoids).

From the results, it was observed that high polar and middle polar extracts contain more polyphenolic contents than non-polar extracts, because water, methanol, and ethyl acetate solvent extracts are higher in polarity than non-polar solvents like dichloromethane and petroleum ether. Hence, there was a close correlation between the bioactivity and the amount of phenolics, tannins, and flavonoids present in the plants, and this plays a significant role in important biological functions. The polyphenolics from medicinal plants have been reported to have multiple biological properties, including anti-oxidant, anti-diabetic, and anti-inflammatory properties. Recent studies suggest that diets rich in polyphenolics play an important role against oxidative stress related disorders because of their anti-oxidant activities (Murugan and Parimelazhagan 2014; Saravanan and Parimelazhagan 2014; Sathyanarayanan et al. 2017; Muniyandi et al. 2017). Therefore, the polyphenolics (phenolic, tannin, and flavonoid) of *E. tectorius* may have the properties to react to oxidative stress related disorders.

TABLE 17.3
Determination of Total Phenolics, Tannins, and Flavonoids Content in *E. tectorius*

Sample	Extracts	Total Phenolics (mg GAE/g extract)	Tannins (mg GAE/g extract)	Total Flavonoids (mg RE/g extract)
Leaf	Petroleum ether	41.80 ± 6.54^{d}	25.33 ± 1.89^{c}	29.300 ± 0.63^{e}
	Dichloromethane	26.73 ± 4.63^{e}	30.53 ± 3.62^{b}	93.050 ± 3.78^{c}
	Ethyl acetate	**658.66 ± 26.06^{a}**	**42.80 ± 7.62^{a}**	117.50 ± 3.00^{b}
	Methanol	324.00 ± 29.22^{b}	**48.00 ± 4.45^{a}**	70.13 ± 1.20^{d}
	Water	133.73 ± 7.06^{c}	**41.86 ± 13.81^{a}**	**226.25 ± 5.12^{a}**

Note: Values are mean of triplicate determination (n = 3) ± SD. GAE: gallic acid equivalents, RE: rutin equivalents. Statistically significant at $p < 0.05$ where $^{a}>^{b}>^{c}>^{d}>^{e}$ in each column.
The numbers in bold indicate that the result is highly significant.

17.3.4 In Vitro Anti-oxidant Assays

The free radical scavenging activity of different extracts of the *E. tectorius* leaf was analyzed and is shown in Table 17.4. By analyzing all the extracts, the ethyl acetate extract of the leaf (9.43 µg/mL) showed better Half Maximal Inhibitory Concentration (IC_{50}) values in the DPPH radical scavenging activity. The minimal activity was seen in the dichloromethane leaf extract (148.21 µg/mL). The lower IC_{50} value indicates the higher DPPH radical scavenging activity. The DPPH radical has been used widely to evaluate the anti-oxidant activity of the plant extracts and foods as free radical scavengers or hydrogen donors (Soare et al. 1997). The highly enhanced DPPH radical scavenging activity of the ethyl acetate extract was observed in the present study. It may be due to the polar nature of the solvent and the extracting ability of the polyphenolics from the *E. tectorius* leaf that was studied.

The ferric reducing ability of the *E. tectorius* extracts was estimated based on their ability to reduce a TPTZ-Fe (III) complex to TPTZ-Fe (II). The ethyl acetate extracts of the leaf (8066.66 mM• (Fe (II) E/mg extract) had the highest ferric reducing power, followed by the methanol leaf extract. However, compared to all the solvent extracts, the petroleum ether extract of the *E. tectorius* leaf showed the minimal scavenging activity. Thus, the ferric reducing power of different extracts of *E. tectorius* revealed that there are phytocompounds in the ethyl acetate leaf extract which have a high affinity to the ferric ions, and thereby quench them through redox reactions. The reducing power of the bioactive phytocompounds, mainly the high molecular polyphenolics, was associated with anti-oxidant activity, specifically scavenging of free radicals (Su et al. 1987; Siddhuraju and Manian 2007). Results have revealed that the *E. tectorius* extracts with the higher amount of phenolics have higher reducing power.

The leaf of the *E. tectorius* was analyzed to determine the total anti-oxidant potential by a phosphomolybdenum assay. Among the different extracts used, the ethyl acetate extract of the leaf (1460.66 mg AAE/g extract) showed the higher reducing ability of Mo (VI) to Mo (V) and the subsequent formation of the green phosphate/Mo (V) complex at an acid pH (Murugan and Parimelazhagan 2014; Thangaraj 2016). The lowest reducing ability was seen in the petroleum ether extract of *E. tectorius*. Thus, the total anti-oxidant capacity observed for the extracts of *E. tectorius* can be correlated with its free radical scavenging activity which is equivalent to that of natural anti-oxidant ascorbic acid. The reduction of Mo (VI) to Mo (V) by the extracts of the *E. tectorius* leaf may be due to the electron transfer or hydrogen ion transfer by the bioactive phytocompounds, specifically the polyphenolics present in the leaf. From the results, the higher radical scavenging activity revealed by the *E. tectorius* leaf extracts points out that the plant is highly potential, and the extracts can also scavenge other free radicals in the body.

17.3.5 Anti-microbial Activity

The in vitro anti-microbial activity against the UTI-responsible pathogenic strains *E. coli*, *S. aureus*, *P. aeruginosa*, and *C. albicans* was examined for successive petroleum ether, dichloromethane, ethyl acetate, methanol, and water extracts from the *E. tectorius* leaf. The anti-microbial activities of these extracts were evaluated by the well diffusion method. The diameter of the inhibitory zones recorded includes the size of the well (6 mm in diameter). The anti-microbial activity of the *E. tectorius* extracts on the different organisms causing UTIs is given in Table 17.5.

The present study indicated the great variation in the anti-microbial activities of the *E. tectorius* leaf extracts against gram-positive, gram-negative, and yeast urinary tract pathogens. The maximum anti-bacterial and anti-candidal activity was shown by the ethyl acetate and methanol extracts of the *E. tectorius* leaf with the inhibitions of a 21.00 ± 1.00 mm to 13.66 ± 0.28 mm zone size at 1 mg concentration (Figure 17.1). Similar results were reported by fruit extracts against the urinary tract pathogens (Manoharan et al. 2019). The ethyl acetate extracts of the *E. tectorius* leaf were found to be active against *E. coli* (15.66 ± 0.57 mm Zone of Inhibition

TABLE 17.4
Determination of Phosphomolybdenum and FRAP Assay of *E. tectorius*

Sample	Extracts	DPPH Radical Scavenging Activity IC50 Value (µg/mL)	FRAP mM (Fe(II)E/g extract)	Phosphomolybdenum (mg AAE/g extract)
Leaf	Petroleum ether	75.29[d]	165.78 ± 11.14[d]	160.00 ± 9.75[e]
	Dichloromethane	148.21[e]	180.19 ± 23.11[d]	335.16 ± 14.33[d]
	Ethyl acetate	**9.43[a]**	**8066.66 ± 887.15[a]**	**1460.66 ± 278.68[a]**
	Methanol	20.66[b]	3717.64 ± 326.82[b]	912.33 ± 196.00[b]
	Water	63.89[c]	1021.56 ± 315.80[c]	377.00 ± 21.79[c]
Rutin (control)		10.83	4705.69 ± 27.66	312.22 ± 30.97

Note: Values are mean of triplicate determination (n = 3) ± SD. AAE: ascorbic acid equivalents. Statistically significant at $p < 0.05$ where $^a>^b>^c>^d>^e$ in each column.
The numbers in bold indicate that the result is highly significant.

TABLE 17.5
Antimicrobial Activity of *E. tectorius* Leaf Extracts

	S. aureus			***E. coli***			***P. aeruginosa***			***C. albicans***		
Extracts	**ZOI[a] (mm)**	**MIC (µg mL)**	**MBC (µg mL)**	**ZOI[a] (mm)**	**MIC (µg mL)**	**MBC (µg mL)**	**ZOI[a] (mm)**	**MIC (µg mL)**	**MBC (µg mL)**	**ZOI[a] (mm)**	**MIC (µg mL)**	**MFC (µg/mL)**
Petroleum ether	00.00 ± 0.00	–	–	00.00 ± 0.00	–	–	00.00 ± 0.00	–	–	07.33 ± 0.57	450	550
Dichloromethane	00.00 ± 0.00	–	–	00.00 ± 0.00	–	–	00.00 ± 0.00	–	–	00.00 ± 0.00	–	–
Ethyl acetate	**16.83 ± 0.76**	**125**	**150**	**15.66 ± 0.57**	**150**	**250**	**15.41 ± 0.52**	**200**	**250**	18.33 ± 1.52	200	350
Methanol	15.58 ± 0.38	150	200	13.66 ± 0.57	200	400	13.66 ± 0.28	300	350	**21.00 ± 1.00**	**150**	**250**
Water	12.33 ± 0.28	150	200	12.83 ± 0.28	200	400	09.50 ± 0.50	400	500	17.16 ± 1.04	250	350
4% DMSO	00.00 ± 0.00	–	–	00.00 ± 0.00	–	–	00.00 ± 0.00	–	–	00.00 ± 0.00	–	–
Ampicillin (15 µg)	22.16 ± 2.56	0.6	1.0	19.00 ± 1.73	0.8	1.0	15.50 ± 0.50	0.7	1.0	NA	NA	NA
Miconazole (30 µg)	NA	NA	NA	NA	NA	NA	NA	NA	NA	22.00 ± 1.00	1.00	1.50

[a] Data are presented in mean values (n = 3) ± Standard deviation at 1 mg concentration of extracts. ZOI: Zone of inhibition, MIC: Minimum inhibitory concentration, MBC: Minimum bactericidal concentration, NA: Not Applicable.
The numbers in bold indicate that the result is highly significant.

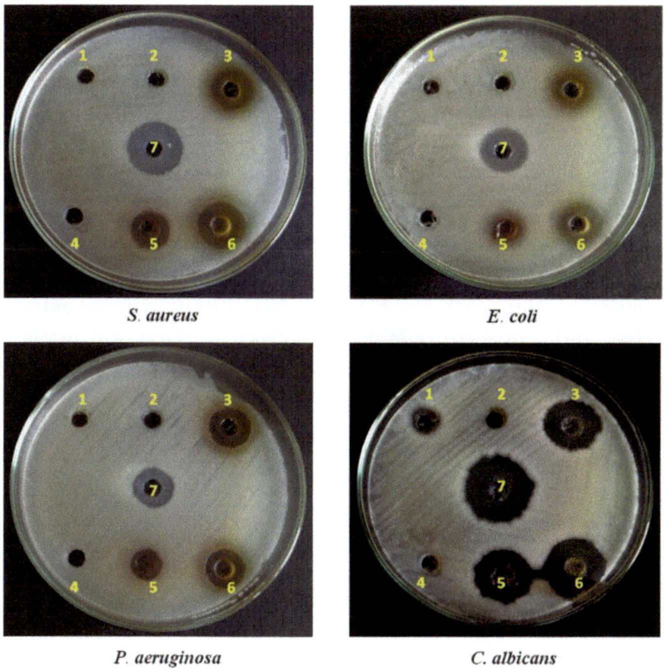

FIGURE 17.1 Antimicrobial activity of *Elaeocarpus tectorius* leaf extracts 1. Petroleum ether leaf extract; 2. dichloromethane leaf extract; 3. ethyl acetate leaf extract; 4. methanol leaf extract; 5. water leaf extract; 6. 4% DMSO control; and 7. positive control (Bacteria-Ampicillin/Candida – Miconazole).

(ZOI), 150 µg mL MIC, 250 µg/mL MBC), *S. aureus* (16.83 ± 0.76 mm ZOI, 125 µg/mL MIC, 150 µg/mL MBC), *P. aeruginosa* (15.41 ± 0.52 mm ZOI, 200 µg/mL MIC, 250 µg/mL MBC), and *C. albicans* (18.33 ± 1.52 mm ZOI, 200 µg/mL MIC, 350 µg/mL MFC), whereas the methanol leaf extract of *E. tectorius* showed major activity with 21.00 ± 1.00 mm of clear inhibition zone, 150 µg/mL MIC, and 250 µg/mL MFC against *C. albicans*. The petroleum ether leaf extract has shown the least activity against only *C. albicans*, but other tested bacteria showed no resistance (Table 17.5). Previous reports from India have revealed that the ethanol extract of *Zingiber officinale*, *Punica granatum*, *Terminalia chebula*, *Ocimum sanctum*, *Cinnamomum cassia*, and *Azadirachta indica* possessed good anti-microbial activity against UTI pathogens (Sharma et al. 2009).

Worldwide, microbial infections are one of the most serious issues in the twenty-first century. The appearance of microbial resistance to antibiotic drugs is a most important health problem, and so it is essential to develop new antibiotic drugs with new mechanisms of action against multidrug resistant microbial strains to defeat these problems (Morris and Masterton 2002; Shaikh et al. 2015; Elangomathavan et al. 2015). Plants have usually provided a source of hope for novel phyto-compounds, as poly herbal mixtures have made great contributions to human health care. The use of plants and their products with known anti-microbial activities can be of great significance for therapeutic treatments (Kumar et al. 2012).

More than 95% of UTIs are caused by *E. coli*, which is the most commonly infecting bacterial pathogen, and others only 5%. About 35% of healthy women suffer the symptoms of a urinary tract infection, and about 5% of women every year suffer with the problem of painful urination (dysuria) (Kumar et al. 2012). UTIs caused by a yeast fungal species are the second most common cause of nosocomial UTIs in children and women. It can spread systemically and can be life threatening. The occurrence of UTIs due to the *Candida* species increased gradually with a prevalence rate 27.2% (Sefton 2000). Plants and their products have been used for several years to control different UTIs. They can affect the UTI pathogens as disinfectants, analgesics, and diuretics (Ubaid et al. 2016). The active components of the plant extracts may display their microbicide potential either by degradation of the cell wall, disruption of the cytoplasmic membrane, affect the synthesis of DNA and RNA, hold up the enzymatic activities inside the cell, and affect nutrient uptake and electron transport (Shan et al. 2007).

From the above results of anti-microbial activity, the ethyl acetate and methanol extracts of the *E. tectorius* leaf extract exhibited broad-spectrum activity against tested UTI pathogens as compared to the petroleum ether and dichloromethane extracts. The susceptibility of the UTI pathogens varied for polar solvent extracts and non-polar solvent extracts. This indicates the contribution of more than one active principle of biological significance for botanicals (Ming et al. 2005). The traditional healers use mainly water as the solvent, but we observed that extracts prepared from the *E. tectorius* leaf in ethyl acetate, methanol, and water as solvents provided a more reliable anti-microbial activity against target pathogens (Parekh 2007). In the present investigation, negligible inhibitory activity was observed in the *E. tectorius* leaf petroleum ether and dichloromethane extracts, which may lack the solubility of the active phyto-constituents. This study may explain the use of *E. tectorius* by indigenous people against various infections for generations. The *E. tectorius* extracts studied here had shown that they are potentially rich in anti-microbial chemical substances and suggest that the extracts may be used for treatment of urinary tract infections.

17.3.6 UV-Vis Spectral Measurement

The ultraviolet/visible spectroscopy can be used for the qualitative analysis of plant extracts because the aromatic molecules in plant extracts are powerful chromophores under ultraviolet/visible range. Natural polyphenolics can be determined by using UV-visible spectroscopy. The phenolic compounds including flavonoids, tannins, anthocyanins, polymer dyes, and phenols form complexes with iron that have been detected by the UV-Vis spectroscopy (Kemp 1991). The UV-Vis absorbance spectra of the ethyl acetate and methanol leaf extract of the *E. tectorius* were recorded in the range of 250–800 nm and have spectral shapes and intensities (Figure 17.2). The methanol *E. tectorius* leaf extract exhibited a high absorbance intensity in the range of 250–350 nm with a maximum absorbance at 278 nm. The absorbance in the range of between 250 and 350 nm is associated with the π-π electronic transitions and usually comes from aromatic phytocompounds, hydroxyl, carbonyl, and other chromophores. These phyto-compounds are natural phenols, and they have poor solubility in water at acidic and physiological pH. Therefore, some natural phenolics are not suitable for use as drugs (Tomren et al. 2007). The UV-Vis spectra of the *E. tectorius* ethyl acetate leaf extract is very different with the *E. tectorius* methanol extract because of the high absorbance intensity in the range of 250–750 nm with a maximum absorbance at 411 nm. In this wavelength, the π-π electronic transitions are responsible for displaying the absorption of the UV-Vis radiation (Kim et al. 2013). These transitions are due to the presence of a high amount of polyphenolics (total phenolic extract (280 nm), flavones (320 nm), phenolic acids (360 nm), flavonoids (510 nm), the total anthokyanids (520 nm) and tannins (710 nm), and other secondary metabolite compounds present in the plant leaf extract (Urbano et al. 2006). From this spectral analysis of both extracts, the spectral region in the range of 250–350 nm will be very useful to discriminate the *E. tectorius* methanol extract from the ethyl acetate extract. Therefore, the spectroscopic UV-Vis analysis gives clear differentiation and information on the composition of the total polyphenolics content and other phyto-compounds present in the successive ethyl acetate and methanol leaf extracts of the *E. tectorius*.

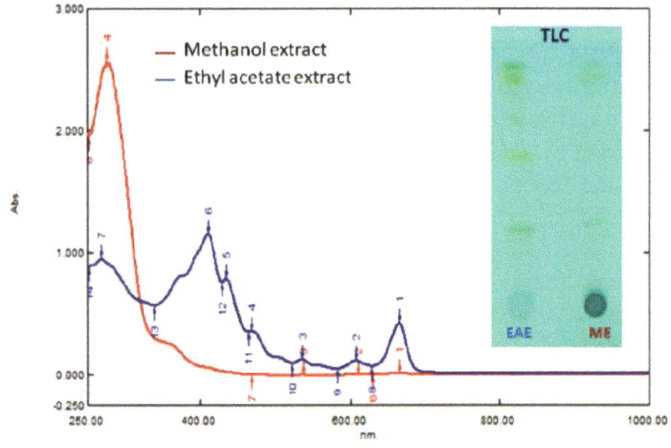

FIGURE 17.2 UV/Vis spectrophotometric and TLC characterization of bioactive *E. tectorius* leaf extracts, EAE: ethyl acetate *E. tectorius* leaf extract; ME: methanol *E. tectorius* leaf extract.

This present study showed that the extraction is an essential step to find active phytocompounds with biological properties from plant extracts. Comparison of these solvent extracts from the *E. tectorius* leaf by successive extraction methods revealed that they produce dissimilar results; however, there are quantitative differences in the total polyphenolics content, anti-oxidant, and anti-microbial activities. Overall, successive ethyl acetate extraction was found to be the best in obtaining anti-oxidant and anti-microbial chemical substances from the *E. tectorius* leaf.

17.3.7 TLC Separation

The phyto-constituents of the extracts were separated by TLC analysis. The TLC profiling of ethyl acetate and methanol extracts of *E. tectorius* gives a distinguished result that expresses the presence of a number of phyto-compounds. The ethyl acetate extract of *E. tectorius* harbors more phyto-compounds, followed by the methanol extract (Figure 17.2). Similar kinds of phyto-compounds were identified from both extracts with various concentrations based on the recovery percentage of the organic solvents used in successive extraction. The major phytochemical constituents were identified from both extracts of *E. tectorius* by UV-visible spectroscopy and thin layer chromatography analysis. These studies revealed that the *E. tectorius* extracts have highly significant anti-oxidant and anti-microbial compounds and are useful in the treatments of several diseases. This information will help in the selection of a suitable solvent system for the further separation of phytocompounds from the *E. tectorius* extracts. The medicinal significance of the plant is due to the fact that each of the bioactive compounds generates characteristic physiological actions on humans (Jagadeesan et al. 2019). It is obvious that the leaf extracts of *E. tectorius* contains various phyto-compounds which may have the appropriate biological properties to be used in the pharmaceutical industry.

17.4 CONCLUSION

This investigation has been done in order to evaluate the effects of different solvents on the successive extraction of bioactive compounds from the *E. tectorius* leaf and the in vitro anti-oxidant activities and anti-microbial effects against target UTI pathogens. The results showed that the polarity of different solvents affect the total phenolics content and anti-oxidant activities of the *E. tectorius* leaf extracts and the ethyl acetate extract exhibited the greatest polyphenolics content, anti-oxidant and anti-microbial activities, followed by the methanol extract. It can be concluded that the ethyl acetate was the most effective solvent for the extraction of soluble phenolic constituents from the *E. tectorius* leaf. The significant anti-microbial activity against UTI pathogenic bacteria and fungi demonstrated the anti-microbial potency of the *E. tectorius* leaf. In conclusion, the *E. tectorius* leaf contains potential anti-oxidant and anti-microbial phyto-constituents that may be of great use in the pharmaceutical and food industries as a remedy against UTIs and other infectious diseases. Further detailed studies are required for the isolation and identification of botanicals from the *E. tectorius* leaf related to anti-oxidant and anti-microbial activity.

REFERENCES

Abdallah, E. M. 2011. Plants: An alternative source for antimicrobials. *Journal of Applied Pharmaceutical Science* 1(6): 16–20.

Abraham, S. N., and Miao, Y. (2015). The nature of immune responses to urinary tract infections. *Nature Reviews Immunology* 15(10): 655.

Ayinde, B. A., Omogbai, E. K., and Amaechina, F. C. 2007. Pharmacognosy and hypotensive evaluation of *Ficus exasperata* Vahl (Moraceae) leaf. *Acta Poloniae Pharmaceutica* 64(6): 543–546.

Azmir, J., Zaidul, I. S. M., Rahman, M. M., Sharif, K. M., Mohamed, A., Sahena, F., and Omar, A. K. M. 2013. Techniques for extraction of bioactive compounds from plant materials: A review. *Journal of Food Engineering* 117(4): 426–436.

Blois, M. S. 1958. Antioxidant determinations by the use of a stable free radical. *Nature* 181(4617): 1199–1200.

Brian, F. H., Thomas-Bigger, J., and Goodman, G. 1985. *The Pharmacological Basis of Therapeutics*. 7th ed. New York: Macmillan.

CLSI. 2008. *Reference Method for Broth Dilution Antifungal Susceptibility Testing of Yeasts*. 3rd ed. Approved standard M27-A3. Wayne, PA: Clinical and Laboratory Standards Institute.

CLSI. 2009. *Methods for Dilution Antimicrobial Susceptibility Tests for Bacteria That Grow Aerobically*. Approved standard-M07-A9. Wayne, PA: Clinical Laboratory Standards Institute.

Cowan, M. M. 1999. Plant products as antimicrobial agents. *Clinical Microbiology Reviews* 12(4): 564–582.

De Castro, M. L., and Priego-Capote, F. 2010. Soxhlet extraction: Past and present panacea. *Journal of Chromatography A* 1217(16): 2383–2389.

Elangomathavan, R., Suman, T., Beaulah, S. N., Ramesh, S., and Patharajan, S. 2015. Comparative analysis of *Cleistanthus collinus* aqueous leaf extract and fractions for its antibacterial potential. *International Journal of Enteric Pathogens* 3(2): e22946.

Jagadeesan, G., Muniyandi, K., Manoharan, A. L., Thamburaj, S., Sathyanarayanan, S., and Thangaraj, P. 2019. Optimization of phenolic compounds extracting conditions from *Ficus racemosa* L. fruit using response surface method. *Journal of Food Measurement and Characterization* 13(1): 312–320.

Kemp, W. 1991. Infrared spectroscopy. In *Organic Spectroscopy*. London, UK: Springer, Palgrave, 19–99.

Kim, H. J., Kim, D. J., Karthick, S. N., Hemalatha, K. V., Raj, C. J., Ok, S., and Choe, Y. 2013. Curcumin dye extracted from *Curcuma longa* L. used as sensitizers for efficient dye-sensitized solar cells. *International Journal of Electrochemical Science* 8(6): 8320–8328.

Kumar, A., Jhadwal, N., Lal, M., and Singh, M. 2012. Antibacterial activity of some medicinal plants used against UTI causing pathogens. *International Journal of Drug Development and Research* 4(2): 278–283.

Mahato, S. B., and Sen, S. 1997. Advances in triterpenoid research, 1990–1994. *Phytochemistry* 44(7): 1185–1236.

Manoharan, A. L., Thamburaj, S., Muniyandi, K., Jagadeesan, G., Sathyanarayanan, S., Nataraj, G., and Thangaraj, P. 2019. Antioxidant and antimicrobial investigations of *Elaeocarpus tectorius* (Lour.) Poir. fruits against urinary tract infection pathogens. *Biocatalysis and Agricultural Biotechnology* 20: 101260.

Ming, D. S., Hillhouse, B. J., Guns, E. S., Eberding, A., Xie, S., Vimalanathan, S., and Towers, G. N. 2005. Bioactive compounds from *Rhodiola rosea* (Crassulaceae). *Phytotherapy Research: An International Journal Devoted to Pharmacological and Toxicological Evaluation of Natural Product Derivatives* 19(9): 740–743.

Morris, A. K., and Masterton, R. G. 2002. Antibiotic resistance surveillance: Action for international studies. *Journal of Antimicrobial Chemotherapy* 49(1): 7–10.

Muniyandi, K., George, E., Mudili, V., Kalagatur, N. K., Anthuvan, A. J., Krishna, K., and Natarajan, G. 2017. Antioxidant and anticancer activities of *Plectranthus stocksii* Hook. f. leaf and stem extracts. *Agriculture and Natural Resources* 51(2): 63–73.

Murugan, R., and Parimelazhagan, T. 2014. Comparative evaluation of different extraction methods for antioxidant and anti-inflammatory properties from *Osbeckia parvifolia* Arn.—An in vitro approach. *Journal of King Saud University-Science* 26(4): 267–275.

Muthuswamya, R. 2014. Pharmacognostical studies on the fruit of *Elaeocarpus oblongus* Gaertn. *Pharmacognosy Journal* 6(3): 72–78.

Nabbugodi, W. F., Gichuhi, J. W., and Mugo, N. W. 2015. Prevalence of urinary tract infection, microbial aetiology, and antibiotic sensitivity pattern among antenatal women presenting with lower abdominal pains at Kenyatta National Hospital, Nairobi, Kenya. *Open Access Journal of Science and Technology* 3 : 1–6.

Nicolle, L. E. 2008. Uncomplicated urinary tract infection in adults including uncomplicated pyelonephritis. *Urologic Clinics of North America* 35(1): 1–12.

Okwu, D. E., and Okwu, M. E. 2004. Chemical composition of *Spondias mombin* Linn plant parts. *Journal of Sustainable Agriculture and the Environment* 6(2): 140–147.

Olurinola, P. F. 1996. A laboratory manual of pharmaceutical microbiology. *Idu, Abuja, Nigeria* 69: 1–105.

Parekh, J. 2007. In vitro screening of antibacterial activity of aqueous and alcoholic extracts of various Indian plant species against selected pathogens from Enterobacteriaceae. *African Journal of Microbiology Research* 1(6): 92–99.

Parivuguna, V. 2008. Antimicrobial properties and phytochemical constituents of *Rheo discolor* hance. *Ethnobotanical Leaflets* 2008(1): 114.

Prieto, P., Pineda, M., and Aguilar, M. 1999. Spectrophotometric quantitation of antioxidant capacity through the formation of a phosphomolybdenum complex: Specific application to the determination of vitamin E. *Analytical Biochemistry* 269(2): 337–341.

Pulido, R., Bravo, L., and Saura-Calixto, F. 2000. Antioxidant activity of dietary polyphenols as determined by a modified ferric reducing/antioxidant power assay. *Journal of Agricultural and Food Chemistry* 48(8): 3396–3402.

Salvatore, S., Salvatore, S., Cattoni, E., Siesto, G., Serati, M., Sorice, P., and Torella, M. 2011. Urinary tract infections in women. *European Journal of Obstetrics & Gynecology and Reproductive Biology* 156(2): 131–136.

Saravanan, S., and Parimelazhagan, T. 2014. In vitro antioxidant, antimicrobial and anti-diabetic properties of polyphenols of *Passiflora ligularis* Juss. fruit pulp. *Food Science and Human Wellness 3*(2): 56–64.

Sathyanarayanan, S., Muniyandi, K., George, E., Sivaraj, D., Sasidharan, S. P., and Thangaraj, P. 2017. Chemical profiling of *Pterolobium hexapetalum* leaves by HPLC analysis and its productive wound healing activities in rats. *Biomedicine & Pharmacotherapy* 95: 287–297.

Sefton, A. M. 2000. The impact of resistance on the management of urinary tract infections. *International Journal of Antimicrobial Agents* 16(4): 489–491.

Shaikh, S., Fatima, J., Shakil, S., Rizvi, S. M. D., and Kamal, M. A. 2015. Antibiotic resistance and extended spectrum beta-lactamases: Types, epidemiology and treatment. *Saudi Journal of Biological Sciences* 22(1): 90–101.

Shan, B., Cai, Y. Z., Brooks, J. D., and Corke, H. 2007. The in vitro antibacterial activity of dietary spice and medicinal herb extracts. *International Journal of Food Microbiology* 117(1): 112–119.

Sharma, A., Verma, R., and Ramteke, P. 2009. Antibacterial activity of some medicinal plants used by tribals against UTI causing pathogens. *World Applied Sciences Journal* 7(3): 332–339.

Siddhuraju, P., and Becker, K. 2003. Studies on antioxidant activities of mucuna seed (*Mucuna pruriens* var *utilis*) extract and various non-protein amino/imino acids through in vitro models. *Journal of the Science of Food and Agriculture* 83(14): 1517–1524.

Siddhuraju, P., and Manian, S. 2007. The antioxidant activity and free radical-scavenging capacity of dietary phenolic extracts from horse gram (*Macrotyloma uniflorum* (Lam.) Verdc.) seeds. *Food Chemistry* 105(3): 950–958.

Soare, J. R., Dinis, T. C., Cunha, A. P., and Almeida, L. 1997. Antioxidant activities of some extracts of *Thymus zygis*. *Free Radical Research* 26(5): 469–478.

Su, J. D., Osawa, T., Kawakishi, S., and Namiki, M. 1987. Antioxidative flavonoids isolated from *Osbeckia chinensis* L. *Agricultural and Biological Chemistry* 51(10): 2801–2803.

Tenney, J., Hudson, N., Alnifaidy, H., Li, J. T. C., and Fung, K. H. 2018. Risk factors for acquiring multidrug-resistant organisms in urinary tract infections: A systematic literature review. *Saudi Pharmaceutical Journal* 26(5): 678–684.

Thangaraj, P. 2016. *Pharmacological Assays of Plant-Based Natural Products*. Geneva, Switzerland: Springer.

Tomren, M. A., Masson, M., Loftsson, T., and Tønnesen, H. H. 2007. Studies on curcumin and curcuminoids: XXXI. Symmetric and asymmetric curcuminoids: Stability, activity and complexation with cyclodextrin. *International Journal of Pharmaceutics* 338(1–2): 27–34.

Toner, L., Papa, N., Aliyu, S. H., Dev, H., Lawrentschuk, N., and Al-Hayek, S. 2016. Extended-spectrum beta-lactamase-producing Enterobacteriaceae in hospital urinary tract infections: Incidence and antibiotic susceptibility profile over 9 years. *World Journal of Urology* 34(7): 1031–1037.

Ubaid, J. M., Hussein, H. M., and Hameed, I. H. 2016. Analysis of bioactive compounds of *Tribolium castaneum* and evaluation of anti-bacterial activity. *International Journal of Pharmaceutical and Clinical Research* 8(7): 655–670.

Urbano, M., De Castro, M. D. L., Pérez, P. M., García-Olmo, J., and Gomez-Nieto, M. A. 2006. Ultraviolet–visible spectroscopy and pattern recognition methods for differentiation and classification of wines. *Food Chemistry 97*(1): 166–175.

World Health Organization (WHO). 2002. *The World Health Report 2002: Reducing Risks, Promoting Healthy Life*. World Health Organization.

Zhishen, J., Mengcheng, T., and Jianming, W. 1999. The determination of flavonoid contents in mulberry and their scavenging effects on superoxide radicals. *Food Chemistry* 64(4): 555–559.

18 Evaluation of Anti-diabetic Activity of *Cleome gynandra* in Alloxan Induced Diabetic Wistar Albino Rats

T. Shanmuganathan and A. V. P. Karthikeyan

CONTENTS

18.1 INTRODUCTION

Cleome gynandra L. (*C. gynandra*), commonly known as cat's whisker and spider flower in English, belongs to the family Cleomaceae (formerly placed in Capparidaceae). It is one of the staple foods in most of the African countries. It has a wide range of medicinal uses and is used in traditional systems of medicine all over the world. The medicinal property of the plant is mainly due to the presence of biologically active substances like flavonoids, polyphenols, saponins, triterpenes, and so on.

Diabetes mellitus is one of the dangerous, fast-growing, and common disorders affecting all types of people around the world. The incidence of this disease among people increases every year. The deficiency or metabolic inactive nature of insulin leading to this disease can be cured by plant-based drugs. It is inevitable that people will need to use plant-based drugs to cure this metabolic disorder. Many researchers ascertained that chemically induced *Diabetes mellitus* in animal models has been cured through plant-based preparations (Syed et al. 2005; Daisy and Rajathi 2009; Ramakrishnan et al. 2011; Rekha et al. 2011; Srilakshmi et al. 2013).

C. gynandra is one such plant with medicinal properties like anti-diabetic (Karimulla et al. 2013), anthelmintic and anti-microbial (Ajaiyeoba 2000), antioxidant (Atawodi 2005; Anbazhagi et al. 2009), anti-inflammatory (Narendhirakannan et al. 2007), anti-nociceptive activity (Ghogare et al. 2009), immunomodulatory (Kori et al. 2009), and anti-tumor activity (Bala et al. 2010). The present study was done with the aim of producing a possible drug for *Diabetes mellitus* from *C. gynandra*. This study explored the anti-diabetic effects of the ethanolic extracts of *C. gynandra* against alloxan-induced diabetes in Wistar albino rats.

18.2 MATERIALS AND METHODS

18.2.1 Chemicals and Reagents

All the solvents and other reagents used in the present study are of analytical grade and purchased from HiMedia Laboratories, India.

18.2.2 Plant Material

C. gynandra L. was collected from the Mayanur and Thanthonimalai Villages of the Karur District, Tamil Nadu, India, and used as an in vivo sample (Figure 18.1). From in vivo plants, in vitro plants (Figure 18.2) were produced through micropropagation (Karthikeyan et al. 2014).

18.2.2.1 Preparation of Extracts

The shade-dried shoots (excluding reproductive parts of in vivo and in vitro shoots) of *C. gynandra* were packed in Soxhlet apparatus for ethanolic extraction (95% ethanol). The yield of ethanolic extracts was used for anti-diabetic studies.

FIGURE 18.1 Morphological characteristics of *Cleome gynandra* L. (a) Habit; (b) inflorescence, (c) flower showing androecium and gynoecium; (d) fruits; and (e) seeds.

FIGURE 18.2 Micropropagation of *Cleome gynandra* L. from shoot tip explants. (a) Shoot initiation on Murashige and Skoog (MS) medium with cytokinins; (b) and (c) shoot multiplication after 20 days; (d) shoot elongation after 30 days; and (e) and (f) shoot proliferation after 45 days.

18.2.3 Preparation of Animals

The experiment was carried out using Wistar albino rats (150–200 g) procured from the Animal House, Nandha College of Pharmacy, Erode, India. The animals were allocated randomly to treatment groups in polypropylene cages with paddy husks as bedding. The animals were housed at a temperature of 24°C ± 2°C, a relative humidity of 30%–70%, and a 12/12 hours light/dark cycle was followed. All animals were allowed free access to water and fed with standard commercial rat chew pellets (Hindustan Lever Ltd, Mumbai). All the experimental procedures and protocols used in this study were reviewed by the Institutional Animal Ethics Committee (Register no: NCP/IAEC/No: 07/2013-14) and were in accordance with the guidelines of the Committee for the Purpose of Control and Supervision of Experiments on Animals (CPCSEA).

18.2.3.1 Induction and Treatment of Diabetes

After fasting for 18 hours, 25 rats were injected intraperitoneally with a single dose of 150 mg/kg of alloxan that had been dissolved in distilled water. After the injection, they were allowed free access to food and water and were given 5% glucose solution to drink overnight to counter the hypoglycemic shock. The development of diabetes was confirmed after 48 hours of the alloxan injection. The animals having fasting blood glucose levels of more than 200 mg/dL were selected for experimentation.

Group I—the normal control animals received distilled water 5 mL/kg PO (per oral) for 21 days. Group II—the alloxan-induced diabetic animals received distilled water 5 mL/kg PO for 21 days. Group III—the alloxan-induced diabetic animals received the standard drug glibenclamide 5 mg/kg PO for 21 days. Group IV—the alloxan-induced diabetic animals received the ethanolic extract of the in vivo plant sample (150 mg/kg) PO for 21 days. Group V—the alloxan-induced diabetic animals received the ethanolic extract of the in vitro plant sample (150 mg/kg) PO for 21 days. The test drugs were administered by suspension in 0.5% carboxy methyl cellulose solution. All the groups of animals received the treatment for 21 days. The blood samples were collected

TABLE 18.1
Anti-diabetic Activity of In Vivo and In Vitro Shoots of *C. gynandra* L. in Alloxan-Induced Wistar Albino Rats

	Blood Glucose Level (mg/dL)				
Animal Model	**Initial**	**4th Day**	**7th Day**	**14th Day**	**21st Day**
Normal control	114.0 ± 3.65	118.00 ± 3.47***	116.33 ± 1.71***	118.00 ± 2.05***	119.1 ± 2.24***
Diabetic control	105.17 ± 2.61	213.33 ± 3.63	213.83 ± 4.29	224.50 ± 3.11	223.83 ± 2.67
Glibenclamide (5 mg/kg)	114.67 ± 2.09	201.50 ± 3.70	188.33 ± 5.00*	146.83 ± 5.74**	126.83 ± 3.67***
In vivo shoots (150 mg/kg)	112.35 ± 3.56	209.23 ± 4.55	186.00 ± 6.75*	145.22 ± 5.36**	124.21 ± 5.21***
In vitro shoots (150 mg/kg)	109.32 ± 2.39	217.66 ± 4.75	189.24 ± 6.26*	144.36 ± 4.24**	121.32 ± 4.41***

Note: Values are in mean ± SEM (n = 5), *P < 0.05, **P < 0.01, and ***P < 0.001 vs. diabetic control.

2 hours after the drug administration on the 4th, 7th, 14th, and 21st days to determine the blood glucose level using a OneTouch Glucometer (Table 18.1).

18.2.4 Statistical Analysis

The values were expressed as mean ± SEM. The statistical analysis was carried out by one-way analysis of variance followed by Dunnett's "t"-test. P values <0.05 were considered significant.

18.3 RESULTS AND DISCUSSION

In the alloxan-induced diabetic rats, there was a significant increase in the mean blood sugar levels on the 4th day. Anti-diabetic activity was observed in the ethanolic extracts (150 mg/kg) of the in vitro and in vivo shoots of *C. gynandra* against the alloxan-induced diabetes in the Wistar albino rats on the 14th and 21st days. The mean blood sugar levels were very near to the glibenclamide group (5 mg/kg), which was used as the reference standard.

The mean blood sugar levels in the normal control animals on the 14th and 21st days were 118.00 ± 2.05 mg/dL and 119.1 ± 2.24 mg/dL, whereas the mean blood sugar levels in the diabetic control animals on the 14th and 21st days were 224.50 ± 3.11 mg/dL and 223.83 ± 2.67 mg/dL (Table 18.1). The ethanolic extracts (150 mg/kg) of the in vivo and in vitro shoots showed a better reduction in the blood sugar levels of 124.21 ± 5.21 mg/dL and 121.32 ± 4.41 mg/dL, respectively, on the 21st day. This reduction in blood sugar level was very close to the standard drug glibenclamide (126.83 ± 3.67 mg/dL).

All the treated animals showed blood sugar levels near to the normal control animals, whereas the alloxan-induced diabetic animals which were not treated showed high blood sugar levels, when compared to the treated ones (Table 18.1). The ethanolic leaf extracts of *C. gynandra* (200 mg/kg) against the alloxan-induced diabetes in the rats showed a significant (p < 0.001) reduction in the blood sugar level on the 15th day (Karimulla et al. 2013).

Similar hypoglycemic effects were observed by different scientists with different plants against alloxan-induced diabetic animal models (Daisy and Rajathi 2009). The ethanolic extract of the *Canthium dicoccum* whole plant at a dose of 200 mg/kg reduced blood glucose levels significantly (p < 0.001) in alloxan-induced diabetes in rats on the 21st day in another study (Santhan et al. 2013). The ethanol, chloroform, and aqueous extracts of *Coccinia indica* fruits were tested against alloxan-induced diabetes in rats, and among the tested extracts, ethanolic 250 mg/kg was the best to reduce blood glucose levels in another study (Ramakrishnan et al. 2011).

18.4 CONCLUSIONS

The ethanolic extracts of in vivo and in vitro shoots of *C. gynandra* (150 mg/kg) showed significant (P < 0.01 and P < 0.001) reductions in blood sugar levels.

ACKNOWLEDGMENT

The authors wish to thank the University Grants Commission, New Delhi for providing financial assistance to carry out the major research project on *C. gynandra* L. (F. No.: 43-143/2014(SR) dated 21.07.2015, University Grants Commission, New Delhi).

REFERENCES

Ajaiyeoba, E. O. 2000. Phytochemical and antimicrobial studies of *Gynandropsis gynandra* and *Buchholzia coriaceae* extracts. *African Journal of Biomedicine* 3: 161–165.

Anbazhagi, T., Kadavul, K., Suguna, G., and Petus, A. J. A. 2009. Studies on the pharmacognostical and *in vitro* antioxidant potential of *Cleome gynandra* Linn. leaves. *Natural Product Radiance* 8(2): 151–157.

Atawodi, S. E. 2005. Antioxidant potential of African medical plants. *African Journal of Biotechnology* 4(2):128–133.

Bala, A., Kar, B., Haldar, P. K., Mazumder, U. K. and Bera, S. 2010. Evaluation of anticancer activity of *Cleome gynandra* on Ehrlich's Ascites carcinoma treated mice. *Journal of Ethnopharmacology* 129: 31–34.

Daisy, P., and Rajathi, M. 2009. Hypoglycemic effects of *Clitoria ternatea* Linn. (Fabaceae) in alloxan-induced diabetes in rats. *Tropical Journal of Pharmaceutical Research* 8(5): 393–398.

Ghogare, U. R., Nirmal, S. A., Patil, R. Y., and Kharya, M. D. 2009. Antinociceptive activity of *Gynandropsis gynandra* leaves. *Natural Product Research* 23(4): 327–333.

Karimulla, S., Afsar S., Deepak, K., and Devi, K. 2013. Evaluation of preliminary phytochemical properties and hypoglycemic activity of *Cleome gynandra* L. *International Journal of Pharmacy and Pharmaceutical Science* 5(3): 824–828.

Karthikeyan, A. V. P., Shanmuganathan, T., and Karthi, C. 2014. *In vitro* propagation of *Cleome gynandra* L. through direct organogenesis. *International Journal of Science and Research* 3(7): 1368–1372.

Kori, M. L., Gaur, K., and Dixit, V. K. 2009. An investigation of immunomodulatory potential of *Cleome gynandra*. *Asian Journal of Pharmaceutical and Clinical Research* 2(1): 35–39.

Narendhirakannan, R. T., Subramanian, S., and Kandaswamy, M. 2007. Anti-inflammatory and lysosomal stability action of *Cleome gynandra L.* studied in adjuvant induced arthritic rats. *Food and Chemical Toxicology* 45(6): 1001–1012.

Ramakrishnan, M., Bhuvaneshwari, R., Duraipandiyan, V., and Dhandapani, R. 2011. Hypoglycaemic activity of *Coccinia indica* wight and arn fruits in alloxan-induced diabetic rats. *Indian Journal of Natural Product Resource* 2(3): 350–353.

Rekha, R., Patil, M. B., Shrivastava, A. K., Singh, G. P., and Shamshath, B. 2011. Antidiabetic effect of the aqueous bark extract of *Ficus benghalensis* on alloxan induced diabetic rats. *International Journal of Pharmaceutical Erudition* 1(1): 72–80.

Santhan, S., Janarthan, M., and Zuber, A. M. 2013. Evaluation of anti-diabetic and nephro protective activity of 95% ethanolic extract of *Canthium dicoccum* whole plant by using albino rats. *Journal of Current Pharmaceutical Science* 6(4): 218–222.

Srilakshmi, P., Janarthan, M., and Zuber, Ali M. 2013. Evaluation of anti-diabetic and hepato protective activity of 95% methanolic extract of *Terminalia tomentosa* bark by using albino rats. *International Journal of Research in Pharmacy and Biosciences* 1(5): 664–667.

Syed, M. A., Vrushabendra Swamy, B. M., Gopkumar, P., Dhanapal, R., and Chandrashekara, V. M. 2005. Anti-diabetic activity of *Terminalia catappa* linn. leaf extracts in alloxan-induced diabetic rats. *International Journal of Pharmacy and Technology* 4(1): 36–39.

19 To Evaluate the Role of a Novel Herbal Formulation by Topical Application in Cancer Therapeutics (Mouse Model)

Umrao Ankita, Rao Gururaj A., Rao S. G. A., and Rao Jyothsna A.

CONTENTS

19.1 INTRODUCTION

Cancer is a group of diseases characterized by the abnormal growth of cells that can eventually spread/invade to other organs. It involves the activation of proto-oncogenes, which leads to uncontrolled cell division (Rang et al. 2011), further leading to imbalance in homeostasis and immune status. Cancer is the second leading cause of death globally, and is responsible for an estimated 9.6 million deaths in 2018 (WHO, 2018). Despite many deaths occur due to side-effects of the treatment, present modalities for cancer treatment include—surgery, radiation, and chemotherapy. These mainstay treatments lead to certain side effects (America Cancer Society 2016), including further loss of immunity (Marina et al. 2016; Kang et al. 2009) and the decrease in quality of life (QoL) (Nayak et al. 2017; Theofilou 2012). Herbal medicines are widely used as complementary and alternative medicines and/or integrative medicine (Klaus et al. 2002), which helps in combating the side effects of the present modalities (Shunsuke and Hiroshi 2015) and boosts the immune system which will overall improve the QoL.

Traditional medicines taken orally get metabolized or go through a first-pass mechanism and get degraded, therefore inhibiting the activity of the active component. In this study, we have developed a topical application of a novel herbal formulation so that it does not get metabolized and stay for long in the blood stream. The effect of the herbal formulation has been evaluated with and without a chemotherapy drug at different doses.

19.2 MATERIALS AND METHODS

19.2.1 Ethical Clearance

The study was approved by the Institutional Animal Ethics Committee (IAEC): Committee for the Purpose of Control and Supervision of Experiments on Animals and ethical clearance was obtained from the Karnataka College of Pharmacy, Bangalore, under the registration number KCP-IAEC/2/17-18/3/15-04-17.

19.2.2 Animal Housing

The in vivo studies were performed on Swiss albino male mice (7–8 weeks old), weighing 20–25 g. The animals were kept in ventilated cages in an airconditioned animal facility with 12 hour light/dark cycles and provided with standard mouse food and tap water ad libitum.

19.2.3 Tumor Development

Mice with growing Ehrlich ascites cells (in the peritoneum region) were obtained from the Amala Cancer Research Centre, Thrissur, Kerala and were grown for 10–15 days in vivo. The fur was removed in groups II–V using a hair removal cream before tumor induction. The ascites was removed, counted, and checked for the viability of cells using Trypan blue. Ninety-eight percent viable, 2×10^6 cells/200 μL were injected intramuscularly at the right hind limb region of each mouse for further experiment (Figure 19.1).

19.2.4 Treatment Groups

The animals were divided in eight groups of six animals each:

- Group I (healthy) animals served as the control and were put on a normal diet and given distilled water (vehicle) by oral gavages daily for the entire study duration.
- Group II (EAC + 20 mg/Kg 5-FU) animals with a tumor were treated with standard 5-Flurouracil (20 mg/Kg), intraperitoneally (Ehrlich Ascites Carcinoma (EAC)).
- Group III (EAC + 20 mg/Kg 5-FU + SR033L) and group IV (EAC + 0.2 mg/Kg 5-FU + SRLape) included a low dose (0.2 mg/Kg) and a high dose (20 mg/Kg) of 5-FU intraperitoneally and SRLape (Swastharakshak Lape (coded as SR033L)) topical application, respectively.
- Group V (EAC + SR033L) received only topical application of SRLape (SR033L) after a tumor induction.
- Group VI (EAC control; without treatment) animals served as the carcinogen control (without treatment).

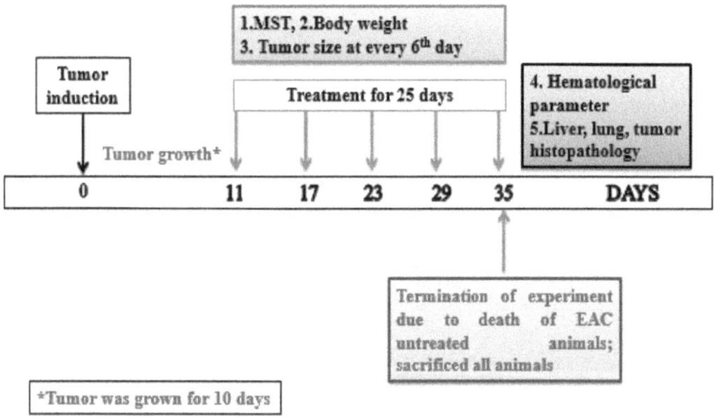

FIGURE 19.1 Work flow for in vivo experiment: Effect of SR033L on Ehrlich ascites solid tumor at right hind limb region.

For all EAC models (groups II–V), treatment was started from day 11 after the tumor induction for 25 days. The experiment was stopped on the 25th day after the tumor induction and parameters were assessed. In groups III–V, the tumor area was cleaned using cotton dipped in warm water, and the SR033L (0.3–0.4 mL/mouse) was applied with the help of a syringe and was spread with a spatula.

19.2.5 Termination of the Experiment and Assessment of Parameters

Animals in all groups were observed for any apparent signs of toxicity and mortality during the entire period of study. The parameters considered for evaluation were survival proportions (%), tumor area (cm^2), and neutrophils:lymphocytes ratio (NLR). The tumor histology was done for animals of all the groups.

19.2.5.1 Differential Blood Count

Blood was collected after euthanizing the animals of the respective groups, and a neutrophils and lymphocytes count was performed.

19.2.5.2 Histology

After the blood collection, the mice were dissected. The tumor was removed and washed with distilled water to remove the outer blood stain and soon was transferred to a 10% neutral formalin fixative. The tissue was kept in a fixative for 48 hours, and then dissected and stained using a hematoxylin and eosin stain.

19.2.6 Statistical Analysis

The data were expressed as mean ± standard deviation values. The statistical significance of the difference between the control and the experimental groups was determined by performing one-way analysis of variance. A *t*-test was used as a post-statistical test. The disease control group is compared with the normal control group, and the treated groups are compared with the disease control group.

19.3 RESULTS

19.3.1 Survival Proportion

The decreased mortality by 33.3% with SR033L (83.3%) as compared to the 20 mg/Kg 5-FU group (50%) and by 16.6% with SR033L + low and high doses of 5-FU (66.6%) as compared to the 20 mg/Kg 5-FU (50%) observed implies the protective effect of topical application, although the tumor burden remains unchanged (Figure 19.2).

19.3.2 Tumor Area

The tumor area of 2 cm^2 was seen in the untreated tumor group (EAC control), which was decreased by 50% by SR033L

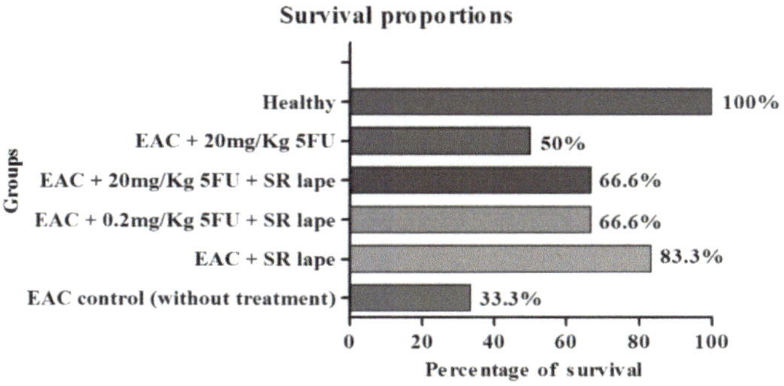

FIGURE 19.2 Survival proportions of different groups.

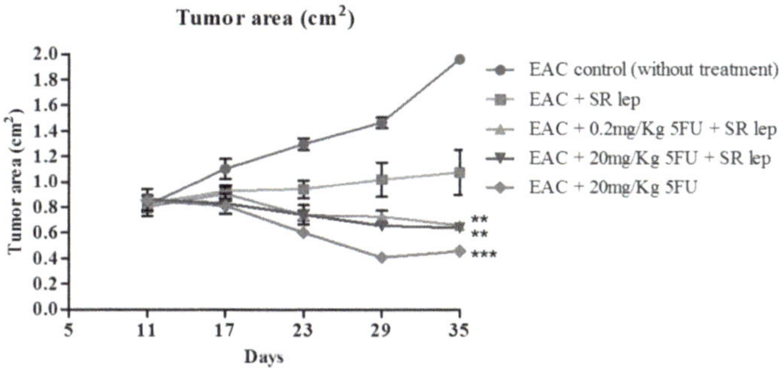

FIGURE 19.3 Tumor area (cm^2) in different groups.

application and further decreased by 80% in SR033L + high and low doses in the 5-FU groups, although the tumor burden was significantly decreased with 20 mg/Kg 5-FU (Figure 19.3).

19.3.3 Immune Status (NLR)

The NLR was decreased by 96% with SR033L as compared to the untreated tumor group and by 90% as compared to the groups treated with 20 mg/Kg 5-FU (Figure 19.4).

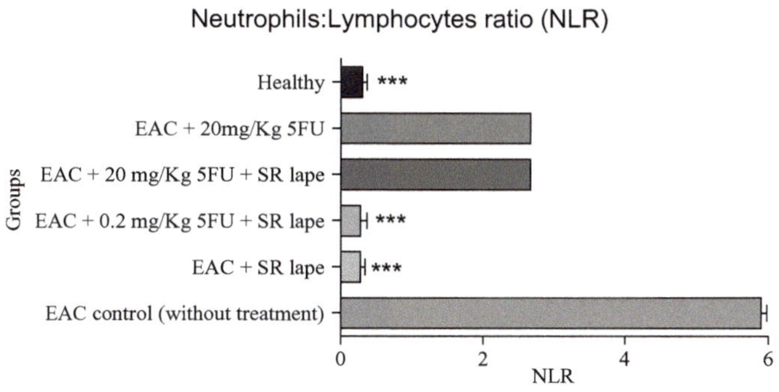

FIGURE 19.4 Neutrophils:lymphocytes ratio in different groups.

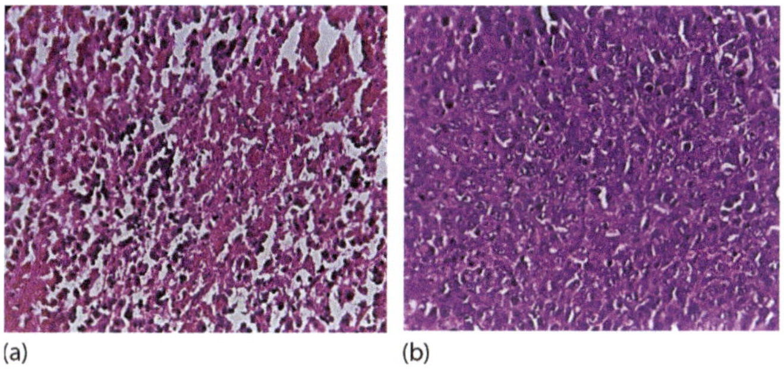

FIGURE 19.5 Tumor biopsy stained with hematoxylin and eosin. (a) EAC control (without treatment) showing dark stained viable tumor cells. (b) EAC treated with SR033L showing mixed lymphocytic cells.

19.3.4 Tumor Histology

No viable tumor was observed in the SR033L group and the tumor cells were replaced by lymphoplasmacytic cells. Tumor infiltrating lymphocytes were absent in the EAC control group (Figure 19.5).

19.4 DISCUSSION

19.4.1 Quality of Life

Quality of life is used to evaluate the general well-being of individuals and societies. QoL is being affected by several illness-related symptoms, as well as treatments, such as chemotherapy in cancer. The SR0333L application has shown to improve the QoL by 33.3%, as compared to the drug and has decreased mortality when used in combination with the drug. The tumor size has decreased significantly with the drug in combination with SR033L, implying that herbal preparation can improve the QoL, although the tumor burden remains unchanged when used alone, nevertheless, it worked equally synergistically with the drug at low and high doses. This suggests the combination can improve the QoL without compromising the drug activity.

19.4.2 Restoration of Homeostasis

The NLR is a valid measurement for chronic stress in the laboratory mouse. A high NLR is a sign of elevated chronic stress (Hickman 2017). Homeostasis was restored by the immunomodulatory effect of the SR033L by a significant decrease in the neutrophil number and increased lymphocyte number (i.e., decreased NLR) in the peripheral blood, which corresponds to a healthy status. Group IV (SR033L + 0.2 mg/Kg 5-FU) and group V (SR033L) restored homeostasis by decreased the NLR ratio in blood.

A high NLR is a better prognosis in a tumor showing the induction of apoptosis in the tumor cells without toxicity. This signifies that the replacement of the tumor by inflammation (lymphoplasmacytic cells), neutrophils at the edges of necrotic modules, and foamy cells at the margins were indications of SR033L application.

19.5 CONCLUSION

This suggests that standard chemotherapeutic drugs can be used in combination with SR033L, but at lower doses, which together will produce an additive anti-cancer effect and also work as a cytoprotective agent. This could potentially decrease the dose of chemotherapy agents, thereby mitigating their oxidative stress in the body.

Thus, a combinational approach can be adopted to reduce the after-effects of chemotherapy (Wei et al. 2017). This is possible by reducing the dose of the chemotherapeutic drugs, but attaining an optimal therapeutic dose for a favorable clinical outcome, keeping the general well-being of the patient.

REFERENCES

America Cancer Society. 2016. Cancer Treatment and Survivorship. Facts and Figures. https://www.cancer.org/content/dam/cancer-org/research/cancer-facts-and-statistics/cancer-treatment-and-survivorship-facts-and-figures/cancer-treatment-and-survivorship-facts-and-figures-2016-2017.pdf. (Accessed January 1, 2016.)

Hickman, D. L. 2017. Evaluation of the neutrophil; lymphocyte ratio as an indicator of chronic distress in the laboratory mouse. *Laboratory Animals (NY)* 46(7):303–307.

Kang, D. H., Weaver, M. T., Park, N. J., et al. 2009. Significant impairment in immune recovery after cancer treatment. *Nursing Research* 58(2):105–114.

Klaus, L., Gerbenter, R., Maria, H., et al. 2002. Systematic reviews of complementary therapies—An annotated bibliography, Part 2: Herbal medicine. *BMC Complementary and Alternative Medicine* 1:5.

Marina, E., Oleg, I. K., Yulia, U., et al. 2016. Assessment of immune status of patients with locally advanced cancer of the upper jaw. *Journal of Clinical Oncology* 34(15):e17562.

Nayak, M. G., George, A., and Vidyasagar, M. S. 2017. Quality of life among cancer patients. *Indian Journal of Palliative Care* 23(4):445–450.

Rang, H. P., Dale, M. M., Ritter, J. M., et al. 2011. *Pharmacology* (7th ed.). London, UK: Churchill Livingstone, Elsevier.

Shunsuke, O., and Hiroshi, T. 2015. Herbal medicines for the treatment of cancer chemotherapy-induced side effects. *Frontiers of Pharmacology* 6:14.

Theofilou, P. 2016. Quality of life outcomes in patients with breast cancer. *Oncology Reviews* 6(1):e2.

Wei, Y., Pu, X., and Zhao. L. 2017. Preclinical studies for the combination of paclitaxel and curcumin in cancer therapy. *Oncology Reports* 37(6):3159–3166.

World Health Organization. Cancer. http://www.who.int/news-room/fact-sheets/detail/cancer, 2018. (Accessed September 12, 2018.)

20 Coconut Sprouts as Phytomedicine
An Approach towards the Development of Nutraceuticals

S. Abiraami Valli and S. Uma Gowrie

CONTENTS

20.1 INTRODUCTION

Medicinal plants are a therapeutic resource for phytomedicine, which are mainly used by the traditional population of the world specifically for health care. Most of the valuable phytochemicals are the secondary metabolites, which may be used as purified compounds or complex mixtures, which serve as medicine, pesticides, herbicides, flavoring agents, food preservatives, etc. The historic and traditional uses of herbal medicine to treat and prevent infectious diseases have been replaced with the emergence of synthetic drugs and anti-microbial agents. However, the worldwide interest in the use of medicinal plants is now increasing, and the beneficial effects of the plant-based medicines and their products, which are known to possess natural anti-oxidants, anti-microbials, cytotoxic properties, and other activities, are being rediscovered, as they serve as a natural approach to treat various diseases (Rochfort and Panozzo 2007).

Arecaceae (Palms) is a very large family, which is of commercial importance, as it is a valuable source of edible fruits, nuts, oil, fiber, and woods. Palm trees include important medicinal and agricultural plants, such as coconut (*Cocos nucifera*), date palm (*Phoenix dactylifera*), oil palm (*Elaeis guineensis*), true sago palm (*Metroxylon sagu*), and rattan (*Calamus* sp.). Palm fruits have been found to contain a number of biological components, namely, carotenes, tocopherols, polyphenols, phenolic acids, and flavonoids with several biological properties (Tan et al. 2007).

Cocos nucifera is the only species of the genus *Cocos*, which includes 27 genera and 600 species (Evans 2002). One of the essential primary natural products from the dry coconut fruit is the coconut oil, which has been used as traditional medicine, functional food, and also in pharmaceuticals. It is referred to as "miracle oil." The coconut tree itself is known as "the Tree of Life," as each of its parts are being used for various purposes. These are highly nutritious and are enriched with bioactive compounds with great medicinal value (Siriphanich et al. 2011). The constituents of *C. nucifera* have some of the biological effects, such as anthelminthic, anti-inflammatory, anti-nociceptive, anti-oxidant, anti-fungal, anti-bacterial, and anti-tumor activities. In addition, other properties, such as anti-hypertensive, cardioprotective, anti-seizure, cytotoxicity, hepatoprotective, vasodilation, nephroprotective, and anti-osteoporosis effects were also reported. Because each part of the *C. nucifera* has different constituents, the pharmacological effects of the plant vary according to the part of the plant evaluated (Lima et al. 2015).

An essential plant product which can be used as a phytomedicine is the sprout of the *Cocos nucifera* L. On the germination of the coconut, the basal part of the embryo, which is embedded in the solid endosperm near the germinating pore, enlarges to form a cotyledonary structure called sprout (haustorium). The coconut sprout is one of the edible parts of the coconut.

These coconut sprouts have been identified as possessing a high nutrient value, including proteins, carbohydrates, low calories, higher content of fibers, vitamins, and enzymes, which boost the metabolic processes as well as the chemical reactions within the body aiding digestion and minimizing the cardiac damage induced by isoproterenol (Chikku and Rajamohan 2012). They are also used to reduce high blood pressure, treat anemia, diarrhea, constipation, lower the cholesterol level, improve digestion, and also help to prevent colorectal cancer. The inhibitory action of the coconut sprout on ghrelin, the hunger hormone, is also identified. The sprouts contain omega-3 fatty acids, which are considered to lower the risk of heart attacks and strokes by reducing the levels of Low-Density Lipoprotein (LDL) cholesterol. The potassium in the sprouts also helps to reduce blood pressure (Anurag and Rajamohan 2003). Apart from all these properties, the coconut sprouts, especially by dilating the blood vessels, increase the circulation and oxygenation, reduce the risk of heart attack, cardiac arrest, strokes, and also exhibit anti-diabetic properties (Preetha et al. 2012). Coconut sprouts are also enriched with various phytoconstituents and anti-oxidants (vitamins), which possess strong anti-bacterial, anti-inflammatory, and anti-oxidant activities.

Hence, the main aim and objective of the present study are to analyze the different medicinal and biological properties by giving scientific validation to the existing bioactive compounds present in the coconut sprouts, and then recommending the coconut sprouts as a natural product and nutraceutical for the treatment of a bacterial infection leading to inflammation.

20.2 MATERIALS AND METHODS

20.2.1 Sample Collection and Preparation of Crude Extracts

The fresh sprouts of a *Cocos nucifera* L. were purchased from local markets in Chennai, Tamil Nadu, India. These fresh sprouts were used for further study. The dried sample analysis was carried out with 100 g of the coconut sprouts using the shade dry method for three weeks. Then they were ground using a blender and stored in airtight containers for further analysis (Figures 20.1 and 20.2). Further, the crude extracts were prepared by the cold percolation method using different solvents like butanol, acetone, methanol, and water (aqueous) separately (1:10) (Eloff 1998).

20.2.2 Qualitative Phytochemical Screening

Butanol, acetone, methanol, and aqueous extracts of the fresh and dried coconut sprouts (CB, fresh coconut sprouts extracted with butanol; CAc, fresh coconut sprouts extracted with acetone; CM, fresh coconut sprouts extracted with methanol; CA,

FIGURE 20.1 Fresh coconut sprouts.

FIGURE 20.2 Dried coconut sprouts.

fresh coconut sprouts extracted with aqueous; CBD, dried coconut sprouts extracted with butanol; CAcD, dried coconut sprouts extracted with acetone; CMD, dried coconut sprouts extracted with methanol; CAD, dried coconut sprouts extracted with aqueous ovalbumin (OVA) induced) were used for the phytochemical analysis. Qualitative phytochemical tests for the identification of alkaloids, saponins, terpenoids, glycosides, steroids and triterpenoids, resin, quinone, gum and mucilage, coumarin, anthroquinone, proteins and amino acids, anthocyanin and betacyanin, carbohydrates, phlobatannins, flavonoids, cardiac glycosides, phenols, tannins, phytosterol, polyphenols, fixed oils and fats, and fatty acids were carried out for all four extracts using standard protocols (Sofowora 1993; Harborne 1998; Raaman 2006).

20.2.3 Quantification of the Phytoconstituents

The methanol and aqueous extracts of the fresh and dried coconut sprouts which showed prominent phytochemicals were taken for further analysis. Using a UV spectrophotometer (UV 1650PC Shimadzu), the quantification of the phytoconstituents, such as the total soluble sugars by the Dinitrosalicylic acid method (Miller 1972; Nelson 1944), proteins by the Lowry method (Lowry et al. 1951), flavonoids by the aluminum chloride method (Patel et al. 2012), and terpenoids by the Ferguson method (Ferguson 1956) were carried out, and the amount of phytic acid was also quantified (Wheeler and Ferrel 1971).

20.2.4 Fourier Transform Infrared Spectrophotometer Analysis

A spectrum Fourier transform infrared spectrophotometer (FTIR) system (Shimadzu, IR Affinity 1, Japan), equipped with a Deuterated Lanthanum α Alanine doped TriGlycine Sulphate (DLATGS) detector with a mirror speed of 2.8 mm/sec, scan range: from 400 to 4000 cm^{-1} with a resolution of 4 cm^{-1} was used for the FTIR analysis. Firstly, the methanol and aqueous extracts of the fresh and dried coconut sprouts were prepared. Then the extracts were evaporated using a flash evaporator, then they were mixed with KBr salt, using a mortar and pestle, and compressed into a thin pellet. Later, infrared spectra were recorded from the KBr pellet on a Shimadzu FTIR spectrometer 4000–5000 cm^{-1}.

20.2.5 Anti-bacterial Assay

The anti-bacterial study was carried out by the agar well diffusion method using Mueller-Hinton agar media with methanol and aqueous extracts of the fresh coconut sprouts. Streptomycin was used as a positive control. Different concentrations of the sprout extracts (50 µg/mL, 75 µg/mL, 100 µg/mL) were assayed against *Staphylococcus aureus*, *Escherichia coli*, *Salmonella typhi*, *Klebsiella pneumoniae*, and *Shigella flexneri*. The bacterial cultures were obtained from the Department of Microbiology, Ethiraj College for Women, Chennai. The triplicate plates of all the samples were incubated at 37°C for bacterial growth, and the zone of inhibition around the well was observed after 24 hours.

20.2.6 In Vitro Anti-inflammatory and Anti-oxidant Assays

The in vitro anti-inflammatory assay (inhibition of the albumin denaturation method) (Mizushima and Kobayashi 1968) and in vitro anti-oxidant assay by a 1,1-diphenyl-2-picrylhydrazyl (DPPH) free radical scavenging activity (Ahmad et al. 2013) using a UV Spectrophotometer (UV 1650PC Shimadzu) of the methanol and aqueous extracts of the fresh and dried coconut sprouts (100–500 µg/mL) were carried out. The experiment was conducted in triplicates, and the values were expressed as equivalents of ascorbic acid in µg/mg of the extract.

20.2.7 Gas Chromatography-Mass Spectrometry Analysis

The gas chromatography-mass spectrometry (GC-MS) analysis of the methanol extract of the fresh coconut sprouts (100 µL) was carried out using the standard method (Adams 1989). A Shimadzu GC-2010 plus gas chromatograph (equipped with a straight deactivated 2 mm direct injector liner and a 15 m Alltech EC-5 column-250 µ I.D., 0.25 µ film thickness) was used. The sample was introduced using a split injection (split ratio-10:1). The oven temperature was programmed to begin at 35°C, then a hold for 2 minutes, then a shift at 20°C per minute to 450°C, and a hold for 5 minutes. Then the helium carrier gas was set at a constant flow mode (2 mL/minute flow rate). The mass spectrum had a direct connection with the capillary column metal quadrupole mass filter pre-rod mass spectrometer operating in an electron ionization mode with software GC-MS solution ver. 2.6 that was used for the analysis. The low-resolution mass spectra were acquired at a resolving power of 1000 (20% height definition) and scanning from m/z 25 to mass to charge ratio (m/z) 1000 at 0.3 seconds per scan with a 0.2 second inter-scan delay. The high resolution mass spectra were acquired at a resolving power of 5000 (20% height definition) and scanning the magnet from m/z 65 to m/z 1000 at 1 second per scan. The identification of the components of the compound was matched with their recorded spectra from the data bank of mass spectra of National Institute of Standards and Technology (NIST) library version 11 provided by the instruments software. The GC/MS metabolomics database was used for the similarity search with a retention index.

20.2.8 In Silico Docking

The compounds identified by the GC-MS analysis of the coconut sprouts were screened against *Helicobacter pylori* (target protein). Then the target molecule was retrieved from the Protein Data Bank, and the basic information about the bioactive compound was retrieved from the PubChem database. Then the bioactive compound was docked against the target protein in the Mcule database. Docking results in interactions between the target and ligand molecules. Diclofenac sodium was used as a standard anti-inflammatory compound (Muhammad and Fatima 2015).

20.2.9 MTT Assay

It is a colorimetric assay that measures the reduction of yellow 3-(4,5-Dimethylthiazol-2-yl)-2,5-diphenyltetrazolium bromide (MTT) by mitochondrial succinate dehydrogenase (Mosmann 1983). An adenocarcinoma gastric cell line was used to determine the cell cytotoxicity activity of the cells obtained from National Centre for Cell Science, Pune, India. The sprout sample concentrations (100–500 μg/mL) were used for the assay.

20.2.10 Statistical Analysis

For each experiment, the data presented are the means of three replicates. The values are expressed as the mean ± Standard Error (SE) of three replicates.

20.3 RESULTS

20.3.1 Qualitative Phytochemical Screening

The phytochemical screening (fresh and dried coconut sprouts) using different solvents, namely, butanol, acetone, methanol, and aqueous (water) revealed the presence of phytoconstituents chiefly in the methanol and aqueous solvents. The fresh and dried coconut sprouts indicated the presence of alkaloids, saponins, terpenoids, steroids, triterpenoids, resin, quinone, proteins, amino acids, carbohydrates, flavonoids, cardiac glycosides, phenols, tannins, fixed oils, fats, and fatty acids. The fresh coconut sprouts only showed the presence of betacyanin.

20.3.2 Quantification of the Phytoconstituents

The methanol extract of the fresh coconut sprouts showed the maximum total soluble sugars (0.60 ± 0.1 mg/g of glucose), protein content (39 ± 0.8 mg/mL of protein), flavonoid content (0.28 ± 0.1 mg QE/g), and terpenoid content (95 ± 1.2 mg/g), with reduced levels of phytic acid (0.12 ± 0.009 mg/g) (Figures 20.3 through 20.5).

20.3.3 FTIR Analysis

The FTIR spectra of the fresh and dried coconut sprouts (methanol and aqueous extracts) showed varied functional groups (Tables 20.1 through 20.4).

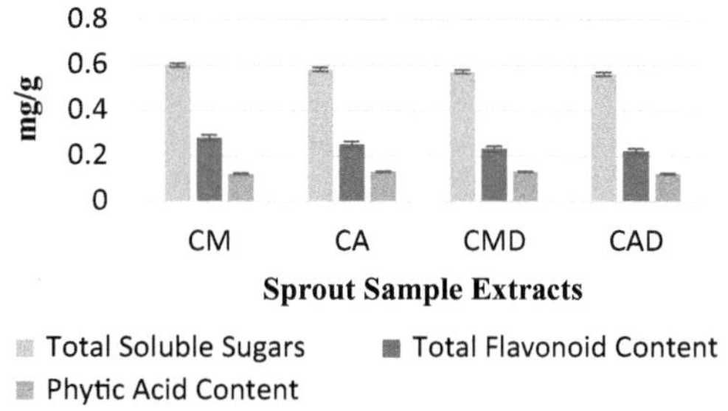

FIGURE 20.3 Total soluble sugars, total flavonoid, and phytic acid content of the coconut sprouts.

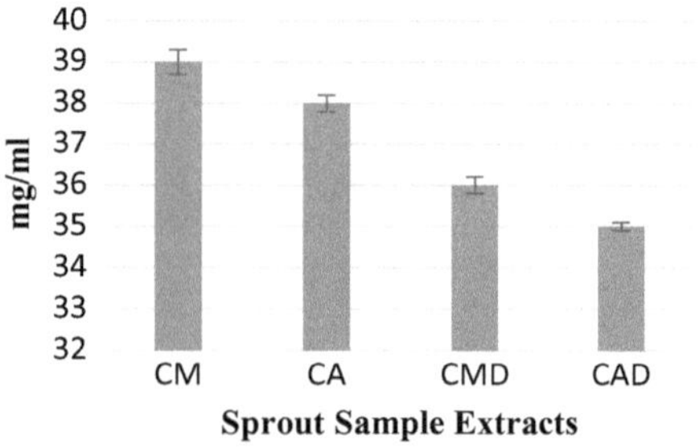

FIGURE 20.4 Total protein content of the coconut sprouts.

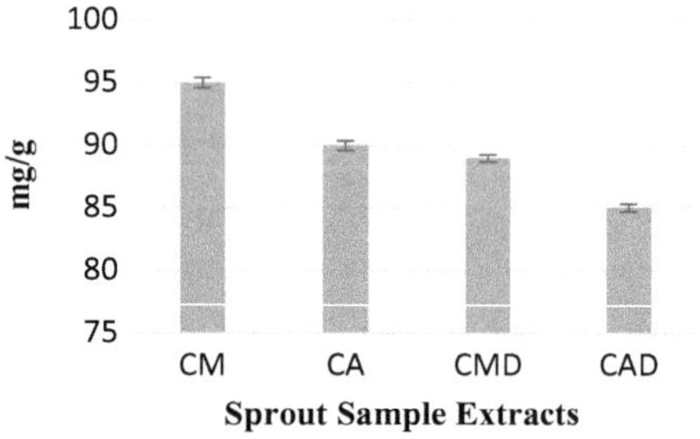

FIGURE 20.5 Total terpenoid content of the coconut sprouts.

20.3.4 Anti-Bacterial Assay

The anti-bacterial assay (fresh methanol and aqueous extracts) showed more prominent anti-bacterial activity than the dried sprouts. The methanol extract of fresh coconut sprouts (CM) at 100 μg/mL showed the maximum zone of inhibition (32 ± 0.6 mm) against *Shigella flexneri* and the minimum zone of inhibition (5 ± 0.1 mm) against *Escherichia coli*. The percent over the control (methanol and aqueous extracts) was found to be 10–25 percent (Figures 20.6 through 20.11).

20.3.5 In Vitro Anti-inflammatory and Anti-oxidant Assays

The results obtained for the anti-inflammatory and anti-oxidant activities revealed percent inhibition in a dose-dependent

TABLE 20.1
FTIR Analysis of the Methanol Extract of Fresh Coconut Sprouts (CM)

Peak Frequencies	Chemical Bond	Functional Group
431.11, 461.97, 470.65, 475.47, 487.05, 518.87, 531.41, 542.02, 559.38, 571.92	C-Br Stretch	Alkyl halides
843.89	C-Cl Stretch	Alkyl halides
932.62	P-OR Esters	Esters
1049.32, 1255.71, 1348.3	C-F Stretch	Alkyl halides
1396.52	C-O Stretch	Carboxylic acids
1415.81, 1432.21	S=O Sulfate ester	Esters
1614.49	C=C Stretch	Alkenes
1636.67	NH out of plane	Amides

TABLE 20.2
FTIR Analysis of the Aqueous Extract of Fresh Coconut Sprouts (CA)

Peak Frequencies	Chemical Bond	Functional Group
433.04, 448.47, 472.58, 537.2, 549.74, 578.67	C-Br Stretch	Alkyl halides
704.05, 839.07	= CH out of plane	Alkenes
848.72	C-Cl Stretch	Alkyl halides
1038.71, 1093.69, 1160.23, 1238.35	C-F Stretch	Alkyl halides
1373.38	CH2 and CH3	Alkanes
1393.63, 1413.88, 1447.64	S=O Sulfate ester	Esters
1641.49	C=C Stretch	Alkenes
1743.72, 1874.99	C=O Stretch doublet	Ketones
2140.11	C=C Stretch	Alkynes
2298.28, 2404.37, 2439.09	P-H Phosphine	Phosphines
2681.17	(O=) PO-H Phosphoric acid	Phosphoric acid
2800.76	Dimer OH	Carboxylic acids
2862.49, 2928.07	CH Stretch	Alkanes
3432.48, 3437.3	O-H Stretch	Alcohols

TABLE 20.3
FTIR Analysis of the Methanol Extract of Dried Coconut Sprouts (CMD)

Peak Frequencies	Chemical Bond	Functional Group
464.86, 563.24, 591.21	C-Br Stretch	Alkyl halides
645.22	C-H Bend	Alkynes
1050.29, 1247.03	C-F Stretch	Alkyl halides
1420.63, 1443.78	S=O Sulfate ester	Esters
1604.84	C-O Stretch	Carboxylic acids
1622.2	NH out of plane	Amides
1741.8	C=O Stretch outlet	Ketones
2102.5	C=C Stretch	Alkynes
2347.47, 2445.84	P-H Phosphine	Phosphines
2866.34	C-H Stretch	Alkanes
2927.1	C-H Stretch	Alkanes
3381.36	Dimer OH	Carboxylic acid

manner. The methanol extract of the fresh coconut sprouts (CM) showed the percent maximum inhibition (anti-inflammatory property) of 85 ± 1.2 at 500 μg/mL with an The biochemical half maximal inhibitory concentration (IC_{50}) value of 211.9 μg/mL (Figure 20.12) and the percent maximum inhibition (DPPH radical scavenging activity) of 88 ± 1.2 at 500 μg/mL with an IC_{50} value of 169.9 μg/mL (Figure 20.13).

20.3.6 GC-MS Analysis

The methanol extract of the fresh coconut sprouts was found to contain a higher amount of phytoconstituents and also showed

TABLE 20.4
FTIR Analysis of the Aqueous Extract of Dried Coconut Sprouts (CAD)

Peak Frequencies	Chemical Bond	Functional Group
584.46	C-Br Stretch	Alkyl halides
667.4	C-H Bend	Alkynes
1043.53, 1150.59, 1248	C-F Stretch	Alkyl halides
1450.53	CH2 and CH3	Alkanes
1602.91	C-O Stretch	Carboxylic acids
1742.76	C=O Stretch doublet	Ketones
2153.62	R-N-C	Isocyanides
2248.13	C=C Stretch	Alkynes
2349.4	P-H Phosphine sharp	Phosphines
2447.77, 2693.71	(O=) PO-H Phosphoric acid	Phosphoric acid
2777.61	RCHO C-H Stretch	Aldehydes
2861.52, 2927.1	CH Stretch	Alkanes
3421.87	O-H Stretch	Alcohols

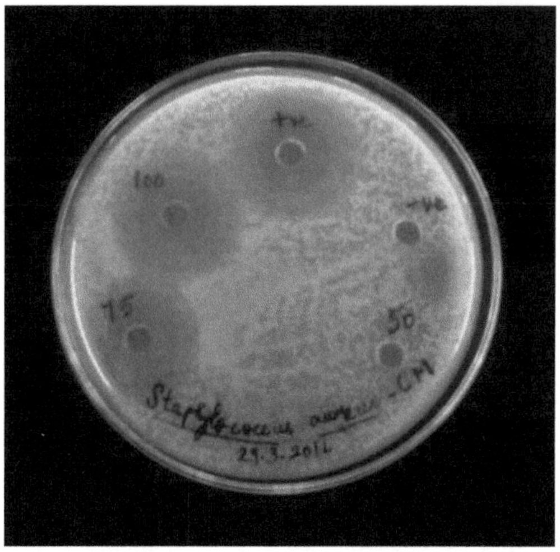

FIGURE 20.6 CM against *Staphylococcus aureus*.

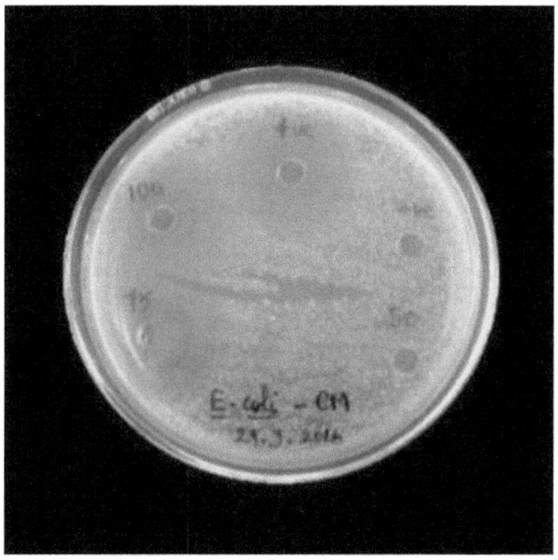

FIGURE 20.7 CM against *Escherichia coli*.

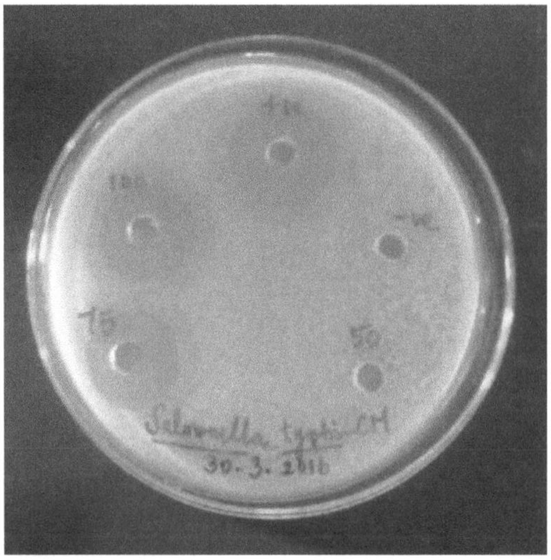

FIGURE 20.8 CM against *Salmonella typhi.*

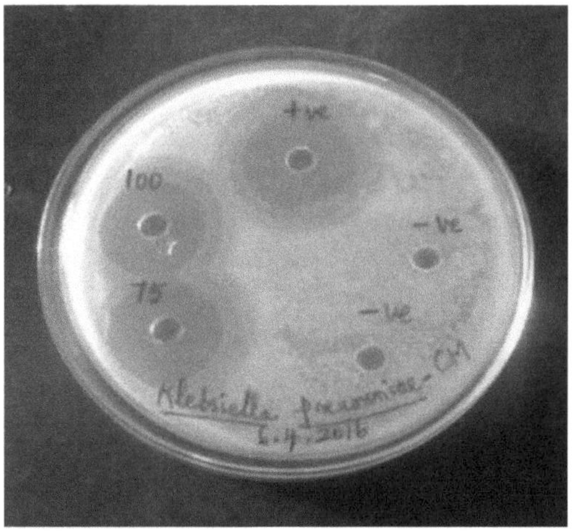

FIGURE 20.9 CM against *Klebsiella pneumoniae.*

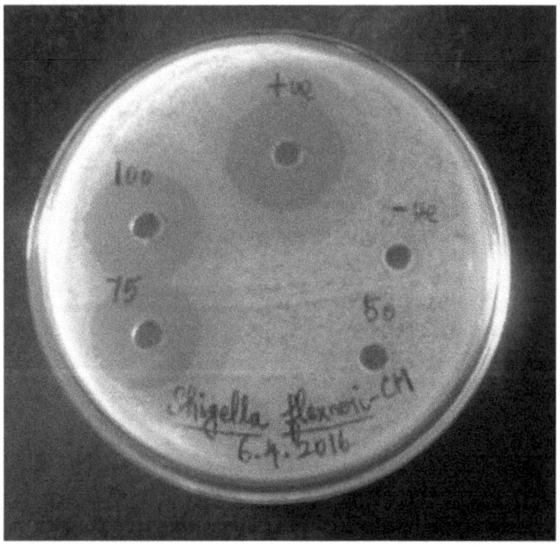

FIGURE 20.10 CM against *Shigella flexneri.*

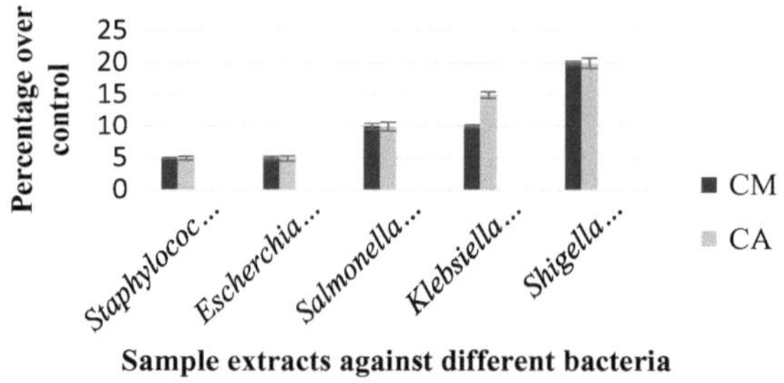

FIGURE 20.11 The percent over control of methanol and aqueous extracts of fresh coconut sprouts at 100 µg/mL concentration against different food borne pathogenic bacteria.

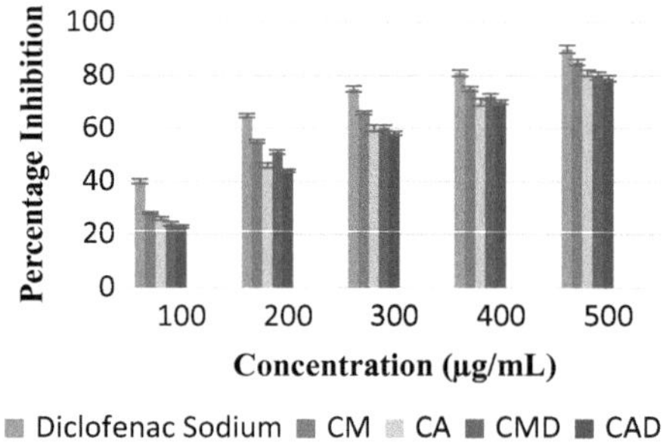

FIGURE 20.12 Anti-inflammatory activity of methanol and aqueous extract of fresh and dried coconut sprouts.

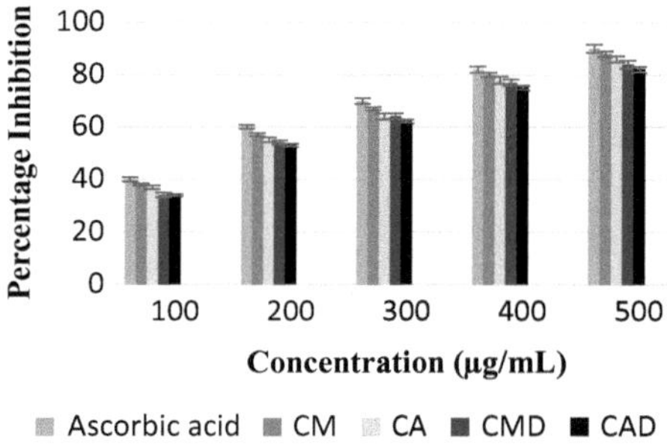

FIGURE 20.13 DPPH radical scavenging activity of methanol and aqueous extracts of fresh and dried coconut sprouts.

more prominent results for anti-bacterial, anti-inflammatory, and anti-oxidant assays, hence, the GC-MS analysis was carried out only in the methanol extract of the coconut sprouts to analyze the presence of specific bioactive compounds. GC-MS studies confirmed the presence of fatty acids, free amino acids, vitamins, terpenoids, and other small functional groups (Table 20.5).

20.3.7 In Silico Docking

From the GC-MS analysis, screening was done for the bioactive compounds related to the anti-ulcer property. Among the compounds screened, squalene was found to possess the anti-ulcer property. Squalene is a triterpenoid, which is an anti-ulcer and anti-cancer agent. Diclofenac sodium was used as the standard. The docking analysis of the squalene from the fresh methanol extract of the coconut sprouts (CM) showed

TABLE 20.5
GC-MS Analysis of Methanol Extract of Fresh Coconut Sprouts

Compound Name	Retention Time	Area Percent	Height Percent
Isosorbide Dinitrate	5.043	8.23	11.79
n-Decanoic acid	6.709	9.18	8.47
Pyrrolidine	6.892	2.71	4.38
1,3-Propanediol	8.542	43.86	15.88
D-Glycero-D-Ido-heptose	10.556	1.72	1.41
Xylitol	11.679	1.06	2.60
Ethylene diacrylate	11.867	0.14	0.31
n-Capric acid	16.429	0.41	0.95
Ascorbic acid	18.435	14.63	22.20
n-Nonadecanol-1	19.195	0.34	0.73
2-Chloroethyl linoleate	22.987	2.26	3.65
Cis-Vaccenic acid	23.179	10.18	15.18
DL-Phenylalanine	23.417	0.68	0.92
Glycerol tricaprylate	28.953	0.65	1.42
Caprylic anhydride	29.361	1.41	2.82
Squalene	43.652	0.80	2.72
4-Nitrophenyl caprylate	45.075	1.75	4.56
		100.00	100.00

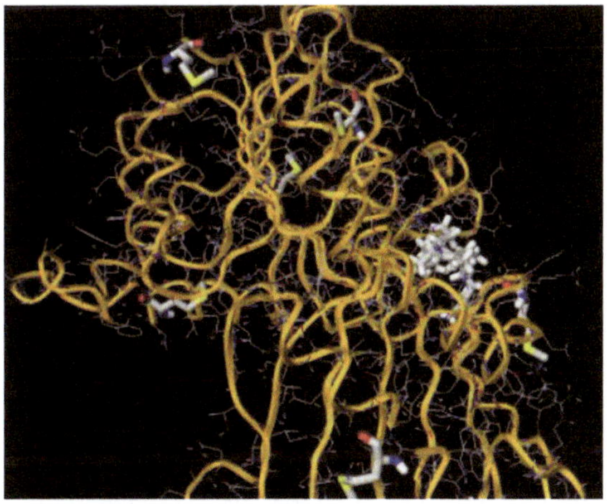

FIGURE 20.14 Squalene docked with target protein 1G60-*Helicobacter pylori* target.

docking scores of –7.7, –7.6, –7.5, and –7.3 (Figure 20.14) and standard diclofenac sodium showed docking scores of –7.5, –7.3, –7.1, and –6.8 (Figure 20.15). Maximum binding affinity was noted for squalene against *Helicobacter pylori* than the standard drug compound diclofenac sodium.

It is a colorimetric assay that measures the reduction of yellow MTT by mitochondrial succinate dehydrogenase. The results obtained were in a concentration-dependant manner. These are a clear evidence for the coconut sprouts having the anti-ulcer property (Figures 20.16 through 20.18).

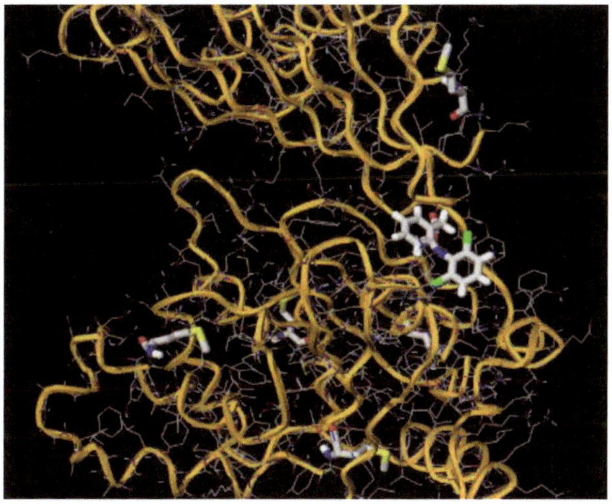

FIGURE 20.15 Diclofenac sodium docked with protein 1G60-*Helicobacter pylori.*

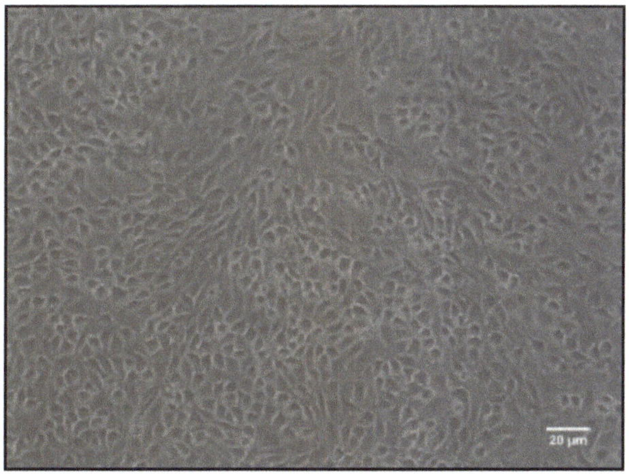

FIGURE 20.16 MTT assay (control).

FIGURE 20.17 MTT assay (500 µg/mL).

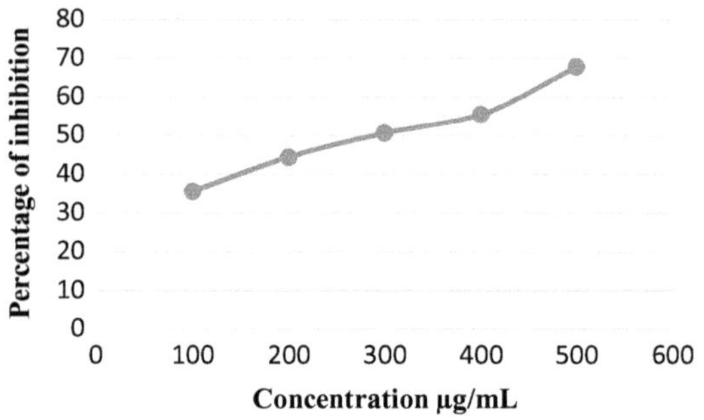

FIGURE 20.18 Cytotoxicity assay (coconut sprouts).

20.4 DISCUSSION

Coconut sprouts are one of the edible parts of the coconut. The present study on the phytoconstituents of coconut sprouts using various bioassays and in silico docking was mainly carried out with an approach toward the development of nutraceuticals.

In the present study, the preliminary phytochemical analysis was performed to find out the different bioactive compounds present in the extracts of coconut sprouts. Variations in the phytoconstituents were observed between different extracts. The methanol and aqueous extracts of the fresh and dried sprouts indicated the maximum amount of phytoconstituents, such as alkaloids, saponins, terpenoids, steroids, triterpenoids, resin, quinone, proteins, amino acids, carbohydrates, flavonoids, cardiac glycosides, phenols, tannins, fixed oils, fats, and fatty acids, whereas the betacyanin was shown only by the fresh coconut sprouts. The phytoconstituents are the dependable sources for the treatment of several health disorders. Phytochemical techniques play a significant role in the discovery of raw materials and resources for the pharmaceutical industry. The tannins and resins can act as astringents both in the gastrointestinal tract and on skin abrasions (Obidoa et al. 2010). In addition to phytochemicals, these are also an excellent source of B complex vitamins, minerals, and fiber (Louise and Susan 1979). These biologically active metabolites present in the sprout samples are essential chemical substances which are involved in several anti-inflammatory, anti-oxidant, anti-diuretic, anti-microbial, anti-cancer, and hepaticidal properties. Similar studies were carried out in the coconut which indicated the presence of different phytoconstituents (Obidoa et al. 2010). Thus, our results were found to be more superficial, as they indicated more amounts of phytoconstituents than the phytochemicals already reported in the coconut. Hence, the coconut sprouts were found to contain a more varied amount of phytoconstituents than that of the coconut, which clearly indicates the significance of the study.

The quantification studies indicated the results, which were found to be prominent when compared to the already existing sprouts, indicating more amounts of phytoconstituents (total soluble sugars, proteins, flavonoids, and terpenoids with reduced levels of phytic acid). The present study is clear evidence that the coconut sprouts have more amounts of phytoconstituents, especially the terpenoids and flavonoids. The terpenoids play an important role in traditional medicine, which is used to cure several diseases. The flavonoids are polyphenolic compounds having a benzo-γ-pyrone structure and are ubiquitously found in plants which are synthesized by the phenyl propanoid pathway, and are found to possess several health promoting properties. Thus, the coconut sprouts are found to have an increased level of terpenoids and flavonoids when compared to the coconut, which was reported to have the least amount of terpenoids (Kumar and Pandey 2013).

The FTIR results revealed the presence of terpenoids (C-H stretch at 2862.49 and 2928.07 cm^{-1}) and saponins (C=O stretch at 1743.72 cm^{-1}) (Skoog et al. 2007), indicating different functional groups which may be attributed to the existence of a variety of potential phytochemicals. The multiple functional groups reflect either the complex structure, or it indicates the nature of the sample as a mixture. The results are clear evidence indicating that the coconut sprouts are enriched with phytoconstituents, especially terpenoids, saponins, and flavonoids.

Prominent results were shown by the extracts of the coconut sprouts in the pharmacological studies, which include anti-bacterial, anti-inflammatory, and anti-oxidant assays. This might be due to the presence of phytoconstituents (terpenoids and flavonoids) present in the sprouts. Several studies have suggested protective effects of flavonoids against many infectious bacterial and viral diseases (Kumar and Pandey 2013). Since ancient times, it has been known that plant polyphenols (terpenoids, flavonoids, etc.) are natural anti-oxidants. These anti-oxidants are biological molecules which reduce the formation of free radicals or react with and neutralize them, thereby protecting the cell from oxidative damage, and they also have a potent activity by inhibiting inflammations. Thus, the results are clear evidence that the coconut sprouts possess strong pharmacological properties which can be further used, leading to the development of nutraceuticals.

GC-MS and docking studies have clearly indicated the presence of an anti-ulcer compound, squalene, a triterpenoid which is a natural anti-inflammatory (anti-ulcer) and anti-cancer agent. The MTT assay measures the cell viability based on the generation of reducing equivalents in metabolic active cells. The higher the absorbance measured, the higher is the cell viability. The cytotoxicity assay clearly confirmed the anti-ulcer property (concentration-dependant manner). The significance of the study is revealed by the presence of an enormous amount of terpenoids and flavonoids in the coconut sprouts, which is an indicator of the sprouts being used as a natural anti-ulcer agent. Thus, the coconut sprouts are enriched with potential and essential phytoconstituents synergized with pharmacological properties that can be used for the development of nutraceuticals.

20.5 CONCLUSION

The coconut sprouts are enriched with phytoconstituents, such as carbohydrates, proteins, terpenoids, and flavonoids which possess strong pharmacological and biological properties. Further, GC-MS and docking studies confirmed the presence of squalene, a triterpenoid showing a strong anti-ulcer property. MTT assays also proved that the coconut sprouts can act against ulcer cells. Thus, the fresh coconut sprouts are a natural, potent food source for human health which can be a phytomedicine and a nutrient supplement with a cost-effective approach. The dried sprouts which are being developed into a nutraceutical product will be certainly beneficial for human health in the management of the diseases related to different types of stomach ulcers.

ACKNOWLEDGMENTS

The authors thank Mrs. Prema Sampathkumar, former associate professor and head, Mrs. V. Manimozhi, associate professor and head, the faculty members, and supporting staff of the Department of Plant Biology and Plant Biotechnology, Ethiraj College for Women (Autonomous), Chennai, and Dr. A. Nirmala, former principal, and Dr. S. Bhuvaneswari, principal i/c, Ethiraj College for Women (Autonomous), for their valuable support and encouragement throughout the entire period of research. We would also like to express our sincere thanks for the facilities extended by the Central Instrumentation Centre of Ethiraj College for Women.

CONFLICTS OF INTERESTS

The authors declared that they had no conflicts of interests.

REFERENCES

Adams, R. P. 1989. *Identification of Essential Oils by Ion Trap Mass Spectroscopy*. Academic Press, New York.

Ahmad, M., Saeed, F., Mehjabeen, M., and Jahan, N. 2013. Evaluation of insecticidal and antioxidant activity of selected medicinal plants. *Journal of Pharmacognosy and Photochemistry* 2:153–158.

Anurag, P., and Rajamohan, T. 2003. Cardioprotective effect of tender coconut water in experimental myocardial infarction. *Plant Foods for Human Nutrition* 58(3):1–12.

Chikku, A. M., and Rajamohan, T. 2012. Dietary coconut sprout beneficially modulates cardiac damage induced by isoproterenol in rats. *Bangladesh Journal of Pharmacology* 7(4):258–265.

Eloff, J. N. 1998. Which extractant should be used for the screening and isolation of antimicrobial components from plants? *Journal of Ethnopharmacology* 60(1):1–8.

Evans, W. C. 2002. *Trease and Evans, Pharmacognosy*, 15th ed. London: W.B. Saunders Company Ltd, pp. 137–139, 230–240.

Ferguson, N. 1956. *A Textbook of Pharmacognosy*. New York: The Macmillan Company, p. 191.

Harborne, A. J. 1998. *Phytochemical Methods a Guide to Modern Techniques of Plant Analysis*. 3rd ed. Houten, Netherlands: Springer Science & Business Media B. V.

Kumar, S., and Pandey, A. K. 2013. Chemistry and biological activities of flavonoids: An overview. *The Scientific World Journal*. 2013:1–16.

Lima, E. B. C., Sousa, C. N. S., Meneses, L. N., Ximenes, N. C., Júnior, S., Vasconcelos, G. S., Lima, N. B. C., Patrocínio, M. C. A., Macedo, D., and Vasconcelos, S. M. M. 2015. *Cocos nucifera* (L.)(Arecaceae): A phytochemical and pharmacological review. *Brazilian Journal of Medical and Biological Research* 48(11):953–964.

Louise, R. B., and Susan, N. K. 1979. Tropical grain legumes, In: Inglett, G.E., and Charalambous, G. (Eds.), *Proceedings of an International Conference on Tropical Foods: Chemistry and Nutrition held in Honolulu*, Hawaii. 547–574.

Lowry, O. H., Rosebrough, N. J., Fan, A. L., and Randall, R. J. 1951. Protein measurement with the Folin-phenol reagent. *Journal of Biological Chemistry* 193:265–275.

Miller, G. L. 1972. Use of dinitrosalicylic acid reagent for determination of reducing sugar. *Analytical Chemistry* 31(3):426–428.

Mizushima, Y., and Kobayashi, M. 1968. Interaction of anti-inflammatory drugs with serum proteins, especially with some biologically active proteins. *Journal of Pharmacy and Pharmacology* 20(3):169–173.

Mosmann, T. 1983. Rapid colorimetric assay for cellular growth and survival: Application to proliferation and cytotoxicity assays. *Immunological Journal of Methods* 65(1–2):55–63.

Muhammad, S. A., and Fatima, N. 2015. In silico analysis and molecular docking studies of potential angiotensin-converting enzyme inhibitor using quercetin glycosides. *Pharmacognosy Magazine* 11(1):123.

Nelson, N. 1944. A photometric adaptation of the Somogyi method for the determination of glucose. *Journal of Biological Chemistry* 153(2):375–380.

Obidoa, O., Joshua, P. E., and Eze, N. J. 2010. Phytochemical analysis of *Cocos nucifera* L. *Journal of Pharmacy Research* 3(2):280–286.

Patel, S., Patel, J., and Patel, R. K. 2012. To study proximate analysis & biological evaluation of *Triphala Guggulu* formulation. *International Journal of Pharmaceutical Sciences and Research* 4:1520–1526.

Preetha, P. P., Devi, V. G., and Rajamohan, T. 2012. Hypoglycemic and antioxidant potential of coconut water in experimental diabetes. *Food & Function* 3(7):753–757.

Raaman, N. 2006. *Phytochemical Analysis*. New Delhi, India: New India Publishing Agency, pp. 19–24.

Rochfort, S., and Panozzo, J. 2007. Phytochemicals for health, the role of pulses. *Journal of Agricultural and Food Chemistry* 55(20):7981–7994.

Siriphanich, J., Saradhuldhat, P., Romphophak, T., Krisanapook, K., Pathaveerat, S., and Tongchipakdee, S. 2011. Coconut (*Cocos nucifera* L.). In: E. Yahia (Ed.), *Postharvest Biology and Technology of Tropical and Subtropical Fruits*. Vol 3: Cocona to Mango. Cambridge, UK: Woodhead Publishing in Food Science Technology and Nutrition, pp. 8–33.

Skoog, A., Holler, E. J., and Crouch, S. R. 2007. *Principles of Instrumental Analysis*.

Sofowora, A. O. 1993. *Medicinal Plants and Traditional Medicine in Africa*. Sunshine House, Nigeria: Spectrum books Ltd., Screening plants for Bioactive agents.

Tan, Y. A., Sambanthamurthi, R., Sundram, K., and Wahid, M. B. 2007. Valorisation of palm by-products as functional components. *European Journal of Lipid Science and Technology* 109(4):380–393.

Wheeler, E. L., and Ferrel, R. E. 1971. A method for phytic acid determination in wheat and wheat fractions. *Cereal Chemistry* 48(3):312–320.

21 Assessment of Secondary Metabolites in the Yams of *Dioscorea oppositifolia* L. & *Dioscorea pentaphylla* L.

S. Vivek, Y. Aron Santhosh Kumar, and M. Palanisamy

CONTENTS

21.1 INTRODUCTION

The forest plays a significant role in the economy as well as in the daily needs of the tribal communities. In times of scarcity, the staple foods are in short supply, and the communities collect wild roots and tubers to supplement their meager food availability at home (Vidyarthi 1987). Yams occupy the third position as the most important food crop in the world, next to cereals and pulses (FAO 2012). Yams have been suggested to have nutritional superiority to tropical root crops (Baquar and Oke 1976). *Dioscorea* is one of the largest and unique genera in Dioscoreaceae under the monocotyledon orders. The members are commonly known as "yams" and widely cultivated for their edible tubers in the tropical and subtropical countries. The genus comprises 850 species worldwide (Mabberley 1997). Among them, 32 species are found in India, of which, 21 are distributed in the Western Ghats (Arackal 2015). The tubers of *Dioscorea* are rich in protein and minerals. In India, the cooked tubers of *Dioscorea* are known to be consumed by the Palliyars and Kanikkars tribes living in the Southeastern slopes of the Western Ghats, Tamil Nadu (Arinathan et al. 2007; Shanthakumari et al. 2008) due to their nutritional value and unique taste. It is suggested that, the dioscorin (storage protein) of the yam may play a vital role as an anti-oxidant and gives some health benefits when people take it as a food additive (Hou et al. 2001; Udensi et al. 2008). Recently, researchers have found that dioscorin might exhibit anti-hypertensive, immuno-modulatory, and lectin activities. Further it can protect airway epithelial cells against dust mite allergen destruction, scavenging 2,2-diphenyl-1-picrylhydrazyl (DPPH), and hydroxyl free radicals (Lu et al. 2012; Hsu et al. 2002). In this respect, the present study investigated the proximate composition, cytotoxicity, and anti-oxidant activities of discorin using the ethanolic tubers extract of *D. Oppositifolia* L. and *D. Pentaphylla* L.

21.2 MATERIALS AND METHODS

21.2.1 Study Area: Gundri

Gundri Village is located between the latitude 11°39′ 1.53″ *N* and longitude 77°24′ 14.98″ *E* in Sathyamangalam Taluk of the Erode District in Tamil Nadu, India. The total area covers 702.78 hectares with the elevation at 329 meters from mean sea level. It covers the lower plains of the Sathyamangalam Range, Bhavanisagar Range, Talavadi Range, and T.N. Palayam Range and exhibits a dry deciduous habitat (Figure 21.1).

FIGURE 21.1 Location of Gundri Village.

FIGURE 21.2 Habit and tubers of *Dioscorea pentaphylla* L. (a & b) and *D. oppositifolia* L. (c & d).

21.2.2 Source of Plants Materials

The tubers of *D. pentaphylla* and *D. oppositifolia* were collected from Gundri hills, Erode District, Tamil Nadu, India (Figure 21.2).

21.2.3 Sample Preparation

The collected tubers were washed thoroughly with tap water, followed by distilled water to remove debris and shade-dried for 3–5 days. The shade-dried tubers were floured separately in a Willey Mill to 60 mesh size for the experiments. The following standard biological assays and laboratory processes were done to evaluate the proximate composition, cytotoxicity, and anti-oxidant properties of *D. oppositifolia* L. and *D. pentaphylla* L. using the ethanolic tubers extract.

21.2.4 Proximate Analysis

The moisture content in the tuber extraction was weighed, and values were estimated before and after incubation in a hot air oven at 80°C for 24 hours, followed by cooling in desiccators. The average loss in weight of the samples was calculated and expressed as the moisture percentage. The nitrogen content was determined by the microKjeldhal method (Humphries 1956). The crude protein was calculated by multiplying the percentage of nitrogen content with the factor 6.25. The crude lipid, crude fiber, and ash were estimated by the Association of Official Analytical Chemists method (AOAC 1970). The nitrogen free extractive (NFE) or crude carbohydrate was calculated as follows 100% of crude protein + % of crude lipid + % of crude fiber + % of ash (Muller and Tobin 1980). The calorific value was determined by multiplying the percentages of crude protein, crude fiber and carbohydrates with the factors 4, 9, and 4 respectively. Amino acids, free sugars, and buffer soluble proteins were determined (Basha et al. 1976).

21.2.5 Cytotoxicity Assay

The A-549 (human lung carcinoma) cell line was the subject of a cytotoxicity study by using a 3-(4,5 Dimethyl thiazol-2-yl)-2,5-diphenyl tetrazolium bromide (MTT) assay at the concentration of 250–62.5 μg/mL.

The percentage of growth inhibition was calculated using the following formula, and the concentration of drug or test samples needed to inhibit cell growth by 50% values were generated from the dose-response curves for each cell line (Francis and Rita 1986).

$$\%\text{ of cell viability} = \frac{\text{Mean OD of individual test group}}{\text{Mean OD of control group}} \times 100$$

21.2.6 Anti-oxidant DPPH Activity

The DPPH radical is obtained through the donation of a proton by antioxidants forming the reduced DPPH (Shimada et al. 1992). The color change from purple to yellow after the reduction can be quantified by its decrease in absorbance at a wavelength of 517 nm. Sample preparation is done by mixing 4ml of sample which is taken from different concentrations of ethanolic extract with 1 ml of ethanolic solution containing 0.2 mM of DPPH. The mixture was shaken vigorously and left to stand for 30 minutes, and the absorbance was measured at 517 nm. Ascorbic acid was used as a control. The percentage of inhibition in the DPPH radical scavenging activity was calculated.

21.2.7 ABTS+ Radical Scavenging Activity

The 2,2'-azino-bis (3-ethylbenzothiazoline-6-sulfonic acid ($ABTS^+$) decolorization assay involves the generation of the $ABTS^+$ chromophore by the oxidation of the $ABTS^+$ with potassium per sulfate. It is applicable for both hydrophilic and lipophilic compounds. The scavenging activity of the extract on the $ABTS^+$ radical cation was measured at 734 nm (Giao et al. 2007). $ABTS^+$ solution is prepared by following methods. An equal volume of 7 mM $ABTS^+$ is mixed with 2.45 mM Potassium per sulphate. The mixture was allowed to stand in the dark place at room temperature for 12–16 hours. The sample is prepared by the addition of 1 ml of diluted $ABTS^+$ with 10 µL of tuber extract in different concentration (50–250 µg/mL) 10 µL of ethanol is served as control. Ascorbic acid was used as the positive control. The absorbance was read at 734 nm after 6 minutes, and the percentage inhibitions were calculated. The inhibition was calculated according to the equation:

$$I = Ao - A1 / Ao \times 100,$$

where Ao is the absorbance of control reaction and A1 is the absorbance of test compound.

21.3 RESULTS AND DISCUSSION

The tubers of *D. oppositifolia* and *D. pentaphylla* were taken to evaluate the proximate components, cytotoxicity and anti-oxidant activities. The results of present study are shown in Tables 21.1, 21.2 and 21.3 respectively.

21.3.1 Proximate Components

The nutritive components' estimated values have been examined under standard assays, and the techniques were calculated in percentages and tabulated in Table 21.1.

The moisture content, crude fiber, ash, buffer protein, and free sugars are found to be more in *D. oppositifolia* than in *D. pentaphylla*. The other components values such as crude proteins, nitrogen, fats and carbohydrates, are comparatively high in *D. pentaphylla* than in *D. oppositifolia*. The calorific values of both tubers show the significant result.

The carbohydrate values of *D. oppositifolia* and *D. pentaphylla* (82.86%) and (82.86%) are quite reasonable, as the most root crops is made up of about 60%–90% carbohydrates. These observations agree with those of reports in *D.rotundata and D. Alata* (Longe 1986).

The tubers of *D. oppositifolia* and *D. pentaphylla* are highly proteinous, with the values of (10.06) and (10.28). These observations agree with those of reports in *D.rotundata* and D. Alata (Longe 1986). The protein content of present study results are also comparable to earlier report of Rajyalakshmi and Geervani (1994). The lipid content of the samples *D. oppositifolia* (1.0%) and *D. pentaphylla* (1.20%) tubers are quite reasonable, as all root crops exhibit very low lipid content Similar results were recorded in *D. bulbifera* (Subhash et al. 2012) and *D. alata* (Saklani et al. 2013). Ekpenyoung (1984). Similar results were recorded in *D. bulbifera* (Subhash et al. 2012), *D. alata* (Saklani et al. 2013).

The ash contents of *D. oppositifolia* (5.0%) and *D. pentaphylla* (4.85%) are quite significant and comparably lower than that of *D. dumentorum* (6.80%) (Ogbuagu 2008). *D. bulbifera* (2.43%) (Celestine and David 2015) and *D. alata* (3.20%) (Saklani et al. 2013) have lower values of ash content than *D. oppositifolia* and *D. pentaphylla*.

21.3.2 Cytotoxicity Activity

According to the MTT assay at the end of 48 hours, 250 µg/mL of ethanolic tubers extract of *D. pentaphylla* and *D. oppositifolia* caused 36.2% and 38% growth viability in A-549 cells, respectively. The observations show that, the

TABLE 21.1
Proximate Analysis of *D. oppositifolia* and *D. pentaphylla*

S. No	Components	*D. oppositifolia* (%)	*D. pentaphylla* (%)
1	Moisture content	9.8	4.1
2	Crude protein	10.06	10.28
3	Nitrogen (Kjeldhal method)	1.61	1.64
4	Crude fiber	1.6	1.03
5	Ash content	5	4.85
6	Fat content	1.0	1.2
7	NFE (Carbohydrate)	82.14	82.86
8	Amino acid	4	4.4
9	Free sugars	19.0	16.0
10	Buffer protein	5.52	5.4

TABLE 21.2
In vitro Cytotoxicity Activity in Ethanolic Extracts of *D. pentaphylla*

S. No	Concentration µg/mL	% viability	Average
1	250	36.29167	0.0871
2	125	37.95833	0.0911
3	62.50	51.29167	0.1231
4	31.25	93.54167	0.2245

IC_{50} value of *Dioscorea pentaphylla:* 105.549.

TABLE 21.3
In vitro Cytotoxicity Activity in Ethanolic Extracts of *D. oppositifolia*

S. No	Concentration µg/mL	% viability	Average
1	250	38.00	0.0912
2	125	38.79167	0.0931
3	62.5	60.45833	0.145
4	31.25	105.875	0.2541

IC_{50} value of *D. oppositifolia:* 123.0152.

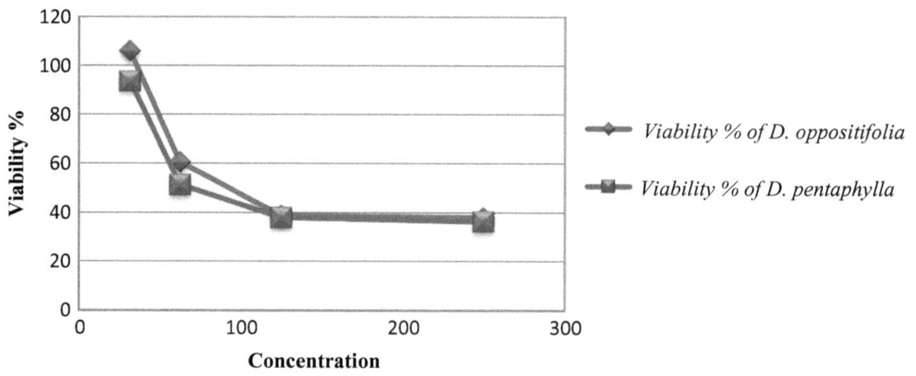

FIGURE 21.3 Graphical representation for in vitro cytotoxicity activity of *D. oppositifolia* and *D. pentaphylla*.

effective viability percentage in *D. pentaphylla* on A-549 cells than in *D. oppositifolia*. The results are presented in Tables 21.2, 21.3 and Figure 21.3.

D. bulbifera rhizome is used for the treatment of leprosy and tumors in Bangladesh within an ethno-botanic frame (Murray et al. 1984). Various impacts at the molecular level were observed regarding different *D. bulbifera* organic extract treatments. Petroleum ether fractions of *D. bulbifera* showed potential effects against Hepatitis A (HepA) cells with microstructure abnormality of the HepA cells surface (Yu et al. 2004). There is a report of (Cui 2012) modulation of the immune response and induction of apoptosis by crude polysaccharides of *D. bulbifera* against cervical carcinoma in U14-bearing mice.

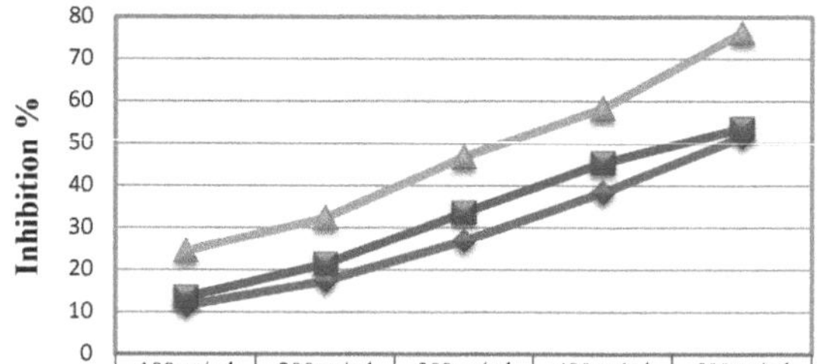

	100µg/ml	200µg/ml	300µg/ml	400µg/ml	500µg/ml
Dioscorea oppositifolia	11.41	17.3	26.8	38.34	51.52
Dioscorea pentaphylla	13.43	21.32	33.57	45.34	53.55
Ascorbic acid	24.45	32.32	46.76	58.17	76.16

FIGURE 21.5 Graphical representation for the ABTS+ activity in different concentrations of *D. oppositifolia* and *D. pentaphylla*.

21.3.3 Anti-oxidant Activities

Anti-oxidant activities of *D. pentaphylla* and *D. oppositifolia* were assessed by using DPPH and ABTS+ assays. In both assays, five different concentrations (100–500 µg/mL) of tuber extracts were used. Vitamin C and ascorbic acid were served as control, and the results are presented in Figures 21.4 and 21.5. Among the five different concentrations, the higher percentages of inhibition (44.35 ± 0.12) and (53.55 ± 0.76) at 500 µg/mL concentration were observed in *D. pentaphylla*. The increase effects of anti-oxidants on the DPPH radical scavenging might be due to presence of hydrogen donating ability reported by Conforti et al. (2005). These observations agree with the reports of Kalaivani and Mathew (2010).

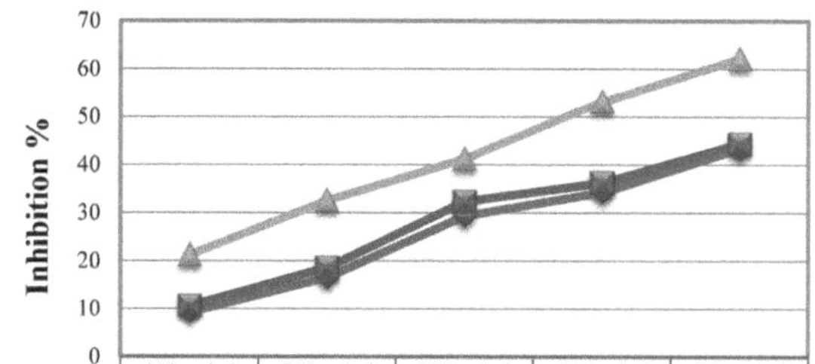

	100µg/ml	200µg/ml	300µg/ml	400µg/ml	500µg/ml
Dioscorea oppositifolia	9.31	16.52	29.32	34.21	43.33
Dioscorea pentaphylla	10.62	18.58	32.32	36.21	44.35
Vitamin C	21.2	32.52	41.2	53.12	62.13

FIGURE 21.4 Graphical representation for the DPPH activity in different concentrations of *D. oppositifolia* and *D. pentaphylla*.

The experiment was conducted in triplicates (n = 3)

IC_{50} value of *Dioscorea oppositifolia*: 459.02 µg/mL
IC_{50} value of *Dioscorea pentaphylla*: 439.17 µg/mL
IC_{50} value of vitamin C (standard): 409.82 µg/mL

The experiment was conducted in triplicates (n = 3).

IC_{50} value of *Dioscorea oppositifolia*: 241.13 µg/mL
IC_{50} value of *Dioscorea pentaphylla*: 239.56 µg/mL
IC_{50} value of ascorbic acid (standard): 184.21 µg/mL

The DPPH is a stable free radical, which has been widely accepted as a tool for estimating free radical scavenging activities of anti-oxidants. The DPPH is a stable free radical and accepts an electron or hydrogen radical to become a stable diamagnetic molecule (Kalaivani and Mathew 2010). The degree of discoloration indicates the potential of the plant extract to scavenge a free radical due to its ability to donate a hydrogen proton. The concentration-dependent curve of the DPPH radical scavenging activity of *D. oppositifolia* and *D. pentaphylla* compared well with the ascorbic acid. The result is indicative of the hydrogen donating ability of *D. pentaphylla*, since the effects of anti-oxidants on the DPPH radical scavenging is thought to be due to their hydrogen donating ability (Conforti et al. 2005).

The ABTS+ radical scavenging activity is a relatively recent one, which involves a more drastic radical, chemically produced, and is often used for screening complex anti-oxidant mixtures, such as plant extracts, beverages, and biological fluids (Huang et al. 2011). The ability in both the organic and aqueous media and the stability in a wide pH range raised the interest in the use of the ABTS for the estimation of anti-oxidant activity (Huang et al. 2011). In vivo studies in both assays proved that the ethanolic extracts of *D. pentaphylla* and *D. oppositifolia* play the most significant role in antioxidant activity. The extract showed potent anti-oxidant activity in the ABTS method, which is comparable to the standard used.

21.4 CONCLUSIONS

Based on the nutritive evaluation studies, the wild yams of *D. oppositifolia* and *D. pentaphylla* could be a good source of proteins, lipids, crude fibers, and minerals. The present study also reveals that, yams of *D. oppositifolia* and *D. pentaphylla* are the good source for cytotoxicity and anti-oxidant activities. The results of the ethanolic extract of tubers also revealed that the cytotoxicity and anti-oxidant activities were present in the both tubers. Further isolation of the particular compounds responsible for the cytotoxicity and anti-oxidant activities is needed.

ACKNOWLEDGMENT

The authors are grateful to the Director, BSI, Kolkata and HoO, BSI, SRC, Coimbatore for providing facilities.

REFERENCES

Arackal, E. J. 2015. Taxonomy, conservation and sustainable utilisation of the genus Dioscorea in Western Ghats Elsamma Joseph Arackal 1 and AG Pandurangan 2. 1. Maharajas College, Ernakulam, Kerala-682011 2. Jawaharlal Nehru Tropical Botanic Garden and Research Institution *Journal of Root Crops* 41(2): 3–16.

Arinathan, V., Mohan, V. R., Britto, A., and Murugan, C. 2007. Wild edibles used by Palliyars of the western Ghats, Tamil Nadu. *Indian Journal of Traditional Knowledge* 6: 163–168.

Association of Official Analytical Chemists and Association of Official Agricultural Chemists (US), 1920. *Official Methods of Analysis*. Washington, DC: Department of Agricultural Division of Chemistry.

Baquar, S. R., and Oke, O. L. 1976. Protein in Nigerian yams (*Dioscorea* spp.) [Nutritive value]. *Nutrition Reports International*.

Basha, S. M. M., Cherry, J. P., and Young, C. T. 1976. Changes in free amino acids, carbohydrates, and proteins of maturing seeds from various peanut (*Arachis hypogaea* L.) cultivars. *Cereal Chemistry* 53: 583–597.

Celestine, A. A., and David, O. I. 2015. Comparative nutritional and phytochemical evaluation of the aerial and underground tubers of air potato (*Dioscorea bulbifera*) Available in Abakaliki, Ebonyi State, Nigeria. *British Journal of Applied Science & Technology* 11(4): 1–7.

Conforti, F., Loizzo, M. R., Statti, G. A., and Menichini, F., 2005. Comparative radical scavenging and antidiabetic activities of methanolic extract and fractions from *Achillea ligustica* A LL. *Biological and Pharmaceutical Bulletin* 28(9): 1791–1794.

Cui, H. X., 2012. Antitumor activity and possible mechanism of crude polysaccharides from *Discorea bulbifera* L. on the mice bearing U14 cervical carcinoma. In *Advanced Materials Research* 560: 374–379. Trans Tech Publications.

Ekpenyong, T. E. 1984. Composition of some tropical tuberous foods. *Food Chemistry* 15(1): 31–36.

FAO (Food and Agricultural Organization). http://www.fao.org/home/en/. (Accessed November 26, 2017.)

Francis, D., and Rita, L. 1986. Rapid colorimetric assay for cell growth and survival. Modifications to the tetrazolium dye procedure giving improved sensitivity and reliability. *Journal of Immunological Methods*. 89(2): 271–277.

Gião, M. S., González-Sanjosé, M. L., Rivero-Pérez, M. D., Pereira, C. I., Pintado, M. E., and Malcata, F. X. 2007. Infusions of Portuguese medicinal plants: Dependence of final antioxidant capacity and phenol content on extraction features. *Journal of the Science of Food and Agriculture* 87(14): 2638–2647.

Harshini, R., and Anitha, M. 2015. Phytochemical analysis of two medicinal plants *Dicoreabulbifera* and *Apium graveooslons*. *International Journal of Pharmaceutical Sciences Review and Research* 33(1): 344–347.

Hou, W. C., Lee, M. H., Chen, H. J., Liang, W. L., Han, C. H., Liu, Y. W., and Lin, Y. H. 2001. Antioxidant activities of dioscorin, the storage protein of yam (*Dioscorea batatas* Decne) tuber. *Journal of Agricultural and Food Chemistry* 49(10): 4956–4960.

Hsu, F. L., Lin, Y. H., Lee, M. H., Lin, C. L., and Hou, W. C. 2002. Both dioscorin, the tuber storage protein of yam (*Dioscorea alata* cv. Tainong No. 1), and its peptic hydrolysates exhibited angiotensin converting enzyme inhibitory activities. *Journal of Agricultural and Food Chemistry* 50(21): 6109–6113.

Huang, M. H., Huang, S. S., Wang, B. S., Wu, C. H., Sheu, M. J., Hou, W. C., Lin, S. S., and Huang, G. J. 2011. Antioxidant and anti-inflammatory properties of *Cardiospermum halicacabum* and its reference compounds ex vivo and in vivo. *Journal of Ethnopharmacology* 133(2): 743–750.

Humphries, E. C., 1956. Mineral components and ash analysis. In *Moderne Methoden der Pflanzenanalyse/Modern Methods of Plant Analysis*. (Eds.), K. Peach and M. V. Tracey. Springer Verlag, Berlin, pp. 468–502.

Kalaivani, T., and Mathew, L. 2010. Free radical scavenging activity from leaves of *Acacia nilotica* (L.) Wild. ex Delile, an Indian medicinal tree. *Food and Chemical Toxicology* 48(1): 298–305.

Longe, O. G., 1986. Energy content of some tropical starchy crops. *The Nigerian Agricultural Journal* 21:136.

Lu, Y. L., Chia, C. Y., Liu, Y. W., and Hou, W. C. 2012. Biological activities and applications of dioscorins, the major tuber storage proteins of yam. *Journal of Traditional and Complementary Medicine* 2(1): 41–46.

Mabberley, D. J. 1997. *The Plant-Book: A Portable Dictionary of the Vascular Plants*. Cambridge University Press, New York.

Mbaebie, B. O., Edeoga, H. O., and Afolayan, A. J. 2012. Phytochemical analysis and antioxidants activities of aqueous stem bark extract of *Schotia latifolia* Jacq. *Asian Pacific Journal of Tropical Biomedicine* 2(2): 118–124.

Muller, H. G., and Tobin, G. 1980. *Nutrition and Food Processing*. Croom Helm, London, UK.

Murray, R. D., Jorge, Z. D., Khan, N. H., Shahjahan, M., and Quaisuddin, M. 1984. Diosbulbin D and 8-epidiosbulbin E acetate, norclerodane diterpenoids from *Dioscorea bulbifera* tubers. *Phytochemistry*, 23(3): 623–625.

Ogbuagu, M. N., 2008. Nutritive and antinutritive composition of the wild (in-edible) species of *Dioscorea bulbifera* (Potato Yam) and *Dioscorea dumentorum* (Bitter Yam). *Journal of Food Technology* 6: 224–226.

Onyenuga, V. A. 1968. *Nigeria's Food and Feeding Stuffs, their Chemistry and Nutritive Value*. Ibadan University Press, Ibadan, Nigeria.

Rajyalakshmi, P., and Geervani, P. 1994. Nutritive value of the foods cultivated and consumed by the tribals of South India. *Plant Foods for Human Nutrition* 46(1): 53–61.

Saklani, S., Chandra, S., Mishrav, A. P. 2013. Nutritional profile, antinutritional profile and phytochemical screening of Garhwal Himalaya medicinal plant *Dioscorea alata* tuber. *International Journal of Pharmaceutical Sciences Review and Research* 23(2): 42–46.

Shanthakumari, S., Mohan, V. R., and de Britto, J. 2008. Nutritional evaluation and elimination of toxic principles in wild yam (*Dioscorea* spp.). *Tropical and Subtropical Agroecosystems* 8(3): 319–325.

Shimada, K., Fujikawa, K., Yahara, K., and Nakamura, T. (1992). Antioxidative properties of xanthan on the autoxidation of soybean oil in cyclodextrin emulsion. *Journal of Agricultural and Food Chemistry* 40(6): 945–948.

Subhash, C., Sarla, S., Abhay, M. P., and Anoop, B. 2012. Nutritional profile and phytochemical screening of Garhwal Himalaya medicinal plant *Dioscorea bulbifera. International Research Journal of Pharmacy* 3(5): 289–294.

Udensi, E. A., Oselebe, H. O., and Iweala, O. O. 2008. The investigation of chemical composition and functional properties of water yam (*Dioscorea alata*): Effect of varietal differences. *Pakistan Journal of Nutrition* 7(2): 342–344.

Vidyarthi, L. P. 1987. Role of forest in tribal life. In *Tribals and Foress.* (Ed.), S.P. Sinha, Bihar Tribal Welfare Research Institute, Ranchi, India, p. 323.

Yu, Z. L., Liu, X. R., McCulloch, M., and Gao, J. 2004. Anticancer effects of various fractions extracted from Dioscorea bulbifera on mice bearing HepA. *China Journal of Chinese Materia Medica* 29: 563–567.

22 Evaluation of Anti-ulcer Potential of *Sphenodesme involucrata* var. *paniculata* (C.B. Clarke) Munir Leaves on Various Gastric Aggressive Factors

P. S. Sreeja, K. Arunachalam, and Parimelazhagan Thangaraj

CONTENTS

22.1 INTRODUCTION

There are different types of ulcers that occur inside and outside the body, such as peptic, stomach, corneal, foot, or leg ulcers, etc. Peptic ulcers are in the lesions/cavities in the mucosal layer and are surrounded by inflammation; they are caused by the imbalance between the protective and aggressive factors in the stomach and duodenum. They are mainly of two types: gastric and duodenum (Singh et al. 2018).

The stomach is an organ of the digestive tract having three parts (fundus, corpus, and antrum), and it is composed of several layers of muscles, smooth and internally covered by a thick layer of mucus. The wall of the stomach has several tissue layers: mucosa, external muscular, and submucosa. The mucosa is the layer of the inner part of the stomach that houses numerous gastric glands; the external muscular layer consists of three layers oblique, internal, and longitudinal circular, and one myoteric plexus; and the submucosa layer is comprised of loose connective tissue and contains a rich vascular network and a plexus, known as the submucosal plexus (Fattini and Dangelo 2007; Schubert and Peura 2008).

The glands in the stomach are composed of three types of cells: (1) the mucosal cells, which are responsible for mucus secretion; (2) peptic or major cells, responsible for the secretion of pepsinogen, which, when in contact with hydrochloric acid is converted into pepsin, exerting the activity of a proteolytic enzyme; and (3) parietal or oxyntic cells, which secrete hydrochloric acid (Schubert and Peura 2008).

Changes in the physiology, tissue structure, and cellular functioning of the stomach may lead to an imbalance in the organ's homeostasis, facilitating the onset of gastric ulcers, the chronic disease that may pass through the mucus layer within tissue cavities (Najm 2011). In addition, other complications may arise, such as bleeding and perforations, which are still important causes of morbidity and mortality in ulcer patients (Sung et al. 2010).

In the last decades, great changes in the epidemiology of this disease have occurred, although having a tendency to reduce it, this is still the disease among the most prevalent gastrointestinal diseases worldwide (Sumbul et al. 2011).

As per the WHO data, up to 2017, 0.66% of the population was affected by peptic ulcers in India (World Life Expectancy 2018). Annually, the incidence of gastric ulcers in the world varies around 0.19% (Sung et al. 2009), and according to Ali et al. (2013), men had a higher incidence of developing gastric ulcers (11%–20% in men; 8%–11% in women) and among females, menopausal women are more likely to have ulcers because of the lower production of hormones (Santin et al. 2013).

22.2 PROTECTIVE FACTORS OF THE GASTRIC MUCOSA

The gastric mucosa is exposed uninterruptedly to harmful agents, such as the acid pH medium ~1.0 and the presence of proteolytic enzymes, such as pepsin, which damage gastric tissues. However, the stomach remains intact and able to perform its physiological functions due to the defense mechanisms inherent to the gastric mucosa (Tulassay and Herszényi 2010). The different mechanisms that maintain the integrity of the gastric mucosa involve several defensive factors that can be divided into: pre-epithelial, the first line of defense represented by the mucus barrier; epithelial cells, being the second line of defense, in which progenitor cell renewal and prostaglandin production occur; and another subepithelial component representing the third line of defense that is related to the continuous blood flow (Tarnawski et al. 2013; Yandrapu and Sarosiek 2015).

The cells of the gastric epithelium are responsible for secreting bicarbonate and mucus ions, which form a barrier that has gelatinous and elastic characteristics, are adherent and viscous, as well as maintain a pH close to neutrality and has the function of protecting the stomach lumen from the action of hydrochloric acid (Ham and Kaunitz 2008).

22.2.1 Mucos, Bicarbonate, and Blood Flow

The production of mucus by the epithelial cells begins from the stomach to the colon. This first gastric mucosal defense block is formed by muco-gel (95% mucin and 5% water), which serves as a prevention of bacterial colonization, in addition to retaining secreted hydrogen ions, and preventing the proteolytic action of pepsin in the gastric epithelium (Gilbert et al. 2015). In addition, mucus prevents back diffusion of H^+ into epithelial cells and removes free radicals that are harmful to gastric mucosa (Allen and Fonström 2005). The phospholipids layer of the epithelial cell is the first line of defense of the mucosa. This layer presents high hydrophobicity, a thin layer of mucus gel, binds the bicarbonate ions, maintains the pH at approximately 7.0, and deposits on the surface of the epithelial cells of the gastric mucosa, protecting it against deleterious effects, mainly from HCl and pepsin (Allen and Fonström 2005).

The bicarbonate secreted and transported by the superficial epithelial cells, the parietal cells, maintains the alkalinity of the gastric mucosa (Tarnawski et al. 2013).

The cellular renewal of the gastric mucosa maintains its structural framework, due to the proliferation of the responsible progenitor cells in a few days to renew the gastric epithelium without loss in the continuity of the functional barrier (Laine et al. 2008).

The continuous blood flow that bathes the gastric mucosa provides a great supply of oxygen and nutrients to the cells of the gastric mucosa and neutralizes the acid secreted by the parietal cells (Tarnawski 2005). This blood flow also removes and dilutes the ions that can backscatter from the gastric lumen. Describing this another way, the gastric microcirculation is responsible in providing nutrients and bicarbonate to the gastric mucosa. Besides, it is worth emphasizing that blood flow is increased in response to the inflammatory processes, thus contributing to the healing process of gastric mucosa due to lesions (Wallace 2008).

22.2.2 Nitric Oxide

Nitric oxide (NO) is a gaseous mediator derived from L-arginine as a consequence of the action of a group of enzymes called nitric oxide synthases (NOS) to form L-citrulline and NO, and it possesses multiple physiological functions. The production of NO has paramount importance in the gastroprotection by preserving the integrity of the gastric epithelium and the mucus barrier. This powerful gas mediator acts as a vasodilator and increases the gastric blood. It also prevents the adhesion of leukocytes to the endothelial cells by acting on the modulation of adhesion molecules and regulates the secretion of hydrochloric acid; thus it assists in the healing of gastric ulcers (Breuillard et al. 2015; Costa et al. 2016).

One of the inducible isoforms of NOS enzymes is induced nitric oxide synthase (iNOS), which is not usually expressed in tissues, but can be induced in inflammatory conditions and is considered as a calcium-independent enzyme (Lopez-Belmonte et al. 1993). In addition, there are other isoforms characterized as constitutive, i.e., expressed in physiological and calcium-dependent conditions, such as endothelial NOS (eNOS-type III; produced from endothelial cells) and neuronal NOS (type I; produced from cerebellum). iNOS (type II; produced from macrophages) is activated by endotoxin or inflammatory cytokines, such as Interleukin (IL)-1β, IL-2, IL-6, TNF-α, and Interferon-γ (Cho 2001).

NO, when generated in physiological quantities by eNOS, has the capacity to reduce the interaction of leukocytes with the endothelial layer in the microcirculation due to the modulation in expression of adhesion molecules, in addition to the induction of vasodilation (Medeiros et al. 2009). Rozza et al. (2011) reported that NO interacts with prostaglandins and neuropeptides via $K^{\pm}$ATP channels, maintaining the integrity of the gastric mucosa.

The NO blockade increases in oxidative stress by activating mast cells, which are found in large quantities in the gastrointestinal tract and release mediators, such as histamine and the platelet-activating factor, causing increased epithelial permeability, and this event is quickly reversed by exogenous NO administration (Arrieta et al. 2003). Besides the various inflammatory cytokines activate the iNOS (type II) (Cho 2001).

In this way, the lesion is intensified with inflammatory infiltrate and edema.

As a result, the balance of expression and activity of iNOS and eNOS in the gastric tissue is very important because the production of NO protects the gastric tissue, in response to harmful agents (Santin et al. 2013). Various studies have shown that NO has thus a controversial role in the ulcerogenic response of the gastrointestinal mucosa, depending on the NOS isoform, i.e., the NO produced by eNOS expresses a protective effect and the NO originated from iNOS has a pro-ulcerogenic effect (Kristjánsson et al. 2005; Nishio et al. 2006). However, studies by Lopez-Belmonte et al. (1993), Andrews et al. (1994), and Calatayud et al. (1999) suggest that NO is a fundamental gaseous mediator in gastric protection. It was proved by many experiments that the actions of various chemical agents that promote lesions in the gastric mucosa can be minimized by the administration of NO, as well as aggravated by its removal. This was confirmed through experiments where the administration of N(G)-nitro-L-arginine methyl ester (non-selective NOS inhibitor) accentuates ethanol-induced gastric lesions and increases H^+, K^+-adenosine triphosphate synthase ATPase activity. In contrast, NO donors (e.g., sodium nitroprusside) repair the ethanol-induced lesions and inhibit the activity of this ATPase in rats (Martin and Motilva 2001). Several studies describe classes of Nonsteroidal Anti-Inflammatory Drugs (NSAIDs) with structural formulations, such as NO donor aspirin, diclofenac, and flurbiprofen release NO despite the inhibition of cyclooxygenase (COX) enzymes (Magierowski et al. 2015). Other NO donors like FK409 and sodium nitroprusside minimize gastric secretion (Lanas 2008).

Hence, the gastroprotective effects that are attributed to the increased blood flow in gastric mucosa are promoted by NO, as well as regulation of gastric mucus secretion.

22.2.3 The Prostaglandins

Prostaglandins are small lipid molecules that have biological actions in various tissues and processes in the body, such as platelet aggregation, renal function, release of neurotransmitters, modulation of the immune response (Tarnawski et al. 2013), and an important role in the defense of the gastric mucosa. The endogenous prostaglandins originate from arachidonic acid through the oxidation by the constitutive COX-1, as well as by the induced isoform (COX-2); these are described as catalysts for the conversion of the arachidonic acid for the endoperoxide H2 (PGH2) and prostaglandin that serve as substrates for the biosynthesis of prostanoids (PGE2, PGF2α, PGD2, PGI2, and Thromboxane A2 (TXA2) (Hata and Breyer 2004).

Prostaglandins, originating from COX-1, the constitutive isoform in the gastrointestinal tract and in several tissues, play a gastroprotective role by increasing the secretion of mucus and bicarbonate, blood flow and motility, as well as protecting epithelial cells against any damage caused by cytokines. Therefore, it is responsible for the healing of gastric ulcers due to plasma exudation and leukocyte recruitment (Laine et al. 2008).

In addition, the prostaglandins have the function of regulating leukocytes with the adhesion to blood vessels, as well as angiogenesis, and PGE2 inhibits the activation of pro-inflammatory mediators (TNF-α, IL-1, and IL-8), activation of mast cells and leukocytes, as well as platelet adhesion (Blackwell et al. 2010). However, prostaglandins from COX-2 are involved in an inflammatory process by the action of cytokines (TNF-α and IL-1β).

22.2.4 Anti-oxidant System

During oxidative stress, redox regulation and signaling occur (Sies 2015). In all biological systems, the reactive oxygen species (ROS) are found, physiologically, however, the increased production of the ROS can affect processes necessary to homeostasis (Lushchak 2014). The main ROS are the superoxide anion and the hydroxyl radical, which are potent and cytotoxic (Ock et al. 2012).

Under physiological conditions, however, during aerobic cellular metabolism, oxygen (O_2) undergoes a reduction to form several reactive intermediates, such as hydroxyl (OH–), hydroperoxyl (HO_2), hydrogen peroxide (H_2O_2), and superoxide radicals (O_2^-) (Semchyshyn and Lozinska 2012). After oxidative stress, the body through its enzymatic defense systems, such as catalase (CAT), superoxide dismutase, glutathione peroxidase, and glutathione reductase, fights with free radicals (Schäfer and Werner 2008). In addition to this enzyme system, there are anti-oxidant compounds, non-enzymatic compounds considered as the second line of defense, represented by the free radical scavengers that react directly with the oxidizing agents. These anti-oxidant substances are produced by the body or acquired by the diet, such as vitamin C (ascorbic acid), vitamin E (tocopherol), carotenoids, and plant-derived polyphenols (Quideau et al. 2011).

Superoxide dismutase is the main anti-oxidant enzyme involved in the reduction of oxidative stress by the constant generation of enzymatic ROS like superoxide radicals formed endogenously by flavoenzymes, such as xanthine oxidase, lipoxygenase, COX, and Nicotinamide Adenine Dinucleotide Phosphate Hydrogen (NADPH) oxidase (Nordberg and Arner 2001).

Another important anti-oxidant, tripeptide γ-glutamyl-cysteinyl-glycine, glutathione, is also involved in cell protection against oxidative stress (Uys et al. 2014). Glutathione acts as a cofactor of several anti-oxidant enzymes, such as glutathione peroxidase and glutathione transferase, and the glutathione peroxidase has been involved in detoxifying hydrogen peroxide and thus avoiding lipid peroxidation (Jozefczak et al. 2012).

The CAT is the enzyme commonly present in all organs and mainly involved in the decomposition of hydrogen peroxide, formed as an intermediate during most of the metabolic processes in the body, and is dangerous as having the property of a free radical, into water and oxygen (Lobo et al. 2010).

Therefore, the presence of these anti-oxidant agents in the gastric mucosa is of paramount importance due to their involvement in the maintenance and production of gastric mucus; the sulfhydryl compounds (homocysteine, cysteine, and glutathione) have glycoprotein subunits joined together by disulfide bonds, which, once reduced, render the mucus water-soluble (Avila et al. 1996).

22.2.5 Physiology of Gastric Secretion

Gastric acid secretion occurs in parietal cells and consist of fluid mucus, electrolytes, HCO_3^-, pepsinogen, intrinsic factor, peptides, and HCl (the main constituent) (Konturek et al. 2004). The gastric secretion is a complex process with three distinct phases: cephalic phase, gastric phase, and intestinal phase. (1) The factors, such as vision, smell, and taste of food, stimulate secretion, which is the cephalic phase and is partially regulated by the hypothalamus and mediated by the vagus nerve; (2) in the gastric phase, the contact of food in the gastric lumen stimulates the secretion, as well as the stomach distension, which entails the activation of the sensory receptors of the body and gastric antrum, with the activation of the vagal reflexes; and (3) the intestinal phase is the mechanical distension and the presence of amino acids and peptides from the chyme stimulate gastric fluid secretion (Goo et al. 2010). HCl has a protective role to the stomach/gastric mucosa through avoiding anti-microbial action, i.e., the growth of microbes from ingested food and water (Schubert and Peura 2008).

Gastric acid secretion is a complex, continuous process in which multiple, central, and peripheral factors contribute to a common end point: the secretion of H^+ by parietal cells. Neuronal (acetylcholine, ACh), paracrine (histamine), and endocrine (gastrin) factors/physiological agonists, all regulate the acid secretion. Their specific receptors (muscarinic, H_2, and cholecystokinin receptor 2, respectively) are on the basolateral membrane of parietal cells in the stomach and also on enterochromaffin-like cells (ECLs; cells seen close to parietal cells and are the source of histamine in the stomach). The receptors seen on ECLs have a role in regulating the release of histamine. The dorsal motor nucleus of the vagal nerve, the hypothalamus, and the solitary tract nucleus are the important structures for the central nervous system stimulation of the gastric acid secretion. Efferent fibers originating in the dorsal motor nuclei descend to the stomach via the vagus nerve and synapse with the ganglion cells of the enteric nervous system (Wallace and Sharkey 2011; Yandrapu and Sarosiek 2015).

The response of the central nervous system to the sight, smell, taste, or anticipation of food alters the activity of the enteric nervous system through ACh to stimulate the gastric acid secretion directly from the parietal cells through its muscarinic receptor. ACh, furthermore, indirectly causes the release of the histamine and gastrin, from the ECL and G cells, respectively, in the stomach. The gastrin, an effective inducer of acid secretion, has a direct effect on the parietal cells and also causes the release of the histamine from the ECL cells. This histamine then activates the H_2 receptors in the nearby parietal cells, which consequently stimulates the G_s–adenylyl cyclase–Cyclic Adenosine Monophosphate–Protein Kinase A (AMP–PKA) pathway, along with the ACh and gastrin receptors with the Gq–PLC-IP_3–Ca^{2+} pathway in parietal cells. Thus, the cyclic AMP and the Ca^{2+}-dependent pathways in the parietal cells trigger the exchanges of hydrogen and potassium ions across the parietal cell membrane, called the proton pump (H^+, K^+-ATPase). This H^+ ion production or gastric acid secretion manages the pH of the lumen, and it may reach up to pH 1.0 at peak secretion, whereas it remains at pH 7.0 near the epithelial cells. This ion transfer or acid secretion into the lumen is terminated by the release of somatostatin from the antral D cells, which suppresses the gastrin production, when the pH at the stomach lumen reaches <3 (Wallace and Sharkey 2011; Yandrapu and Sarosiek 2015). Thus, the two gastric endocrine cells: gastric G and somatostatin D cells regulate the gastric secretion. The somatostatin exerts paracrine control on the gastrin release.

Therefore, the gastric acid secretion is controlled through the interaction of three processes: (1) neural stimulation via the vagus nerve with the release of acetylcholine; (2) endocrine stimulation by the gastrin; and (3) paracrine regulation, promoted by the local release of histamine (Jain et al. 2007).

22.3 ETIOLOGY OF GASTRIC ULCER

The etiology of the gastric ulcer is pluricausal. An imbalance between antagonistic and protective factors starts the ulcerogenic process, and the gastric ulcers arise due to the exposure of the gastric mucosa constantly to harmful stimuli, such as acid-peptic juice. The presence of the excess gastric acid secretion and pepsin are one among the endogenous effects of the ulcer. The gastric acid gets secreted from the parietal cells of the epithelial layer into the lumen, which is activated by many factor receptors, like histamine, acetylcholine, and gastrin that receive the stimulus from the central nervous system by stress, smell, taste, etc. of various substances. This gastric acid activates the conversion of the pepsinogen into pepsin optimally at $pH < 3.5$, whereas at $pH > 5$, this gets inactivated/denatured. Hence, at low pH (high H^+ ions), the highest secretion of acid and pepsin causes mucus and cellular damage. At this situation, the mucosal lining is breached and the epithelial layer gets perforated, which eventually results in the penetration of the stomach content, including acidic fluid into the abdominal cavity, causing intense pain, local peritonitis, that can become generalized, and finally leads to a systemic inflammatory response syndrome and sepsis, with the risk of multiorgan failure and death. In the presence of stress, there is an increase in the secretion of gastric acid and pepsin, with the involvement of two pathways, the hypothalamic-pituitary-adrenal axis and the sympathoadrenal system. Besides, the presence of glucocorticoids and catecholamines in the microcirculation of gastric tissue also provokes ischemia and tissue necrosis in the stomach, consequently, the appearance of gastric ulcers (Saavedra et al. 2011; Yandrapu and Sarosiek 2015).

Among the exogenous factors that attack the mucosa, which predispose the formation of gastric ulcers through exposing the gastric epithelium to acid, we can mention smoking, the ingestion of alcohol and diets with food of spicy content, *Helicobacter pylori* infection, and the continued use of drugs (NSAIDs) that cause adverse effects. NSAIDs are the group of drugs that are used worldwide for pain, inflammation, and fever (Bhandari et al. 2010). About 60% of people who use NSAIDs are continuously prone to develop gastric lesions (Santin et al. 2013). It is known that NSAIDs promote a blockade in the biosynthesis of PGE2 due to the inhibition of COX-1. PGE2 is responsible for secreting mucus, the release of the bicarbonate, and the increase of blood flow of the gastric mucosa, which are the factors essential for gastric protection, and thus their absence/or decrease in production leads to the formation of lesions in stomach (Vane and Botting 1995; Yandrapu and Sarosiek 2015). According to Berstad et al. (2002), NSAIDs have direct effects on the degradation of phospholipids and adhered mucus, reducing hydrophobicity and facilitating the H-ion back diffusion in the gastric mucosa. This ultimately helps the other gastric aggressive agents, such as coffee, smoking, and alcohol to damage the stomach effortlessly by irritating the stomach cells to secrete more gastric acid (Burkitt et al. 2017).

It is also possible to emphasize alcohol as an aggressive agent to the gastric mucosa, since this reduces gastric blood flow and induces oxidative stress. Alcohol raises the activity of xanthine oxidase and reduces the expression of glutathione peroxidase, reductase, and transferase, that promotes the lipid peroxidation mediated by ROS, as evidenced by the increase in the concentration of malondialdehyde (Antonisamy et al. 2014). Thus, excessive consumption of alcohol stimulates severe instability in the mucosa, implementing damaging effects directly into the epithelium, causing the formation of necrotic lesions (Massignani et al. 2009).

According to Lee et al. (2004), the bacterium *Helicobacter pylori*, which is in the form of a rod Gram-negative, flagellate, and spiral, requires a microaerophilic atmosphere for its growth. Even though the gastric microenvironment, with the presence of high acidic pH and digestive enzymes and low partial oxygen pressure, is unfriendly for the bacterium, it colonizes in the stomach and human duodenum by developing a suitable microenvironment through intense urease activity that elevates the pH of this site through generating ammonia and acts as an etiological factor to cause a stomach ulcer and chronic gastritis (Takashima et al. 2001; Ustün et al. 2006; Burkitt et al. 2017). Its infection causes chronic gastritis and peptic ulcers in humans. It was estimated that 50% of the world's population have been affected with *H. pylori*, while only a small percentage have the symptoms of related diseases (Gubina and Ihan 2014).

Primarily, this bacterium inhabits the antrum with the help of the flagella that helps its motility and the mucus layer for survival and colonization, then over time migrates toward the proximal segments of the stomach, *H. pylori* infection accompanied with gastritis. The *H. pylori* infection causes the increase of interferon-γ and other inflammatory cells in the antrum (stomach), and thereby the gastrin release and inhibition of the somatostatin production, resulting in increased acid secretion (McColl and El-Omar 1994; Singh et al. 2018). The *pylori* infection affects the normal physiological control on the gastric secretion. Also, the bacterium results in the upregulation of pro-inflammatory cytokines like IL-6, IL-8, TNF-α, etc., thus, it induces inflammation and oxidative stress. There occurs the activation of pro-inflammatory cells (mainly neutrophils and persistent lymphocytes) to remove the bacterium, and thereby DNA damage and apoptosis of the epithelial cells by the ROS is generated, to result in an ulcer (Malfertheiner et al. 2009; Burkitt et al. 2017).

Thus, the *H. pylori* infection mainly causes the gastritis, while there occur changes in the gastric acid secretion due to the inhibition of the somatostatin release and over secretion of gastrin. Some patients are reported with an increased acid secretion, while some with low, these changes were caused by the inflammatory cytokine IL-1β in the gastric mucosa (Takashima et al. 2001).

This infectious agent is present in approximately 80% of the patients with gastric ulcers.

The *H. pylori* infection is seen in association with other ulcers, and the reduction in acid secretion increases the epithelial damage or ulceration. Hence, *H. pylori* is considered as the main causal factor of chronic gastritis present in 95% of cases of duodenal ulcers, for that reason, this bacterium is regarded as being the first bacterium associated with the risk of developing cancer, gastric cancer (Kumar et al. 2005).

Foods that are processed or overcooked, spicy, and those that have low level/lack of digestive enzymes, and fibers, stimulate the stomach to secrete more such enzymes and fluctuate levels of acid, which adversely affect the stomach by helping the inhabited bacteria to thrive in the stomach (Burkitt et al. 2017). Long-term emotional stress has also played a similar role in producing ulcers through altering the factors maintaining the mucosal integrity including the prostaglandins production by COX-1 activity (Jain 2016). Stress results in an ischemic condition in the stomach via both sympathetic (results in arteriolar vasoconstriction) and parasympathetic (causes muscular contraction by increasing the gastrointestinal motility) ways, which results in the increase of the generation of hydroxyl radicals to cause oxidative damage in the area. The stress increases the histamine release, thereby, to increase the acid secretion and reduce mucus production and blood flow (Tandon et al. 2004; Adinortey et al. 2013).

Another important aggressive factor in ulcers and many other degenerative diseases like strokes, inflammatory processes, inflammatory bowel disease, and gastritis is ROS. The NSAID-, ethanol-, and *H. pylori*-like ulcerogenic agents diminish the endogenous anti-oxidant enzyme levels in the gastric tissues, which creates oxidative stress in the area, and its effect is seen as damage to proteins, lipids, and nucleic acids in gastric mucosa and epithelial cells (Suzuki et al. 2011).

22.4 CURRENT PHARMACOTHERAPY FOR GASTRIC ULCERS

Mainly, ulcer treatments are followed in two ways: prophylactic and therapeutic types. The prophylactic treatment mainly depends on strengthening the mucosal defensive factors, such as stimulating the somatostatin and prostaglandin synthesis and inhibiting the secretion of gastrin, to provide gastroprotection through protecting the mucus layer and reducing acid secretion. Apart from this, the cytoprotection of the stomach is achieved by reducing the oxidative damage through increasing the catalase activity and suppressing the peroxidation of lipids, and also the anti-inflammatory effect and the involvement of the NO-synthase pathway played gastroprotective roles, whereas, the therapeutic way of treatment involves the usage of anti-secretory or ulcer-healing drugs. Anti-secretory drugs put forth inhibitory effects on the histaminergic and cholinergic effects on the proton pump mechanism, and the ulcer healing is satisfied through the mucosal enhancement by the anti-ulcer agents (Singh et al. 2018).

For decades, the treatment for ulcer disease aimed at suppressing the gastric acid secretion in order to re-establish the balance of gastrointestinal mucosa (Kangwan et al. 2014). However, this type of therapy continuously caused side effects. In the middle of the twentieth century, the treatment of gastric ulcers was to remove the vagus nerve (vagotomy) and to use drugs that would minimize gastric acidity, in addition to limiting dietary intake; however, there were high morbidity and mortality rates at that time. Soon afterwards, in the 1970s, new drugs were introduced into the market.

H2 receptor antagonists are reversible competitive blockers for histamine receptors (CYP450) in the parietal cells. Representatives of this class of drugs are cimetidine, famotidine, and ranitidine. Cimetidine also has an affinity for androgen receptors. These inhibitory effects, when activated, lead to side effects such as headaches, diarrhea, renal impairment, and impotence gynecomastia, nausea, and drowsiness in the elderly (Kang et al. 2016; Yuan et al. 2006; Singh et al. 2018).

In addition, another class of drugs that was introduced include the proton pump inhibitors (inhibits H^+, K^+-ATPase pathway in parietal cells) represented by omeprazole, lansoprazole, rabeprazole, pantoprazole, and omeprazole. These drugs were considered as effective in minimizing acid secretion and cellular restitution during the treatment of gastric ulcers (Palileo and Kaunitz 2011). The process for decreasing acid in the gastric environment by these drugs occurs through the acid secretory canaliculi of the parietal cell. By this action, the inactivation of the proton pump occurs via the formation of disulfide bonds between the structure of the drugs and the protein structure of said pump. Its active form, cyclic sulfenamide or sulfenic acid, reacts by covalently binding to the sulfhydryl group of the proton pump extracellular domain cysteine, which enables the inhibition of the hydrochloric acid secretion. However, this type of drug therapy causes side effects, such as lower vitamin B12 and iron absorption, hypergastrinemia, thrombocytopenia, risk of pneumonia, headaches, nausea, weakness, diarrhea, and gastric cancer, as well as a greater susceptibility to bone fractures (Dacha et al. 2015; Singh et al. 2018).

Sodium nitroprusside- and FK409-like NO donors were also used in the ulcer treatment, which was to exert their activity through minimizing the gastric secretion by acting on the Cyclic Guanosine Monophosphate Pathway (cGMP) (Lanas 2008). Anti-muscarinie is another class of anti-ulcer drug with an anti-secretary property by inhibiting the vagal stimulation and blocking the M1 muscarinic receptor. Pirenzepine is a popularly used anti-muscarinie drug, but this also coupled with various side effects to humans, including, dry mouth, blurred vision, tachycardia, and photobia (Singh et al. 2018).

Other medicines developed with gastroprotective action through mucosal protection are still to be mentioned; sulcralfate, bismuth, and subsalicylate are drugs that forms a gel-like mucus layer to inhibit the action of pepsin and also have inhibitory action against *H. pylori*; carbenoxolone, which increases gastric mucus production; and misoprostol, a synthetic analog of PGE1 that facilitates inhibition of gastric acid secretion, increased blood flow, enhanced mucus and bicarbonate secretion and, consequently, regulates gastric motility. Nonetheless, these medicines are accompanied with several side effects also, the usage of bismuth, subsalicylate, and sulcralfate may be hazardous to some of the patients in the form of dry mouth, headache, darkening of stools, and severe constipation. While, the misoprostol reported with after effects such as abdominal pain, vomiting and nausea, headache, and diarrhea during the course of treatment (Sung et al. 2010; Singh et al. 2018).

In addition to the drug inhibitors of the acid secretion and cytoprotectors, several antibiotics and combinations have been used in the pharmacological therapy to eradicate the *H. pylori* bacterium infection in the stomach. Among them are metronidazole and clarithromycin, which act on salivary and gastric secretions, and amoxiline and tetracycline, which act on the lumen. The therapeutic schemes proposed for the eradication of *H. pylori* to date have shown some drawbacks, especially regarding the upcoming antibiotic-resistant strains of *H. pylori*, other adverse effects, and cost of treatment, factors that have been contributing to the lack of adherence to the treatment (Eisig et al. 2003; Lin et al. 2010; Burkitt et al. 2017).

Therefore, the ideal ulcer treatment relies on maintaining the proteolytic factor (gastric acid and digestive enzymes) and protective factor (mucus layer), thus, the treatment stimulates the stomach to secrete sufficient mucus to coat the stomach lining in critical situations and to reduce the secretion, while maintaining the acid environment in the stomach that is essential for maintaining the gastric function (Burkitt et al. 2017). However, despite the drug variability in the pharmaceutical market to inhibit gastric acid secretion and provide gastroprotection, attention has been paid to the treatment of the gastric ulcer (Sung et al. 2010). It is known that the gastric ulcer has a recurrent profile, the drugs used in this therapy are expensive, not totally effective, and cause adverse side effects such as: vomiting and nausea, gynecomastia, and loss of libido (Lin et al. 2010). Another fact in ulcer treatment is its slow healing and recurrence of the disease condition on breaking/not completing the treatment (Singh et al. 2018). Therefore, considering these facts, the side effects and expenses of treatment, it is necessary to expand the research in order to develop a treatment product with an anti-ulcerogenic and fewer side effects (Arrieta et al. 2003).

In this context, many natural products and isolated compounds derived from plants with promising results in the treatment of gastric ulcers show promise as alternatives to the clinic (Zayachkivska et al. 2005). Still, there is no drug on the market that leads to complete remission of gastro-duodenal ulcers. Hence, an important point to be analyzed when deciding on the work performed with a given therapeutic class is the incidence of morbidity in the population; this type of data reflect the binomial: importance in public health and local and global markets for potentially useful substances in pathology (Hiruma-Lima et al. 2002). Hence, the demand for medication with plants is increasing among the people nowadays. In this context, medicinal plants are presented as a viable alternative for all types of diseases, particularly multifactorial illnesses like ulcers, since the substances that compose them may present synergism or antagonism between them (Ghasemian et al. 2016).

22.5 ALTERNATIVE DRUGS FROM NATURAL PRODUCTS

From the earliest times, therapeutic agents from nature including plants, animals, and minerals have been used to treat human disorders/health and, consequently, humans mainly made use of plants as remedies (Fabricant and Farnsworth 2001) as revealed by the Egyptian papyri (2000 BC). According to the studies by Quer and Davit (1962), the Greeks Hippocrates and Dioscorides contributed much to the legacy about the herbal drugs used to heal the ills/problems of mankind at the time, showing that the plants possessed therapeutic properties by the action, including synergistic action, of the active chemical constituents in it. The traditional medication advises the usage of crude extracts to utilize the multifactorial activity of multicompounds in plants instead of considering one disease-one target-one drug observations. Traditional practitioners believe in the synergistic actions and multitarget properties of plant extracts, which contain a mixture of compounds that are able to minimize the multiple factors observed associated with a disease (Leonti and Casu 2013). Research on herbal products took on a greater proportion from the discovery of morphine, isolated from the plant *Papaver somniferum*, which was used in visceral pain. Another example of a drug is salicylic acid, isolated in 1828 from the *Salix alba* plant, and used to combat fever and pain.

In this way, plants are a source of therapy and are assisting in the discovery of new molecules with pharmacological properties and therapeutic usefulness (Di Stasi 1996). One of the characteristics that seem to be striking in many medicinal plants is their anti-oxidant activity, which could also contribute to their anti-inflammatory and many other interlinked effects.

22.5.1 Anti-ulcer Activity of Medicinal Plants

Several medicinal plants have been used for the treatment of gastrointestinal disorders. The research on medicinal plants with actions in the gastrointestinal tract, especially with anti-ulcerogenic activity and that stimulate the protective factors, has intensified in the last few years (Farzaei et al. 2013). Many plants of various families have been reported with anti-ulcer properties either through providing gastroprotection by augmenting the gastric defensive factors or therapeutically healing the ulcer. The anti-oxidative and anti-secretory properties of herbal medicines with H^+, K^+-ATPase activity as well as the stimulation of the mucous cell proliferation mechanism by phytomedicines provide gastroprotection, thus the anti-ulcer properties of plants lie on the multiple action/mechanism of phyto-ingredients (Bi et al. 2014). Thus, the plants exhibit gastroprotection through multiple ways of mechanism. Arunachalam et al. (2017) reported the gastroprotective activity of *Gallesia integrifolia* by exerting its anti-secretory property, increasing the production of mucus and NO, augmenting the anti-oxidant enzyme (CAT) activity, and decreasing the TNF-α and IL-1β levels.

The extracts of various plants viz. *Zizyphus oenoplia*, *Zingiber officinale*, *Anacardium humile*, *Coccinia grandis*, *Zataria multiflora*, etc. provided their anti-ulcer property through cytoprotection in the gastric tissues. The other plants that provide gastroprotection through enhancing the anti-oxidant enzyme activity include *Encholirium spectabile*, *Glycyrrhiza glabra*, *Carica papaya*, etc. Plants that provide the anti-ulcer property through the healing activity include *Terminalia chebula*, *Pausinystalia macroceras*, *Quassia amara*, etc., and the extract of *Rhizophora mangle* forms a coating over the mucosal layer by binding with it to protect from gastric offensive factors (Singh et al. 2018). In addition, some plants provide both prophylactic and therapeutic ways of anti-ulcer therapy; *Polyalthia longifolia, Strychnos potatorum*, *Syngonanthus bisulcatus*, *Pausinystalia macroceras*, etc. are some of them (Singh et al. 2018). Anti-microbial drugs play a more important role in ulcer treatment since the bacterium *Helicobacter pylori* was reported to have a close association with ulcers. The extract of *Croton lechleri* has shown an inhibitory effect on *H. pylori* growth in an acetic acid-induced gastric ulcer model. Similar to *H. pylori*, inflammation is also interlinked with ulcers by the neutrophil migration to the stomach (Bi et al. 2014). Yang et al. (2017) reported the anti-ulcer property of *Taraxacum coreanum* by inhibiting the inflammatory mediators' migration to the stomach and by increasing the endogenous anti-oxidant enzymes activity. The anti-ulcer properties demonstrated by several plants are presented in Table 22.1.

Researches on the phytochemical profiling of plant extracts proved that the various chemical constituents present in plants for their own defensive mechanisms have many biological properties for human illness. According to Bi et al. (2014), many herbal drugs produced a high efficacy rate (57%–63%) in ulcer treatments and a low recurrence rate after one year of treatment as compared to the popularly used

TABLE 22.1
The Anti-Ulcer Property of Several Medicinal Plants

Sl. No.	Plant Name	Mode of Action	References
1	*Myristica malabarica, Piper betel, Phyllanthus emblica, Alchornea glandulosa, Plantago lanceolata*	Gastroprotective activity through anti-oxidant action, augmenting mucus content proliferation as well as inhibition of gastric inflammation process and gastric & acid secretion, in NSAID and acetic acid induced ulcer models	Bi et al. (2014)
2	*Cinnamomi Cortex, Zingiberis rhizoma, Zizyphi fructus, Scutellariae radix, Azadirachta indica* leaf	Anti-ulcer property in cold restraint stress ulcer model via anti-oxidant action	Vimala and Gricilda Shoba (2014); Bi et al. (2014)
3	*Tabebuia avellanedae, Centella asiatica*	Anti-ulcer activity through mucosal cell proliferation that could be by the upregulation of epidermal growth factor and the expression of its receptors	Bi et al. (2014)
4	*Solanum nigrum* fruits extract	Inhibitory effect on acid secretion through inhibition of H^+/K^+ ATPase activity and increase in PGE2 production	Vimala and Gricilda Shoba (2014)
5	*Aegle marmelos, Azadirachta indica* leaf	Inhibit acid secretion in pylorus ligation ulcer model	Vimala and Gricilda Shoba (2014)
6	*Bauhinia variegate* root	Protect mucus from damage by inhibiting the acid secretion	Vimala and Gricilda Shoba (2014)
7	*Careya arborea, Momordica charantia*	Healing of gastric ulcer property was demonstrated in ethanol, cold restraint stress, and pylorus ligation induced ulcer models	Vimala and Gricilda Shoba (2014)
8	*Carica papaya* seed, *Mangifera indica*	Protected the gastric mucosa and reduced the gastric juice volume and acidity	Vimala and Gricilda Shoba (2014)
9	*Ficus religiosa*	Exhibited gastroprotection against ethanol and aspirin aggressive effect	Vimala and Gricilda Shoba (2014)
10	*Mimosa pudica* leaf	Gastroprotection via anti-oxidant property	Vimala and Gricilda Shoba (2014)
11	*Tamarindus indica*	Reduces the total volume of gastric juice and free and total acidity of gastric secretion in pylorus ligation induced ulcer model	Vimala and Gricilda Shoba (2014)

synthetic drugs omeprazole and ranitidine, where the recurrence rate recorded is >50%. Therefore, it was well accepted that the phytomedicines are more effective, while special caution should be given for the patients' internal conditions and a check for the safety of the herbal products. Recently, the research on the active ingredients obtained from medicinal plants has increased, as it may lead to the discovery of new compounds that are useful for the development of new drugs at more affordable prices for the majority of the population (Farzaei et al. 2013). The first systematically effective drug against gastric ulcers, carbenoxolone, was isolated from the plant *Glycyrrhiza glabra* (Liquorice) based on the report on medicinal usage by indigenous people (Gohar and Zaki 2014; Dastagir and Rizvi 2016). D-002 is a therapeutic agent with gastroprotective activity that is formed by mixing the aliphatic primary alcohols isolated from beeswax (Singh et al. 2018)

The compounds isolated from medicinal plants that have anti-ulcerogenic activity are characterized by several chemical structures belonging to the classes such as triterpenes, flavonoids, alkaloids, glycosides, saponins, and polysaccharides (Jain 2016). In addition, some medicinal plants with pharmacological properties can be partially conferred to volatile constituents, which are part of the essential oils (Takayama et al. 2011). Singh et al. (2018) have reported concise literature on the anti-ulcer properties of many plants like *Abrus cantoniensis* and *Saussurea lappa* with an inhibitory effect on *Helicobacter pylori* growth that have secondary metabolites like saponins, anthraquinones, alkaloids, flavonoids, and other aromatic compounds, sterols, alkaloids, etc. Also, several compounds identified in plant extracts with an anti-ulcer effect, e.g., betulinic acid and oleanolic acid from the plant *Melaleuca bracteata* stem bark extract have been reported (Singh et al. 2018). Recent data in the literature demonstrate the wide variety of plant-isolated chemical substances that have anti-ulcerogenic activity and are effective against gastrointestinal tract disorders, such as the work of Hiruma-Lima et al. (2002), who report the isolated *trans*-dehydrocrotonin and *trans*-crotonin from *Croton cajucara* Benth, and the work of Farzaei et al. (2013) reported about ternatin isolated from *Egletes viscose*. Bi et al. (2014) have reported the potent anti-ulcer properties of several isolated compounds, such as curcumin, glycyrrhetic acid, b-sitosterol, berberine, baicalin, ginsenoside, and bisdemethoxycurcumin by decreasing the acid secretion and inflammation, and these compounds furnished anti-oxidant action to protect the gastric tissues in the ulcer-induced animal models. Vimala and Gricilda Shoba (2014) have presented many active constituents isolated from various medicinal plants possessing anti-ulcer properties; namely, Mangiferin (*Mangifera indica*), Nimbidin (*Azadirachta indica*), Naringenin (*Ficus religiosa*), Chymopapain & Papain (*Carica papaya*), Luvangetin (*Aegle marmelos*), and Barbaloin & Isobarbolin (*Aloe vera*). Breviglieri et al. (2017) have reported the isolated compound 2-phenylquinoline, the main alkaloid of the plant *Galipea longiflora*, produced anti-ulcer activity as a result of regulating the histaminergic and gastrinergic pathways to reduce the acid secretion and also through providing cytoprotection from oxidative damage.

In this context, the researches with medicinal plants are in continuous expansion with the intention of preventing and treating different pathologies, such as those related to the gastrointestinal tract, being valued by the pharmaceutical industry as a source of discovery for new bioactive molecules (Doughari 2012).

Since the plants produce a wide variety of compounds, most of the plants possess a therapeutic effect for one or more diseases. *Eugenia punicifolia* has experimentally demonstrated a beneficial effect on pain and inflammation through the inhibition of the glutamatergic system, the production of NO, and also against ulcers by improving the gastroprotective factors, such as mucus production (Basting et al. 2014). Likewise, *Piper aleyreanum* demonstrated analgesic and anti-inflammatory properties with therapeutic activity on gastroprotection by means of enhancing the mucus production and glutathione activity (Lima et al. 2012).

Even though there are hundreds of compounds isolated from plants with medicinal properties, some compounds, like taspine, isolated from *Croton lechleri*, has both anti-inflammatory property by reducing the inflammatory process, as well as gastroprotection property by reducing the *H. pylori* bacteria content in the stomach. Thus Taspine demonstrates that it has multitargets of the therapeutic property for curing diseases and, hence, it is able to be used for more than one illness, both anti-inflammatory, as well as gastroprotection. In same way, the compound colchicine derived from *Colchicum luteum* possesses anti-inflammatory activity, pain relieving effects on joints and relief from bronchial diseases; salicin from the plant *Salix alba* proves to have analgesic and anti-rheumatic properties, and is also a proven febrifuge; marrubiin obtained from the plant *Marrubium vulgare* demonstrated anti-nociceptive and vasorelaxant activity; and glycirrhizin derived from the plant *Glycyrrhiza glabra* can be used as an expectorant and also for duodenal ulcer and inflammation treatment (Sultan et al. 2013; Jain 2016; Uritu et al. 2018).

Likewise, Breviglieri et al. (2017) have reported the gastroprotective effect of 2-phenylquinoline, alkaloid from *Galipea longiflora*, through inhibiting the acid secretion and oxidative stress; also Costa et al. (2018) have reported the anti-ulcer property of four compounds, 3-prenyl-4-dihydroxycinnamoiloxy cinnamic acid (baccharin), 3-prenyl-4-hydroxycinnamic acid (drupanin), aromadendrin-4′-O-methyl-ether, and kaempferide, isolated from Brazilian green propolis.

22.6 LAMIACEAE

Lamiaceae, generally called the "mint" family, is a multinational botanical family and considered as the 6th largest family among the angiosperms with 236 genera and approximately 7000 species of plants distributed worldwide. It includes herbs, shrubs, lianas, and trees with cosmopolitan distribution (Li et al. 2016) and is found in most of the folklore medicine for a variety of health problems like throat infections, stomach disorders, ulcers, spasms, colds, hemorrhages, and skin diseases (Ciocarlan 2016). Thyme, mint, basil, sage, savory, rosemary, lemon balm, etc., are the most known members of this family with many uses in food seasonings and as medicines, and it also presents some more plants with only limited usage in public health care (Uritu et al. 2018). Generally, the plants of this family contain a variety of phenolic compounds, polyphenols, tannins, coumarins, steroids, alkaloids, and saponins showing anti-oxidant and anti-microbial activity, and hence, several lamiaceae species are used as food seasonings and as medicines all over the world (El-Gharbaoui et al. 2017). The presence of essential oils is also characteristic for these members. These chemical constituents also have demonstrated many therapeutic properties like anti-inflammatory, hemostatic, cicatrizing, stomachic, sedative, spasmolytic, diuretic, expectorant, cardiac, hypotensive, etc. (Ciocarlan 2016). On considering its wide variety of curative properties and its usage in traditional/folklore medicines, researchers have paid attention to research on this family for developing new and efficient drugs based on the ethnobotanical report on it.

22.7 SYMPHOREMATACEAE

Symphoremataceae is considered as a synonym of one of the largest angiosperm families, the lamiaceae (Gbif 2018; Npgs 2018). According to Kadereit (2012) and Li et al. (2016), among the seven subfamilies of lamiaceae, the symphorematoideae consists of three genera (*Congea* Roxb., *Sphenodesme* Jack, and *Symphorema* Roxb.), and on occasion, this subfamily is considered as a distinct family, symphoremataceae (Thorne 2000; The taxonomicon 2018). The members of the family symphoremataceae are seen as endemic to continental Asia, including India, Sri Lanka, Southeast China, Peninsular Malaysia, Sumatra, Borneo, Java, Tanimbar Islands, and the Philippines (Li et al. 2016).

22.8 *SPHENODESME INVOLUCRATA* VAR. *PANICULATA*

The genus *Sphenodesme*, native to the Indian subcontinent and South China (Ghosh 2014), has spread to South Asia including Vietnam, Myanmar, Sri Lanka, Bangladesh, Sumatra, Cambodia, Thailand, Guangdong, Hainan, and Taiwan, and consists of 15 species classified in the family symphoremataceae (Revolvy 2018; The Plant List 2018a).

Sphenodesme involucrata var. *paniculata*, endemic to South India, is a climber/liana with white flowers arranged in panicles and with involucral bracts, commonly known as arambodal or Njarambodal (Malayalam) (Keralaplants 2018).

22.8.1 Ethanobotanical Report

Sphenodesme involucrata has been used by the local people or/in folklore medication for many diseases including pain, rheumatism, abdominal disorders, wounds, etc. According to Srivastava and Choudhary (2008), the plant is used for abdominal disorders, worms, and burns, and its stem bark ground with a small amount of water is taken for abdominal pain in a Thailand medicinal system of treatment (Saralamp et al. 1996). Moreover, the root and stem of *Sphenodesme* in the form of decoction is being used against human illnesses, fever, and rheumatism (Johnson 1999).

In addition, several researchers have reported the therapeutic value in folklore medicine. The plant *Sphenodesme* is considered as a promising Asian medicinal plant currently used in traditional herbal formulations for treating mouth ulcer, malaria, neurological, and neuralgia disorders (Oudhia 2012). Wiart (2006) and Quattrocchi (2012) have reported the usage of this plant for fever, rheumatism, central nervous system disorders, and as a galactogogue, vermifuge and anti-bacterial agent.

Binu (2011) has reported its usage by the indigenous communities in Kerala, India, such as Malappandaran, Urali, Mala-arayan, Ulladan, Malavedan, and Malakurava for curing body pain in the form of a paste of fresh mature leaves in water, and had two times per day with rice gruel (Afroz et al. 2014).

22.8.2 Pharmacological Properties

The *S. involucrate* var. *paniculata* leaf methanol extract was proven as free of toxicity in an acute toxicity animal model. Recently, *Sphenodesme involucrata* var. *paniculata* leaf methanol extract was reported to have analgesic activity through central and peripheral pain mechanisms in hot plate and acetic acid-induced mice models, it also demonstrated to have an anti-inflammatory property in carrageenan and cotton pellet-induced animal models. The lipopolysaccharide-induced model proved the anti-inflammatory property of *S. involucrate* var. *paniculata* leaf methanol extract through an immunomodulatory mechanism by inhibiting the infiltration of neutrophil migration and NO production (Sreeja et al. 2018a).

Also, the plant was reported to have gastroprotection in various experimental animal models such as an absolute ethanol, acidified ethanol, and non-steroidal anti-inflammatory drug (indomethacin)-induced. Three doses of the plant extract were tested, and all doses significantly reduced the ulceration in the animals. The mode of anti-ulcer action studies in the plant leaves methanol extract has demonstrated the anti-ulcer action was through the anti-secretory property and the cytoprotective action was through attenuating the enzymatic anti-oxidant system and reducing the accumulation of inflammatory infiltrates in stomach tissues (Sreeja et al. 2018b).

The *S. involucrate* var. *paniculata* demonstrated anti-*helicobacter pylori* activity in a broth microdilution assay with minimum inhibitory concentration of 100 μg/mL (Sreeja et al. 2018b).

22.8.3 Phytochemical Studies

The quantitative phytochemical studies in *Sphenodesme involucrate* var. *paniculata* leaf and bark extracts showed the presence of high phenolic and flavonoid compounds in the leaf extracts (Sreeja et al. 2018a, 2018b).

The presence of a variety of 21 phytoconstituents, such as phenol, 2,4-bis(1,1-dimethylethyl), hexadecanoic acid, (9E,12E)-9,12-Octadecadienoyl chloride, palmitic acid, phytol, methyl stearate, 9,12-Octadecadienoiccid, etc., was found through Gas chromatography–mass spectrometry (GC-MS) study in methanol leaf extracts. Also, the Liquid chromatography–mass spectrometry (LC-MS) analysis in the leaf methanol extracts observed the presence of various compounds, such as austricine, benzylglucosinolate, gossypin, justicidin B, and cirsimarin.

These compounds were reported with various medicinal properties like anti-inflammatory, analgesic, anti-oxidant, gastroprotective, anti-microbial, hepatoprotective, anti-arthritic, and anti-diabetic by many researchers, and according to Sreeja et al. (2018a, 2018b), the presence of these compounds could be the reason behind the pharmacological properties of *S. involucrata.*

22.9 CONCLUSION

Sphenodesme involucrate var. *paniculata* was traditionally used for its many medicinal properties to fight against pain, inflammation, rheumatism, stomach pain, and abdominal disorders. Also, the medicinal properties of leaf methanol extracts against free radicals, pain, inflammation, and ulcers were proved in various pharmacological experiments. Thus, the plant showed gastroprotective effects by maintaining the balance between various protective and aggressive stomach factors, which may be through its anti-oxidant and anti-inflammatory properties. This could be due to the presence and synergistic or individual action of the various phytocompounds in it.

REFERENCES

Adinortey, M. B., Ansah, C., Galyuon, I., and Nyarko, A. 2013. In vivo models used for evaluation of potential antigastroduodenal ulcer agents. *Ulcers.* 2013: 1–12.

Afroz, S., Hossain, I., Eusufzai, T. K., Islam, S., Jabin, D., Rahman, S., and Rahmatullah, M. 2014. Antinociceptive activity evaluation of an Indonesian herbal preparation Cleng Marem. *Advances in Natural and Applied Sciences* 8(2): 75–82.

Ali, N., Ullah, A., Akhtar, S., Shah, A., Wadood, S., and Junaid, M. 2013. Factors associated with peptic ulcer: A single centre experience at tertiary care hospital of Khyber Pakhtunkhwa. *Khyber Medical University Journal* 5(1): 18–21.

Allen, A., and Flemström, G. 2005. Gastroduodenal mucus bicarbonate barrier: Protection against acid and pepsin. *American Journal of Physiology-Cell Physiology* 288(1): C1–C19.

Andrews, F. J., Malcontenti-Wilson, C., and O'Brien, P. E. 1994. Protection against gastric ischemia-reperfusion injury by nitric oxide generators. *Digestive Diseases and Sciences* 39(2): 366–373.

Antonisamy, P., Subash-Babu, P., Alshatwi, A. A., Aravinthan, A., Ignacimuthu, S., Choi, K. C., and Kim, J. H. 2014. Gastroprotective effect of nymphayol isolated from *Nymphaea stellata* (Willd.) flowers: Contribution of antioxidant, anti-inflammatory and anti-apoptotic activities. *Chemico-Biological Interactions* 224: 157–163.

Arrieta, J., Benitez, J., Flores, E., Castillo, C., and Navarrete, A. 2003. Purification of gastroprotective triterpenoids from the stem bark of *Amphipterygium adstringens*; role of prostaglandins, sulfhydryls, nitric oxide and capsaicin-sensitive neurons. *Planta Medica* 69(10): 905–909.

Arunachalam, K., Balogun, S. O., Pavan, E., de Almeida, G. V. B., de Oliveira, R. G., Wagner, T., Cechinel Filho, V. and de Oliveira Martins, D. T. 2017. Chemical characterization, toxicology and mechanism of gastric antiulcer action of essential oil from *Gallesia integrifolia* (Spreng.) Harms in the in vitro and in vivo experimental models. *Biomedicine and Pharmacotherapy* 94: 292–306.

Avila, J. R., De La Lastra, C. A., Martin, M. J., Motilva, V., Luque, I., Delgado, D., Esteban, J. and Herrerias, J. 1996. Role of endogenous sulphydryls and neutrophil infiltration in the pathogenesis of gastric mucosal injury induced by piroxicam in rats. *Inflammation Research* 45(2): 83–88.

Basting, R. T., Nishijima, C. M., Lopes, J. A., Santos, R. C., Périco, L. L., Laufer, S., Bauer, S., Costa, M. F., Santos, L. C., Rocha, L. R. and Vilegas, W. 2014. Antinociceptive, anti-inflammatory and gastroprotective effects of a hydroalcoholic extract from the leaves of *Eugenia punicifolia* (Kunth) DC. in rodents. *Journal of Ethnopharmacology* 157: 257–267.

Berstad, A. E., Berstad, K., and Berstad, A. 2002. pH-activated phospholipase A 2: An important mucosal barrier breaker in peptic ulcer disease. *Scandinavian Journal of Gastroenterology* 37(6): 738–742.

Bhandari, S. V., Parikh, J. K., Bothara, K. G., Chitre, T. S., Lokwani, D. K., Devale, T. L., Modhave, N. S., Pawar, V. S., and Panda, S. 2010. Design, synthesis, and evaluation of anti-inflammatory, analgesic, ulcerogenicity, and nitric oxide releasing studies of novel indomethacin analogs as non-ulcerogenic derivatives. *Journal of Enzyme Inhibition and Medicinal Chemistry* 25(4): 520–530.

Bi, W. P., Man, H. B., and Man, M. Q. 2014. Efficacy and safety of herbal medicines in treating gastric ulcer: A review. *World Journal of Gastroenterology: WJG* 20(45): 17020.

Binu, S. 2011. Medicinal plants used for treating body pain by the tribals in. *Indian Journal of Traditional Knowledge* 10: 547–549.

Blackwell, K. A., Raisz, L. G., and Pilbeam, C. C. 2010. Prostaglandins in bone: Bad cop, good cop? *Trends in Endocrinology and Metabolism* 21(5): 294–301.

Breuillard, C., Cynober, L., and Moinard, C. 2015. Citrulline and nitrogen homeostasis: An overview. *Amino Acids* 47(4): 685–691.

Breviglieri, E., da Silva, L. M., Boeing, T., Somensi, L. B., Cury, B. J., Gimenez, A., Cechinel Filho, V. and de Andrade, S. F. 2017. Gastroprotective and anti-secretory mechanisms of 2-phenylquinoline, an alkaloid isolated from *Galipea longiflora*. *Phytomedicine* 25: 61–70.

Burkitt, M. D., Duckworth, C. A., Williams, J. M., and Pritchard, D. M. 2017. *Helicobacter pylori*-induced gastric pathology: Insights from *in vivo* and *ex vivo* models. *Disease Models and Mechanisms* 10(2): 89–104.

Calatayud, S., Sanz, M. J., Canet, A., Bello, R., De Rojas, F. D., and Esplugues, J. V. 1999. Mechanisms of gastroprotection by transdermal nitroglycerin in the rat. *British Journal of Pharmacology* 127(5): 1111–1118.

Cho, C. H. 2001. Current roles of nitric oxide in gastrointestinal disorders. *Journal of Physiology-Paris* 95(1–6): 253–256.

Ciocarlan, N. 2016. Family Lamiaceae: Main important spontaneous medicinal and aromatic species in the republic of Moldova. *Journal of Botany* 8: 86–91.

Costa, E. D., Rezende, B. A., Cortes, S. F., and Lemos, V. S. 2016. Neuronal nitric oxide synthase in vascular physiology and diseases. *Frontiers in Physiology* 7: 206.

Costa, P., Almeida, M. O., Lemos, M., Arruda, C., Casoti, R., Somensi, L. B., Boeing, T., et al. 2018. Artepillin C, drupanin, aromadendrin-4'-O-methyl-ether and kaempferide from Brazilian green propolis promote gastroprotective action by diversified mode of action. *Journal of Ethnopharmacology* 226: 82–89.

Dacha, S., Razvi, M., Massaad, J., Cai, Q., and Wehbi, M. 2015. Hypergastrinemia. *Gastroenterology Report* 3(3): 201–208.

Dastagir, G., and Rizvi, M. A. 2016. *Glycyrrhiza glabra* L. (Liquorice). *Pakistan Journal of Pharmaceutical Sciences* 29(5): 1727–1733.

Di Stasi, L. C. 1996. Medicinal plants: Art and science. An interdisciplinary study guide-Publisher of Universidade Estadual Paulista-São Paulo-SP.

Doughari, J. H. 2012. Phytochemicals: Extraction methods, basic structures and mode of action as potential chemotherapeutic agents. In *Phytochemicals-A Global Perspective of Their Role in Nutrition and Health*. InTech. 1–32.

Eisig, J. N., André, S. B., Silva, F. M., Hashimoto, C., Moraes-Filho, J. P. P., and Laudanna, A. A. 2003. The impact of *Helicobacter pylori* resistance on the efficacy of a short course pantoprazole based triple therapy. *Arquivos de gastroenterologia* 40(1): 55–60.

El-Gharbaoui, A., Benítez, G., González-Tejero, M. R., Molero-Mesa, J., and Merzouki, A. 2017. Comparison of Lamiaceae medicinal uses in eastern Morocco and eastern Andalusia and in Ibn al-Baytar's Compendium of Simple Medicaments (13th century CE). *Journal of Ethnopharmacology* 202: 208–224.

Fabricant, D. S., and Farnsworth, N. R. 2001. The value of plants used in traditional medicine for drug discovery. *Environmental Health Perspectives* 109(Suppl 1): 69.

Farzaei, M. H., Khazaei, M., Abbasabadei, Z., Feyzmahdavi, M., and Mohseni, G. R. 2013. Protective effect of *Tragopogon graminifolius* DC against ethanol induced gastric ulcer. *Iranian Red Crescent Medical Journal* 15(9): 813.

Fattini, C. A., and Dangelo, J. G. (2007). *Systemic and Segmental Human Anatomy*. São Paulo, Brazil, Atheneu.

Gbif. 2018 (https://www.gbif.org/species/4077952). (Retrieved on April 12, 2018.)

Ghasemian, M., Owlia, S., and Owlia, M. B. 2016. Review of anti-inflammatory herbal medicines. *Advances in Pharmacological Sciences* 2016: 11 pages.

Ghosh, A. 2014. Endemic climbing genera of North Andaman Islands, India. *International Journal of Pharmacy and Life Sciences* 5(6): 3611–3616.

Gilbert, A., Hauberline, M. D. M., Breuil, D. F. R., William, Y. N., and Albert, K. 2015. Antiulcerogenic and antiulcer activities of *Dissotis thollonii* (Melastomataceae) leaves in rats. *International Journal Pharmacy Pharmaceutical Sciences* 7 (Suppl. 8): 562–569.

Gohar, A. A., and Zaki, A. A. 2014. Assessment of some herbal drugs for prophylaxis of peptic ulcer. *Iranian Journal of Pharmaceutical Research: IJPR* 13(3): 1081.

Goo, T., Akiba, Y., and Kaunitz, J. D. 2010. Mechanisms of intragastric pH sensing. *Current Gastroenterology Reports* 12(6): 465–470.

Gubina, M., and Ihan, A. 2014. The immune response to helicobacter pylori. *Food Technology and Biotechnology* 52(2): 204–209.

Ham, M., and Kaunitz, J. D. 2008. Gastroduodenal mucosal defense. *Current Opinion in Gastroenterology* 24(6): 665–673.

Hata, A. N., and Breyer, R. M. 2004. Pharmacology and signaling of prostaglandin receptors: Multiple roles in inflammation and immune modulation. *Pharmacology and Therapeutics* 103(2): 147–166.

Hiruma-Lima, C. A., Toma, W., de Souza Gracioso, J., de Almeida, A. B. A., Batista, L. M., Magri, L., de Paula, A. C. B., Soares, F. R., Nunes, D. S., and Brito, A. R. M. S. 2002. Natural trans-Crotonin: The antiulcerogenic effect of another diterpene isolated from the bark of *Croton cajucara* B ENTH. *Biological and Pharmaceutical Bulletin* 25(4): 452–456.

Jain, K. S., Shah, A. K., Bariwal, J., Shelke, S. M., Kale, A. P., Jagtap, J. R., and Bhosale, A. V. 2007. Recent advances in proton pump inhibitors and management of acid-peptic disorders. *Bioorganic and Medicinal Chemistry* 15(3): 1181–1205.

Jain, P. 2016. Secondary metabolites for antiulcer activity. *Natural Product Research* 30(6): 640–656.

Johnson, T. 1999. *CRC Ethnobotany Desk Reference.* CRC Press, Taylor & Francis Group: New York.

Jozefczak, M., Remans, T., Vangronsveld, J., and Cuypers, A. 2012. Glutathione is a key player in metal-induced oxidative stress defenses. *International Journal of Molecular Sciences* 13(3): 3145–3175.

Kadereit, J. W. (Ed.). 2012. *Flowering Plants Dicotyledons: Lamiales (except Acanthaceae Including Avicenniaceae)* (Vol. 7). Springer Science & Business Media, Berlin, Germany.

Kang, S. H., Kim, Y. S., Kim, E. K., Hwang, J. W., Jeong, J. H., Dong, X., Lee, J. W., Moon, S. H., Jeon, B. T., and Park, P. J. 2016. Anticancer effect of thymol on AGS human gastric carcinoma cells. *Journal of Microbiology and Biotechnology* 26(1): 28–37.

Kangwan, N., Park, J. M., Kim, E. H., and Hahm, K. B. 2014. Quality of healing of gastric ulcers: Natural products beyond acid suppression. *World Journal of Gastrointestinal Pathophysiology* 5(1): 40.

Keralaplants, 2018 (http://keralaplants.in/keralaplantsdetails.aspx?id=Sphenodesme_ paniculata), Accessed on: May 2018.

Konturek, P. C., Konturek, S. J., and Ochmański, W. 2004. Neuroendocrinology of gastric H+ and duodenal $HCO3^-$ secretion: The role of brain–gut axis. *European Journal of Pharmacology* 499(1–2): 15–27.

Kristjánsson, G., Högman, M., Venge, P., and Hällgren, R. 2005. Gut mucosal granulocyte activation precedes nitric oxide production: Studies in coeliac patients challenged with gluten and corn. *Gut* 54(6): 769–774.

Kumar, V., Abbas, A. K., and Fausto, N. 2005. Philadelphia, PA: Elsevier Saunders. *Robbins and Cotran—Pathologic Basis of Disease.* Philadelphia, PA: Elsevier Saunders, p. 22.

Laine, L., Takeuchi, K., and Tarnawski, A. 2008. Gastric mucosal defense and cytoprotection: Bench to bedside. *Gastroenterology* 135(1): 41–60.

Lanas, A. 2008. Role of nitric oxide in the gastrointestinal tract. *Arthritis Research & Therapy* 10(2): S4.

Lee, K. M., Yeo, M., Choue, J. S., Jin, J. H., Park, S. J., Cheong, J. Y., Lee, K. J., Kim, J. H., and Hahm, K. B. 2004. Protective mechanism of epigallocatechin-3-gallate against *Helicobacter pylori*-induced gastric epithelial cytotoxicity via the blockage of TLR-4 signaling. *Helicobacter* 9(6): 632–642.

Leonti M., and Casu L. 2013. Traditional medicines and globalization: Current and future perspectives in ethnopharmacology. *Frontiers in Pharmacology* 4: 92.

Li, B., Cantino, P. D., Olmstead, R. G., Bramley, G. L., Xiang, C. L., Ma, Z. H., Tan, Y. H., and Zhang, D. X. 2016. A large-scale chloroplast phylogeny of the Lamiaceae sheds new light on its subfamilial classification. *Scientific Reports* 6(1): 1-18.

Lima, D. K., Ballico, L. J., Lapa, F. R., Gonçalves, H. P., de Souza, L. M., Iacomini, M., de Paula Werner, M. F. et al. 2012. Evaluation of the antinociceptive, anti-inflammatory and gastric antiulcer activities of the essential oil from *Piper aleyreanum* C. DC in rodents. *Journal of Ethnopharmacology* 142(1): 274–282.

Lin, P. C., Chang, C. H., Hsu, P. I., Tseng, P. L., and Huang, Y. B. 2010. The efficacy and safety of proton pump inhibitors vs histamine-2 receptor antagonists for stress ulcer bleeding prophylaxis among critical care patients: A meta-analysis. *Critical Care Medicine* 38(4): 1197–1205.

Lobo, V., Patil, A., Phatak, A., and Chandra, N. 2010. Free radicals, antioxidants and functional foods: Impact on human health. *Pharmacognosy Reviews* 4(8): 118.

Lopez-Belmonte, J., Whittle, B. J., and Moncada, S. 1993. The actions of nitric oxide donors in the prevention or induction of injury to the rat gastric mucosa. *British Journal of Pharmacology* 108(1): 73–78.

Lushchak, V. I. 2014. Free radicals, reactive oxygen species, oxidative stress and its classification. *Chemico-Biological Interactions* 224: 164–175.

Magierowski, M., Magierowska, K., Kwiecien, S., and Brzozowski, T. 2015. Gaseous mediators nitric oxide and hydrogen sulfide in the mechanism of gastrointestinal integrity, protection and ulcer healing. *Molecules* 20(5): 9099–9123.

Malfertheiner, P., Chan, F. K., and McColl, K. E. 2009. Peptic ulcer disease. *The Lancet* 374(9699): 1449–1461.

Martin, M. J., and Motilva, V. 2001. New issues about nitric oxide and its effects on the gastrointestinal tract. *Current Pharmaceutical Design* 7(10): 881–908.

Massignani, J. J., Lemos, M., Maistro, E. L., Schaphauser, H. P., Jorge, R. F., Sousa, J. P. B., Bastos, J. K., and de Andrade, S. F. 2009. Antiulcerogenic activity of the essential oil of *Baccharis dracunculifolia* on different experimental models in rats. *Phytotherapy Research: An International Journal Devoted to Pharmacological and Toxicological Evaluation of Natural Product Derivatives* 23(10): 1355–1360.

McColl, K. E. L., and El-Omar, E. 1994. Effect of H. pylori infection on gastrin and gastric acid secretion. In Hunt R. H., Tytgat G. N. J. (Eds.), *Helicobacter pylori* (pp. 245–256). Springer, Dordrecht.

Medeiros, J. V. R., Bezerra, V. H., Gomes, A. S., Barbosa, A. L. R., Lima-Júnior, R. C. P., Soares, P. M. G., Brito, G. A. C., Ribeiro, R. A., Cunha, F. Q., and Souza, M. H. 2009. Hydrogen sulfide prevents ethanol-induced gastric damage in mice: Role of ATP-sensitive potassium channels and capsaicin-sensitive primary afferent neurons. *Journal of Pharmacology and Experimental Therapeutics* 330(3) 764–770.

Najm, W. I. 2011. Peptic ulcer disease. *Primary Care: Clinics in Office Practice* 38(3): 383–394.

Nishio, H., Hayashi, Y., Terashima, S., and Takeuchi, K. 2006. Role of endogenous nitric oxide in mucosal defense of inflamed rat stomach following iodoacetamide treatment. *Life sciences* 79(16): 1523–1530.

Nordberg, J., and Arner, E. S. 2001. Reactive oxygen species, antioxidants, and the mammalian thioredoxin system. *Free Radical Biology and Medicine* 31(11): 1287–1312.

Npgs 2018 (https://npgsweb.ars-grin.gov/gringlobal/taxonomyfamily.aspx?id=619). (Retrieved on June 21, 2018.)

Ock, C. Y., Kim, E. H., Choi, D. J., Lee, H. J., Hahm, K. B., and Chung, M. H. 2012. 8-Hydroxydeoxyguanosine: Not mere biomarker for oxidative stress, but remedy for oxidative stress-implicated gastrointestinal diseases. *World Journal of Gastroenterology: WJG* 18(4): 302.

Oudhia, P. 2012. Medicinal Plants of India used in treatment of neurological disorders from Pankaj Oudhia's Medicinal Plant Database. Pankaj Oudhia's Ethnobotanical Survey from year 1990–2012, India.

Palileo, C., and Kaunitz, J. D. 2011. Gastrointestinal defense mechanisms. *Current Opinion in Gastroenterology* 27(6): 543–548.

The Plant List. 2018a (http://www.theplantlist.org/browse/A/Lamiaceae/ Sphenodesme/). (Accessed on May 24, 2018.)

Quattrocchi, U. 2012. *CRC World Dictionary of Medicinal and Poisonous Plants: Common Names, Scientific Names, Eponyms, Synonyms, and Etymology.* CRC Press: Boca Raton, FL.

Quer, P. F., and Davit, S. 1962. *Medicinal Plants: The Renewed Dioscorides* (p. 543). Work. Peninsular Publishing Company, Barcelona, BCN, Spain.

Quideau, S., Deffieux, D., Douat-Casassus, C., and Pouysegu, L. 2011. Plant polyphenols: Chemical properties, biological activities, and synthesis. *Angewandte Chemie International Edition* 50(3): 586–621.

Revolvy. 2018 (https://www.revolvy.com/main/index.php?s= Sphenodesme). (Retrieved on May 24, 2018.)

Rozza, A. L., de Mello Moraes, T., Kushima, H., Tanimoto, A., Marques, M. O. M., Bauab, T. M., Hiruma-Lima, C. A., and Pellizzon, C. H. 2011. Gastroprotective mechanisms of citrus lemon (Rutaceae) essential oil and its majority compounds limonene and β-pinene: Involvement of heat-shock protein-70, vasoactive intestinal peptide, glutathione, sulfhydryl compounds, nitric oxide and prostaglandin E2. *Chemico-Biological Interactions* 189(1–2): 82–89.

Saavedra, J. M., Sánchez-Lemus, E., and Benicky, J. 2011. Blockade of brain angiotensin II AT1 receptors ameliorates stress, anxiety, brain inflammation and ischemia: Therapeutic implications. *Psychoneuroendocrinology* 36(1): 1–18.

Santin, J. R., Machado, I. D., Rodrigues, S. F., Teixeira, S., Muscará, M. N., Galdino, S. L., da Rocha Pitta, I., and Farsky, S. H. 2013. Role of an indole-thiazolidine molecule PPAR pan-agonist and COX inhibitor on inflammation and microcirculatory damage in acute gastric lesions. *PLoS One* 8(10): e76894.

Santin, J. R., Machado, I. D., Rodrigues, S. F., Teixeira, S., Muscará, M. N., Galdino, S. L., da Rocha Pitta, I., and Farsky, S. H. 2013. Role of an indole-thiazolidine molecule PPAR pan-agonist and COX inhibitor on inflammation and microcirculatory damage in acute gastric lesions. *PLoS One* 8(10): e76894.

Saralamp, P., Mahidon, M., and Phesatchaphruksasat, P. W. 1996. *Medicinal Plants in Thailand Department of Pharmaceutical Botany, Faculty of Pharmacy*, Mahidol University, Bangkok, Thailand, p. 218.

Schäfer, M., and Werner, S. 2008. Oxidative stress in normal and impaired wound repair. *Pharmacological Research* 58(2): 165–171.

Schubert, M. L., and Peura, D. A. 2008. Control of gastric acid secretion in health and disease. *Gastroenterology* 134(7): 1842–1860.

Semchyshyn, H. M., and Lozinska, L. M. 2012. Fructose protects baker's yeast against peroxide stress: Potential role of catalase and superoxide dismutase. *FEMS Yeast Research* 12(7): 761–773.

Sies, H. 2015. Oxidative stress: A concept in redox biology and medicine. *Redox biology* 4: 180–183.

Singh, A. K., Singh, S. K., Singh, P. P., Srivastava, A. K., Pandey, K. D., Kumar, A., and Yadav, H. 2018. Biotechnological aspects of plants metabolites in the treatment of ulcer: A new prospective. *Biotechnology Reports* 1, 18:e00256.

Sreeja, P. S., Arunachalam, K., de Oliveira Martins, D. T., da Silva Lima, J. C., Balogun, S. O., Pavan, E., Saikumar, S., Dhivya, S., Kasipandi, M., and Parimelazhagan, T. 2018b. *Sphenodesme involucrata* var. *paniculata* (CB Clarke) Munir.: Chemical characterization, anti-nociceptive and anti-inflammatory activities of methanol extract of leaves. *Journal of* Ethnopharmacology 225: 71–80.

Sreeja, P. S., Arunachalam, K., Saikumar, S., Kasipandi, M., Dhivya, S., Murugan, R., and Parimelazhagan, T. 2018a. Gastroprotective effect and mode of action of methanol extract of *Sphenodesme involucrata* var. *paniculata* (CB Clarke) Munir (Lamiaceae) leaves on experimental gastric ulcer models. *Biomedicine and Pharmacotherapy* 97: 1109–1118.

Srivastava, R. C., and Choudhary, R. K. 2008. Species diversity and economic importance of the family Verbenaceae in Arunachal Pradesh. *Bulletin of Arunachal Forest Research* 24(1&2): 1–21.

Sultan, R., Wani, M. A., and Irshad, A. N. 2013. Herbal drugs-Current status and future prospects. *International Journal of Medicinal Plant and Alternative Medicine* 1(2): 020–029.

Sumbul, S., Ahmad, M. A., Mohd, A., and Mohd, A. 2011. Role of phenolic compounds in peptic ulcer: An overview. *Journal of Pharmacy and Bioallied Sciences* 3(3): 361.

Sung, J. J. Y., Kuipers, E. J., and El-Serag, H. B. 2009. Systematic review: The global incidence and prevalence of peptic ulcer disease. *Alimentary Pharmacology and Therapeutics* 29(9): 938–946.

Sung, J. J., Tsoi, K. K., Ma, T. K., Yung, M. Y., Lau, J. Y., and Chiu, P. W. 2010. Causes of mortality in patients with peptic ulcer bleeding: A prospective cohort study of 10,428 cases. *The American Journal of Gastroenterology* 105(1): 84.

Suzuki, H., Nishizawa, T., Tsugawa, H., Mogami, S., and Hibi, T. 2011. Roles of oxidative stress in stomach disorders. *Journal of Clinical Biochemistry and Nutrition* 50(1): 35–39.

Takashima, M., Furuta, T., Hanai, H., Sugimura, H., and Kaneko, E. 2001. Effects of *Helicobacter pylori* infection on gastric acid secretion and serum gastrin levels in Mongolian gerbils. *Gut* 48(6): 765–773.

Takayama, C., de-Faria, F. M., de Almeida, A. C. A., Rehen, C. S., Dunder, R. J., Socca, E. A. R., Manzo, L. P. et al. 2011. Gastroprotective and ulcer healing effects of essential oil from *Hyptis spicigera* Lam. (Lamiaceae). *Journal of Ethnopharmacology* 135(1): 147–155.

Tandon, R., Khanna, R. D., Dorababu, M. and Goel, R. K. 2004. Oxidative stress and antioxidants status in peptic ulcer and gastric carcinoma. *Indian Journal of Physiology and Pharmacology* 48(1): 115–118.

Tarnawski, A. S. 2005. Cellular and molecular mechanisms of gastrointestinal ulcer healing. *Digestive Diseases and Sciences* 50(1): S24–S33.

Tarnawski, A., Ahluwalia, A., and Jones, M. K. 2013. Gastric cytoprotection beyond prostaglandins: Cellular and molecular mechanisms of gastroprotective and ulcer healing actions of antacids. *Current Pharmaceutical Design* 19(1): 126–132.

The Taxonomicon. 2018 (http://taxonomicon.taxonomy.nl/ TaxonTree.aspx?id =1137204&src=0). (Retrieved on July 15, 2018.)

Thorne, R. F. 2000. The classification and geography of the flowering plants: Dicotyledons of the class Angiospermae. *The Botanical Review* 66(4): 441–647.

Tulassay, Z. and Herszényi, L. 2010. RETRACTED: Gastric mucosal defense and cytoprotection. *Best Practice and Research Clinical Gastroenterology* 24: 99–108.

Uritu, C. M., Mihai, C. T., Stanciu, G. D., Dodi, G., Alexa-Stratulat, T., Luca, A., Leon-Constantin, M. M. et al. 2018. Medicinal plants of the family Lamiaceae in pain therapy: A review. *Pain Research and Management* 2018: 44 pages.

Ustün, O., Ozçelik, B., Akyön, Y., Abbasoglu, U., and Yesilada, E. 2006. Flavonoids with anti-*Helicobacter pylori* activity from *Cistus laurifolius* leaves. *Journal of Ethnopharmacology* 108(3): 457–461.

Uys, J. D., Mulholland, P. J., and Townsend, D. M. 2014. Glutathione and redox signaling in substance abuse. *Biomedicine & Pharmacotherapy* 68(6): 799–807.

Vane, J. R., and Botting, R. M. 1995. New insights into the mode of action of anti-inflammatory drugs. *Inflammation Research* 44(1): 1–10.

Vimala, G., and Gricilda Shoba, F. 2014. A review on antiulcer activity of few Indian medicinal plants. *International Journal of Microbiology* 2014: 14 pages.

Wallace, J. L. 2008. Prostaglandins, NSAIDs, and gastric mucosal protection: Why doesn't the stomach digest itself? *Physiological Reviews* 88(4): 1547–1565.

Wallace, J. L., and Sharkey, K. A. 2011. Pharmacotherapy of gastric acidity, peptic ulcers and gastroesophageal reflux disease. *Goodman & Gilman's. The Pharmacological Basis of Therapeutics.* 12th ed. McGraw-Hill Companies, San Diego, CA, pp. 1309–1322.

Wiart, C. 2006. *Medicinal Plants of Asia and the Pacific.* CRC Press, Taylor & Francis Group, New York, pp. 336. doi:10.1021/np0781460.

World Life Expectancy. 2018 (http://www.worldlifeexpectancy.com/india-peptic-ulcer-disease). (Retrieved on July 4, 2018.)

Yandrapu, H., and Sarosiek, J. 2015. Protective factors of the gastric and duodenal mucosa: An overview. *Current Gastroenterology Reports* 17(6): 24.

Yang, H. J., Kim, M. J., Kwon, D. Y., Kang, E. S., Kang, S. and Park, S. 2017. Gastroprotective actions of *Taraxacum coreanum* Nakai water extracts in ethanol-induced rat models of acute and chronic gastritis. *Journal of Ethnopharmacology* 208: 84–93.

Yuan, Y., Padol, I. T., and Hunt, R. H. 2006. Peptic ulcer disease today. *Nature Reviews Gastroenterology and Hepatology* 3(2): 80.

Zayachkivska, O. S., Konturek, S. J., Drozdowicz, D., Konturek, P. C., Brzozowski, T., and Ghegotsky, M. R. 2005. Gastroprotective effects of flavonoids in plant extracts. *Journal of Physiology and Pharmacology Supplement*, 56(1): 219–231.

23 Seaweeds
A Potent Source of Bioactive Compounds

L. Stanley Abraham, Vasantharaja Raguraman, R. Thirugnanasambandam, K. M. Smitha, D. Inbakandan, and P. Premasudha

CONTENTS

23.1 INTRODUCTION

Seaweeds are widely accepted as food and biologically active value added products by the countries of Southeast Asia and Europe, as they have significant potent biologically active metabolites. The practice of utilization of seaweeds as a food source dates back as early as 3000 BC (Doty 1979), and they are frequently used in medical preparations as prescribed in Chinese materia medica around 600 BC. In modern lifestyle it has gained importance as they nourish important nutrients without increasing the calories (Aikaterini et al. 2018). On the other hand, seaweed is also termed as sea vegetables, resembling higher terrestrial plants since seaweeds illustrate specialization and differentiation of their vegetative cells (Barsanti and Gualtieri 2014; Khan et al. 2018; Pereira 2018). They are comprised of several species which differ from one another by size, morphology, pigment, habitat, reserve food products, flagellation, cell wall, and cell structure.

23.2 MACRO ALGAE

Macro algae play a vital role in the ecosystem, as they are large primary producers in the ocean with a diverse group of algal communities. They are mostly benthic organisms and lack specific physical structures like roots, stems, and leaves that are found in the terrestrial plant. Instead of right plant parts, they possess advanced structures resembling the plant, such as holdfast-root like, stipe-stem like, lamina or blades-leaf like, thallus-simple body part, and float-helps in absorbing more sunlight. They usually reproduce by asexual mode, i.e., vegetative growth (Nisbet and Sleep 2001). Seaweeds are found under inter and subtidal regions of the ocean with varied species. Some of them are perennials, whereas other types of seaweed are annuals, and they can withstand some extreme conditions of high salinity, light, and temperature. The cultivation and harvesting are more simple for these seaweeds than for micro algae. Thus, biomass production can be measured with less difficulty. Also, due to their longevity and seasonal adaptation, there is a rapid increase in the area of ocean science. Moreover, seaweeds are a source of food with high active beneficial nutrients and they contain various biochemicals predominated by polysaccharides, which have therapeutic potential (Buggeln 1983).

23.3 COMPOSITION AND CLASSIFICATION OF SEAWEEDS

Seaweeds are comprised of 90% water, 25%–50% carbohydrates, 7%–15% proteins, 1%–5% low lipids, pigments, and other components (Vasantharaja et al. 2019). The higher percentage of carbohydrates provides the research platform for the isolation of sulfated polysaccharides with potential health benefits. The major distinguishing characteristics used for classification are from the biochemical characterizations of pigments, morphology, physiological functions, and electron microscopic studies to differentiate the macro algal species based on Dhargalkar and Devanand (2004).

1. Photosynthetic pigments
2. Storage food products
3. Cell wall component
4. Flagella

In terms of photosynthetic pigment, the macro algae are divided into three diversified classes, such as green (Chlorophyceae), red (Rhodophyceae), and brown (Phaeophyceae) (Hong et al. 2012). The different pigments of each class of seaweeds are tabulated in Table 23.1.

The seaweeds are classified into three major classes, viz:

1. Brown seaweeds (Phaeophyceae)
2. Red seaweeds (Rhodophyceae)
3. Green seaweeds (Chlorophyceae)

23.3.1 Phaeophyceae

Phaeophyceae is a class of algae grouped under the division of Heterokontophyta and commonly known as brown algae. They originated from branched filament to kelp, thus, are multicellular and possess a higher differentiation at the morphological and anatomical levels. The majority of the algal species are lithophytes, and others possibly will grow epiphytically. Few members may be free-floating, for example, *Sargassum* is found as floating algal masses. They possess certain accessory pigments such as chlorophyll a and c, α-, β-, and ε-carotene and xanthophylls like fucoxanthin, which imparts brown color to the macro algae. The cell wall of Phaeophyceae species is comprised of the inner cellulose, and the outer pectic layers containing alginic acid, which can be extracted and used as stabilizers in various food and other industrial products.

The macro algal cells tend to have high metabolic activity due to the presence of abundant refractive vesicles called fucosan vesicles. The species of brown algae reproduce by vegetative, asexual, and sexual means. The ecological importance of brown algae is food, bioactive compounds, fertilizers, and soil conditioners. The sulfated polysaccharides producing major Phaeophyceae were grouped under the order of Fucales, such as *Sargassum*, and Dictyotales such as *Padina* (Barsanti and Gualtieri 2014; Buggeln 1983).

TABLE 23.1
Characteristics of Broadly Classified Marine Seaweeds

Class	Common Name	Pigments
Phaeophyceae	Brown algae	Chlorophyll a, c, Carotenoids-α-, β- and ε- Carotene, Xanthophylls-Fucoxanthin
Rhodophyceae	Red algae	Chlorophyll a, Phycobilins, Carotenoids-α- and β-Carotene, Xanthophylls-Lutein
Chlorophyceae	Green algae	Chlorophyll a, b, Carotenoids-α-, β- and γ-Carotene, Xanthophylls-Lutein, Prasinoxanthin

Source: Vasantharaja, R. et al., *Carbohydr. Polym.*, 203, 441–449, 2019.

23.3.2 Rhodophyceae

The red algae consist of several species with no flagellated stages, and they are grouped under the Rhodophyta division. They are marine organisms and commonly found in abundance in warmer waters with greater depth. The majority of the other species grow on rocks or another substratum, and mostly, they are photoautotrophs. The thallus structure differs among the species from unicellular to filamentous, and its form can be from that of an ancestor, organized from parenchymatous filaments. The accessory pigments are chlorophyll a, phycobilins as r,b-phycoerythrin and r-phycocyanin, carotenoids as α- and β-carotene, and xanthophylls as lutein. The cell wall of these algae contains pectin, cellulose, and polysulphide esters. The microfibrillar fraction of cellulose is significantly lower than the other algal groups.

In some species, the cell wall contains pits to maintain the protoplasmic continuity between the adjacent cells. The reserve foods for the red algae are in the form of floridean starch (α-1,4-glucan), and their substances are located in the cytoplasm of algal cells other than chloroplast. The economical importance of the red algae is used to extract stabilizing agents such as phyco-colloids widely utilized in food flavoring, gelling, cosmetics, paints, rubber, and textile industries. The sulfated polysaccharides of red algae include sulfated galactans and carrageenan from commercially important species like *Gracilaria*, *Laurencia*, *Hypnea*, and *Gigartina* (Barsanti and Gualtieri 2014; Buggeln 1983).

23.3.3 Chlorophyceae

The Chlorophyceae is a class of green algae that is placed under the division of Chlorophyta. The majority of these species are freshwater organisms; hence, only a few are reported as a marine organism. The marine green algae grow along the coastal areas and stay attached to the substratum. They are grouped under the orders of Ulvales and Siphonales. Most of the species are photoautotrophs and some are heterotrophs and most species like *Ulva* and *Codium* possess cell layers attached to the substrate and also contain siphonous filaments that make use of them. The major pigments responsible for their green coloration are chlorophyll a and b, xanthophylls like lutein and prasinoxanthin, and α-, β-, and γ-carotene. The green algae store food in the form of starch (α-1,4-glucan), and its granules are found inside the chloroplast itself. The cell walls of these algae are rich in cellulose and pectin. The species like *Ulva, Codium, Enteromorpha, Caulerpa*, and *Monostroma* result in the production of the sulfated polysaccharide. There is only minor importance of these algae, such as food and few bioactive compounds (Barsanti and Gualtieri 2014; Buggeln 1983).

23.4 BIOACTIVE COMPOUNDS FOUND IN SEAWEED

Seaweeds are excellent reserves of bioactive compounds including polysaccharides, pigments, proteins, minerals, vitamins, and dietary fibers (Koutsaviti 2018; Holdt and Kraan 2011). Specifically, brown seaweed contains numerous bioactivity compounds that have anti-oxidant, anti-bacteria, anti-coagulant, anti-hypertension, and anti-tumor activities (Holdt and Kraan 2011; Heo et al. 2005; Mayer and Hamann 2004). In recent years, sulfated polysaccharides and other natural compounds from brown seaweed exhibited various bioactivities including neuroprotective, anti-viral, anti-cancer, anti-inflammatory, anti-coagulant activities. Figure 23.1 illustrates the bioactive compounds of seaweeds (Chen et al. 2012; Ina et al. 2007). The seaweeds of Pheophyceae have highly potent bioactive compounds with varied neuroprotective properties, such as anti-viral, anti-cancer, anti-inflammatory, anti-coagulant activity and anti-aging (Chen et al. 2012).

23.4.1 Bioactivity of Carbohydrates from Seaweeds

Carbohydrates are the major source of energy for carrying out most of the physiological functions, which are present as three forms in the macro algal biomass as detailed below,

1. Monosaccharide
2. Oligosaccharide
3. Polysaccharide

The biological properties, such as antioxidant, anticoagulant, anti-inflammatory, anti-arthritic, immunomodulatory, hepatoprotective, neuroprotective, chemopreventive, anti-diabetic, anti-aging, and antiobesity are being exhibited by algal polysaccharides and oligomers (Mayer and Hamann 2004).

23.4.1.1 Bioactivity of Oligosaccharides from Seaweeds

Oligosaccharides are low molecular weight carbohydrates with three or more monosaccharides generally resistant to acid hydrolysis, and they are not vulnerable to digestion by human gastric enzymes, but they can act as probiotic agents and are considered food supplements (Patel and Goyal 2011). Oligosaccharides that are derived from polysaccharides by enzyme digestion have shown effective anti-oxidant potential (Fleita et al. 2015). These oligosaccharides possess several bioactivities, such as anti-diabetic, anti-obesity, anti-oxidant, and immunomodulating agents.

The anti-diabetic activity of oligomannuronate from the *Laminaria* spp., after forming chromium (III) complexes, has been examined to study the insulin activity in C_2C_{12} cell lines

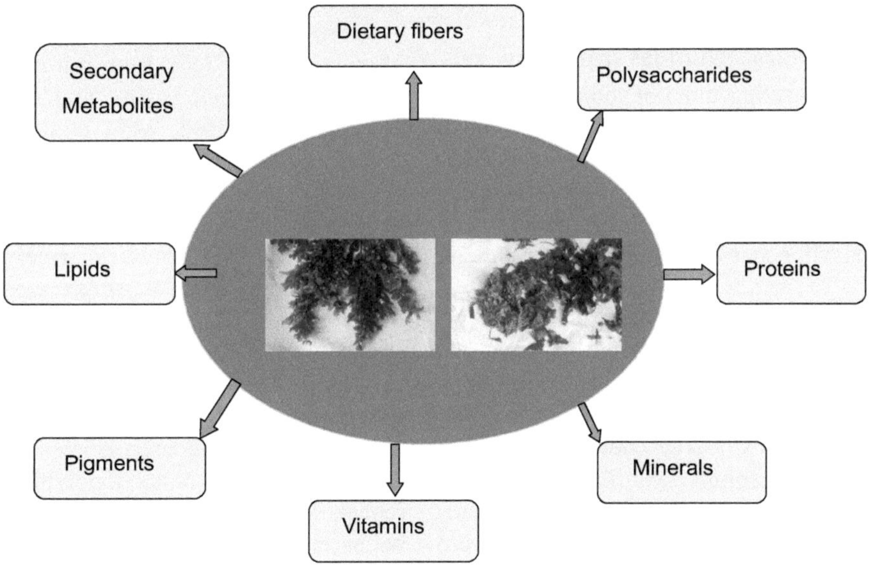

FIGURE 23.1 Bioactive compounds of seaweeds.

(Hao et al. 2011). There was considerable activity from the complex, but comparatively less to that of anti-diabetic drug metformin. But their bio assimilation and compatibility prove that these compounds are suitable for the treatment of diabetes mellitus.

Certain oligomers have effectively shown anti-obesity properties. Nakazono et al. (2016) compared the anti-obesity activity of acid and enzyme-digested alginate oligomers. The oral administration of enzymatically treated derivatives of alginate oligomers have improved the biochemical profile of the serum of obese mice more than the other form of oligomers. In vitro assays also infer the inhibition of lipid deposition in 3T3-L1 adipocytes cell lines by enzymatically synthesized oligomers. This study also proves the suitability of digested oligomers as anti-obesity agents when compared to their corresponding parental polysaccharide.

Similarly, Fucooligomers produced by enzymatic degradation exhibit increased thrombin time (Kim et al. 2010). These oligomers also exhibited lesser anti-coagulant activity than the whole fucoidan polysaccharide. Kalitnik et al. (2016) explored the oligomers from rhodophycean algae *Tichocarpus crinitus* for enhancement of interleukin-10 production in murine and human cells as dose-dependent at higher concentrations.

These oligosaccharides also possess the immunomodulation and immunostimulation activities, which were investigated by Yao et al. (2014) by evaluating the immunomodulating activity of carrageenan-derived oligosaccharides for the treatment of non-communicable neurodegenerative diseases. The authors proved the immunomodulatory activity of sulfated oligomers which inhibits the microglial cells.

The anti-oxidant activity of oligosaccharides are due to not only by the type and structure of the hydrolysates, but also by the functional group, such as sulfate to the core derivative. Dun et al. (2015) investigated if the guluronate of *F. vesiculosus* synthesized by microwave extraction method exhibited the defensive mechanism on the Tumour Necrosis Factor-α-induced oxidative damage and mitochondrial dysfunction in C_2C_{12} skeletal muscle cells.

23.4.1.2 Bioactivity of Polysaccharides from Seaweeds

Polysaccharides are the repeating units of monosaccharides linked by a glycosidic bond. They are structural storage metabolites in plants and algae. The anti-microbial properties of algal polysaccharides are due to the presence of surface glycoprotein receptors. As a result of this, the permeability of the cellular membranes will occur for cell lysis (Vasantharaja et al. 2019).

Polysaccharides from phycocolloids, such as agar, alginates, and carrageenan were widely produced by seaweeds. In addition, they have been proven to have anti-cancer, anti-microbial, as well as anti-viral properties (Ely et al. 2004; Newman et al. 2003). Fucoidans, fucose containing sulfated polysaccharides, isolated from *Sargassum cristaefolium* have proven to possess anti-oxidant, anti-cancer, anti-viral, anti-coagulant, and immunomodulating properties. They were found to have anti-cancer properties against colon cancer cells (Cheng-Yuan Wang et al. 2015). *Sargassum tenerrium*, collected from the coastal region of Mandapam, is having a wide range of pharmaceutical applications. The sulfated polysaccharide from *Sargassum tenerrium* was found to exhibit anti-cancer properties due to its free radical scavenging potential (70.33 ± 2.33%), DDPH activity (64.66 ± 2.08%), H_2O_2 (61.56 ± 2.05%), and TAC (62.55 ± 1.40) (Vijayabaskar and Vaseela 2012). Breast cancer is a predominant cancer among women, and its prevalence increases every year. The toxicity to normal cells and the acquired resistance of tumors to drugs seem to be barriers for the treatment of cancer by chemotherapy. Drug combinations with natural dietary supplements decrease the toxicity to normal cells, as well as the amount of dose administration. The bioactive compounds that form brown seaweed; phloroglucinol, fucoxanthin, and fucoidan have been proven to have potent treatment effects against breast cancer (Pádua et al. 2015). For nearly 2000 years, the *Sargassum* spp. was found to be utilized as a traditional Chinese medicine for treating various diseases since it is a rich source of bioactive compounds. They used two *Sargassum* species that have been listed in Chinese pharmacopeia (2010 version), namely, *Sargassum pallidum and Sargassum fusiforme*. Several studies have proven that the *Sargassum* species has been found to have anti-cancer, anti-microbial, anti-viral, anti-inflammatory, liver protective, and anti-oxidant activities. In recent years, the synthesis of nanoparticles using biological extracts has attracted the attention of many researchers because of better uptake, the rate of reaction is high, particles are more stable and more biocompatible. The biosynthesis gold nanoparticles using *Sargassum swartzii* is having a proven cytotoxic effect against human cervical carcinoma cells and possessing heparin-like activity, which can be used for the treatment of cardiovascular problems (Stalin Dhas et al. 2014).

The polysaccharides in their sulfated form were abundantly found in the cell wall of seaweeds (Pereira 2018). The sulfated polysaccharides and their derivatives are the complex polymers that differ by their occurrence, structure, and function. They can be extracted from all the classes of macro algae, such as green (Chlorophyceae), red (Rhodophyceae), and brown (Phaeophyceae). The Chlorophyceae such as *Monostroma* and *Ulva* sp. result in the production of sulfated polysaccharides *viz.* ulvan and rhamnan (Liu et al. 2018; Tziveleka et al. 2018; Wang et al. 2018). The Rhodophyceae species of *Gracilaria*, *Hypnea*, *Laurencia*, and *Gigartina* contribute to the production of galactans and carrageenan (De Almeida et al. 2011). Fucoidans and fucans are brown seaweed's sulfated polysaccharides, which are abundant in several species such as *Sargassum, Padina, Ascophyllum, Nizamuddina, Dictyopteris, Dictyota*, and *Kjellmaniella* (Li et al. 2008). Polysaccharides are biopolymers comprising of same or different monomers to form the essential components for the living system. They have a wide diversity among their composition of monomers, such as direct, interrupted repeat, alternating repeat, block copolymer, branched, and complex repeat. The analysis made on these structures may lead to a better understanding of their function. The polysaccharide is composed of individual sugar units linked with glycosidic linkages, and functional groups. Based on monomeric constituents, it can be divided into a homopolysaccharide, acquired from the same type of sugar residues, and heteropolysaccharide, made from the linkage of different monomers as tabulated in Table 3. The higher and diverse structure of various kinds of polysaccharides tends to possess certain unique biological activities. Specificity augments among these polysaccharides due to their interchain interactions (Eric 2005; Harding et al. 2017). Moreover, the sulfated polysaccharides possess high therapeutic potential as compared to other forms of polysaccharides.

23.4.2 Bioactivity of Proteins from Seaweeds

The protein content is the second largest type of bioactive compound, and it varies from species to species of macro algae. Rhodophyceaen algae species possess high protein content to about 47% of dry weight and as low as among phaeophycean algae species (3% ± 15% of dry weight), whereas in Chlorophycean algae, the protein content was around 9% ± 26% of dry weight. But certain species of brown algae like *Undaria pinnatifida* (wakame) have abundant content of protein to about 11%–24% of dry biomass, when compared to other edible brown seaweeds which have less than 15% of dry weight. Similarly, the green seaweed *Ulva spp.* possess high protein content of about 9%–33% of dry biomass, particularly, the *Ulva pertusa* which possess 26% of dry weight. The protein content of the *Ulva lactuca* was found to be containing around 32% of dry biomass which is a seasonal high value. Among Rhodophycean macro algae, the *Porphyra yezoensi* has protein content upto 47% of dry biomass (Fujiwara-Arasaki et al. 1984).

The essential amino acids, such as glycine, arginine, alanine, and glutamic acid are found in compliance with the Food and Food Agriculture Organization (FAO)/World Health Organization (WHO) regulations for requirements of nutritional proteins. Unfortunately, they have less content of lysine, threonine, tryptophan, cysteine, and methionine than other protein-rich foods. In terms of amino acid profile, the red seaweeds have higher amino acid content than other types, such as green and brown seaweeds (FAO 1981).

The proteins of seaweeds exhibit diversified health benefits, which includes anti-hypertensive, anti-oxidative, antithrombotic, hypocholesterolemic, mineral binding, appetite suppression, anti-microbial, immunomodulatory, and cytomodulatory properties (Pangestuti and Kim 2015). Heo et al. (2008) reported the anti-oxidative effects of edible brown seaweeds collected from South Korea. Carnosine and glutathione are well known anti-oxidant peptides that are usually found in muscles, but they are also seen in seaweeds. Phycobiliproteins of seaweeds are used widely as food colorants and in cosmetics, and they are also used as anti-oxidant agents in fermented milk products and ice (Pangestuti and Kim 2015; Sekar and Chandramohan 2007).

23.4.3 Bioactivity of Lipids from Seaweeds

The third major class of compounds present in seaweeds are lipids, which are approximately estimated to about 3% (Aikaterini et al. 2018), and include non-polar (e.g., monounsaturated fatty acids and Poly unsaturated Fatty acids (PUFAs) and polar lipids (e.g., glycolipids and phospholipids). These lipids exhibit diversified bio activities, such as anti-obesity, anti-inflammatory, and anti-tumor, which are underexploited in food applications (Aikaterini et al. 2018).

23.4.3.1 Bioactivity of Omega-3 Fatty Acids

Omega-3 fatty acids are long chain PUFAs and are considered to be essential nutrients and acquired mostly through diet, as mammals cannot synthesize these on their own.

Eicosapentaenoic and docosahexaenoic acids are the predominant omega-3 fatty acids, found abundantly in fish oils and micro algae, and these are also seen in seaweeds (Aikaterini et al. 2018; Polat and Ozogul 2013; Skulas-Ray et al. 2008; Dawczynski et al. 2007). Pereira et al. (2012) investigated the prevalence of PUFAs among all the classes of seaweeds, and among them, brown and red seaweeds are more important sources of non-polar fatty acids than the chlorophyceae except for *U. lactuca* (sea lettuce). Eicosapentaenoic acid was the common constituent among total lipid fractions in *Cystoseira humilis*, which shows low atherogenicity and thrombogenic indices (Belattmania et al. 2016; Aikaterini et al. 2018).

The lipid extracts of seaweeds also exhibit anti-inflammatory potential to combat cardiovascular diseases and diabetes mellitus type 2 (Aikaterini et al. 2018; Robertson et al. 2015; Kathiravan et al. 2015; Jaswir and Monsur 2011).

23.4.3.2 Bioactivity of Glycolipids

Glycolipids of seaweeds exhibit anti-inflammatory and antiproliferative activity, as well as biofunctional effects on lipid metabolism (Aikaterini et al. 2018; Costa et al. 2017; Kellogg et al. 2015; Kendel et al. 2015; Banskota et al. 2014; Lopes et al. 2014; Ma et al. 2014).

Several researchers report the anti-inflammatory potential of a monoacylglycerol and other two mixture of monogalactosyl diacylglycerols from *Fucus spiralis* exhibiting the inhibition of NO production in RAW 264.7 macrophage cells (Aikaterini et al. 2018; Lopes et al. 2014). Similarly, Kellogg et al. (2015) studied the anti-inflammatory potential of six coastal Alaskan seaweeds in RAW 264.7 cells by suppressing the mRNA expression for the production of COX-2, TNF-α, interleukin-10, nitric oxide synthase, and monocyte chemoattractant protein (Aikaterini et al. 2018).

23.4.3.3 Bioactivity of Phospholipids

Phospholipids are the elementary units of cellular membranes consisting of a polar phosphate group with non-polar fatty acids through glycerol linkage. Generally, these lipids possess anti-oxidant, anti-inflammatory, and anti-proliferative activities (Costa et al. 2017; Banskota et al. 2014; Yuan 2007; Bhaskar et al. 2004).

Phospholipids also show their anti-oxidant capacity by its scavenging potential due to the presence of phosphoric functional groups and enhance their anti-oxidant potential (Yuan 2007).

Banskota et al. (2014) extensively studied about the anti-inflammatory effect of phospholipids from *P. palmata* in RAW 264.7 macrophage cells by proving the downregulation of nitric oxide synthase and NO, a signaling factor that plays an important role in inflammation. In certain studies, the Fucoxantin-rich lipids extracted from *Undaria pinnatifida* exhibit enhanced anti-obesity properties (Maeda et al. 2009).

23.5 BIOACTIVITY OF PIGMENTS FROM SEAWEEDS

Pigments facilitate the phototrophs to trap light as energy, one such abundant pigment is chlorophyll, which has a very close relation with the common green pigment found commonly in all the autotrophs. The next major seaweed pigment is an orange colored pigment viz. carotenoids that play a harmonizing role in photosynthesis by transferring electrons onto chlorophyll. The chlorophyll and carotenoid content in macro algae differs according to the varying levels of UV intensity (Yuan 2007). These pigments exhibit anti-oxidant properties (Aikaterini et al. 2018; Lordan et al. 2011; Yuan 2007).

Carotenoids are isoprenoid compounds and are categorized as carotenes (unsaturated hydrocarbons) and xanthophylls (Pangestuti and Kim 2015; Christaki et al. 2013). β-carotene is the predominant pigment found in Phaeophyceae and Chlorophycaean seaweeds, whereas in red seaweed, they are found in lesser levels. The most abundant xanthophylls of brown seaweeds are fucoxanthin, along with violaxanthin and zeaxanthin, whereas, in the red macroalgal species, the major pigment is the zeaxanthin along with lutein, and in green seaweeds, violaxanthin and neoxanthin are the predominant xanthophylls (Aikaterini et al. 2018; Takaichi 2011).

In most of the brown algae fucoxanthin, orange colored anti-oxidant pigments, along with chlorophyll a, c, and β-carotene altogether form a peculiar structure with allenic bonds and functional groups as carbonyl, hydroxyl, and carboxyl moieties with diversified properties (Nomura et al. 1997). These compounds are strong in oxygen scavenging properties (Fung et al. 2013; D'Orazio et al. 2012; Yan et al. 1999). Particularly, they show high DPPH radical scavenging activity, under anoxic conditions (Yan et al. 1999; Nomura et al. 1997). Zhang et al. (2014) investigated the formation of double bonds of fucoxanthin for its radical scavenging potential. In addition, Heo et al. (2008) investigated the H_2O_2-induced oxidative stress are being protected by fucoxanthin in Vero cells (monkey kidney fibroblasts). As a result, fucoxanthin augmented the cell viability by reducing the production of the reactive oxygen species. Vasantharaja et al. (2019) clearly explain the detoxification of the reactive oxygen species in Figure 23.2.

Furthermore, Zheng et al. (2014) showed that fucoxanthin enhances the cellular anti-oxidant property by elevating the levels of reduced Glutathione (GSH) in human keratinocytes through the stimulation by expression of enzymes necessary for GSH synthesis.

In a recent study, the anti-inflammatory properties of the fucoxanthin were evaluated by showing their inhibitory activities against pro-inflammatory agents, such as NO, prostaglandin E2, TNF-α, interleukin-1β, and interleukin-6 (Aikaterini et al. 2018).

Maeda et al. (2009) studied the fucoxanthin and fucoxanthinol downregulate the proliferator-activated receptor gamma and, in turn, inhibit adipocyte differentiation and lipid accumulation in 3T3-L1 cells. Jeon et al. (2010) experimented to find that the ethanolic extract of the fucoxanthin extract from *U. pinnatifida* improved the lipid profile in mice by lowering the triglyceride and cholesterol levels.

Chlorophylls are porphyrin-related chlorin-like compounds with a magnesium ion core performing photosynthesis and protecting against excessive UV radiation and, in addition, they safeguard the algal cells from oxidative stress (Yuan 2007). Osuna-Ruiz et al. (2016) evaluated the anti-oxidant activity along with the chemoprotective property of several macro algae from Mexico. They found the strong affinity of chlorophylls towards polycyclic aromatic hydrocarbons, heterocyclic amines, and aflatoxins and neutralizing these carcinogenic hydrophobic compounds (Lordan et al. 2011; Chernomorsky et al. 1999).

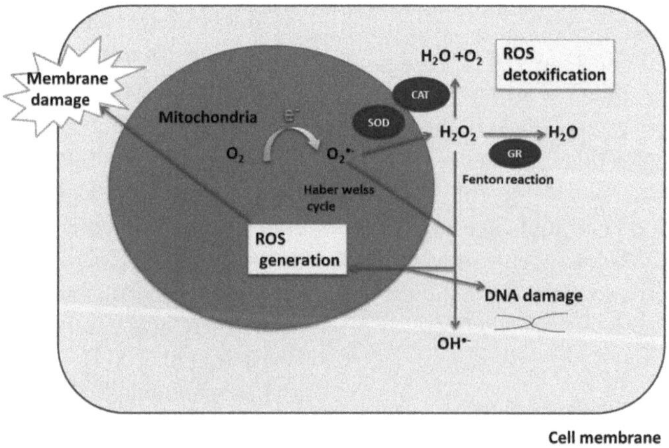

FIGURE 23.2 Detoxification of reactive oxygen species in a cell.

23.6 BIOACTIVITY OF SECONDARY METABOLITES FROM SEAWEEDS

Secondary metabolites are novel compounds that are produced as a part of normal metabolism and during certain adverse conditions. Aikaterini et al. (2018) reported in detail about the diversified bioactivity of secondary metabolites from seaweeds. These compounds have been extracted by different methods using solvent extraction, ultra sonication, rotary evaporation, and ultra filtration since the 1960s (MarinLit 2017). These natural products possess varied vulnerable potent bioactivities, such as anti-oxidant, anti-coagulant, anti-inflammatory, anti-cancer, anti-diabetic, and anti-obesity, providing complete health benefits for human well-being.

The major bioactive metabolites of seaweeds are alkaloids, terpenoids, quinone, sterols, and polyphenols, which are found in all the classes of seaweeds with the majority being found in brown seaweeds (Padua et al. 2015).

Caulerpin, an important alkaloid of the green seaweed Caulerpa, which exhibits anti-inflammatory and anti-nociceptive properties, and it was attributed to the structural resemblance of the indomethacin inspite of having less activity than the indomethacin (De Souza et al. 2009). The most abundant diterpenoid, phytol, compounds are found in seaweeds, exhibits anti-oxidant potential and also shows effective healing benefits against neurodegeneration diseases (Syad et al. 2016). Sargaquinoic acid and sargahydroquinoic acid are common meroterpenoids, found in brown algae, *Sargassum*. In addition, sargachromenol has exhibited in vitro anti-oxidant activity and several other health benefits against metabolic disorders (Liu et al. 2012). In addition, these compounds also showed anti-inflammatory properties proving to possess beneficial effects against anti-inflammatory ailments (Joung et al. 2017).

Plastoquinones are quinone class of compounds mostly isolated from brown seaweeds, involved in light-dependent reactions of photosynthesis (Aikaterini et al. 2018). Mori et al. (2006) extracted three different types of plastoquinones from *Sargassum micracanthum* and evaluated their anti-ulcer effect. Gil et al. (1995) observed the anti-inflammatory activity of epitaondiol, and Areche et al. (2015) evaluated the gastroprotective property of the meroterpenoid seco-taondiol from the brown seaweed *Stypopodium flabelliforme* (Aikaterini et al. 2018).

Fucosterol is a predominant sterol found in marine macro algae and elucidated for several biological activities, such as anti-oxidant, anti-diabetic, anti-inflammatory, and anti-obesity. In addition, the above compound was also found to regulate the levels of cholesterol (Han et al. 2015; Jung et al. 2013, 2014; Hoang et al. 2012; Liu et al. 2012; Lee et al. 2003).

Phenolic acids are well known anti-oxidant agents found in seaweeds exerting high DPPH scavenging activity with high Fe^{2+} chelating potential (Shanura Fernando et al. 2016). Gallic, protocatechuic, vanillic, caffeic, and p-hydroxybenzoic acids are some of the bioactive compounds identified in several algal species (Shanura Fernando et al. 2016). These phenolic acids exhibit hepatoprotective activity by inducing the anti-oxidant enzymes and increasing the oxidative status of the liver by inhibiting the lipid peroxidation in iron-induced liposomes (Aikaterini et al. 2018; Mancini-Filho et al. 2009).

Phlorotannins, polymer of phloroglucinol, are the most important polyphenols found in brown seaweeds exhibit high anti-oxidant potential with DPPH, superoxide scavenging properties, and metal bioadsorption activities (Koutsaviti et al. 2018; Karthik et al. 2016). Osuna-Ruiz et al. (2016) extensively studied on secondary metabolites from marine macroalgal extracts from *Spyridia filamentosa*, *Rhizoclonium riparium*, and *Caulerpa sertularioides* that exerted chemo-preventive properties mainly due to the phenolic compounds. In addition, phloroglucinol has also shown hepatoprotective properties (Kim et al. 2005).

23.7 CONCLUSION

Seaweeds are a potent source of versatile bioactive compounds ranging from macromolecules to small molecules. Several works have been carried out to prove their efficacy by in vitro and in vivo studies. These results from bioactive compounds may provide a sustainable platform to develop several promising drugs for treating life threatening diseases like cancer, diabetes, multidrug resistant microbial diseases, and anti-aging. Further dose-dependent in vivo and clinical studies have to be carried out to substantiate the development of promising beneficial drugs for the contribution to increased life expectancy and life quality of human kind without any undesirable side effects. These marine life forms envisage for the development of a new arena for hopes from sea, the cradle of life.

ACKNOWLEDGMENTS

The authors sincerely thank the management for providing all the facilities and support for carrying out the successful research. The authors from the Sathyabama Institute of Science and Technology, Chennai, sincerely thank Department of Science and Technology-Fund for Improvement of Science and Technology infrastructure (SR/FST/ESI-145/2016) for the financial support for establishing the advanced research facility at Centre for Ocean Research.

REFERENCES

Areche, C., Benites, J., Cornejo, A., et al. 2015. Seco-taondiol, an unusual meroterpenoid from the chilean seaweed *Stypopodium flabelliforme* and its gastroprotective effect in mouse model. *Mar Drugs.* 13:1726–1738. doi:10.3390/md13041726.

Banskota, A. H., Stefanova, R., Sperker, S., et al. 2014. Polar lipids from the marine macroalga *Palmaria palmata* inhibit lipopolysaccharide-induced nitric oxide production in RAW264.7 macrophage cells. *Phytochemistry.* 101:101–108. doi:10.1016/j.phytochem.2014.02.004.

Barsanti, L., and Gualtieri, P. 2014. *Algae: Anatomy, Biochemistry, and Biotechnology*. 2nd ed., Boca Raton, FL: Taylor & Francis Group.

Belattmania, Z., Engelen, A. H., Pereira, H., et al. 2016. Potential uses of the brown seaweed *Cystoseira humilis* biomass: 2-fatty acid composition, antioxidant and antibacterial activities. *J Mater Environ Sci.* 7(6):2074–2081.

Bhaskar, N., Hosakawa, M., and Miyashita, K. 2004. Growth inhibition of human pro-myelocytic leukemia (HL-60) cells by lipid extracts of marine alga *Sargassum marginatum* (Fucales, Phaeophyta) harvested off Goa (west coast of India) with special reference to fatty acid composition. *Indian J Geo-Mar Sci.* 33(4): 355–360.

Buggeln, R. G. 1983. Algal biology: A physiological approach. *Phycologia.* 22(4): 457–458. doi:10.2216/i0031-8884-22-4-457.1.

Chen, X., Nie, W., Yu, G., et al. 2012. Antitumor and immunomodulatory activity of polysaccharides from *Sargassum fusiforme. Food Chem Toxicol.* 50(3–4):695–700. doi:10.1016/j.fct.2011.11.015.

Cheng-Yuan Wang, Tien-Chau Wu, Shu-ling Hsieh, Yung-Hsiang Tsai, Chia-Wen Yeh, Chun-Yung Huang. 2015. Antioxidant activity and growth inhibition of human colon cancer cells by crude and purified fucoidan preparations extracted from *Sragassum critaefollium. Journal of Food and Drug Analysis* 1–12.

Chernomorsky, S., Segelman, A., and Poretz, R. D. 1999. Effect of dietary chlorophyll derivatives on mutagenesis and tumor cell growth. *Teratog Carcinog Mutagen.* 19(5): 313–322.

Christaki, E., Bonos, E., Giannenasa, I., and Florou-Paneria, P. 2013. Functional properties of carotenoids originating from algae. *J Sci Food Agric.* 93(1):5–11. doi:10.1002/jsfa.5902.

Costa, D. E., Melo, T., Moreira, A. S. P., et al. 2017. Valorization of lipids from *Gracilaria* sp. through lipidomics and decoding of antiproliferative and anti-inflammatory activity. *Mar Drugs.* 15(3):E62. 1–17. doi:10.3390/md15030062.

D'Orazio, N., Gemello, E., Gammone, M. A., De Girolamo, M., Ficoneri, C., and Riccioni, G. 2012. Fucoxantin: A treasure from the sea. *Mar Drugs.* 10(3):604–616. doi:10.3390/md10030604.

Dawczynski, C., Schubert, R., and Jahreis, G. 2007. Amino acids, fatty acids, and dietary fibre in edible seaweed products. *Food Chem.* 103(3):891–899. doi:10.1016/j.foodchem.2006.09.041.

De Almeida, C. L. F., De S. Falcão H., De M. Lima. G. R., et al. 2011. Bioactivities from Marine Algae of the Genus Gracilaria. *International Journal of Molecular Sciences* 12(7):4550–4573.

De Souza, E. T., De Lira, D. P., De Queiroz, A. C., et al. 2009. The antinociceptive and anti-inflammatory activities of caulerpin, a bisindole alkaloid isolated from seaweeds of the genus Caulerpa. *Mar Drugs.* 7(4):689–704. doi:10.3390/md7040689.

Dhargalkar, V. K., and Devanand, K. 2004. In: *Seaweeds—A Field Manual*, Verlecar, X.N., and Vijaykumar, R. (Eds.), pp. 1–36. National Institute of Oceanography, Dona Paula, Goa.

Doty, M. S. 1979. Status of marine agronomy with special reference to the tropics. *Proceedings of the International Seaweed Symposia.* 9:35–58.

Dun, Y. L., Zhou, X. L., Guan, H. S., et al. 2015. Low molecular weight guluronate prevents TNF-α-induced oxidative damage and mitochondrial dysfunction in C2C12 skeletal muscle cells. *Food Funct.* 9:3056–3064. doi:10.1039/c5fo00533g.

Edens, R. E. 2005. *Polysaccharides: Structural Diversity and Functional Versatility.* ed. Severian Dumitriu, University of Sherbrooke, Quebec, Marcel Dekker: New York.

Ely, R., Supriya, T., and Naik, C. G. 2004. Antimicrobial activity of marine organisms collected off the coast of South East India. *J Exp Mar Biol Ecol.* 309(1):121–127. doi:10.1016/j.jembe.2004.03.010.

Fleita, D., El-Sayed, M., and Rifaat, D. 2015. Evaluation of the antioxidant activity of enzymatically hydrolyzed sulfated polysaccharides extracted from red algae Pterocladiacapillacea. *LWT-Food Sci Technol.* 63(2):1236–1244. doi:10.1016/j.lwt.2015.04.024.

Fujiwara-Arasaki, T., Mino, N., and Kuroda, M. 1984. The protein value in human nutrition of edible marine algae in Japan. In: Bird, C.J., Ragan, M.A. (Eds.), *Eleventh International Seaweed Symposium. Developments in Hydrobiology*, p. 22. Springer, Dordrecht, the Netherlands. doi:10.1007/978-94-009-6560-7_104.

Fung, A., Hamid, N., and Lu, J. 2013. Fucoxanthin content and antioxidant properties of *Undaria pinnatifida. Food Chem.* 136(2):1055–1062. doi:10.1016/j.foodchem.2012.09.024.

Gil, B., Ferr Bndizl, M. L., Sanzl, M. J., et al. 1995. Inhibition of inflammatory responses by epitaondiol and other marine natural products. *Life Sci.* 57(2):25–30. doi:10.1016/0024-3205(95)00260-d.

Han, Y. R., Ali, M. Y., Woo, M.-H., Jung, H. A., and Choi, J. S. 2015. Anti-diabetic and anti-inflammatory potential of the edible brown alga *Hizikia fusiformis. J Food Biochem.* 39(4):417–428. doi:10.1111/jfbc.12138.

Hao, C., Hao, J., Wang, W., et al. 2011. Insulin sensitizing effects of oligomannuronate-chromium (III) complexes in C2C12 skeletal muscle cells. *PLoS One.* 6(9):e24598. doi:10.1371/journal.pone.0024598.

Harding, S. E., Tombs, M. P., Adams, G. G., Paulsen, B. S., Inngjerdingen, K. T., and Barsett, H. 2017. *An Introduction to Polysaccharide Biotechnology.* 2nd ed., CRC Press, New York, 153–190.

Harper, A. 1981. FAO, Amino acid scoring patterns. FAO/WHO/UNU EPR/81/31. http://www.fao.org/3/M3013E/M3013E00.htm.

Heo, S. J., Ko, S. C., Kang, S. M., et al. 2008. Cytoprotective effect of fucoxanthin isolated from brown algae *Sargassum siliquastrum* against H_2O_2-induced cell damage. *Eur Food Res Technol.* 228(1):145–151. doi:10.1007/s00217-008-0918-7.

Heo, S. J., Park, E. J., Lee, K. W., and Jeon, Y. J. 2005. Antioxidant activities of enzymatic extracts from brown seaweeds. *Bioresour Technol.*14:1613–1623. doi:10.1016/j.biortech.2004.07.013.

Hoang, M. H., Jia, Y., Jun, H. J., Lee, J. H., Lee, B. Y., and Lee, S. J. 2012. Fucosterol is a selective liver X receptor modulator that regulates the expression of key genes in cholesterol homeostasis in macrophages, hepatocytes, and intestinal cells. *J Agric Food Chem.* 60(46):11567–11575. doi:10.1021/jf3019084.

Holdt, S. L., and Kraan, S. 2011. Bioactive compounds in seaweed: Functional food applications and legislation. *J Appl Phycol.* 23(3):543–597. doi:10.1007/s10811-010-9632-5.

Hong, S. W., Lee, H. S., Jung, K. H., Lee, H., and Hong, S. S. 2012. Protective effect of fucoidan against acetaminophen induced liver injury. *Arch Pharm Res.* 35(6):1099–1105.

Ina, A., Hayashi, K. I., Nozaki, H., and Kamei, Y. 2007. Pheophytin a low molecular weight compound found in the marine brown alga *Sargassum fulvellum*, promotes the differentiation of PC12 cells. *Int J Dev Neurosci.* 25(1):63–68. doi:10.1016/j.ijdevneu.2006.09.323.

Jaswir, I., and Monsur, H. A. 2011. Anti-inflammatory compounds of macro algae origin: A review. *J Med Plant Res.* 5(33):7146–7154. doi:10.5897/JMPR11.018.

Jeon, S. M., Kim, H. J., Woo, M. N., et al. 2010. Fucoxanthin rich seaweed extract suppresses body weight gain and improves lipid metabolism in high-fat-fed C57BL/6J mice. *Biotechnol J* 5(9):961–969. doi:10.1002/biot.201000215.

Joung, E. J., Gwon, W. G., Shin, T., Jung, B. K., Choi, J. S., and Kim, H. R. 2017. Anti-inflammatory action of the ethanolic extract from *Sargassum serratifolium* on lipopolysaccharide-stimulated mouse peritoneal macrophages and identification of active components. *J Appl Phycol.* 29(1):563–573.

Jung, H. A., Jin, S. E., Ahn, B. R., Lee, C. M., and Choi, J. S. 2013. Anti-inflammatory activity of edible brown alga *Eisenia bicyclis* and its constituents fucosterol and phlorotannins in LPS-stimulated RAW264.7 macrophages. *Food Chem Toxicol.* 59:199–206. doi:10.1016/j.fct.2013.05.061.

Jung, H. A., Jung, H. J., Jeong, H. Y., Kwon, H. J., Kim, M. S., and Choi, J. S. 2014. Anti-adipogenic activity of the edible brown alga *Ecklonia stolonifera* and its constituent fucosterol in 3 T3-L1 adipocytes. *Arch Pharm Res.* 37(6):713–720. doi:10.1007/s12272-013-0237-9.

Kalitnik, A. A., Anastyuk, S. D., Sokolova, E. V., Kravchenko, A. O., Khasina, E. I., and Yermak, I. M. 2016. Oligosaccharides of κ/β-carrageenan from the red alga *Tichocarpus crinitus* and their ability to induce interleukin 10. *J Appl Phycol.* 28(1):545–553. doi:10.1007/s10811-015-0577-6.

Karthik, R., Manigandan, V., Sheeba, R., Saravanan, R., and Rajesh, P. R. 2016. Structural characterization and comparative biomedical properties of phloroglucinol from Indian brown seaweeds. *J Appl Phycol.* 28(6):3561–3573. doi:10.1007/s10811-016-0851-2.

Kathiravan, V., Panneerselvam, N., and Palanikumar, L. 2015. An untapped resource for natural anti-inflammatory compounds from marine macroalgae. *Int J Pharm Bio Sci.* 6(2):579–595.

Kellogg, J., Esposito, D., Grace, M. H., Komarnytsky, S., and Lila, M.A. 2015. Alaskan seaweeds lower inflammation in RAW 264.7 macrophages and decrease lipid accumulation in 3T3-L1 adipocytes. *J Funct Foods.* 15:396–407. doi:10.1016/j.jff.2015.03.049.

Kendel, M., Wielgosz-Collin, G., Bertrand, S., Roussakis, C., Bourgougnon, N., and Bedoux, G. 2015. Lipid composition, fatty acids and sterols in the seaweeds *Ulva armoricana*, and *Solieria chordalis* from Brittany (France): An analysis from nutritional, chemotaxonomic, and antiproliferative activity perspectives. *Mar Drugs.* 13(9):5606–5628. doi:10.3390/md13095606.

Khan, M. I., Shin, J. H., and Kim, J. D. 2018. The promising future of microalgae: Current status, challenges, and optimization of a sustainable and renewable industry for biofuels, feed, and other products. *Microbial Cell Factories* 17(1):36.

Kim, W. J., Koo, Y. K., Jung, M. K., et al. 2010. Anticoagulating activities of low-molecular weight fuco-oligosaccharides prepared by enzymatic Digestion of fucoidan from the sporophyll of Korean *Undaria pinnatifida*. *Arch Pharm Res.* 33(1):125–131. doi:10.1007/s12272-010-2234-6.

Kim, Y. C., Ren, B. A., Na, Y. Y., Taek, J. N., and Jae, S. C. 2005. Hepatoprotective constituents of the edible brown alga *Ecklonia stolonifera* on tacrine-induced cytotoxicity in Hep G2 cells, *Arch Pharm Res.* 28 (12):1376–1380. doi:10.1007/bf02977904.

Koutsaviti, A., Loannou, E., and Roussis, V. 2018. Bioactive seaweed substances, In: *Bioactive Seaweeds for Food Applications*, Yimin Qin (Ed.), pp. 25–51. Elsevier London, UK. doi:10.1016/B978-0-12-813312-5.00002-9.

Lee, S., Lee, Y. S., Jung, S. H., Kang, S. S., and Shin, K. H. 2003. Anti-oxidant activities of fucosterol from the marine algae *Pelvetia siliquosa*. *Arch Pharm Res.* 26(9):719–722. doi:10.1007/bf02976680.

Li, B., Lu, F., Wei, X., and Zhao, R. 2008. Fucoidan: Structure and bioactivity. *Molecules.* 13(8):1671–1695. doi:10.3390/molecules13081671.

Liu, L., Heinrich, M., Myers, S., and Dworjanyn, S. A. 2012. Towards a better understanding of medicinal uses of the brown seaweed *Sargassum* in Traditional Chinese Medicine: A phytochemical and pharmacological review. *J Ethnopharmacol.* 142(3):591–619. doi:10.1016/j.jep.2012.05.046.

Liu, S., Wang, Q., Song, Y., et al. 2018. Studies on the hepatoprotective effect of fucoidans from brown algae *Kjellmaniella crassifolia*. Carbohydr. *Polym.* 193:298–306. doi:10.1016/j.carbpol.2018.03.077.

Lopes, G., Daletos, G., Proksch, P., Andrade, P. B., and Valentão, P. 2014. Anti-inflammatory potential of monogalactosyl diacylglycerols and a monoacylglycerol from the edible brown seaweed *Fucus spiralis* Linnaeus. *Mar Drugs.* 12(3):1406–1418. doi:10.3390/md12031406.

Lordan, S., Ross, R. P., and Stanton, C. 2011. Marine bioactives as functional food ingredients: Potential to reduce the incidence of chronic diseases. *Mar Drugs.* 9(6):1056–1100. doi:10.3390/md9061056.

Ma, A. C., Chen, Z., Wang, T., et al. 2014. Isolation of the molecular species of monogalactosyl diacylglycerols from brown edible seaweed *Sargassum horneri* and their inhibitory effects on triglyceride accumulation in 3T3-L1 adipocytes. *J Agric Food Chem.* 62(46):11157–11162. doi:10.1021/jf503068n.

Maeda, H., Hosokawa, M., Sashima, T., Murakami-Funayama, K., and Miyashita, K. 2009. Anti-obesity and anti-diabetic effects of fucoxanthin on diet-induced obesity conditions in a murine model. *Mol Med Rep.* 2(6):897–902. doi:10.3892/mmr_00000189.

Mancini-Filho, J., Novoa, A. V., González, A. E. B., et al. 2009. Free phenolic acids from the seaweed *Halimeda monile* with antioxidant effect protecting against liver injury. *Z Naturforsch C J Biosci.* 64(9–10):657–663. doi:10.1515/znc-2009-9-1009.

MarinLit, *A Database of the Marine Natural Products Literature*. Royal Society of Chemistry, London. 2017. http://pubs.rsc.org/marinlit/.

Mayer, A. M. S., and Hamann, M. T. 2004. Marine pharmacology in 2000: Marine compounds with antibacterial, anticoagulant, antifungal, anti-inflammatory, antimalarial, antiplatelet, antituberculosis, and antiviral activities; affecting the cardiovascular, immune, and nervous systems and other miscellaneous mechanisms of action. *Mar Biotechnol* (NY). 6(1):37–52. doi:10.1007/s10126-003-0007-7.

Mori, J., Hayashi, T., Iwashima, M., Matsunaga, T., and Saito, H. 2006. Effects of plastoquinones from the brown alga *Sargassum micracanthum* and a new chromene derivative converted from the plastoquinones on acute gastric lesions in rats. *Biol Pharm Bull.* 29(6):1197–1201. doi:10.1248/bpb.29.1197.

Nakazono, S., Cho, K., Isaka, S., et al. 2016. Anti-obesity effects of enzymatically digested alginate oligomer in mice model fed a high-fat-diet. *Bioact Carbohydr Diet Fibre.* 7(2):1–8. doi:10.1016/j.bcdf.2016.02.001.

Newman, D. J., Cragg, G. M., and Snader, K. M. 2003. Natural products as source of new drugs over the period 1981–2002, *J Nat Prod.* 66(7):1022–1037. doi:10.1021/np030096l.

Nomura, T., Kikuchi, M., Kubodera, A., and Kawakami, Y. 1997. Proton-donative antioxidant activity of fucoxanthin with 1,1-diphenyl-2-picrylhydrazyl (DPPH), *Biochemistry and Molecular Biology International.* 42 (2):361–370. doi:10.1080/15216549700202761.

Osuna-Ruiz, I., López-Saiz, C.-M., Burgos-Hernández, A., Velázquez, C., Nieves-Soto, M., and Hurtado-Oliva, M. A. 2016. Antioxidant, antimutagenic and antiproliferative activities in selected seaweed species from Sinaloa, Mexico. *Pharm Biol.* 54(10):2196–2210. doi:10.3109/13880209.2016.1150305.

Pádua, D., Rocha, E., Gargiulo, D., and Ramos, A. A. 2015. Bioactive compounds from brown seaweeds: Phloroglucinol, fucoxanthin and fucoidan as promising therapeutic agents against breast cancer. *Phytochem Lett.* 14:91–98. doi:10.1016/j.phytol.2015.09.007.

Pangestuti, R., and Kim S. K. 2015. Carotenoids, bioactive metabolites derived from seaweeds, In: Se-Kwon Kim ed., pp. 816–821. *Springer Handbook of Marine Biotechnology*, Springer.

Pangestuti, R., and Kim, S. K. 2014. Biological activities of carrageenan. *Adv Food Nutri Res.* 72:113–124. doi:10.1016/B978-0-12-800269-8.00007-5.

Patel, S., and Goyal, A. 2011. Functional oligosaccharides: Production, properties and applications. *World J Microbiol Biotechnol.* 27(5):1119–1128. doi:10.1007/s11274-010-0558-5.

Pereira, H., Barreira, L., Figueiredo, F., et al. 2012. Polyunsaturated fatty acids of marine macroalgae: Potential for nutritional and pharmaceutical applications. *Mar Drugs.* 10(9):1920–1935. doi:10.3390/md10091920.

Pereira, L. 2018. Biological and therapeutic properties of the seaweed polysaccharides. *Inter Bio Rev.* 2(2):1–41. doi:10.18103/ibr.v2i2.1762.

Polat, S., and Ozogul, Y. 2013. Seasonal proximate and fatty acid variations of some seaweeds from the northeastern Mediterranean coast. *Oceanologia.* 55(2):375–391. doi:10.5697/oc.55-2.375.

Robertson, R. C., Guihéneuf, F., Bahar, B., et al. 2015. The anti-inflammatory effect of algae-derived lipid extracts on lipopolysaccharide (LPS)-stimulated human THP-1 macrophages. *Mar Drugs.* 13(8):5402–5424. doi:10.3390/md13085402.

Sekar, S., and Chandramohan, M. 2007. Phycobiliproteins as a commodity: Trends in applied research, patents and commercialization. *J Appl Phycol.* 20(2):113–136. doi:10.1007/s10811-007-9188-1.

Shanura Fernando, I. P., Kim, M., Son, K. T., Jeong, Y., and Jeon, Y. J. 2016. Antioxidant activity of marine algal polyphenolic compounds: A mechanistic approach. *J Med Food.* 19(7): 615–628. doi:10.1089/jmf.2016.3706.

Skulas-Ray, A. C., West, S. G., Davidson, M. H., Kris-Etherton, P. M. 2008. Omega-3 fatty acid concentrates in the treatment of moderate hyper triglyceridemia. *Expert Opin Pharmacother.* 9(7):1237–1248. doi:10.1517/14656566.9.7.1237.

Stalin Dhas, T., Ganesh Kumar, V., Karthick, V., Govindaraju, K., and Shankara Narayana, T. 2014. Biosynthesis of gold nanoparticle using *Sargassum swartzii* and its cytotoxicity effect on Hela cells. *Spectrochim Acta A Mol Biomol Spectrosc.* 133: doi:10.1016/j.saa.2014.05.042.102-106.

Syad, A. N., Rajamohamed, B. S., Shunmugaiah, K. P., and Kasi, P. D. 2016. Neuroprotective effect of the marine macroalga *Gelidiella acerosa*: Identification of active compounds through bioactivity-guided fractionation, *Pharm Biol.* 54(10):2073–2081. doi:10.3109/13880209.2016.1145700.

Takaichi, S. 2011. Carotenoids in algae: Distributions, biosyntheses and functions. *Mar Drugs.* 9(6):1101–1118. doi:10.3390/md9061101.

Tziveleka, L. A, Pippa, N., Georgantea, P., Ioannou, E., Demetzos, C., and Roussis, V. 2018. Marine sulfated polysaccharides as versatile polyelectrolytes for the development of drug delivery nanoplatforms: Complexation of ulvan with lysozyme. *International Journal of Biological Macromolecules*, 118:Part A:69–75. doi:10.1016/j.ijbiomac.2018.06.050.

Vasantharaja, R., Stanley Abraham, L., Jyotsna, J., et al. 2019. Sulfated polysaccharide from *Sargassum tenerrimum* attenuates oxidative stress induced reactive oxygen species production in in vitro and in zebrafish model. *Carbohydr Polym.* 203:441–449. doi:10.1016/j.carbpol.2018.09.056.

Vijayabaskar, P., and Vaseela, N. 2012. In vitro antioxidant properties of sulphated polysaccharides from brown marine algae *Sargassum tenerrimum. Asian Pac J Trop Dis.* 2(2):S890–S896. doi:10.1016/S2222-1808(12)60287-4.

Wang, C. Y., Wu, T. C., Hsieh, H. L., Tsai, Y. H., Yeh, C. W., and Huang, C. Y. 2015. Antioxidant activity and growth inhibition of human colon cancer cells by crude and purified fucoidan preparations extracted from *Sargassum cristaefollium. J Food Drug Anal.* 23:766–777. doi:10.1016/j.jfda.2015.07.002.

Wang, S., Wang, W., Hao, C., et al. 2018. Antiviral activity against enterovirus 71 of sulfated rhamnan isolated from the green alga *Monostroma latissimum. Carbohydr Polym.* 200:43–53. doi:10.1016/j.carbpol.2018.07.067.

Yan, X., Chuda, Y., Suzuki, M., and Nagata, T. 1999. Fucoxanthin as the major antioxidant in *Hijikia fusiformis*, a common edible seaweed. *Biosci Biotechnol Biochem.* 63(3):605–607. doi:10.1271/bbb.63.605.

Yao, Z., Xu, L., and Wu, H. 2014. Immunomodulatory function of κ-carrageenan oligosaccharides acting on LPS-activated microglial cells. *Neurochem Res.* 39(2):333–343.doi:10.1007/s11064-013-1228-4.

Yuan, Y. V. 2007. Antioxidants from edible seaweeds. *ACS Symposium Series.* 956:268–301. doi:10.1021/bk-2007-0956.ch019.

Zhang, Y., Fang, H., Xie, Q., et al. 2014. Comparative evaluation of the radical scavenging activities of fucoxanthin and its stereoisomers. *Molecules.* 19(2): 2100–2113. doi:10.3390/molecules19022100.

Zheng, J., Piao, M. J., Kim, K. C., Yao, C. W., Cha, J. W., and Hyun, J. W. 2014. Fucoxanthin enhances the level of reduced glutathione via the Nrf2-mediated pathway in human keratinocytes. *Mar Drugs.* 12(7):4214–4230. doi:10.3390/md12074214.

24 Bioactive Compounds in Marine Macro Algae and Their Role in Pharmacological Applications

Subramaniam Kalidass, Lakshmanan Ranjith, Palavesam Arunachalam, Amarnath Mathan Babu, and Karuppasamy Kaviarasan

CONTENTS

24.1 INTRODUCTION

Marine organisms are a great source for the discovery of novel bioactive molecules with commercially valuable pharmacological applications (Heo and Jeon 2009; Díaz-Rubio et al., 2009). Macro algae or seaweeds are the most important sustainable resources in the marine environment that can be used for industrial and therapeutic applications (Manilal et al. 2009), these produce numerous bioactive molecules with various structures and functional properties (Choi et al. 2002; Amarowicz et al. 2004; Shibata et al. 2008; Kong et al. 2009; Kim and Bae 2010). The bioactive molecules include sulfated polysaccharides, proteins, polyunsaturated fatty acids, minerals, and plant growth hormones (Chojnacka et al. 2012). Some of them are thought to be valuable with functional health benefits, and the extracted components of seaweeds will increase the value of the potential pharmacological and industrial aspects in a broad extent.

Brown algae, the Phaeophyceae, are the second largest and complex group of marine macro algae, representing 18 orders and more than 2000 species (Guiry and Guiry 2011), that are available in the intertidal zone. Most of the brown algae have the pigment fucoxanthins that are responsible for the greenish-brown color, and hence the name. Brown algae produce a rich source of bioactive molecules, such as polysaccharides, proteins, pigments, and phlorotannins, and most of them are responsible for the specific bioactivities which give potential for their economic utilization. They produce a large amount of sulfated polysaccharides (e.g., fucans, laminarans, galactans, and sargassans) and polyphenols (phlorotannin, bromophenol, and meroditerpenoids) that are used for commercially important and global industrial applications. Seaweed polysaccharides are abundantly used for food, pharmaceutical, and industrial applications. Therefore, the seaweed-derived polysaccharide industries operate in extremely regulated surroundings.

For more than a decade, the isolated bioactive molecules from brown algae have been the subject of more attention in the fields of pharmacology and biochemistry. Bioactive compounds from brown algal extracts are known to reveal many biological properties, such as anti-viral, anti-coagulant, anti-inflammatory, and anti-tumoral actions (Costa et al. 2010; Shibata et al. 2008; Lee et al. 2008). In the last few years, sulfated polysaccharides extracted from brown algae can be used in pharmacological applications (Yang and Zhang 2009). Hence, the main objective of this review paper is to find the present results on the biological properties of brown algae with their pharmacological applications. Moreover, additional efforts have also been made to update the information covered in new research articles on this subject. This communication aims toward the appraisal of bioactive molecules from brown algae with regards to pharmacological applications.

24.2 BIOACTIVE COMPOUND IN MACRO ALGAE

24.2.1 Polysaccharides

Seaweeds consist of different types of polysaccharides that have chemical structures that are related to their cell structure and algal taxonomic classification (Ferreira et al. 2012; Wijesinghe and Jeon 2012). The polysaccharide contents exhibit seasonal changes from one place to another, and 76% of dry weight is present in the total percentage of seaweeds. In prebiotics, seaweed polysaccharides enhance the growth of beneficial bacteria in the digestive system and also produce health-improving effects (Vidanarachchi et al. 2009). Bioactive-sulfated polysaccharides slow down the bacterial activity of many species other than viruses (Leonard et al. 2010). Various brown algal polysaccharides are found, but the most important are fucoidan, alginates, laminarin, and galactans (Ferreira et al. 2012), and most of them contain soluble dietary fibers that have a good effect on the digestive systems of animals.

24.2.2 Proteins

Proteins isolated from seaweeds have biologically active components, but there is not a lot of documentation on polysaccharides because of their different biological properties and structures. Generally, the protein content in seaweeds is lower than five percentage (Leonard et al. 2010), and the red and green algae have a higher protein content when compared to the brown algae. Proteins can be isolated from macro algae, and the most essential biologically active components are lectins, which attach to carbohydrates and contribute the most to the biological system, for instance, intercellular communication.

24.2.3 Poly Unsaturated Fatty Acids

The two major classes of lipids that are present in seaweeds are phospholipids and glycolipids. If the environmental temperatures are decreased, polyunsaturated fatty acids will occur in seaweeds. In the marine environment, polyunsaturated fatty acids containing the species in cold waters live better than the species living in hot temperatures (Holdt and Kraan 2011). Long chain polyunsaturated fatty acids play a vital role in regulating human health, and they are synthesized only in plants (Pulz and Gross 2004). Twenty carbon atoms as a minimum with two double bonds are present in these lipids. If the first bond is situated in the third carbon atom, then the lipid molecule is considered as omega-3.

24.2.4 Pigments

There are three major forms of pigments, i.e., carotenoids, chlorophylls, and phycobiliproteins that are found in seaweed pigments. Carotenoids, in the form of organic pigments, are present in chromoplasts and chloroplasts (Wijesinghe and Jeon 2012). These pigments are common in nature and available in the marine algae, plants, fungi, and some of the bacteria (Liaua et al. 2010; Li and Kim 2011). The three main groups, such as β-carotene, fucoxanthin, and tocopherol play a significant part in carotenoids. The fucoxanthin has the total carotenoid content of about 70% and β-carotene substance, which are present in the values of 36 to 4500 mg/kg from the algal dry mass (Holdt and Kraan 2011). The carotenoid pigment is mainly composed of polyenes that are soluble in lipids, and many types of carotenoids that are available in different algal species are powerful anti-oxidants. The properties of these pigments have the capacity to reduce singlet oxygen and scavenge free radicals (Li and Kim 2011). The water soluble pigments are found in the cyanobacteria that produce the phycobiliproteins and crypto-monads (Mihova et al. 1996), and they hold a total algal dry weight of a few percent.

24.2.5 Polyphenols

Polyphenols are one of the bioactive components that are found mainly in plants, including seaweeds (Antonisamy and Raj 2011). Macro algal extracts have a considerable amount of polyphenols, and their substance is powerful depending on the extraction method. The polyphenols are produced by means of the polymerization of phloroglucinol on the acetate-malonate pathway (Li et al. 2011; Wijesinghe and Jeon 2012). These polymers consist of various biological activities in the organisms that occur in the host defense mechanisms (Swanson and Druehl, 2002). *Ascophyllum* spp. has the highest amount of polyphenols compared to other seaweeds, whereas the lowest substances of these compounds are present in the *Ulva* spp. (Keyrouz et al. 2011; Craigie 2011). Tannins are a compound of polyphenols that are present in both marine and terrestrial plants. The polyphenolic substances like phlorotannin (eckol or dieckol) are a group of tannin compounds that are abundant in brown algae (Antonisamy and Raj 2011; Gupta and Abu-Ghannam 2011). The content of phlorotannin varies in the phlorotannin skeleton, which is composed of eight phenol rings (O'Sullivan et al. 2011), whereas in terrestrial plants, tannins generate about 3 to 4 rings (Antonisamy and Raj 2011). Phenol rings can be used as electron traps for the free radicals (Gupta and Abu-Ghannam 2011).

24.2.6 Minerals

Minerals are rich in seaweeds (Nwosu et al. 2011), and their biomass content is sometimes as high as 40% (Kumar et al. 2011) because seaweeds can occur in metal ions from salt water and give attention to those substances as carbonate salts (Aslam et al. 2010). The researchers reported the mineral substances were found in concentrates on harvested seaweeds on Japanese beaches. The highest concentrations were present in potassium (2.71 g/L), magnesium (0.19 g/L), and calcium (0.16 g/L) ions in the *Sargassum ringgoldianum* subsp. *Coreanum* extract. The *Codium fragile* have a good source of sodium ions that are found in 1.21 g/L (Kuda and Ikemori 2009). The high levels of magnesium and calcium ions are present in the *Kapaphycus alvarezii* (Rathore et al. 2009), and their concentrations in extracts were 581.20 mg/L and 460.11 mg/L, respectively.

24.2.7 Plant Growth Hormones

Plant growth hormones are available in seaweed extracts, which are used to induce plant growth and to improve the photosynthesis. Cytokinins are plant growth regulators that protect plants from temperature variations (Tarakhovskaya et al. 2007; Zhang et al. 2010), and these are synthesized by means of the bio-chemical modification of adenine. Other plant hormones are auxin, abscisic acid, and betaines that are found in macro algal extracts. Auxin functions to start the root formation and reduce its elongation, their concentration may vary, and it depends up on the species. Gibberellins play a major role to start seed germination and are formed in developing seeds from glyceraldehydes-3-phosphate. They were first identified in two brown algal extracts, such as a *Fucus vesiculosus* and *Fucus spiralis* (Tarakhovskaya et al. 2007). Abscisic acid is formed from carotenoids by more than 60 species of algae, and betaines are not usual plant hormones, which are also found in seaweed extracts (MacKinnon et al. 2010), and their role is to guard the plants from drought and frost (Craigie 2011). The brown algal extract, *Ascophyllum nodosum*, has a rich source of betaines (Khan et al. 2009; Craigie 2011).

24.3 TYPES OF SEAWEED-DERIVED POLYSACCHARIDES

24.3.1 Alginates

Alginate is an anionic polysaccharide that is composed of two hexuronic acids, namely, mannuronic and guluronic acids (Andriamanantoanina and Rinaudo 2010). They are present in the cell wall of about an 18%–40% dry weight basis of brown algae, such as *Ascophyllum nodosum* and *Laminaria* sp. (Rioux and Turgeon 2015). They are used as by-products in the production of food, such as a gelling agent, thickening agent, stabilizer, and encapsulation that are also useful for human health (Brownlee et al. 2009). In addition, alginate-based composite biomaterials are used for bone-tissue regeneration (Venkatesan et al. 2015a, 2015b).

24.3.2 Fucoidan

Fucoidan is the most important and dominant sulfated polysaccharide, which mainly is composed of fucose linked in (1, 3) and (1, 4) glycosidic linkage (Daniel et al. 2001). In the two brown algal families, such as Fucaceae and Laminariaceae, the cell walls are composed of 2%–10% dry weight of the fucoidans. Fucoidans are derived from numerous species of BA, such as *Ascophyllum nodosum, Cladosiphon okamuranus, Fucus vesiculosis*, and *Undaria pinnatifida*. Fucoidan has a higher bioactive potential, and pharmacological activities, such as anti-oxidant, anti-bacterial, anti-virus, anti-tumor, anti-coagulation, immunomodulatory, and anti-inflammatory have been performed (Smit 2004; Cumashi et al. 2007; Wijesinghe and Jeon 2012). Fucoidan will show potential in bone-tissue engineering during analyses for the biomedical field (Jeong et al. 2013; Lowe et al. 2016).

24.3.3 Laminarin

Laminarin is primarily formed of (1,3)-β-d-glucan with β (1, 6) branching (Rioux et al. 2007). Moreover, it is classified into two types of laminarin chains with mannitol and glucose, and their molecular weight is about 5000 Da. In brown algae, the laminarin content is up to 35% on a dry basis that varies depending on the species. Potential pharmacological activities are found in laminarin, for instance, anti-tumor, anti-oxidant, and anti-inflammatory activity (Ren et al. 1994; Hoffman et al. 1995; Miao et al. 1999).

24.3.4 Ulvan

Ulvan is composed of water-soluble-sulfated polysaccharides, which are abundant in green algae and primarily consists of l-rhamnose, d-xylose, d-glucose, and d-glucuronic acid (Rioux and Turgeon 2015). The cell wall of ulvan is formed of 8%–29% of dry weight (Lahaye and Robic 2007). They can be isolated with ammonium oxalate and precipitates of ethyl alcohol with water at 80°C–90°C (Lahaye and Robic 2007).

24.3.5 Agar

Agar is present in red algae formed of Ca-, Mg-, K-, and Na-sulfated esters of d- and l-galactose units (Wijesekara and Karunarathna 2017). The cell wall consists of 20% of dry weight and is isolated from the *Gelidium* and *Gracilaria*s genera. Moreover, *Pterocladia* and *Gelidiella* species are used to isolate agar and play a vital role in food and industrial applications.

24.3.6 Carrageenans

Carrageenans are sulfated linear polysaccharides formed of ammonium, Ca-, Mg-, K-, and Na-sulfated esters of d-galactose and (3,6)-anhydro-d-galactose (Wijesekara and Karunarathna 2017) that are naturally present in red algae. Three types of carrageenan are formed, i.e., kappa carrageenan (k), iota carrageenan (ι), and lambda carrageenan (λ). The cell walls of carrageenan consist of 30%–80% dry weight and are extracted from the *Kappaphycus alvarezii* and *Chondrus crispus* species. In carrageenan, the biological properties, chemical alteration, and its structural analysis have been reviewed (Campo et al. 2009). Tofu is a fermented food that mainly contains k/ι-hybrid carrageenans, and it shows the maximum rheological properties, which could be a food additive to change its food textures (Shen and Kuo 2017).

24.3.7 Floridean Starch

Red algae contain the starch granules called floridean starch that are found on the outer surface of the plastids and lack amylase (Wijesekara and Karunarathna 2017).

24.4 BIOLOGICAL ACTIVITIES OF SEAWEEDS WITH SPECIAL FOCUSES ON PHARMACOLOGICAL APPLICATIONS

Seaweeds offer a good source of bioactive molecules with a broad range of pharmacological, therapeutic, and food applications (Table 24.1). The bioactive compounds, such as phenolics, fatty acids, steroids, and several acids are found in many structures of seaweeds that have the potential source of antibiotic or anti-fungal activities (Bhagavathy et al. 2011). Most of the seaweed extracts have been found to have biological activities, such as antibiotics, anti-viral, anti-tumor, anti-oxidant, and anti-inflammatory (Cumashi et al. 2007; Costa et al. 2010; Bhagavathy et al. 2011; Wijesinghe and Jeon 2012). Several halogenated compounds contain bromine, chlorine, and iodine metabolites, and some of them like diterpenes and triterpenes are the potential source of various biological activities such as anti-bacterial, ichtyotoxic, anti-oxidant, anti-malarian, insecticidal, and cytotoxic (Shui-Chun and Yue-Wei 2010).

Polysaccharides have the immunological properties that range from non-specific stimulation of the host immune system as a result of containing biological activities, such as anti-viral, anti-infection, and anti-tumor effects to anti-mutagenic, hematopoietic, or anti-oxidant activity (Nwosu et al. 2011; Athukorala et al. 2006b; Wijesekara et al. 2011). Seaweed polysaccharides exhibit good immunomodulatory properties that are linked with anti-tumor effects. The sulfated polysaccharides are a source of pharmacological activities, such as anti-proliferative activity in cancer cell lines and inhibitory activity against tumors (Suganthy et al. 2010). The brown seaweed *Sargassum stenophyllum* contains the sulfated polysaccharides that are extracted to reduce vasculogenesis, inhibit the developmental angiogenesis in chick embryos (Dias et al. 2008), and in the brown seaweed *Ecklonia cava*, the anti-coagulant activity was investigated (Wijesinghe et al. 2011). Many of the studies were carried out in fucoidans because those have potential biological activities, such as anti-coagulant, anti-thrombotic, immunomodulatory, anti-proliferative, and anti-cancer (Wijesinghe and Jeon 2012; Takahashi 1983; Rioux et al. 2010). This molecule contains a high number of sulfate polysaccharides and less than 20% sulfate content that exhibits the potential of anti-tumor and anti-angiogenic activity (Koyanagi et al. 2003). The anti-viral properties are also found in fucoidan toward viruses, such as HIV and human cytomegalovirus (Gupta and Abu-Ghannam 2011). Fucoidan polymers are recommended due to the stimulation of the host immune system against tumors (You et al. 2010).

Protein extracts have the potential source of anti-oxidant properties that are found in some of the phycobiliproteins, such as C-phycocyanin and allophycocyanin. The role of protein in pharmacological applications is less compared to the polysaccharides and polyphenols. Natural pigments are one of the bioactive compounds in macro algal extracts that have been found in advantageous activities, such as anti-cancer (Moreau et al. 2006), anti-obesity, anti-inflammatory, neuroprotective, and anti-angiogenic (Pangestuti and Kim 2011). In recent years, the anti-oxidant molecules were identified as some of the pigments, such as fucoxanthin, astaxantin, and carotenoids (Heo et al. 2005). Several seaweed pigment products are recommended for use with anti-obesity, immunity booster, and many others that are available in the market (Dufossé et al. 2005).

In seaweed extracts, most of the polyphenolic compounds are present with photoprotective and anti-photo aging activities, which can prevent oxidative stress and damage from UV radiation (Shui-Chun and Yue-Wei 2010; Ryu et al. 2009; Jung et al. 2009; Kumar et al. 2008). They have the properties to suppress the lipid peroxidation in the anti-oxidant activity (Ryu et al. 2009). Phenol-rich extracts from seaweeds have the capacity to stop the digestive enzymes, to treat anti-diabetic effects (Nwosu et al. 2011), and have the therapeutic potential for fighting bronchial asthma and allergic diseases. These compounds that are injected intraperitoneally resulted in a reduction of all asthmatic reactions (Jung et al. 2009). Tannins are plant growth hormones that are documented to exhibit their HIV-1 inhibitory mode of action by slowing down HIV-1RT (polymerase and ribonuclease activities) (Gupta and Abu-Ghannam 2011).

Phlorotannins (polyphenols) are synthesized by algae that have the capacity to absorb UV radiation and perform as photo-protective cells against photo-damage (Li et al. 2011; Kang et al. 2010). These extracts can stop the dose-dependent relationship of α-amylase and α-glucosidase, whereas some other organisms are used to slow down the normal increase in postprandial blood glucose observed for about 20 minutes after a meal by 90% and repeatedly decrease the peak insulin secretion by 40% (Roy et al. 2011). Fucosterol is a compound of phlorotannins that is used to decline the serum glucose concentration, and it showed a reduction of sorbitol substances in rat lenses (Gupta and Abu-Ghannam 2011). Such compounds are used to guard the animals against ultraviolet B radiation (UVB)-induced skin carcinogenesis (Ryu et al. 2009) and natural anti-oxidants in various pharmaceutical products (Guinea et al. 2012). *Ecklonia cava* is a brown algal species that contains a large amount of phlorotannin derivatives, such as eckol and dieckol to decrease the intracellular reactive oxygen species, which are formed

TABLE 24.1
Bioactive Compounds from Brown Algae with Various Functional Properties as Pharmacological Applications

Species	Compound	Biological Activities	References
Adenocystis utricularis	Fucoidan	Inhibitory against HSV and Anti-HIV and Anti-viral	Ponce et al. (2003) and Trinchero et al. (2009)
Alaria marginata	Fucoidan	Anti-cancer	Usoltseva et al. (2016)
Analipus japonicus	Fucoidan	Anti-coagulant	Ushakova et al. (2009)
Ascophyllum nodosum	Fucoidan and Laminaran	Immunomodulatory, Anti-inflammatory, Anti-coagulant, Anti-thrombotic, Anti-metastatic, Anti-tumor, Anti-adhesive, Restenosis preventive; and Anti-angiogenic	Percival (1968); Ren et al. (1994); Hoffman et al. (1995); Miao et al. (1999); Chevolot et al. (2001); (1999); Anastase-Ravion et al. (2002); Matou et al. (2002); Colliec-Jouault et al. (2003); Luyt et al. (2003); Smitt (2004); Cumashi et al. (2007); Ushakova et al. (2009); Foley et al. (2011)
Bifurcaria bifurcata	Diterpenes	Cytotoxicity	Smyrniotopoulos et al. (2017)
C. novae-caledoniae	Fucoidan	Anti-tumor	Zhang et al. (2011)
Canistrocarpus cervicornis	Fucoidan	Anti-coagulant, Anti-oxidant and Anti-proliferative	Costa et al. (2010); Camara et al. (2011)
Chnoospora bicanaliculata	Fucoidan	Anti-tumor, Anti-proliferation, and Anti-cancer	Haneji et al. (2005)
Chorda filum	Fucoidan	Anti-coagulant	Ushakova et al. (2009)
Cladosiphon okamuranus	Fucoidan	Anti-proliferative, Anti-viral, Anti-inflammatory, Anti-adhesive, Anti-tumor, Immunomodulator; Angiogenic, Gastroprotective, Cardioprotective, Restenosis preventive	Takahashi et al. (1998); Ushakova et al. (2009); Cumashi et al. (2007); Wijesinghe and Jeon (2012); Hidari et al. (2007); Nagaoka et al. (1999); Shimizu et al. (2005); Teruya et al. (2007); Kawamoto et al. (2006); Shibata et al. (2003); Shibata et al. (2000); Thomes et al. (2010)
Coccophora langsdorfii	Fucoidan	Anti-cancer	Imbs et al. (2016)
Colpomenia sinuosa	Fucoidan	Anti-coagulant and Anti-cancer	Shora et al. (2018); Yousef et al. (2018)
Costaria costata	Fucoidan	Anti-tumor	Matou et al. (2002)
Cystoseira sedoides	Fucoidan	Anti-inflammatory	Ammar et al. (2015)
Dictyopteris delicatula	Fucoidan	Anti-coagulant, Anti-oxidant, Anti-tumor, and Anti-proliferative	Magalhaes et al. (2011); Costa et al. (2010); Bilan and Usov (2008)
Dictyota menstrualis	Fucoidan	Peripheral anti-nociceptive, Anti-inflammatory, Anti-oxidant, Anti-coagulant, and Anti-proliferative	Albuquerque et al. (2004, 2013); Costa et al. (2010)
	Diterpenes	Anti-retroviral	Pereira et al. (2004)
Dictyota mertensis	Fucoidan	Peripheral anti-nociceptive, Anti-inflammatory, Anti-oxidant, Anti-coagulant, and Anti-proliferative	Costa et al. (2010)
Dictyota pfaffii	Diterpenes	Inhibitory against HSV-1 and Anti-HIV	Abrantes et al. (2010); Barbosa et al. (2007)
Dictyota polypodioides	Fucoidan	Anti-tumor	Sokolova et al. (2011)
Dictyota sp.	Fucoidan	Blood Clotting	Devi et al. (2016)
	Diterpenes	Cytotoxic	Jongaramruong and Kongkam (2007)
Dictyota. Indica	Fucoidan	Anti-cancer	Yousef et al. (2018)
Ecklonia cava	Phlorotannin	Anti-diabetic, Anti-hypertension, MMP inhibition, Immunomodulatory, Anti-HIV, Anti-oxidant, Anti-allergy inhibition of melanin formation, Anti-proliferative, Anti-cancer, Anti-tumor Whitening effect, Anti-photoaging	Okada et al. (2004); Lee et al. (2009); Jung et al. (2006); Ryu et al. (2009); Lee et al. (2010); Park et al. (2010); Shibata et al. (2008); Li et al. (2009); Yoon et al. (2009); Artan et al. (2008); Ahn et al. (2004); Heo (2009); Kong et al. (2009); Ahn et al. (2007); Kang et al. (2005); Park et al. (2008); Ahn et al. (2010); Kim et al. (2006); Wijesinghe et al. (2011); Ryu et al. (2008); Lee et al. (2008); Ah Kang et al. (2006, 2007); Joe et al. (2006)
	Fucoidan	Anti-proliferative, Anti-cancer, Anti-tumor, Anti-coagulative, Anti-inflammatory, Anti-allergy, and Anti-thrombotic	Athukorala et al. (2009); Jung et al. (2007); Kang et al. (2011); Athukorala et al. (2006); Ermakova et al. (2011); Yamamoto et al. (1984); Nishino et al. (1989, 1991)

(*Continued*)

TABLE 24.1 (*Continued*)
Bioactive Compounds from Brown Algae with Various Functional Properties as Pharmacological Applications

Species	Compound	Biological Activities	References
Ecklonia arborea	Phlorotannin	Anti-allergy	Sugiura et al. (2006)
Ecklonia bicyclis	Phlorotannin	Anti-HIV, Anti-oxidation, Anti-diabetic, and MMP inhibition	Okada et al. (2004); Shibata et al. (2008); Lee et al. (2009, 2010)
Ecklonia kurome	Phlorotannin	Anti-HIV, Anti-oxidation, Bactericidal, Algicidal, Anti-allergy, Anti-diabetic, Tyrosinase inhibitor, Anti-photoaging, and Anti-hypertension	Shibata et al. (2008, 2003); Lee et al. (1996); Jung et al. (2006, 2008)
	Fucoidan	Anti-proliferative, Anti-tumor, Anti-coagulant, Anti-oxidant, Anti-thrombotic, Anti-inflammatory, Anti-cancer	Ermakova et al. (2011); Yamamoto et al. (1984); Nishino et al. (1989); (1991); Hu et al. (2001); Kang et al. (2011)
Eisenia arborea	Pholorotannin	Anti-diabetic and Anti-allergy	Sugiura et al. (2007)
Eisenia bicyclis	Fucoidan	Anti-cancer	Vischuk et al. (2009)
	Fucoidan; laminaran	Anti-proliferative, Anti-tumor, Anti-coagulant, Anti-tumor, and Anti-cancer	Yamamoto et al. (1984); Takahashi (1983); Usui et al. (1980); Rioux et al. (2010)
Enteromorpha prolifera	Fucoidan	Immunomodulatory	Kim et al. (2011)
Epinephelus bruneus	Sodium alginate	Immunomodulatory	Harikrishnan et al. (2011)
Epinephelus coioides	Sodium alginate	Immunomodulatory	Cheng et al. (2007)
Fucus distichus	Fucoidan	Anti-coagulant	Ushakova et al. (2009)
Fucus evanescens	Fucoidan	Anti-tumor, Anti-metastatic, Anti-viral, Anti-coagulant, Thrombolytic, Hepatoprotective, Immunomodulatory, and Anti-cancer	Alekseyenko et al. (2007); menshova et al. (2016); Kuznetsova et al. (2003); Ushakova et al. (2009)
Fucus serratus	Polyphenol	Anti-oxidant	O'Doherty et al. (2011)
	Fucoidan	Anti-coagulant	Ushakova et al. (2009)
Fucus sp.	Fucoidan	Anti-cancer activity and Immunomodulation	Xue et al. (2012); Kim et al. (2010)
	Laminaran	Anti-tumor, Anti-bacterial, immunomodulator, Anti-coagulant, Decreases liver triglyceride, cholesterol and phospholipid levels; Serum hypocholesterolemic, Hypotensive	Hoffman et al. (1995); Miao et al. (1999); Percival (1968); Ren et al. (1994)
Fucus spiralis	Fucoidan	Anti-coagulant	Ushakova et al. (2009)
Fucus vesiculosus	Fucoidan	Inhibitor of avian RT, Anti-HIV, Immunomodulatory, Anti-viral, Anti-tumor, Anti-proliferative, Anti-adhesive, Anti-coagulant, Anti-oxidant, Anti-metastatic, Anti-inflammatory, Anti-angiogenic, Anti-thrombotic	Queiroz et al. (2008); Mourão (2004); Kim and Joo (2008); Yang et al. (2008); Choi et al. (2005); Costa et al. (2010); Cumashi et al. (2007); Ale et al. (2011); Beress et al. (1993); Bilan et al. (2002); Mourão and Pereira (1999); Pereira et al. (1999); Nakamura et al. (2006); Park et al. (2011); Synytsya et al. (2010); Yang et al. (2006, 2008); Kim et al. (2008); Do et al. (2010); Jintang et al. (2010); De Souza et al. (2007); Dockal et al., 2011; O'Doberty et al. (2011); Ushakova et al. (2009)
Hizikia fusiforme	Fucoidan	Immunomodulatory Anti-coagulant and Anti-thrombotic	Jeong et al. (2015); Dobashi et al. (1989)
Ishige okamurae	Phlorotannin	Whitening effect, Anti-diabetic, Anti-oxidant, and Anti-HIV	Heo et al. (2009, 2010); Ahn et al. (2006)
Iyngaria stellate	Fucoidan	Anti-cancer	Yousef et al. (2018)
Laminaria digitata	Fucoidan	Anti-coagulant and Anti-thrombic	Ushakova et al. (2009); Grauffel et al. (1989)
Laminaria japonica	Laminaran	Anti-oxidant, Hypolipidemic activities, Anti-apoptotic	Cheng et al. (2011); Zha et al. (2012); Kim et al. (2006); Li et al. (2011)
	Fucoidan	Anti-lipidemic, Increases HDL, Anti-viral, Anti-tumor, Anti-oxidant, Neuroprotective, Immunomodulatory, and Anti-inflammation	Fedorov et al. (2013); Cumashi et al. (2007); Vishchuk et al. (2011); Xiaolin et al. (1995); Dockal et al. (2011).

(*Continued*)

TABLE 24.1 (*Continued*)
Bioactive Compounds from Brown Algae with Various Functional Properties as Pharmacological Applications

Species	Compound	Biological Activities	References
Laminaria saccharina	Fucoidan	Anti-coagulant	Ushakova et al. (2009)
	Laminarin	Skin treatment	Yvin et al. (1999)
Laminaria sp.	Laminaran	Anti-tumor, Anti-coagulant, Decreases Liver triglyceride, Cholesterol and Phospholipid levels, Serum Hypocholesterolemic, Hypotensive, Anti-bacterial, Immunomodulatory	Hoffman et al. (1995); Miao et al. (1999); Ren et al. (1994); Park et al. (2012, 2013); Cheng et al. (2011)
	Fucoidan	Anti-oxidant, Anti-coagulant, Anti-thrombotic, Anti-adhesive, Anti-proliferative, Anti-inflammatory, Anti-angiogenic, Anti-metastatic	Cumashi et al. (2007); Chevolot et al. (1999); Yamamoto et al. (1984); Usov et al. (1998); Maruyama and Yamamoto (1984); Kitamura et al. (1992); Huang et al. (2010); Xue et al. (2004)
Laminaria uryanovae	Fucoidan	Anti-tumor, Anti-proliferation, Anti-cancer	Lee et al. (2008)
Leathesia difformis	Fucoidan	Anti-viral	Feldman et al. (1999)
Lessonia vadose	Fucoidan	Anti-coagulant, Anti-tumor, Anti-thrombotic	Li et al. (2008)
Lobophora variegata	Fucoidan	Anti-oxidant, Anti-coagulant, Anti-inflammatory	Medeiros et al. (2008); Paiva et al. (2011)
Monocystis pyrifera	Sodium alginate	Immunomodulatory	Fujiki and Yano (1997)
Okinawa mozuku	Fucoidan	Anti-proliferative, Antiviral, Anti-inflammatory, Anti-adhesive, Anti-tumor, Immunomodulator, Angiogenic, Gastroprotective, Cardioprotective, Restenosis preventive	Cumashi et al. (2007); Wijesinghe and Jeon (2012); Hidari et al. (2008); Nagaoka et al. (1999); Shimizu et al. (2005); Teruya et al. (2007); Kawamoto et al. (2006); Shibata et al. (2000, 2003); Thomes et al. (2010)
Padina boryana	Fucoidan	Anti-cancer	Usoltseva et al. (2018)
Padina gymnospora	Fucoidan	Anti-oxidant, Anti-coagulant, Anti-thrombotic, Anti-viral	Costa et al. (2010); Silva et al. (2005); Zhu et al. (2003); De Souza et al. (2007)
Padina pavonica	Fucoidan	Anti-coagulant and Anti-HIV	Shora et al. (2018); Mohamed and Agili (2013)
Padina sp.	Fucoidan	Cytotoxicity and Anti-oxidant	Isnansetyo et al. (2017); Dang et al. (2018)
Padina tetrastomatica	Fucoidan	Immunomodulatory, Anti- bacterial, Anti-coagulant, Blood Clotting	Rani et al. (2017); Chandini et al. (2008)
Pelvetia canaliculata	Fucoidan	Anti-viral	Klarzynski et al. (2003)
Pelvetia fastigiata	Fucoidan	Anti-viral	Venkateswaran et al. (1989)
Pelvetia siliquosa	Phlorotannin	Anti-diabetic	Lee et al. (2004)
Phyllospora comosa	Fucoidan	Anti-oxidant	Dang et al. (2018)
Punctaria plantaginea	Fucoidan	Anti-coagulant and Anti-thrombotic	Ustyuzhanina et al. (2016)
Saccharina cichorioides	Fucoidan	Anti-cancer	Vischuk et al. (2009)
Sargassm fusiforme	Fucoidan	Immunomodulatory	Huang et al. (2010)
Sargassum cristaefolium	Fucoidan	Anti-inflammatory	Wu et al. (2015)
Sargassum duplicatum	Laminaran, Fucoidan and Alginic acid	Anti-diabetic, Anti-cancer	Hardoko et al. (2014); Usoltseva et al. (2017)
Sargassum filipendula	Fucoidan	Anti-oxidant, Anti-proliferative	Costa et al. (2010, 2011)
Sargassum fulvellum	Fucoidan	Anti-coagulant, Anti-thrombotic	Zoysa et al. (2008)
Sargassum hemiphyllum	Fucoidan	Immunomodulatory	Hwang et al. (2011)
Sargassum henslowianum	Fucoidan	Anti-proliferative, Anti-tumor	Ale et al. (2011)
Sargassum horneri	Fucoidan	Anti-tumor, Anti-viral, Anti-cancer	Ale et al. (2011); Hoshino et al. (1998); Ermakova et al. (2011)
Sargassum kjelimanianum	Fucoidan	Anti-cancer	Yamamoto et al. (1981)
Sargassum latifolium	Fucoidan	Anti-coagulant	Shora et al. (2018)
Sargassum linearifolium	Fucoidan	Anti-oxidant	Dang et al. (2018)
Sargassum marginatum	Polyphenol	Anti-oxidant	Chandini et al. (2008)
Sargassum mcclurei	fucoidan	Anti-cancer, Anti-HIV	Thuy et al. (2015)
Sargassum muticum	Polyphenol	Anti-cancer	Namvar et al. (2013)
Sargassum oligocystum	Fucoidan	Anti-bacterial	Rani et al. (2017)
Sargassum patens	Fucoidan	Anti-viral	Zhu et al. (2003)

(*Continued*)

TABLE 24.1 (*Continued*)
Bioactive Compounds from Brown Algae with Various Functional Properties as Pharmacological Applications

Species	Compound	Biological Activities	References
Sargassum podocanthum	Fucoidan	Anti-oxidant	Dang et al. (2018)
Sargassum polycystum	Fucoidan	Anti-HIV, Immunomodulatory, Anti-oxidant, and Anti-cancer	Thuy et al. (2015); Chotigeat et al. (2004); Palanisamy et al. (2017)
Sargassum siliquastrum	Fucoidan	Anti-oxidant	Heo et al. (2008); Heo and Jeon (2009)
Sargassum sp.	Fucoidan	Anti-tumor, Anti-oxidant, Emulsification, Cytotoxicity, Prevent hyperlipidemia, Normalize dislipidemia	Sokolova et al. (2011); Ale et al. (2011); Duarte et al. (2001); Hoshino et al. (1998); Isnansetyo et al. (2017); Raghavendran et al. (2005); Josephine et al. (2007)
Sargassum tenerrimum	Fucoidan	Anti-coagulant	Guru et al. (2013)
Sargassum vestitum	Fucoidan	Anti-oxidant	Dang et al. (2018)
Sargassum vulgare	Alginic acid	Anti-tumor	De Souza et al. (2007)
Sargassum wighti	Alginic acid	Immunomodulatory	Kudus et al. (2017)
	Fucoidan	Immunomodulatory, Hepatoprotective, Anti-coagulant, Blood Clotting	Immanuel et al. (2012); Kanimozhi et al. (2013); Prabu et al. (2016); Guru et al. (2013); Jamuna Devi et al. (2016)
	Sodium alginate	Immunomodulatory	Immanuel et al. (2012)
Scytothamnus australis	Fucoidan	Anti-HSV1	Wozniak et al. (2015)
Spatoglossum schroederi	Fucoidan	Anti-coagulant, Anti-thrombotic	Almeida-Lima et al. (2011)
	Fucoidan and Alginic acid	Anti-viral	Queiroz et al. (2008)
Spatoglossum schröederi	Fucoidan	Anti-thrombotic, Peripheral anti-nociceptive, Anti-proliferative, Anti-adhesive, Anti-oxidant	Costa et al. (2010); Almeida-Lima et al. (2010); (2011); Farias et al. (2011); Rocha et al. (2005); Alwarsamy et al. (2016)
Turbinaria conoides	Polyphenol	Anti-oxidant	Chandini et al. (2008)
	Fucoidan	Anti-coagulant activity	Guru et al. (2013)
	Alginic acid	Anti-oxidant	Chattopadhyay et al. (2010)
Turbinaria decurens	Laminaran, Fucoidan and Alginic acid	Anti-diabetic	Hardoko et al. (2014)
Turbinaria ornate	Fucoidan	Anti-coagulant, Anti-HIV, immunomodulatory, Anti-bacterial	Guru et al. (2013); Thuy et al. (2015); Rani et al. (2017)
Turbinaria sp.	Fucoidan	Cytotoxicity	Isnansetyo et al. (2017)
Undaria pinnatifida	Fucoidan	Anti-coagulant, Anti-tumor, Serum hypocholesterolemic, Hypotensive, Anti-bacterial, Immunomodulatory, Anti-proliferative, Induced osteoblastic differentiation	Hemmingson et al. (2006); Yang et al. (2008); Yoo et al. (2007); Hayashi et al. (2008); Faggio et al. (2015); Hoffman et al. (1995); Miao et al. (1999); Ren et al. (1994) Preobrazhenskaya et al. (1997); Fedorov et al. (2013); Chevolot et al. (1999); Synytsya et al. (2010); Thompson and Dragar (2004); Maruyama et al. (2003); Cho et al. (2009, 2011); Maruyama et al. (2005)

by gamma-ray radiation (Li et al. 2011). In pharmaceutical applications, bioactive compounds, such as phlorotannin and dieckol have the potential of a whitening effect (Gupta and Abu-Ghannam 2011). Hence, phlorotannins have the potential to cure skin diseases and be used in pharmaceuticals.

Rich extracts of polyphenol and isolated phlorotannin have the potential to stop the multiplication of cancer cells and to manipulate anti-inflammatory responses (Nwosu et al. 2011). For example, *Eucheuma cottonii* contains rich polyphenol extracts that have the potential of anti-proliferative activity against human breast cancer cells while they are non-toxic to normal cell lines (Namvar et al. 2012). The extracts of the *P. palmate* species possess the anti-oxidant activities that are associated with the occurrence of secondary metabolites and function as the UV-absorbing sunscreen molecule (Yuan and Walsh 2006). Most of the phlorotannins that were extracted from *Eisenia arborea* were documented to have anti-allergic properties. The seaweed extracts of *Sargasum hemiphyllum* (brown algae) are used in the treatment of various allergic diseases and Korean folk medicine (Vo et al. 2012). Phloroglucinol is a derivative of phlorotannin that released a maximum anti-proliferative activity in a human breast cancer cell and stimulates a major proliferative inhibition and apoptosis (Gupta and Abu-Ghannam 2011).

24.5 CONCLUSION

Biologically active compounds from seaweeds are a key product for the manufacturing of bio-based materials in food, pharmaceutics, therapeutics, and cosmetics. Different types of bioactive molecules can be either formed or released by marine macroalgae. Seaweeds are a rich source of polysaccharides; many of them contain sulfated polysaccharides. Even so, few similarities can be made among the polysaccharides from every group of organisms, and these can be diverse and morphologically varied. Several advantageous bioactivities have been demonstrated by the polysaccharides and polyphenols, both in vitro and in vivo.

Brown algae are well known owing to simply being available producers of sulfated polysaccharides. These produce a variety of significant polyphenolic compounds (phlorotannins), polysaccharides, and carotenoids, together called bioactive compounds. This review describes the brown algal extracts and chemically defined bioactive compounds of natural sources that exhibit a variety of biological effects. The functional metabolites of brown algae show a potential approach toward the pharmaceutical preparations. In conclusion, the brown algal bioactive compounds have the potential sources of bioactivities that can be used in search of novel functional applications in the relevant pharmaceuticals.

REFERENCES

Abrantes, J. L., Barbosa, J., Cavalcanti, D., Pereira, R. C., Fontes, C. L. F., Teixeira, V. L., Souza, T. L. M., and Paixão, I. C. 2010. The effects of the diterpenes isolated from the Brazilian brown algae *Dictyota pfaffii* and *Dictyota menstrualis* against the herpes simplex type-1 replicative cycle. *Planta Medica*, 76(04): 339–344.

Ah Kang, K., Zhang, R., Hwa Lee, K., Chae, S., Joon Kim, B., Sook Kwak, Y., Woo Park, J., Ho Lee, N., and Won Hyun, J. 2006. Protective effect of triphlorethol-A from *Ecklonia cava* against ionizing radiation in vitro. *Journal of Radiation Research*, 47(1): 61–68.

Ahn, G., Park, E., Park, H. J., Jeon, Y. J., Lee, J., Park, J. W., and Jee, Y., 2010. The classical NFκB pathway is required for phloroglucinol-induced activation of murine lymphocytes. *Biochimica et Biophysica Acta (BBA)-General Subjects*, 1800(7): 639–645.

Ahn, G. N., Kim, K. N., Cha, S. H., Song, C. B., Lee, J., Heo, M. S., Yeo, I. K., Lee, N. H., Jee, Y. H., Kim, J. S., and Heu, M. S. 2007. Antioxidant activities of phlorotannins purified from *Ecklonia cava* on free radical scavenging using ESR and H_2O_2-mediated DNA damage. *European Food Research and Technology*, 226(1–2): 71–79.

Albuquerque, I. R. L., Queiroz, K. C. S., Alves, L. G., Santos, E. A. D., Leite, E. L., and Rocha, H. A. O. 2004. Heterofucans from *Dictyota menstrualis* have anticoagulant activity. *Brazilian Journal of Medical and Biological Research*, 37(2): 167–171.

Albuquerque, I. R. L., Cordeiro, S. L., Gomes, D. L., Dreyfuss, J. L., Filgueira, L. G. A., Leite, E. L., Nader, H. B., and Rocha, H. A. O. 2013. Evaluation of anti-nociceptive and anti-inflammatory activities of a heterofucan from *Dictyota menstrualis*. *Marine Drugs*, 11: 2722–2740.

Ale, M. T., Mikkelsen, J. D., and Meyer, A. S. 2011. Important determinants for fucoidan bioactivity: A critical review of structure-function relations and extraction methods for fucose-containing sulfated polysaccharides from brown seaweeds. *Marine Drugs*, 9(10): 2106–2130.

Alekseyenko, T. V., Zhanayeva, S. Y., Venediktova, A. A., Zvyagintseva, T. N., Kuznetsova, T. A., Besednova, N. N., and Korolenko, T. A. 2007. Antitumor and antimetastatic activity of fucoidan, a sulfated polysaccharide isolated from the Okhotsk Sea *Fucus evanescens* brown alga. *Bulletin of Experimental Biology and Medicine*, 143(6): 730–732.

Almeida-Lima, J., Costa, L. S., Silva, N. B., Melo-Silveira, R. F., Silva, F. V., Cansancao Felipe, M. B. M., Batistuzzo Medeiros, S. R., Leite, E. L., and Rocha, H. A. O. 2010. Evaluating the possible genotoxic, mutagenic and tumor cell proliferation-inhibition effects of a non-anticoagulant, but antithrombotic algal heterofucan. *Journal of Applied Toxicology*, 30(7): 708–715.

Almeida-Lima, J., Dantas-Santos, N., Gomes, D. L., Cordeiro, S. L., Sabry, D. A., Costa, L. S., Freitas, M. D. L., Silva, N. B., Moura, C. E., Lemos, T. M., and Leite, E. L. 2011. Evaluation of acute and subchronic toxicity of a non-anticoagulant, but antithrombotic algal heterofucan from the *Spatoglossum schröederi* in Wistar rats. *Revista Brasileira de Farmacognosia*, 21(4): 674–679.

Alwarsamy, M., Gooneratne, R., and Ravichandran, R. 2016. Effect of fucoidan from *Turbinaria conoides* on human lung adenocarcinoma epithelial (A549) cells. *Carbohydrate Polymers*, 152: 207–213.

Amarowicz, R., Pegg, R. B., Rahimi-Moghaddam, P., Barl, B., and Weil, J. A. 2004. Free-radical scavenging capacity and antioxidant activity of selected plant species from the Canadian prairies. *Food Chemistry*, 84(4): 551–562.

Ammar, H. H., Lajili, S., Said, R. B., Le Cerf, D., Bouraoui, A., and Majdoub, H. 2015. Physico-chemical characterization and pharmacological evaluation of sulfated polysaccharides from three species of Mediterranean brown algae of the genus *Cystoseira*. *DARU Journal of Pharmaceutical Sciences*, 23(1): 1.

Anastase-Ravion, S., Carreno, M. P., Blondin, C., Ravion, O., Champion, J., Chaubet, F., Haeffner-Cavaillon, N., and Letourneur, D. 2002. Heparin-like polymers modulate proinflammatory cytokine production by lipopolysaccharide-stimulated human monocytes. *Journal of Biomedical Materials Research*, 60(3), pp. 375–383.

Andriamanantoanina, H., and Rinaudo, M. 2010. Characterization of the alginates from five Madagascan brown algae. *Carbohydrate Polymers*, 82(3), pp. 555–560.

Antonisamy, J. M., and Raj, E. D. S. 2016. UV–VIS and HPLC studies on *Amphiroa anceps* (Lamarck) Decaisne. *Arabian Journal of Chemistry*, 9, pp. S907–S913.

Artan, M., Li, Y., Karadeniz, F., Lee, S. H., Kim, M. M., and Kim, S. K. 2008. Anti-HIV-1 activity of phloroglucinol derivative, 6, 6′-bieckol, from Ecklonia cava. *Bioorganic & Medicinal Chemistry*, 16(17), pp. 7921–7926.

Aslam, M. N., Kreider, J. M., Paruchuri, T., Bhagavathula, N., DaSilva, M., Zernicke, R. F., Goldstein, S. A., and Varani, J. 2010. A mineral-rich extract from the red marine algae *Lithothamnion calcareum* preserves bone structure and function in female mice on a Western-style diet. *Calcified Tissue International*, 86(4), pp. 313–324.

Athukorala, Y., Ahn, G. N., Jee, Y. H., Kim, G. Y., Kim, S. H., Ha, J. H., Kang, J. S., Lee, K. W., and Jeon, Y. J. 2009. Antiproliferative activity of sulfated polysaccharide isolated from an enzymatic digest of *Ecklonia cava* on the U-937 cell line. *Journal of Applied Phycology*, 21(3), pp. 307–314.

Athukorala, Y., Jung, W. K., Vasanthan, T., and Jeon, Y. J. 2006a. An anticoagulative polysaccharide from an enzymatic hydrolysate of *Ecklonia cava*. *Carbohydrate Polymers*, 66(2), pp. 184–191.

Athukorala, Y., Kim, K. N., and Jeon, Y. J. 2006b. Antiproliferative and antioxidant properties of an enzymatic hydrolysate from brown alga, *Ecklonia cava*. *Food and Chemical Toxicology*, 44(7), pp. 1065–1074.

Barbosa, J. P., Fleury, B. G., da Gama, B. A., Teixeira, V. L., and Pereira, R. C. 2007. Natural products as antifoulants in the Brazilian brown alga *Dictyota pfaffii* (Phaeophyta, Dictyotales). *Biochemical Systematics and Ecology*, 35(8), pp. 549–553.

Camara, R. B.G., Silva Costa, L., Pereira Fidelis, G., Duarte Barreto Nobre, L. T., Dantas-Santos, N., Lima Cordeiro, S., Santana Santos Pereira Costa, M., Guimaraes Alves, L., and Oliveira Rocha, H. A. 2011. Heterofucans from the brown seaweed *Canistrocarpus cervicornis* with anticoagulant and antioxidant activities. *Marine Drugs*, 9(1), pp. 124–138.

Béress, A., Wassermann, O., Bruhn, T., Béress, L., Kraiselburd, E. N., Gonzalez, L. V., de Motta, G. E., and Chavez, P. I. 1993. A new procedure for the isolation of anti-HIV compounds (polysaccharides and polyphenols) from the marine alga *Fucus vesiculosus*. *Journal of Natural Products*, 56(4), pp. 478–488.

Bhagavathy, S., Sumathi, P., and Bell, I. J. S. 2011. Green algae *Chlorococcum humicolaa* new source of bioactive compounds with antimicrobial activity. *Asian Pacific Journal of Tropical Biomedicine*. 1(1), pp. S1–S7.

Bilan, M. I., and Usov, A. I. 2008. Structural analysis of fucoidans. *Natural Product Communications*, 3, 1639–1648.

Bilan, M. I., Grachev, A. A., Ustuzhanina, N. E., Shashkov, A. S., Nifantiev, N. E., and Usov, A. I. 2002. Structure of a fucoidan from the brown seaweed *Fucus evanescens* C. Ag. *Carbohydrate Research*, 337(8), pp. 719–730.

Brownlee, I. A., Seal, C. J., Wilcox, M., Dettmar, P. W., and Pearson, J. P. 2009. Applications of alginates in food. In *Alginates: Biology and Applications* edited by B.H.A. Rehm (Vol. 13, pp. 211–228). Springer, Berlin, Germany.

Campo, V. L., Kawano, D. F., da Silva Jr, D. B., and Carvalho, I. 2009. Carrageenans: Biological properties, chemical modifications and structural analysis–A review. *Carbohydrate Polymers*, 77(2), pp. 167–180.

Chandini, S. K., Ganesan, P., and Bhaskar, N. 2008. In vitro antioxidant activities of three selected brown seaweeds of India. *Food Chemistry*, 107(2), pp. 707–713.

Chattopadhyay, N., Ghosh, T., Sinha, S., Chattopadhyay, K., Karmakar, P., and Ray, B. 2010. Polysaccharides from Turbinaria conoides: Structural features and antioxidant capacity. *Food Chemistry*, 118(3), pp. 823–829.

Cheng, A. C., Tu, C. W., Chen, Y. Y., Nan, F. H., and Chen, J. C. 2007. The immunostimulatory effects of sodium alginate and iota-carrageenan on orange-spotted grouper *Epinephelus coicoides* and its resistance against *Vibrio alginolyticus*. *Fish & Shellfish Immunology*, 22(3), pp. 197–205.

Cheng, X., Sun, K., Li, W., Zhang, T., and Li, C. 2011. A new species of Hansfordia isolated from the marine brown alga, *Colpomenia sinuosa*. *Mycotaxon*, 116, pp. 431–436.

Chevolot, L., Foucault, A., Chaubet, F., Kervarec, N., Sinquin, C., Fisher, A. M., and Boisson-Vidal, C. 1999. Further data on the structure of brown seaweed fucans: Relationships with anticoagulant activity. *Carbohydrate Research*, 319(1–4), pp. 154–165.

Chevolot, L., Mulloy, B., Ratiskol, J., Foucault, A., and Colliec-Jouault, S. 2001. A disaccharide repeat unit is the major structure in fucoidans from two species of brown algae. *Carbohydrate Research*, 330(4), pp. 529–535.

Cho, M. L., Lee, B. Y., and You, S. G. 2011. Relationship between oversulfation and conformation of low and high molecular weight fucoidans and evaluation of their in vitro anticancer activity. *Molecules*, 16(1), pp. 291–297.

Cho, Y. S., Jung, W. K., Kim, J. A., Choi, I. W., and Kim, S. K. 2009. Beneficial effects of fucoidan on osteoblastic MG-63 cell differentiation. *Food Chemistry*, 116(4), pp. 990–994.

Choi, C. W., Kim, S. C., Hwang, S. S., Choi, B. K., Ahn, H. J., Lee, M. Y., Park, S. H., and Kim, S. K. 2002. Antioxidant activity and free radical scavenging capacity between Korean medicinal plants and flavonoids by assay-guided comparison. *Plant Science*, 163(6), pp. 1161–1168.

Choi, E. M., Kim, A. J., Kim, Y. O., and Hwang, J. K. 2005. Immunomodulating activity of arabinogalactan and fucoidan in vitro. *Journal of Medicinal Food*, 8(4), pp. 446–453.

Chojnacka, K., Saeid, A., Witkowska, Z., and Tuhy, L. 2012. August. Biologically active compounds in seaweed extracts-the prospects for the application. *In The Open Conference Proceedings Journal*, 3(1), pp. 20–25.

Chotigeat, W., Tongsupa, S., Supamataya, K., and Phongdara, A. 2004. Effect of fucoidan on disease resistance of black tiger shrimp. *Aquaculture*, 233(1–4), pp. 23–30.

Colliec-Jouault, S., Millet, J., Helley, D., Sinquin, C., and Fischer, A. M. 2003. Effect of low-molecular-weight fucoidan on experimental arterial thrombosis in the rabbit and rat. *Journal of Thrombosis and Haemostasis*, 1(5), pp. 1114–1115.

Costa, L. S., Fidelis, G. P., Cordeiro, S. L., Oliveira, R. M., Sabry, D. D. A., Câmara, R. B. G., Nobre, L. T. D. B., Costa, M. S. S. P., Almeida-Lima, J., Farias, E. H. C., and Leite, E. L. 2010. Biological activities of sulfated polysaccharides from tropical seaweeds. *Biomedicine & Pharmacotherapy*, 64(1), pp. 21–28.

Costa, L. S., Fidelis, G. P., Telles, C. B. S., Dantas-Santos, N., Camara, R. B. G., Cordeiro, S. L., Costa, M. S. S. P., Almeida-Lima, J., Melo-Silveira, R. F., Oliveira, R. M., and Albuquerque, I. R. L. 2011. Antioxidant and antiproliferative activities of heterofucans from the seaweed *Sargassum filipendula*. *Marine Drugs*, 9(6), pp. 952–966.

Craigie, J. S. 2011. Seaweed extract stimuli in plant science and agriculture. *Journal of Applied Phycology*, 23(3), pp. 371–393.

Cumashi, A., Ushakova, N. A., Preobrazhenskaya, M. E., D'Incecco, A., Piccoli, A., Totani, L., Tinari, N., Morozevich, G. E., Berman, A. E., Bilan, M. I., and Usov, A. I. 2007. A comparative study of the anti-inflammatory, anticoagulant, antiangiogenic, and antiadhesive activities of nine different fucoidans from brown seaweeds. *Glycobiology*, 17(5), pp. 541–552.

Dang, T. T., Van Vuong, Q., Schreider, M. J., Bowyer, M. C., Van Altena, I. A., and Scarlett, C. J. 2017. Optimisation of ultrasound-assisted extraction conditions for phenolic content and antioxidant activities of the alga *Hormosira banksii* using response surface methodology. *Journal of Applied Phycology*, 29(6), pp. 3161–3173.

Daniel, R., Berteau, O., Chevolot, L., Varenne, A., Gareil, P., and Goasdoue, N. 2001. Regioselective desulfation of sulfated l-fucopyranoside by a new sulfoesterase from the marine mollusk *Pecten maximus*. Application to the structural study of algal fucoidan (Ascophyllum nodosum). *European Journal of Biochemistry*, 268(21), pp. 5617–5626.

De Souza, A. P. A., Torres, M. R., Pessoa, C., de Moraes, M. O., Rocha Filho, F. D., Alves, A. P. N. N., and Costa-Lotufo, L. V. 2007. In vivo growth-inhibition of Sarcoma 180 tumor by alginates from brown seaweed Sargassum vulgare. *Carbohydrate polymers*, 69(1), pp. 7–13.

De Zoysa, M., Nikapitiya, C., Jeon, Y. J., Jee, Y., and Lee, J. 2008. Anticoagulant activity of sulfated polysaccharide isolated from fermented brown seaweed *Sargassum fulvellum*. *Journal of Applied Phycology*, 20(1), pp. 67–74.

Dias, P. F., Siqueira Jr, J. M., Maraschin, M., Ferreira, A. G., Gagliardi, A. R., and Ribeiro-do-Valle, R. M. 2008. A polysaccharide isolated from the brown seaweed *Sargassum stenophyllum* exerts antivasculogenic effects evidenced by modified morphogenesis. *Microvascular Research*, 75(1), pp. 34–44.

Díaz-Rubio, M. E., Pérez-Jiménez, J., and Saura-Calixto, F. 2009. Dietary fiber and antioxidant capacity in *Fucus vesiculosus* products. *International Journal of Food Sciences and Nutrition*, 60(suppl2), pp. 23–34.

Dobashi, K., Nishino, T., Fujihara, M., and Nagumo, T. 1989. Isolation and preliminary characterization of fucose-containing sulfated polysaccharides with blood-anticoagulant activity from the brown seaweed *Hizikia fusiforme*. *Carbohydrate Research*, 194, pp. 315–320.

Dockal, M., Till, S., Knappe, S., Ehrlich, H. J., and Scheiflinger, F. 2011. Anticoagulant activity and mechanism of non-anticoagulant sulfated polysaccharides. *Blood*, 118, p. 1208

Duarte, M. E., Cardoso, M. A., Noseda, M. D., and Cerezo, A. S. 2001. Structural studies on fucoidans from the brown seaweed *Sargassum stenophyllum*. *Carbohydrate Research*, 333(4), pp. 281–293.

Dufossé, L., Galaup, P., Yaron, A., Arad, S. M., Blanc, P., Murthy, K. N. C., and Ravishankar, G. A. 2005. Microorganisms and microalgae as sources of pigments for food use: A scientific oddity or an industrial reality? *Trends in Food Science & Technology*, 16(9), pp. 389–406.

El-Shora, H. M., Abou-El-Wafa, G. S., and Abu-Eftouh, N. M. 2018. Fucoidan and fucoidanase from brown seaweeds and applications. *International Journal of Current Microbiology and Applied Science*, 7(2): 3707–3715.

Ermakova, S., Sokolova, R., Kim, S. M., Um, B. H., Isakov, V., and Zvyagintseva, T. 2011. Fucoidans from brown seaweeds Sargassum hornery, Eclonia cava, Costaria costata: Structural characteristics and anticancer activity. *Applied Biochemistry and Biotechnology*, 164(6), pp. 841–850.

Faggio, C., Morabito, M., Minicante, S. A., Piano, G. L., Pagano, M., and Genovese, G. 2015. Potential use of polysaccharides from the brown alga *Undaria pinnatifida* as anticoagulants. *Brazilian Archives of Biology and Technology*, 58(5), pp. 798–804.

Farias, W. R., Lima, P. C. W., Rodrigues, N. V. F., Siqueira, R. C. L., Amorim, R. M., Pereira, M. G., and Assreuy, A. M. S. 2011. A novel antinociceptive sulphated polysaccharide of the brown marine alga *Spatoglossum schroederi*. *Natural Product Communications*, 6(6), p. 1934578 × 1100600626.

Fedorov, S., Ermakova, S., Zvyagintseva, T., and Stonik, V. 2013. Anticancer and cancer preventive properties of marine polysaccharides: Some results and prospects. *Marine Drugs*, 11(12), pp. 4876–4901.

Feldman, S. C., Reynaldi, S., Stortz, C. A., Cerezo, A. S., and Damonte, E. B. 1999. Antiviral properties of fucoidan fractions from *Leathesia difformis*. *Phytomedicine*, 6(5), pp. 335–340.

Ferreira, L. G., Noseda, M. D., Gonçalves, A. G., Ducatti, D. R., Fujii, M. T., and Duarte, M. E. 2012. Chemical structure of the complex pyruvylated and sulfated agaran from the red seaweed *Palisada flagellifera* (Ceramiales, Rhodophyta). *Carbohydrate Research*, 347(1), pp. 83–94.

Foley, S. A., Szegezdi, E., Mulloy, B., Samali, A., and Tuohy, M. G. 2011. An unfractionated fucoidan from *Ascophyllum nodosum*: Extraction, characterization, and apoptotic effects in vitro. *Journal of Natural Products*, 74(9), pp. 1851–1861.

Fujiki, K., Matsuyama, H., and Yano, T. 1994. Protective effect of sodium alginates against bacterial infection in common carp, *Cyprinus carpio* L. *Journal of Fish diseases*, 17(4), pp. 349–355.

Fujiki, K., and Yano, T. 1997. Effects of sodium alginate on the non-specific defence system of the common carp (Cyprinus carpio L.). *Fish & Shellfish Immunology*, 7(6), pp. 417–427.

Grauffel, V., Kloareg, B., Mabeau, S., Durand, P., and Jozefonvicz, J. 1989. New natural polysaccharides with potent antithrombic activity: Fucans from brown algae. *Biomaterials*, 10(6), pp. 363–368.

Guinea, M., Franco, V., Araujo-Bazán, L., Rodríguez-Martín, I., and González, S. 2012. In vivo UVB-photoprotective activity of extracts from commercial marine macroalgae. *Food and Chemical Toxicology*, 50(3–4), pp. 1109–1117.

Guiry, M. D., and Guiry, G. M., 2011. *AlgaeBase. World-Wide Electronic Publication.* National Institute Ireland, Galway, Ireland.

Gupta, S., and Abu-Ghannam, N. 2011. Recent developments in the application of seaweeds or seaweed extracts as a means for enhancing the safety and quality attributes of foods. *Innovative Food Science & Emerging Technologies*, 12(4), pp. 600–609.

Guru, S. M., Mahesh, K. P., Vasanthi, M., and Anant, A. 2013. Anticoagulant property of sulphated polysaccharides extracted from marine brown algae collected from Mandapam Island, India. *African Journal of Biotechnology*, 12(16), pp. 1937–1945.

Haneji, K., Matsuda, T., Tomita, M., Kawakami, H., Ohshiro, K., Uchihara, J. N., Masuda, M., Takasu, N., Tanaka, Y., Ohta, T., and Mori, N. 2005. Fucoidan extracted from *Cladosiphon okamuranus* Tokida induces apoptosis of human T-cell leukemia virus type 1-infected T-cell lines and primary adult T-cell leukemia cells. *Nutrition and Cancer*, 52(2), pp. 189–201.

Hardoko, S. T., Eveline, Y. M., and Olivia, S. 2014. An in vitro of antidiabetic activity of *Sargassum duplicatum* and *Turbinaria decurens* seaweed. *International Journal of Pharmaceutical Science Invention*, 3, pp. 13–18.

Harikrishnan, R., Kim, M. C., Kim, J. S., Han, Y. J., Jang, I. S., Balasundaram, C., and Heo, M. S. 2011. Immunomodulatory effect of sodium alginate enriched diet in kelp grouper *Epinephelus bruneus* against *Streptococcus iniae*. *Fish & Shellfish Immunology*, 30(2), pp. 543–549.

Hayashi, K., Nakano, T., Hashimoto, M., Kanekiyo, K., and Hayashi, T. 2008. Defensive effects of a fucoidan from brown alga *Undaria pinnatifida* against herpes simplex virus infection. *International Immunopharmacology*, 8(1), pp. 109–116.

Hemmingson, J. A., Falshaw, R., Furneaux, R. H., and Thompson, K. 2006. Structure and antiviral activity of the galactofucan sulfates extracted from *Undariapinnatifida* (Phaeophyta). *Journal of Applied Phycology*, 18(2), p. 185.

Heo, S. J., and Jeon, Y. J. 2009. Protective effect of fucoxanthin isolated from *Sargassum siliquastrum* on UV-B induced cell damage. *Journal of Photochemistry and Photobiology B: Biology*, 95(2), pp. 101–107.

Heo, S. J., Ko, S. C., Kang, S. M., Kang, H. S., Kim, J. P., Kim, S. H., Lee, K. W., Cho, M. G., and Jeon, Y. J. 2008. Cytoprotective effect of fucoxanthin isolated from brown algae Sargassum siliquastrum against H 2 O 2-induced cell damage. *European Food Research and Technology*, 228(1), pp. 145–151.

Heo, S. J., Park, E. J., Lee, K. W., and Jeon, Y. J. 2005. Antioxidant activities of enzymatic extracts from brown seaweeds. *Bioresource Technology*, 96(14), pp. 1613–1623.

Heo, S. J., Yoon, W. J., Kim, K. N., Ahn, G. N., Kang, S. M., Kang, D. H., Oh, C., Jung, W. K., and Jeon, Y. J. 2010. Evaluation of anti-inflammatory effect of fucoxanthin isolated from brown algae in lipopolysaccharide-stimulated RAW 264.7 macrophages. *Food and Chemical Toxicology*, 48(8–9), pp. 2045–2051.

Hidari, K. I., Takahashi, N., Arihara, M., Nagaoka, M., Morita, K., and Suzuki, T. 2008. Structure and anti-dengue virus activity of sulfated polysaccharide from a marine alga. *Biochemical and Biophysical Research Communications*, 376(1), pp. 91–95.

Hoffman, R., Donaldson, J., Alban, S., and Franz, G. 1995. Characterisation of a laminarin sulphate which inhibits basic fibroblast growth factor binding and endothelial cell proliferation. *Journal of Cell Science*, 108(11), pp. 3591–3598.

Holdt, S. L., and Kraan, S. 2011. Bioactive compounds in seaweed: Functional food applications and legislation. *Journal of Applied Phycology*, 23(3), pp. 543–597.

Hoshino, T., Hayashi, T., Hayashi, K., Hamada, J., Lee, J. B., and Sankawa, U. 1998. An antivirally active sulfated polysaccharide from *Sargassum horneri* (TURNER) C. AGARDH. *Biological and Pharmaceutical Bulletin*, 21(7), pp. 730–734.

Hu, J. F., Geng, M. Y., Zhang, J. T., and Jiang, H. D. 2001. An in vitro study of the structure-activity relationships of sulfated polysaccharide from brown algae to its antioxidant effect. *Journal of Asian Natural Products Research*, 3(4), pp. 353–358.

Huang, L., Wen, K., Gao, X., and Liu, Y. 2010. Hypolipidemic effect of fucoidan from *Laminaria japonica* in hyperlipidemic rats. *Pharmaceutical Biology*, 48(4), pp. 422–426.

Hwang, P. A., Chien, S. Y., Chan, Y. L., Lu, M. K., Wu, C. H., Kong, Z. L., and Wu, C. J. 2011. Inhibition of lipopolysaccharide (LPS)-induced inflammatory responses by *Sargassum hemiphyllum* sulfated polysaccharide extract in RAW 264.7 macrophage cells. *Journal of Agricultural and Food Chemistry*, 59(5), pp. 2062–2068.

Imbs, T. I., Ermakova, S. P., Malyarenko, O. S., Isakov, V. V., and Zvyagintseva, T.N. 2016. Structural elucidation of polysaccharide fractions from the brown alga *Coccophora langsdorfii* and in vitro investigation of their anticancer activity. *Carbohydrate Polymers*, 135, pp. 162–168.

Immanuel, G., Sivagnanavelmurugan, M., Marudhupandi, T., Radhakrishnan, S., and Palavesam, A. 2012. The effect of fucoidan from brown seaweed *Sargassum wightii* on WSSV resistance and immune activity in shrimp *Penaeus monodon* (Fab). *Fish & Shellfish Immunology*, 32(4), pp. 551–564.

Isnansetyo, A., Lutfia, F. N. L., Nursid, M., and Susidarti, R. A. 2017. Cytotoxicity of fucoidan from three tropical brown algae against breast and colon cancer cell lines. *Pharmacognosy Journal*, 9(1), pp. 14–20.

Jamuna, J. D., Santhosh, S., Hemalatha, V., and Dhandapani, R. 2016. Extraction and purification of sulfated polysaccharide from brown algae and its efficacy in preventing blood clotting. *Asian Journal of Biological and Life Sciences*. 5(3), pp. 237–244.

Jeong, H. S., Venkatesan, J., and Kim, S. K., 2013. Hydroxyapatite-fucoidan nanocomposites for bone tissue engineering. *International Journal of Biological Macromolecules*, 57, pp. 138–141.

Jeong, S. C., Jeong, Y. T., Lee, S. M., and Kim, J. H. 2015. Immune-modulating activities of polysaccharides extracted from brown algae Hizikia fusiforme. *Bioscience, Biotechnology, and Biochemistry*, 79(8), pp. 1362–1365.

Jeong, S. I., Jeon, O., Krebs, M. D., Hill, M. C., and Alsberg, E. 2012. Biodegradable photo-crosslinked alginate nanofibre scaffolds with tuneable physical properties, cell adhesivity and growth factor release. *European Cells & Materials*, 24, p. 331.

Jintang, S., Alei, F., Yun, Z., Shanzhen, S., Weixu, H., Meixiang, Y., Fengcai, W., and Xun, Q. 2010. Fucoidan increases TNF-α-induced MMP-9 secretion in monocytic cell line U937. *Inflammation Research*, 59(4), pp. 271–276.

Jongaramruong, J., and Kongkam, N. 2007. Novel diterpenes with cytotoxic, anti-malarial and anti-tuberculosis activities from a brown alga *Dictyota* sp. *Journal of Asian Natural Products Research*, 9(8), pp. 743–751.

Josephine, A., Veena, C. K., Amudha, G., Preetha, S. P., and Varalakshmi, P. 2007. Protective role of sulphated polysaccharides in abating the hyperlipidemic nephropathy provoked by cyclosporine A. *Archives of Toxicology*, 81(5), pp. 371–379.

Jung, H. A., Hyun, S. K., Kim, H. R., and Choi, J. S. 2006. Angiotensin-converting enzyme I inhibitory activity of phlorotannins from Ecklonia stolonifera. *Fisheries Science*, 72(6), pp. 1292–1299.

Jung, W. K., Athukorala, Y., Lee, Y. J., Cha, S. H., Lee, C. H., Vasanthan, T., Choi, K. S., Yoo, S. H., Kim, S. K., and Jeon, Y. J. 2007. Sulfated polysaccharide purified from *Ecklonia cava* accelerates antithrombin III-mediated plasma proteinase inhibition. *Journal of Applied Phycology*, 19(5), pp. 425–430.

Jung, W. K., Choi, I., Oh, S., Park, S. G., Seo, S. K., Lee, S. W., Lee, D. S., Heo, S. J., Jeon, Y. J., Je, J. Y., and Ahn, C. B. 2009. Anti-asthmatic effect of marine red alga (*Laurencia undulata*) polyphenolic extracts in a murine model of asthma. *Food and Chemical Toxicology*, 47(2), pp. 293–297.

Kang, C., Jin, Y. B., Lee, H., Cha, M., Sohn, E. T., Moon, J., Park, C., Chun, S., Jung, E. S., Hong, J. S., and Kim, S. B. 2010. Brown alga *Ecklonia cava* attenuates type 1 diabetes by activating AMPK and Akt signaling pathways. *Food and Chemical Toxicology*, 48(2), pp. 509–516.

Kang, K. A., Lee, K. H., Park, J. W., Lee, N. H., Na, H. K., Surh, Y. J., You, H. J., Chung, M. H., and Hyun, J. W. 2007. Triphlorethol-A induces heme oxygenase-1 via activation of ERK and NF-E2 related factor 2 transcription factor. *FEBS Letters*, 581(10), pp. 2000–2008.

Kang, S. M., Kim, K. N., Lee, S. H., Ahn, G., Cha, S. H., Kim, A. D., Yang, X. D., Kang, M. C., and Jeon, Y. J. 2011. Anti-inflammatory activity of polysaccharide purified from AMG-assistant extract of *Ecklonia cava* in LPS-stimulated RAW 264.7 macrophages. *Carbohydrate Polymers*, 85(1), pp. 80–85.

Kanimozhi, S., Krishnaveni, M., Deivasigmani, B., Rajasekar, T., and Priyadarshni, P. 2013. Immunomostimulation effects of *Sargassum whitti* on *Mugil cephalus* against *Pseudomonas fluorescence. International Journal of Current Microbiology and Applied Science*, 7, pp. 93–103.

Kawamoto, H., Miki, Y., Kimura, T., Tanaka, K., Nakagawa, T., Kawamukai, M., and Matsuda, H. 2006. Effects of fucoidan from Mozuku on human stomach cell lines. *Food Science and Technology Research*, 12(3), pp. 218–222.

Keyrouz, R., Abasq, M. L., Le Bourvellec, C., Blanc, N., Audibert, L., ArGall, E., and Hauchard, D. 2011. Total phenolic contents, radical scavenging and cyclic voltammetry of seaweeds from Brittany. *Food Chemistry*, 126(3), pp. 831–836.

Khan, W., Rayirath, U. P., Subramanian, S., Jithesh, M. N., Rayorath, P., Hodges, D. M., Critchley, A. T., Craigie, J. S., Norrie, J., and Prithiviraj, B. 2009. Seaweed extracts as biostimulants of plant growth and development. *Journal of Plant Growth Regulation*, 28(4), pp. 386–399.

Kim, E. J., Park, S. Y., Lee, J. Y., and Park, J. H. Y. 2010. Fucoidan present in brown algae induces apoptosis of human colon cancer cells. *BMC Gastroenterology*, 10(1), p. 96.

Kim, J. K., Cho, M. L., Karnjanapratum, S., Shin, I. S., and You, S. G. 2011. In vitro and in vivo immunomodulatory activity of sulfated polysaccharides from *Enteromorpha prolifera. International Journal of Biological Macromolecules*, 49(5), pp. 1051–1058.

Kim, K. N., Heo, S. J., Song, C. B., Lee, J., Heo, M. S., Yeo, I. K., Kang, K. A., Hyun, J. W., and Jeon, Y. J. 2006. Protective effect of *Ecklonia cava* enzymatic extracts on hydrogen peroxide-induced cell damage. *Process Biochemistry*, 41(12), pp. 2393–2401.

Kim, M. H., and Joo, H. G. 2008. Immunostimulatory effects of fucoidan on bone marrow-derived dendritic cells. *Immunology Letters*, 115(2), pp. 138–143.

Kim, T. H., and Bae, J. S. 2010. Ecklonia cava extracts inhibit lipopolysaccharide induced inflammatory responses in human endothelial cells. *Food and Chemical Toxicology*, 48(6), pp. 1682–1687.

Kitamura, K., Matsuo, M., and Tsuneo, Y. 1992. Enzymic degradation of fucoidan by fucoidanase from the hepatopancreas of Patinopecten yessoensis. *Bioscience, Biotechnology, and Biochemistry*, 56(3), pp. 490–494.

Klarzynski, O., Descamps, V., Plesse, B., Yvin, J. C., Kloareg, B., and Fritig, B. 2003. Sulfated fucan oligosaccharides elicit defense responses in tobacco and local and systemic resistance against tobacco mosaic virus. *Molecular Plant-Microbe Interactions*, 16(2), pp. 115–122.

Kong, C. S., Kim, J. A., Yoon, N. Y., and Kim, S. K. 2009. Induction of apoptosis by phloroglucinol derivative from *Ecklonia cava* in MCF-7 human breast cancer cells. *Food and Chemical Toxicology*, 47(7), pp. 1653–1658.

Koyanagi, S., Tanigawa, N., Nakagawa, H., Soeda, S., and Shimeno, H. 2003. Oversulfation of fucoidan enhances its anti-angiogenic and antitumor activities. *Biochemical Pharmacology*, 65(2), pp. 173–179.

Kuda, T., and Ikemori, T. 2009. Minerals, polysaccharides and antioxidant properties of aqueous solutions obtained from macroalgal beach-casts in the Noto Peninsula, Ishikawa, Japan. *Food Chemistry*, 112(3), pp. 575–581.

Kudus, K., Grasian, I., Madasamy, S., and John, A. 2017. Immunomodulatory effect of alginic acid from brown seaweed *Sargassum wightii* on disease resistance in *Penaeus monodon. Journal Clean WAS*, 1(1), pp. 26–29.

Kumar, K. S., Ganesan, K., and Rao, P. S. 2008. Antioxidant potential of solvent extracts of *Kappaphycus alvarezii* (Doty) Doty–An edible seaweed. *Food Chemistry*, 107(1), pp. 289–295.

Kumar, M., Kumari, P., Trivedi, N., Shukla, M. K., Gupta, V., Reddy, C. R. K., and Jha, B. 2011. Minerals, PUFAs and antioxidant properties of some tropical seaweeds from Saurashtra coast of India. *Journal of Applied Phycology*, 23(5), pp. 797–810.

Kuznetsova, T. A., Besednova, N. N., Mamaev, A. N., Momot, A. P., Shevchenko, N. M., and Zvyagintseva, T. N. 2003. Anticoagulant activity of fucoidan from brown algae *Fucus evanescens* of the Okhotsk Sea. *Bulletin of Experimental Biology and Medicine*, 136(5), pp. 471–473.

Lahaye, M., and Robic, A. 2007. Structure and functional properties of ulvan, a polysaccharide from green seaweeds. *Biomacromolecules*, 8(6), pp. 1765–1774.

Lee, N. Y., Ermakova, S. P., Zvyagintseva, T. N., Kang, K. W., Dong, Z., and Choi, H. S. 2008. Inhibitory effects of fucoidan on activation of epidermal growth factor receptor and cell transformation in JB6 Cl41 cells. *Food and Chemical Toxicology*, 46(5), pp. 1793–1800.

Lee, S. H., Karadeniz, F., Kim, M. M., and Kim, S. K. 2009. α-Glucosidase and α-amylase inhibitory activities of phloroglucinal derivatives from edible marine brown alga, *Ecklonia cava. Journal of the Science of Food and Agriculture*, 89(9), pp. 1552–1558.

Lee, S. H., Park, M. H., Heo, S. J., Kang, S. M., Ko, S. C., Han, J. S., and Jeon, Y. J. 2010. Dieckol isolated from Ecklonia cava inhibits α-glucosidase and α-amylase in vitro and alleviates postprandial hyperglycemia in streptozotocin-induced diabetic mice. *Food and Chemical Toxicology*, 48(10), pp. 2633–2637.

Lee, Y. S., Shin, K. H., Kim, B. K., and Lee, S. 2004. Anti-diabetic activities of fucosterol from *Pelvetia siliquosa. Archives of Pharmacal Research*, 27(11), pp. 1120–1122.

Leonard, S. G., Sweeney, T., Pierce, K. M., Bahar, B., Lynch, B. P., and O'Doherty, J.V. 2010. The effects of supplementing the diet of the sow with seaweed extracts and fish oil on aspects of gastrointestinal health and performance of the weaned piglet. *Livestock Science*, 134(1–3), pp. 135–138.

Li, B., Lu, F., Wei, X., and Zhao, R. 2008. Fucoidan: Structure and bioactivity. *Molecules*, 13(8), pp. 1671–1695.

Li, Y., Qian, Z. J., Ryu, B., Lee, S. H., Kim, M. M., and Kim, S. K. 2009. Chemical components and its antioxidant properties in vitro: An edible marine brown alga, *Ecklonia cava. Bioorganic & Medicinal Chemistry*, 17(5), pp. 1963–1973.

Li, Y. X., and Kim, S. K. 2011. Utilization of seaweed derived ingredients as potential antioxidants and functional ingredients in the food industry: An overview. *Food Science and Biotechnology*, 20(6), pp. 1461–1466.

Li, Y. X., Wijesekara, I., Li, Y., and Kim, S. K. 2011. Phlorotannins as bioactive agents from brown algae. *Process Biochemistry*, 46(12), pp. 2219–2224.

Liaua, B. C., Shena, C. T., and Lianga, F. P. 2010. Supercritical fluids extraction and antisolvent purification of carotenoids from microalgae and associated bioactivity. *Journal of Supercritical Fluids*, 55, pp. 169–175.

Lowe, B., Venkatesan, J., Anil, S., Shim, M. S., and Kim, S. K. 2016. Preparation and characterization of chitosan-natural nano hydroxyapatite-fucoidan nanocomposites for bone tissue engineering. *International Journal of Biological Macromolecules*, 93, pp. 1479–1487.

Luyt, C. E., Meddahi-Pellé, A., Ho-Tin-Noe, B., Colliec-Jouault, S., Guezennec, J., Louedec, L., Prats, H., Jacob, M. P., Osborne-Pellegrin, M., Letourneur, D., and Michel, J. B. 2003. Low-molecular-weight fucoidan promotes therapeutic revascularization in a rat model of critical hindlimb ischemia. *Journal of Pharmacology and Experimental Therapeutics*, 305(1), pp. 24–30.

MacKinnon, S. L., Hiltz, D., Ugarte, R., and Craft, C. A. 2010. Improved methods of analysis for betaines in *Ascophyllum nodosum* and its commercial seaweed extracts. *Journal of Applied Phycology*, 22(4), pp. 489–494.

Magalhaes, K. D., Costa, L. S., Fidelis, G. P., Oliveira, R. M., Nobre, L. T. D. B., Dantas-Santos, N., Camara, R. B. G., Albuquerque, I. R. L., Cordeiro, S. L., Sabry, D. A., and Costa, M. S. S. P. 2011. Anticoagulant, antioxidant and antitumor activities of heterofucans from the seaweed *Dictyopteris delicatula. International Journal of Molecular Sciences*, 12(5), pp. 3352–3365.

Manilal, A., Sujith, S., Kiran, G. S., Selvin, J., Shakir, C., Gandhimathi, R., and Panikkar, M. V. N. 2009. Biopotentials of seaweeds collected from southwest coast of India. *Journal of Marine Science and Technology*, 17(1), pp. 67–73.

Maruyama, H., and Yamamoto, I. 1984. An antitumor fucoidan fraction from an edible brown seaweed, *Laminaria religiosa*. In Eleventh International Seaweed Symposium (pp. 534–536). Springer, Dordrecht, the Netherlands.

Maruyama, H., Tamauchi, H., Hashimoto, M., and Nakano, T. 2003. Antitumor activity and immune response of Mekabu fucoidan extracted from Sporophyll of *Undaria pinnatifida. In Vivo*, 17(3), pp. 245–249.

Maruyama, H., Tamauchi, H., Hashimoto, M., and Nakano, T. 2005. Suppression of Th2 immune responses by mekabu fucoidan from *Undaria pinnatifida* sporophylls. *International Archives of Allergy and Immunology*, 137(4), pp. 289–294.

Matou, S., Helley, D., Chabut, D., Bros, A., and Fischer, A. M. 2002. Effect of fucoidan on fibroblast growth factor-2-induced angiogenesis in vitro. *Thrombosis Research*, 106(4–5), pp. 213–221.

Medeiros, V. P., Queiroz, K. C. S., Cardoso, M. L., Monteiro, G. R. G., Oliveira, F. W., Chavante, S. F., Guimaraes, L. A., Rocha, H. A. O., and Leite, E. L. 2008. Sulfated galactofucan from *Lobophora variegata*: Anticoagulant and anti-inflammatory properties. *Biochemistry* (Moscow), 73(9), pp. 1018–1024.

Menshova, R. V., Shevchenko, N. M., Imbs, T. I., Zvyagintseva, T. N., Malyarenko, O. S., Zaporoshets, T. S., Besednova, N. N. and Ermakova, S. P. 2016. Fucoidans from brown alga *Fucus evanescens*: Structure and biological activity. *Frontiers in Marine Science*, 3, p. 129.

Miao, H. Q., Elkin, M., Aingorn, E., Ishai-Michaeli, R., Stein, C. A., and Vlodavsky, I. 1999. Inhibition of heparanase activity and tumor metastasis by laminarin sulfate and synthetic phosphorothioate oligodeoxynucleotides. *International Journal of Cancer*, 83(3), pp. 424–431.

Mihova, S. G., Georgiev, D. I., Minkova, K. M., and Tchernov, A. A. 1996. Phycobiliproteins in *Rhodella reticulata* and photoregulatory effects on their content. *Journal of Biotechnology*, 48(3), pp. 251–257.

Mohamed, S. F., and Agili, F. A. 2013. Antiviral sulphated polysaccharide from brown algae *Padina pavonia* characterization and structure elucidation. *International Journal of ChemTech Research*, 5(4), pp. 1469–1476.

Moreau, D., Tomasoni, C., Jacquot, C., Kaas, R., Le Guedes, R., Cadoret, J. P., Muller-Feuga, A., Kontiza, I., Vagias, C., Roussis, V., and Roussakis, C. 2006. Cultivated microalgae and the carotenoid fucoxanthin from *Odontella aurita* as potent anti-proliferative agents in bronchopulmonary and epithelial cell lines. *Environmental Toxicology and Pharmacology*, 22(1), pp. 97–103.

Mourão, P. A., and Pereira, M. S. 1999. Searching for alternatives to heparin: Sulfated fucans from marine invertebrates. *Trends in Cardiovascular Medicine*, 9(8), pp. 225–232.

Mourão, P. A. 2004. Use of sulfated fucans as anticoagulant and antithrombotic agents: Future perspectives. *Current Pharmaceutical Design*, 10(9), pp. 967–981.

Nagaoka, M., Shibata, H., Kimura-Takagi, I., Hashimoto, S., Kimura, K., Makino, T., Aiyama, R., Ueyama, S., and Yokokura, T. 1999. Structural study of fucoidan from *Cladosiphon okamuranus* Tokida. *Glycoconjugate Journal*, 16(1), pp. 19–26.

Nakamura, T., Suzuki, H., Wada, Y., Kodama, T., and Doi, T. 2006. Fucoidan induces nitric oxide production via p38 mitogen-activated protein kinase and NF-κB-dependent signaling pathways through macrophage scavenger receptors. *Biochemical and Biophysical Research Communications*, 343(1), pp. 286–294.

Namvar, F., Mohamad, R., Baharara, J., Zafar-Balanejad, S., Fargahi, F., and Rahman, H. S. 2013. Antioxidant, antiproliferative, and antiangiogenesis effects of polyphenol-rich seaweed (*Sargassum muticum*). *BioMed Research International*, 2013, pp. 604787.

Namvar, F., Mohamed, S., Fard, S. G., Behravan, J., Mustapha, N. M., Alitheen, N. B. M., and Othman, F. 2012. Polyphenolrich seaweed (Eucheuma cottonii) extract suppresses breast tumour via hormone modulation and apoptosis induction. *Food Chemistry*, 130(2), pp. 376–382.

Nishino, T., Kiyohara, H., Yamada, H., and Nagumo, T. 1991. An anticoagulant fucoidan from the brown seaweed Ecklonia kurome. *Phytochemistry*, 30(2), pp. 535–539.

Nishino, T., Yokoyama, G., Dobashi, K., Fujihara, M., and Nagumo, T. 1989. Isolation, purification, and characterization of fucose-containing sulfated polysaccharides from the brown seaweed Ecklonia kurome and their blood-anticoagulant activities. *Carbohydrate Research*, 186(1), pp. 119–129.

Nwosu, F., Morris, J., Lund, V. A., Stewart, D., Ross, H. A., and McDougall, G. J. 2011. Anti-proliferative and potential anti-diabetic effects of phenolic-rich extracts from edible marine algae. *Food Chemistry*, 126(3), pp. 1006–1012.

O'Doherty, J. V., McDonnell, P., and Figat, S. 2011. The effect of dietary laminarin and fucoidan in the diet of the weanling piglet on performance and selected faecal microbial populations. *Livestock Science*, 134, pp. 208–210.

O'Sullivan, A. M., O'Callaghan, Y. C., O'Grady, M. N., Queguineur, B., Hanniffy, D., Troy, D. J., Kerry, J. P., and O'Brien, N. M. 2011. In vitro and cellular antioxidant activities of seaweed extracts prepared from five brown seaweeds harvested in spring from the west coast of Ireland. *Food Chemistry*, 126(3), pp. 1064–1070.

Okada, Y., Ishimaru, A., Suzuki, R., and Okuyama, T. 2004. A new phloroglucinol derivative from the brown alga *Eisenia bicyclis*: Potential for the effective treatment of diabetic complications. *Journal of Natural Products*, 67(1), pp. 103–105.

Paiva, A. A., Castro, A. J., Nascimento, M. S., Will, L. S., Santos, N. D., Araújo, R. M., Xavier, C. A., Rocha, F. A., and Leite, E. L. 2011. Antioxidant and anti-inflammatory effect of polysaccharides from *Lobophora variegata* on zymosan-induced arthritis in rats. *International Immunopharmacology*, 11, pp. 1241–1250.

Palanisamy, S., Vinosha, M., Marudhupandi, T., Rajasekar, P., and Prabhu, N. M. 2017. Isolation of fucoidan from *Sargassum polycystum* brown algae: Structural characterization, in vitro antioxidant and anticancer activity. *International Journal of Biological Macromolecules*, 102, pp. 405–412.

Pangestuti, R., and Kim, S. K. 2011. Biological activities and health benefit effects of natural pigments derived from marine algae. *Journal of Functional Foods*, 3(4), pp. 255–266.

Park, E., Ahn, G. N., Lee, N. H., Kim, J. M., Yun, J. S., Hyun, J. W., Jeon, Y. J., Wie, M. B., Lee, Y. J., Park, J. W., and Jee, Y. 2008. Radioprotective properties of eckol against ionizing radiation in mice. *FEBS Letters*, 582(6), pp. 925–930.

Park, H. S., Hwang, H. J., Kim, G. Y., Cha, H. J., Kim, W. J., Kim, N. D., Yoo, Y. H., and Choi, Y. H. 2013. Induction of apoptosis by fucoidan in human leukemia U937 cells through activation of p38 MAPK and modulation of Bcl-2 family. *Marine Drugs*, 11(7), pp. 2347–2364.

Park, H. S., Kim, G. Y., Nam, T. J., Deuk Kim, N., and Hyun Choi, Y. 2011. Antiproliferative activity of fucoidan was associated with the induction of apoptosis and autophagy in AGS human gastric cancer cells. *Journal of Food Science*, 76(3), pp. T77–T83.

Park, H. Y., Han, M. H., Park, C., Jin, C. Y., Kim, G. Y., Choi, I. W., Kim, N. D., Nam, T. J., Kwon, T. K., and Choi, Y. H. 2011. Anti-inflammatory effects of fucoidan through inhibition of NF-κB, MAPK and Akt activation in lipopolysaccharide-induced BV2 microglia cells. *Food and Chemical Toxicology*, 49(8), pp. 1745–1752.

Park, S. B., Chun, K. R., Kim, J. K., Suk, K., Jung, Y. M., and Lee, W. H. 2010. The differential effect of high and low molecular weight fucoidans on the severity of collagen-induced arthritis in mice. *Phytotherapy Research*, 24(9), pp. 1384–1391.

Park, S. J., Lee, K. W., Lim, D. S., and Lee, S. 2012. The sulfated polysaccharide fucoidan stimulates osteogenic differentiation of human adipose-derived stem cells. *Stem Cells and Development*, 21(12), pp. 2204–2211.

Percival, E. 1968. Glucuronoxylofucan, a cell-wall component of *Ascophyllum nodosum*. Part I. *Carbohydrate Research*, 7(3), pp. 272–283.

Pereira, H., Leão-Ferreira, L. R., Moussatché, N., Teixeira, V. L., Cavalcanti, D. N., Costa, L. J., Diaz, R., and Frugulhetti, I. C. P. P. 2004. Antiviral activity of diterpenes isolated from the Brazilian marine alga *Dictyota menstrualis* against human immunodeficiency virus type 1 (HIV-1). *Antiviral Research*, 64(1), pp. 69–76.

Pereira, M. S., Mulloy, B., and Mourão, P. A. 1999. Structure and anticoagulant activity of sulfated fucans comparison between the regular, repetitive, and linear fucans from echinoderms with the more heterogeneous and branched polymers from brown algae. *Journal of Biological Chemistry*, 274(12), pp. 7656–7667.

Ponce, N. M., Pujol, C. A., Damonte, E. B., Flores, M. L., and Stortz, C. A. 2003. Fucoidans from the brown seaweed *Adenocystis utricularis*: Extraction methods, antiviral activity and structural studies. *Carbohydrate Research*, 338(2), pp. 153–165.

Prabu, D. L., Sahu, N. P., Pal, A. K., Dasgupta, S., and Narendra, A. 2016. Immunomodulation and interferon gamma gene expression in sutchi cat fish, *Pangasianodon hypophthalmus*: Effect of dietary fucoidan rich seaweed extract (FRSE) on pre and post challenge period. *Aquaculture Research*, 47(1), pp. 199–218.

Preobrazhenskaya, M. E., Berman, A. E., Mikhailov, V. I., Ushakova, N. A., Mazurov, A. V., Semenov, A. V., Usov, A. I., Nifant'Ev, N. E., and Bovin, N. V. 1997. Fucoidan inhibits leukocyte recruitment in a model peritonial inflammation in rat and blocks interaction of P-selectin with its carbohydrate ligand. *IUBMB Life*, 43(2), pp. 443–451.

Pulz, O., and Gross, W. 2004. Valuable products from biotechnology of microalgae. *Applied Microbiology and Biotechnology*, 65(6), pp. 635–648.

Queiroz, K. C. S., Medeiros, V. P., Queiroz, L. S., Abreu, L. R. D., Rocha, H. A. O., Ferreira, C. V., Jucá, M. B., Aoyama, H., and Leite, E. L. 2008. Inhibition of reverse transcriptase activity of HIV by polysaccharides of brown algae. *Biomedicine & Pharmacotherapy*, 62, pp. 303–307.

Raghavendran, H. R. B., Sathivel, A., and Devaki, T. 2005. Antioxidant effect of Sargassum polycystum (Phaeophyceae) against acetaminophen induced changes in hepatic mitochondrial enzymes during toxic hepatitis. *Chemosphere*, 61(2), pp. 276–281.

Rani, V., Jawahar, P., Shakila, R. J., and Srinivasan, A. 2017. Antibacterial activity of some brown seaweeds of gulf of Mannar, south east coast of India. *Journal of Pharmaceutical and BioSciences*, 4, pp. 14–21.

Rathore, S. S., Chaudhary, D. R., Boricha, G. N., Ghosh, A., Bhatt, B. P., Zodape, S. T., and Patolia, J. S. 2009. Effect of seaweed extract on the growth, yield and nutrient uptake of soybean (Glycine max) under rainfed conditions. *South African Journal of Botany*, 75(2), pp. 351–355.

Ren, D., Noda, H., Amano, H., and Nisizawa, K. 1994. Antihypertensive and antihyperlipidemic effects of funoran. *Fisheries Science*, 60(4), pp. 423–427.

Rioux, L.-E., and Turgeon, S. L. 2015. Seaweed carbohydrates. In: Tiwari, B.K., Troy, D. (Eds.), *Seaweed Sustainability*. Elsevier, San Diego, CA, pp. 141–192.

Rioux, L. E., Turgeon, S. L., and Beaulieu, M. 2007. Characterization of polysaccharides extracted from brown seaweeds. *Carbohydrate Polymers*, 69(3), pp. 530–537.

Rioux, L. E., Turgeon, S. L., and Beaulieu, M. 2010. Structural characterization of laminaran and galactofucan extracted from the brown seaweed *Saccharina longicruris*. *Phytochemistry*, 71(13), pp. 1586–1595.

Rocha, H. A., Moraes, F. A., Trindade, E. S., Franco, C. R., Torquato, R. J., Veiga, S. S., Valente, A. P., Mourão, P. A., Leite, E. L., Nader, H. B., and Dietrich, C. P. 2005. Structural and hemostatic activities of a sulfated galactofucan from the brown alga *Spatoglossum schroederi* an ideal antithrombotic agent? *Journal of Biological Chemistry*, 280(50), pp. 41278–41288.

Roy, M. C., Anguenot, R., Fillion, C., Beaulieu, M., Bérubé, J., and Richard, D. 2011. Effect of a commercially-available algal phlorotannins extract on digestive enzymes and carbohydrate absorption in vivo. *Food Research International*, 44(9), pp. 3026–3029.

Ryu, B., Li, Y., Qian, Z. J., Kim, M. M., and Kim, S. K. 2008. October. Exhibitory effects of compounds from brown alga *Ecklonia cava* on the human osteoblasts. In *13th International Biotechnology Symposium and Exhibition* (第 13 届 IUPAC 国际生物工程会议) (pp. 588). 大连理工大学.

Ryu, B., Qian, Z. J., Kim, M. M., Nam, K. W., and Kim, S. K. 2009. Anti-photoaging activity and inhibition of matrix metalloproteinase (MMP) by marine red alga, *Corallina pilulifera* methanol extract. *Radiation Physics and Chemistry*, 78(2), pp. 98–105.

Shen, Y. R., and Kuo, M. I. 2017. Effects of different carrageenan types on the rheological and water-holding properties of tofu. *LWT-Food Science and Technology*, 78, pp. 122–128.

Shibata, H., Iimuro, M., Uchiya, N., Kawamori, T., Nagaoka, M., Ueyama, S., Hashimoto, S., Yokokura, T., Sugimura, T., and Wakabayashi, K. 2003. Preventive effects of *Cladosiphon* fucoidan against Helicobacter pylori infection in *Mongolian gerbils*. *Helicobacter*, 8(1), pp. 59–65.

Shibata, H., Kimura-Takagi, I., Nagaoka, M., Hashimoto, S., Aiyama, R., Iha, M., Ueyama, S., and Yokokura, T. 2000. Properties of fucoidan from *Cladosiphon okamuranus* tokida in gastric mucosal protection. *Biofactors*, 11(4), pp. 235–245.

Shibata, T., Ishimaru, K., Kawaguchi, S., Yoshikawa, H., and Hama, Y. 2008. Antioxidant activities of phlorotannins isolated from Japanese Laminariaceae. *Journal of Applied Phycology* , 20, pp. 705–711.

Shimizu, J., Wada-Funada, U., Mano, H., Matahira, Y., Kawaguchi, M., and Wada, M. 2005. Proportion of murine cytotoxic T cells is increased by high molecular-weight fucoidan extracted from Okinawa mozuku (*Cladosiphon okamuranus*). *Journal of Health Science*, 51(3), pp. 394–397.

Shui-Chun, M. A. O., and Yue-Wei, G. U. O. 2010. Sesquiterpenes from Chinese red alga *Laurencia okamurai*. *Chinese Journal of Natural Medicines*, 8(5), pp. 321–325.

Silva, T. M. A., Alves, L. G., De Queiroz, K. C. S., Santos, M. G. L., Marques, C. T., Chavante, S. F., Rocha, H. A. O., and Leite, E. L. 2005. Partial characterization and anticoagulant activity of a heterofucan from the brown seaweed Padina gymnospora. *Brazilian Journal of Medical and Biological Research*, 38(4), pp. 523–533.

Smit, A. J. 2004. Medicinal and pharmaceutical uses of seaweed natural products: A review. *Journal of Applied Phycology*, 16(4), pp. 245–262.

Smyrniotopoulos, V., Merten, C., Kaiser, M., and Tasdemir, D. 2017. Bifurcatriol, a new antiprotozoal acyclic diterpene from the brown alga *Bifurcaria bifurcata*. *Marine Drugs*, 15(8), p. 245.

Sokolova, R. V., Ermakova, S. P., Awada, S. M., Zvyagintseva, T. N., and Kanaan, H. M. 2011. Composition, structural characteristics, and antitumor properties of polysaccharides from the brown algae *Dictyopteris polypodioides* and Sargassum sp. *Chemistry of Natural Compounds*, 47(3), p. 329.

Suganthy, N., Pandian, S. K., and Devi, K. P. 2010. Neuroprotective effect of seaweeds inhabiting South Indian coastal area (Hare Island, Gulf of Mannar Marine Biosphere Reserve): Cholinesterase inhibitory effect of *Hypnea valentiae* and *Ulva reticulata*. *Neuroscience letters*, 468(3), pp. 216–219.

Sugiura, Y., Matsuda, K., Yamada, Y., Nishikawa, M., Shioya, K., Katsuzaki, H., Imai, K., and Amano, H. 2006. Isolation of a new anti-allergic phlorotannin, phlorofucofuroeckol-B, from an edible brown alga, *Eisenia arborea*. *Bioscience, Biotechnology, and Biochemistry*, 70(11), pp. 2807–2811.

Sugiura, Y., Matsuda, K., Yamada, Y., Nishikawa, M., Shioya, K., Katsuzaki, H., Imai, K., and Amano, H. 2007. Anti-allergic phlorotannins from the edible brown alga, *Eisenia arborea*. *Food Science and Technology Research*, 13(1), pp. 54–60.

Swanson, A. K., and Druehl, L. D. 2002. Induction, exudation and the UV protective role of kelp phlorotannins. *Aquatic Botany*, 73(3), pp. 241–253.

Synytsya, A., Kim, W. J., Kim, S. M., Pohl, R., Synytsya, A., Kvasnička, F., Čopíková, J., and Park, Y. I. 2010. Structure and antitumour activity of fucoidan isolated from sporophyll of Korean brown seaweed *Undaria pinnatifida*. *Carbohydrate Polymers*, 81(1), pp. 41–48.

Takahashi, M. 1983. Studies on the mechanism of host mediated antitumor action of crude fucoidan from a brown marine alga *Eisenia bicyclis*. *Journal-Japanese Society for Horticultural Science*, 22, pp. 269–283.

Takahashi, T., Kuniyasu, Y., Toda, M., Sakaguchi, N., Itoh, M., Iwata, M., Shimizu, J., and Sakaguchi, S. 1998. Immunologic self-tolerance maintained by CD25+ CD4+ naturally anergic and suppressive T cells: Induction of autoimmune disease by breaking their anergic/suppressive state. *International Immunology*, 10(12), pp. 1969–1980.

Tarakhovskaya, E. R., Maslov, Y. I., and Shishova, M. F. 2007. Phytohormones in algae. *Russian Journal of Plant Physiology*, 54(2), pp. 163–170.

Teruya, T., Konishi, T., Uechi, S., Tamaki, H., and Tako, M. 2007. Anti-proliferative activity of oversulfated fucoidan from commercially cultured *Cladosiphon okamuranus* TOKIDA in U937 cells. *International Journal of Biological Macromolecules*, 41(3), pp. 221–226.

Thomes, P., Rajendran, M., Pasanban, B., and Rengasamy, R. 2010. Cardioprotective activity of *Cladosiphon okamuranus* fucoidan against isoproterenol induced myocardial infarction in rats. *Phytomedicine*, 18(1), pp. 52–57.

Thompson, K. D., and Dragar, C. 2004. Antiviral activity of *Undaria pinnatifida* against herpes simplex virus. *Phytotherapy Research: An International Journal Devoted to Pharmacological and Toxicological Evaluation of Natural Product Derivatives*, 18(7), pp. 551–555.

Thuy, T. T. T., Ly, B. M., Van, T. T. T., Van Quang, N., Tu, H. C., Zheng, Y., Seguin-Devaux, C., Mi, B., and Ai, U. 2015. Anti-HIV activity of fucoidans from three brown seaweed species. *Carbohydrate Polymers*, 115, pp. 122–128.

Trinchero, J., Ponce, N. M., Córdoba, O. L., Flores, M. L., Pampuro, S., Stortz, C. A., Salomón, H., and Turk, G. 2009. Antiretroviral activity of fucoidans extracted from the brown seaweed *Adenocystis utricularis*. *Phytotherapy Research: An International Journal Devoted to Pharmacological and Toxicological Evaluation of Natural Product Derivatives*, 23(5), pp. 707–712.

Ushakova, N. A., Morozevich, G. E., Ustyuzhanina, N. E., Bilan, M. I., Usov, A. I., Nifantiev, N. E., and Preobrazhenskaya, M. E. 2009. Anticoagulant activity of fucoidans from brown algae. *Biochemistry (Moscow) Supplement Series B: Biomedical Chemistry*, 3(1), pp. 77–83.

Usoltseva, R. V., Anastyuk, S. D., Ishina, I. A., Isakov, V. V., Zvyagintseva, T. N., Thinh, P. D., Zadorozhny, P. A., Dmitrenok, P. S., and Ermakova, S. P. 2018. Structural characteristics and anticancer activity in vitro of fucoidan from brown alga Padina boryana. *Carbohydrate Polymers*, 184, pp. 260–268.

Usoltseva, R. V., Anastyuk, S. D., Shevchenko, N. M., Surits, V. V., Silchenko, A. S., Isakov, V. V., Zvyagintseva, T. N., Thinh, P. D., and Ermakova, S.P. 2017. Polysaccharides from brown algae *Sargassum duplicatum*: The structure and anticancer activity in vitro. *Carbohydrate Polymers*, 175, pp. 547–556.

Usoltseva, R. V., Anastyuk, S. D., Shevchenko, N. M., Zvyagintseva, T. N., and Ermakova, S. P. 2016. The comparison of structure and anticancer activity in vitro of polysaccharides from brown algae *Alaria marginata* and *A. angusta*. *Carbohydrate Polymers*, 153, pp. 258–265.

Usov, A. I., Smirnova, G. P., Bilan, M. I., and Shashkov, A.S. 1998. Polysaccharides of algae. 53. brown alga *Laminaria saccharina* (l.) lam. asa source of fucoidan. *Russian Journal of Bioorganic Chemistry*, 24, pp. 437–445.

Ustyuzhanina, N. E., Bilan, M. I., Gerbst, A. G., Ushakova, N. A., Tsvetkova, E. A., Dmitrenok, A. S., Usov, A. I., and Nifantiev, N. E. 2016. Anticoagulant and antithrombotic activities of modified xylofucan sulfate from the brown alga *Punctaria plantaginea*. *Carbohydrate Polymers*, 136, pp. 826–833.

Usui, T., Asari, K., and Mizuno, T. 1980. Isolation of highly purified "fucoidan" from *Eisenia bicyclis* and its anticoagulant and antitumor activities. *Agricultural and Biological Chemistry*, 44(8), pp. 1965–1966.

Venkatesan, J., Bhatnagar, I., Manivasagan, P., Kang, K. H., and Kim, S. K. 2015a. Alginate composites for bone tissue engineering: A review. *International Journal of Biological Macromolecules*, 72, pp. 269–281.

Venkatesan, J., Lowe, B., Anil, S., Manivasagan, P., Kheraif, A. A. A., Kang, K. H., and Kim, S. K. 2015b. Seaweed polysaccharides and their potential biomedical applications. *Starch-Stärke*, 67(5–6), pp. 381–390.

Venkateswaran, P. S., Millman, I., and Blumberg, B. S. 1989. Interaction of fucoidan from *Pelvetia fastigiata* with surface antigens of hepatitis B and woodchuck hepatitis viruses. *Planta Medica*, 55(03), pp. 265–270.

Vidanarachchi, J. K., Iji, P. A., Mikkelsen, L. L., Sims, I., and Choct, M. 2009. Isolation and characterization of water-soluble prebiotic compounds from Australian and New Zealand plants. *Carbohydrate Polymers*, 77(3), pp. 670–676.

Vischuk, O. S., Ermakova, S. P., Tin, F. D., Shevchenko, N. M., Li, B. M., and Zvyagintseva, T.N. 2009. Anticancer activity of brown algae-derived fucoidans. *Pacific Medical Journal*, 3, pp. 86–89.

Vishchuk, O. S., Ermakova, S. P., and Zvyagintseva, T. N. 2011. Sulfated polysaccharides from brown seaweeds Saccharina japonica and Undaria pinnatifida: Isolation, structural characteristics, and antitumor activity. *Carbohydrate Research*, 346(17), pp. 2769–2776.

Vo, T. S., Ngo, D. H., and Kim, S. K., 2012. Marine algae as a potential pharmaceutical source for anti-allergic therapeutics. *Process Biochemistry*, 47(3), pp. 386–394.

Wijesekara, I., and Karunarathna, W. K. D. S. 2017. Chapter 18 –Usage of seaweed polysaccharides as nutraceuticals. In *Seaweed Polysaccharides* (pp. 341–348). Elsevier, Saint Louis.

Wijesekara, I., Pangestuti, R., and Kim, S. K. 2011. Biological activities and potential health benefits of sulfated polysaccharides derived from marine algae. *Carbohydrate Polymers*, 84(1), pp. 14–21.

Wijesinghe, W. A. J. P., and Jeon, Y. J. 2012. Biological activities and potential industrial applications of fucose rich sulfated polysaccharides and fucoidans isolated from brown seaweeds: A review. *Carbohydrate Polymers*, 88(1), pp. 13–20.

Wijesinghe, W. A. J. P., and Jeon, Y. J. 2012. Enzyme-assistant extraction (EAE) of bioactive components: A useful approach for recovery of industrially important metabolites from seaweeds: A review. *Fitoterapia*, 83(1), pp. 6–12.

Wozniak, M., Bell, T., Dénes, Á., Falshaw, R., and Itzhaki, R. 2015. Anti-HSV1 activity of brown algal polysaccharides and possible relevance to the treatment of Alzheimer's disease. *International Journal of Biological Macromolecules*, 74, pp. 530–540.

Wu, X., Song, M., Rakariyatham, K., Zheng, J., Guo, S., Tang, Z., Zhou, S., and Xiao, H. 2015. Anti-inflammatory effects of 4′-demethylnobiletin, a major metabolite of nobiletin. *Journal of Functional Foods*, 19, pp. 278–287.

Xiaolin, Y., Juyun, S., Hannian, X., Xiaohui, L., Yishen, Z., and Shaolun, Z. 1995. An experimental study on immunoregulatory effect of fucoidan [j]. *Chinese Journal of Marine Drugs*, 3, p. 002.

Xue, C., Chen, L. E. I., Li, Z., Cai, Y., Lin, H., and Fang, Y. 2004. Antioxidative activities of low molecular fucoidans from kelp *Laminaria japonica*. In *Developments in Food Science*, 42, pp. 139–145.

Xue, M., Ge, Y., Zhang, J., Wang, Q., Hou, L., Liu, Y., Sun, L., and Li, Q. 2012. Anticancer properties and mechanisms of fucoidan on mouse breast cancer in vitro and in vivo. *PLoS One*, 7(8), p. e43483.

Yamamoto, I., Nagumo, T., Takahashi, M., Fujihara, M., Suzuki, Y., and Iizima, N. 1981. Antitumor effect of seaweeds. III. Antitumor effect of an extract from *Sargassum kjellmanianum*. *The Japanese Journal of Experimental Medicine*, 51(3), pp. 187–189.

Yamamoto, I., Takahashi, M., Tamura, E., Maruyama, H., and Mori, H. 1984. Antitumor activity of edible marine algae: Effect of crude fucoidan fractions prepared from edible brown seaweeds against L-1210 leukemia. *Hydrobiologia*, 116, pp. 145–148.

Yang, C., Chung, D., Shin, I. S., Lee, H., Kim, J., Lee, Y., and You, S. 2008. Effects of molecular weight and hydrolysis conditions on anticancer activity of fucoidans from sporophyll of *Undaria pinnatifida*. *International Journal of Biological Macromolecules*, 43(5), pp. 433–437.

Yang, J. W., Yoon, S. Y., Oh, S. J., Kim, S. K., and Kang, K. W. 2006. Bifunctional effects of fucoidan on the expression of inducible nitric oxide synthase. *Biochemical and Biophysical Research Communications*, 346(1), pp. 345–350.

Yang, L., and Zhang, L.M. 2009. Chemical structural and chain conformational characterization of some bioactive polysaccharides isolated from natural sources. *Carbohydrate Polymers*, 76(3), pp. 349–361.

Yoo, Y. C., Kim, W. J., Kim, S. Y., Kim, S. M., Chung, M. K., Park, J. W., Suh, H. H., Lee, K. B., and Park, Y. I. 2007. Immunomodulating activity of a fucoidan isolated from Korean *Undaria pinnatifida* sporophyll. *Algae*, 22(4), pp. 333–338.

Yoon, S. J., Pyun, Y. R., Hwang, J. K., and Mourão, P. A. 2007. A sulfated fucan from the brown alga *Laminaria cichorioides* has mainly heparin cofactor II-dependent anticoagulant activity. *Carbohydrate Research*, 342(15), pp. 2326–2330.

You, S., Yang, C., Lee, H., and Lee, B. Y. 2010. Molecular characteristics of partially hydrolyzed fucoidans from sporophyll of *Undaria Pinnatifida* and their in vitro anticancer activity. *Food Chemistry*, 119(2), pp. 554–559.

Yuan, Y. V., and Walsh, N. A. 2006. Antioxidant and antiproliferative activities of extracts from a variety of edible seaweeds. *Food and Chemical Toxicology*, 44(7), pp. 1144–1150.

Yvin, J. C., Levasseur, F., and Hud'Homme, F., Laboratoires Goemar SA. 1999. Use of laminarin and oligosaccharides derived therefrom in cosmetics and for preparing a skin treatment drug. U.S. Patent 5,980,916.

Zha, X. Q., Xiao, J. J., Zhang, H. N., Wang, J. H., Pan, L. H., Yang, X. F., and Luo, J. P. 2012. Polysaccharides in Laminaria japonica (LP): Extraction, physicochemical properties and their hypolipidemic activities in diet-induced mouse model of atherosclerosis. *Food Chemistry*, 134(1), pp. 244–252.

Zhang, X., Wang, K., and Ervin, E. H. 2010. Optimizing dosages of seaweed extract-based cytokinins and zeatin riboside for improving creeping bentgrass heat tolerance. *Crop Science*, 50(1), pp. 316–320.

Zhang, Z., Teruya, K., Eto, H., and Shirahata, S. 2011. Fucoidan extract induces apoptosis in MCF-7 cells via a mechanism involving the ROS-dependent JNK activation and mitochondria-mediated pathways. *PloS one*, 6(11), p. e27441.

Zhao, X., Xue, C. H., Li, B. F. 2008. Study of antioxidant activities of sulfated polysaccharides from *Laminaria japonica*. *Journal of Applied Phycology*, 20, pp. 431–436.

Zhu, W., Ooi, V. E. C., Chan, P. K. S., and Ang Jr, P. O. 2003. Inhibitory effect of extracts of marine algae from Hong Kong against Herpes simplex viruses. In Chapman ARO, Anderson RJ, Vreeland VJ, Davison IR (eds), In *Proceedings of the 17th International Seaweed Symposium*, Oxford University Press, Oxford (pp. 159–164).

25 Fisher Folks Usage of Medicinal Plants against Gynecological Disorders

Vijayakumar Dhaarani and Arumugam Rajendran

CONTENTS

25.1 INTRODUCTION

The reproductive problems such as syphilis, gonorrhea, chlamydia, and trichomoniasis occur annually throughout the world in adults aged 15–49 years (WHO 1999). Infections of the reproductive tract, complications after childbirth, and reproductive problems continue to be major health challenges worldwide. An impressive number of plant species is traditionally used to remedy such afflictions, and some have been investigated for their efficacy with positive results. In many Asian countries, phytomedicine is widely used even though Western medicine is often readily available (Bussmann and Glenn 2010).

In developing countries, traditional medicines are still the main sources of the health care system, and it has been estimated that 80% of the population in such countries are dependent on traditional medicines (WHO 2002; Mahmood et al. 2013). According to Schippman et al. (2002), more than 50,000 flowering plants have been used for medicinal purposes throughout the world. Ethnomedicinal studies are significant for the discovery of new crude drugs from reported indigenous medicinal plants. Right from the beginning, ethnobotany with special emphasis on the documentation of traditional medicinal knowledge of plants, has directed a number of key modern drugs (Flaster; Balick and Cox 1996, 2000).

Nowadays, gynecological disorders are often treated with allopathic medicine, surgery, and non-steroidal, anti-inflammatory drugs, though the efficacy of these treatments is great, numerous potential side effects are common (Safeghi and Mahmood 2014). Medicinal approaches or knowledge incorporating plant or any other natural/spiritual therapies used to treat gynecological problems such as abortion, delivery procedures, among others are taken into consideration (Patel 2012; Lawal et al. 2013). Keeping the above point in mind, the current study was undertaken to record the ethno-gynecological indigenous medicinal knowledge from fisher folklore residing in the coastal regions of Tamil Nadu.

25.2 MATERIALS AND METHODS

25.2.1 Study Area

The study area, Nagapattinam, Tamil Nadu is a coastal district situated on the eastern side of Tamil Nadu. The district is surrounded by Karaikal, Cuddalore, and Tiruvarur Districts. It lies between 10.7906°N latitude and 79. 8428°E longitude. It covers about 2715.83 sq. km. The district enjoys a coastal maritime climate; it receives 60% of its total annual rainfall during the Northeast monsoon that begins in October and ends in December. The district has sandy coastal alluvial soil and black soil. The district is rich with its own kind of vegetation, and the forest cover is estimated to about 5037.21 ha, which involves important forests and wildlife areas like point Calimere Wildlife Sanctuary, Karaivetti Bird Sanctuary, and Muthupet Mangrove Forest.

25.2.2 Plant Identification and Data Collection

The study was conducted during 2016–2017. The data were recorded along with the botanical name of the plant, family, vernacular name, plant parts used, and mode of preparation. About 250 informants were interviewed in their local language, and the data were recorded on questionnaires. The plant materials were collected with the help of the informants, and the plants collected were preserved and made into standard herbarium (Jain and Rao 1977). The correct identity of the plants was verified using various local, regional floras and monographs (Balasubramaniam and Ramachandran 1991; Umamaheswari and Daniel 1999; Gamble and Fisher 1957; Mathew, 1983; Ravishankar and Shukla 2007; Hooker

1875; Sambandan et al. 2015). The herbarium specimens were submitted to the herbarium in the Department of Botany, Bharathiar University.

25.3 RESULTS

From the study a total of 50 medicinal plant species belonging to 47 genera and 33 families were identified and documented as herbal remedies for gynecological health care issues in the coastal areas of Tamil Nadu. The results of the present study are presented in Table 25.1. The plants are arranged in alphabetical scientific names. For each species, the correct botanical identity, vernacular name(s), family, part(s) used, mode of administration, and uses/ailments treated are enlisted. The common dosage form includes decoction, paste, powder, and vegetables. Most of the herbal remedies are taken orally. A decoction is prepared by boiling water or the herbs can be taken along with honey, milk, and ghee as adjuvants. Many remedies were prepared as a mixture of multiple ingredients.

TABLE 25.1
Medicinal Plants Used by the Fisher Folks

S. No	Binomial Name	Vernacular Name	Family	Diseases Treated	Parts Used	Mode of Preparation & Administration
1.	*Tephrosia purpurea* (L.) Pers.	Kozhinji	Fabaceae	Painful menses after delivery	Root	Dried root powder with honey administered orally
2.	*Gymnema sylvestre* (Retz.) R.Br. ex Sm.	Sarkaraikolli	Apocyanaceae	Pregnancy sugar	Leaves	Fresh juice administered orally
3.	*Tribulus terrestris* L.	Yanainerunji	Zygophyllaceae	Used to treat sexually transmitted diseases	Root	Fresh juice administered orally
4.	*Punica granatum* L.	Madhulai	Punicaceae	Controls blood loss	Rind of the fruit	Fresh juice administered orally
5.	*Glycyrrhiza glabra* L.					
6.	*Curcuma longa* L.	Manjal	Zingeberaceae	Heals wounds after delivery	Rhizome	Dried rhizome made into paste and applied topically
7.	*Cocos nucifera* L.	Thengaai	Araceae	Heals ulcer	Coconut milk, tender coconut	Coconut milk taken orally
8.	*Ocimum tenuiflorum* L.		Lamiaceae	Relieves cold	Leaves	Decoction taken orally
9.	*Allium sativum* L.	Poondu	Lilliaceae	Increases milk secretion	Bulb	Cooked and taken along with food
10.	*Sesamum indicum* L.	Ellu	Pedaliaceae	Removes wastes from uterus	Seeds	Powdered seeds with jaggery administered orally
11.	*Zingiber officinale* Roscoe	Ingi, sukku	Zingeberaceae	Indigestion	Rhizome	Decoction of rhizome is given orally
12.	*Senna alexandrina* Mill.	Aavarai	Caesalpiniaceae	Pregnancy sugar	Flowers	Fresh flowers can be eaten as such
13.	*Vigna mungo* (L.) Hepper	Ulundhu	Fabaceae	Strengthens backbone	Seeds	Fried and powdered seeds are cooked taken as food
14.	*Ocimum basilicum* L.	Tulasi	Lamiaceae	Treat cold	Leaves	Decoction of the leaves along with honey administered orally
15.	*Chrysopogon zizanioides* (L.) Roberty	Vettiver	Poaceae	Regulates body heat, used to treat vaginal discharge	Root	Decoction of the root administered orally
16.	*Cucumis sativus* L.	Vellari	Cucurbitaceae	Regulates periods	Seed	Powdered seeds are taken orally along with milk
17.	*Benincasa hispida* (Thunb.) Cogn.	Poosani	Cucurbitaceae	Regulates periods	Seed	Seeds are powdered and administered orally along with milk
18.	*Cucurbita maxima* Duchesne	Parangikaai	Cucurbitaceae	Regulates irregular periods	Seeds	Dried seeds along with the seed powder of cucumber and ash guard and drank along with milk

(Continued)

TABLE 25.1 (*Continued*)
Medicinal Plants Used by the Fisher Folks

S. No	Binomial Name	Vernacular Name	Family	Diseases Treated	Parts Used	Mode of Preparation & Administration
19.	*Abrus precatorius* L.	Kundumani	Fabaceae	Used as contraceptive	Seeds	Seeds along with cow milk is administered orally
20.	*Calotropis gigantea* (L.) Dryand.	Erukku	Asclepiadaceae	Uterus contraction	Root	Dried root powder applied topically
21.	*Abutilon indicum* (L.) Sweet	Thuthi	Malvaceae	Regulates blood level and bowel movements	Leaves	Fresh juice administered orally
22.	*Achyranthes aspera* L.	Naayuruvi	Amaranthaceae	Reduced delivery pain	Leaves	Fresh juice given orally
23.	*Citrullus lanatus* (Thunb.) Matsum. & Nakai	Tharpoosani	Cucurbitaceae	Increases water content in the uterus	Fruits	Eaten as such or made into juice and administered orally
24.	*Narthex asafoetida* Falc. ex Lindl.	Perungayam	Apiaceae	Treats cyst in ovaries	Root resin	Pinch of powdered resin along with salt and water administered orally
25.	*Trigonella foenum-graecum* L.	Vendhayam	Fabaceae	Reduces body heat	Seeds	Soaked in water and given orally
26.	*Solanum americanum* Mill.	Manithakkli	Solanaceae	Used to treat vitamin deficiency and ulcer	Leaves, fruits	Cooked and taken orally
27.	*Hemidesmus indicus* (L.) R. Br. ex Schult.	Thelkodukku	Boraginaceae	Used to treat vitamin deficiency	Leaves	Juice along with honey administered orally
28.	*Sesuvium portulacastrum* (L.) L.		Aizoaceae	Used to treat vitamin deficiency	Stem	Cooked and taken as food
29.	*Melia azedarach* L.	Malaivembu	Meliaceae	Used for cleaning uterus, treats infertility	Leaves	Fresh juice given orally
30.	*Hybanthus enneaspermus* (L.) F. Muell.	Oridalthamarai	Violaceae	Treats infertility	Whole plant	Decoction administered orally
31.	*Portulaca oleracea* L.		Chenopodiaceae	Vitamin substitute	Whole plant	Cooked and taken as food
32.	*Talinum portulacifolium* (Forssk.) Asch. ex Schweinf.	Pasalaikeerai	Chenopodiaceae	Eases delivery	Leaves	Make into soup and taken regularly during pregnancy
33.	*Daucus carota* L.	Carrot	Apiaceae	Maintains calcium level	Root	To avoid joint pain during pregnancy, carrot is included in daily diet
34.	*Citrus limon* (L.) Osbeck	Elumichai	Rutaceae	Shrinks fibroids during pregnancy	Fruit	Administered in the form of juice
35.	*Citrus maxima* (Burm.) Merr.	Bumblimass	Rutaceae	Shrinks fibroids	Fruit	Taken as such
36.	*Ananas comosus* (L.) Merr.	Annasi	Bromeliaceae	Gets rid of clogged milk	Fruit	Fresh juice taken orally
37.	*Carum carvi* L.					
38.	*Phyllanthus emblica* L.	Nellikkai	Phyllanthaceae	Increases vitamin C and helps in preventing clogging of milk	Fruit	Fruits are edible
39.	*Terminalia chebula* Retz.	Kadukkai	Combretaceae	Decreases vomiting sensation	Fruit	Dried fruit power mixed with water and administered orally
40.	*Justicia adhatoda* L.	Adathodai	Acanthaceae	For painless delivery	Leaves	Leaves boiled in water and given orally
41.	*Hibiscus rosasinensis*					
42.	*Cyclea peltata* (Lam.) Hook.f. & Thomson	Seendhilkodi	Menispermaceae	Increases lactation	Leaves	Leaves boiled in water and taken orally
43.	*Cynodon dactylon* (L.) Pers.	Arugampul	Poaceae	Cleans uterus and regularizes periods	Leaves	Leaves made into soup and taken orally

(*Continued*)

TABLE 25.1 (*Continued*)
Medicinal Plants Used by the Fisher Folks

S. No	Binomial Name	Vernacular Name	Family	Diseases Treated	Parts Used	Mode of Preparation & Administration
44.	*Psidium guajava* L.	Koyya	Myrtaceae	Used to stop excessive bleeding	Leaves	Decoction of the leaves administered orally
45.	*Momordica charantia* L.	Pagarkaai	Cucurbitaceae	Increases lactation	Fruits	Cooked and taken along with food
46.	*Moringa oleifera* Lam.	Murungai	Moringaceae	Nutritious and increases lactation	Leaves	Used as food
47.	*Phyla nodiflora* (L.) Greene	Poduthalai	Asteraceae	Treats vaginal infection	Flowers	Dried flowers powdered and taken along with ghee
48.	*Madhuca longifolia* (J. Koenig ex L.) J.F.Macbr.	Illuppai	Sapotaceae	Increases lactation	Leaves	Leaf paste applied topically
49.	*Mimosa pudica* L.	Thottalcinungi	Mimosaceae	Treats vaginal infection	Whole plant	Fresh juice taken along with milk
50.	*Musa paradisiaca* L.	Vaazhai	Musaceae	Used to stop excessive bleeding during periods	Stem	Stem juice administered orally

The important medicinal plant species are represented by their usage in the study area and in other regions. Majority of the herbal preparations were prepared from the leaves of the plants (33%), the whole plant (7%), stem (5%), and roots (13%), while other plant parts were used much less frequently. Mostly in all of the medical preparation, fresh plant material was used to prepare the remedies (Figure 25.1).

The mode of the administration of the medicine involves oral administration and sometimes the drug is applied topically. The study reveals that the fisher folks use the oral mode of administration in treating most of the ailments (Figure 25.2).

A list of plants have been shown to be used against menopausal syndrome, including (*Tephrosia purpurea* (L.) Pers.), *Cucumis sativus* L., *Benincasa hispida* (Thunb.) Cogn., *Cucurbita maxima* Duchesne, *Cynodon dactylon* (L.) Pers., *Psidium guajava* L., and *Musa paradisiaca* L. (Figure 25.3).

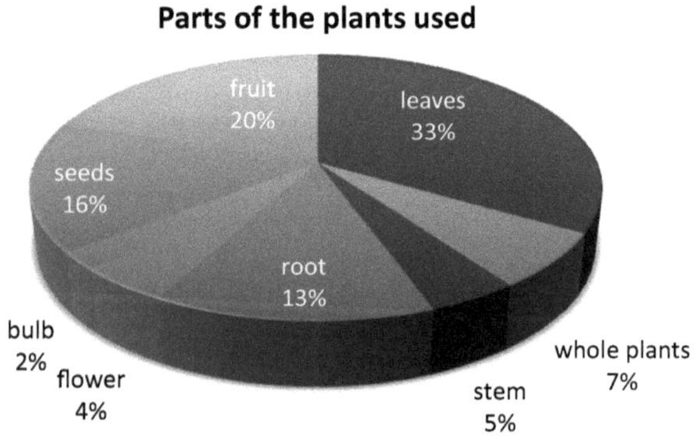

FIGURE 25.1 Showing % of the plant parts used against various ailments.

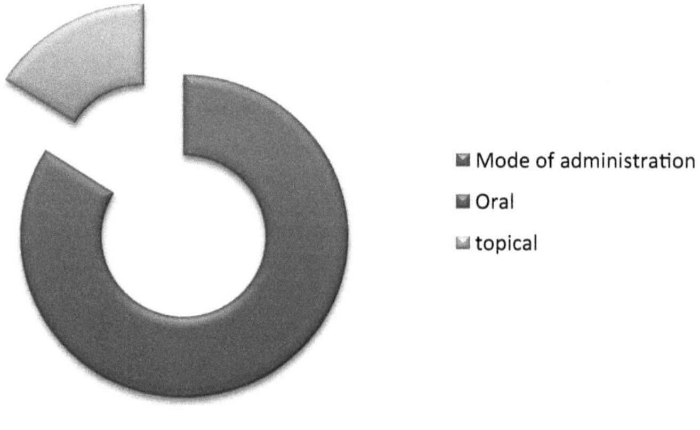

FIGURE 25.2 Showing the mode of administration.

Ocimum spp. are known for their usage in treating colds. Some of the medicinal plants, including *Calotropis gigantea* (L.) Dryand, *Achyranthes aspera* L., *Talinum portulacifolium* (Forssk.) Asch. exSchweinf., *Justicia adhatoda* L., *Allium sativum* L., *Phyllanthus emblica* L., *Momordica charantia* L., and *Madhuca longifolia* (J. Koenig ex L.) J.F.Macbr., were found to be the most effective indigenous remedies for easy delivery and lactation.

Plants like *Gymnema sylvestre* (Retz.) R.Br. *ex* Sm. and *Senna alexandrina* Mill. are effective in treating pregnancy sugar.

Plants like *Trigonella foenum-graecum* L., *Solanum americanum* Mill., and *Hemidesmus indicus* (L.) R. Br. ex Schult. *Sesuvium portulacastrum* (L.) L. are widely used among the fisher folks as a source to treat vitamin deficiencies.

It is noteworthy that a few of the plant species like *Tribulus terrestris* L., *Senna alexandrina* Mill., *Portulaca oleracea* L., and *Mimosa pudica* L. were not native species of the studied area, but still they are used as medicines.

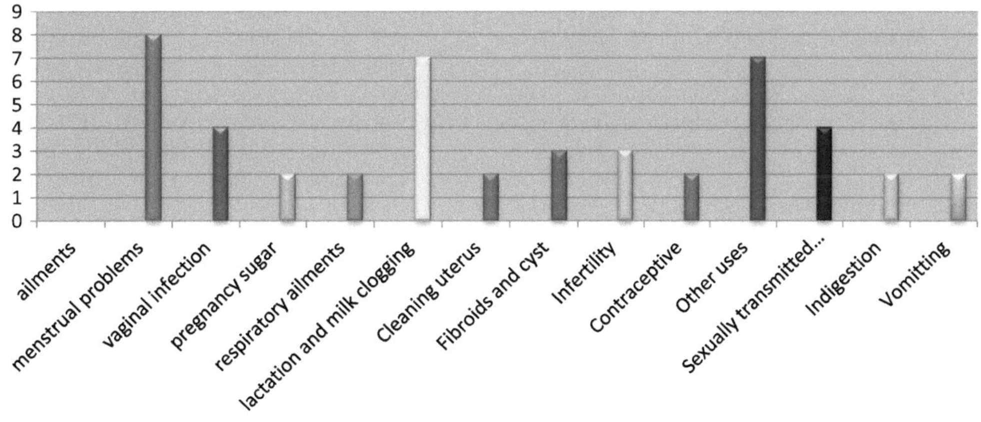

FIGURE 25.3 Ailments and number of plant genera used for the ailments.

25.4 DISCUSSION

India is country rich with plant medicinal resources. Though many studies have been conducted so far regarding phytomedicines, a detailed work on the phytomedicine usage against gynecological disorders has not been documented so far. Some studies have been distributed worldwide. Olanipekun et al. (2013) reported the different bioactive ingredients, such as alkaloids, saponins, tannins, and cardiac glycoside as natural agents responsible for the effectiveness of some of the identified plants. The present study has revealed the herbal knowledge of the fisher folks of the coastal regions of Tamil Nadu. This tends to suggest that the level of awareness, importance, and acceptability of the medicinal species was high among the respondents (Kayode et al. 2015). The people in the study area believe that the phytomedicines have lesser or no side effects, are cheaper to cure various ailments than orthodox medicine, and could save the nation a large amount of money that can be converted for other uses which will help to further national development (Arowosegbe et al. 2015, Kayode et al. 2015; Lewis 2003; Olapade 2002).

25.5 CONCLUSION

Female reproductive disorders can cause domestic issues in many countries of the world. It is necessary to search for solutions to female disorders in all possible ways including the herbal approach. The plants documented in this study should be further researched to confirm the claims of the fisher folks.

ACKNOWLEDGMENT

I am thankful to the professor and head of the Department of Botany, Bharathiar University, for providing the necessary guidance and facilities to carry out the studies.

REFERENCES

Arowosegbe, S., Olanipekun, M. K., and Kayode, J. 2015. Ethnobotanical survey of medicinal plants used for the treatment of Diabetes mellitus in Ekiti-State Senatorial districts, Nigeria. *European Journal of Botany and Plant Science* 2(4), 1–8.

Balasubramaniam, V., and Ramachandran, V. S. 1991. *Flora of Ramanathapuram District Tamil Nadu.*

Balick, M. J., and Cox, P. A. 1996. *Plants, People, and Culture: The Science of Ethnobotany.* Scientific American Library, New York, 17–53.

Bussmann, R. W., and Glenn, A. 2010. Medicinal plants used in northern Peru for reproductive problems and female health. *Journal of Ethnobiology and Ethnomedicine* 6, 1–12.

Cox, P. A. 2000. Will tribal knowledge survive the millennium. *Science* 287, 44–45.

Flaster, T. 1995. Ethnobotanical approaches to the discovery of bioactive compounds. *Progress in new crops: Proceedings of the third national symposium.* New crops: New opportunities, new technologies. Indianapolis Alexandria: ASHS Press Janick E. pp. 561–656.

Flaster, T. 2010. Ethnobotanical approaches to the discovery of bioactive compounds. Progress in new crops. In *Proceedings of the third National Symposium*, ASHS Press, Alexandria, Egypt, 561–565.

Gamble, J. S., and Fisher, C. E. C. 1957. *Flora of the Presidency of Madras.* Vols I–III. Adlard & Co., London, UK. O9Reprinted, Botanical Survey of India, Calcutta. 1915–1936.

Hooker, J. D. 1875–2006. *Flora of British India.* Vols I–VII, L. Reeva and Company Ltd., eds. The East House Book. Ashford, Kent, UK.

Jain, S. K., and Rao, R. R. *A Handbook of Field and Herbarium Methods.* Today and Tomorrow Publishers, New Delhi, 1977.

Kayode, J., Michael, A. O., Modupe, J. A., and Ayodele, A. O. 2015. Stem barks and roots extravitism in Ekiti state Nigeria: Need for conservation as a sustainable innovation in healthcare management in rural areas. *American Journal of Biosciences* 3(2), 28–33.

Lawal, I., Amao A. K., Alamu, T., and Savunmi, L. 2013. Phytotherapy approach for the treatment of gynaecological disorder among women in Ido Local Government Area of Ibadan, Oyo state, Nigeria. *Journal of Advanced Scientific Research* 4, 41–44.

Lewis, W. H. 2003. Pharmaceutical discoveries based on ethnomedicinal plants: 1985 to 2000 and beyond. *Economic Botany* 57(1), 126–134.

Mahmood, A., Mahmood, A., Malik, R. N., Shinwari, Z. K. 2013. Indigenous Knowledge of medicinal plants from Gujranwala district, Pakistan. *Journal of Ethnopharmacology* 148, 714–723.

Mathew, K. M. 1983. *The Flora of Tamil Nadu Carnatic Part (1–3).* Rapinet herbarium, Tiruchirapalli.

Olanipekun, M. K., Kayode, J., and Akinyemi, M. O. 2013. Ethnoveterinary survey and proximate analysis of plants used in managing Newcastle disease of poultry animals in Ekiti State, Southwestern, Nigeria. *Bulletin of Pure Applied Sciences* 32(1), 29–33.

Olapade, E. O. 2002. *The Herbs for Good Health: The 50th Anniversary Lecture of the University of Ibadan.* NARL Specialist Clinic.

Patel, P. K. 2012. Ethnogynaecological uses of plants from Gujarat, India. *Bangladesh Journal of Plant Taxonomy* 19, 93–94.

Ravishankar, B., and Shukla, V. J. 2007. Indian system of medicine: A brief profile. *African Journal of Traditional, Complementary and Alternative Medicines AJTCAM* 4(3), 319–337.

Schippman, D., Leaman, J., and Cunningham, A. B. 2002. Biodiversity and the ecosystem approach in Agriculture, Forestry and Fisheries. *Impact of Cultivation and Gathering of Medicinal Plants on Biodiversity: Global Trends and Issues* 1, 1–21.

Safeghi, Z., and Mahmood, A. 2014. Ethno-gynaecological knowledge of medicinal plants used by Baluchj tribes, Southeast of Baluchistan, Iran. *Brazilian Journal of Pharmacognosy* 706–715.

Sambandan, K., Datchanamoorthy, N., and Jagadeesan, J. 2015. E-Flora of Karaikal District. U.T. of Puducherry India. Retrieved from http:// www.eflorakkl.

Umamaheswari, P., and Daniel, P. 1991. Studies on the Flora of the Gulf of Mannar Biosphere Reserve, Southern India. Botanical Survey of India, Southern Regional Circle, Coimbatore. World Health Organization: Traditional, Complementary and Alternative Medicines and Therapies. Washington, DC

World Health Organization. 1999. Traditional, Complementary and Alternative Medicines and Therapies. Washington DC, WHO Regional Office for the Americas/Pan American Health Organization, Geneva, Switzerland.

World Health Organization: WHO TM Strategy 2002–2005. World Health Organization, Geneva, Switzerland.

26 *In Vitro* Anticancer Activity of *Syzygium calophyllifolium* on A549 Lung Cancer Cells

Rahul Chandran and Heidi Abrahamse

CONTENTS

26.1 INTRODUCTION

Cancer has become one of the most studied type of diseases, and the reason for this being the high mortality and incidence reported every year. The condition is characterized by the irregular proliferation of cells that leads to life threatening events (Roy A et al. 2017). According to recent cancer statistics, 18.1 million cancer cases were reported with an estimated mortality of 9.6 million in 2018 (Bray et al. 2018). The statistics also reported a high incidence of lung cancer in both sexes (11.6%), which led to the highest number of cancer death cases (18.4%) as compared to other cancer types. Small cell and non-small cell are the histologic classifications of lung cancer, with both having common symptoms like cough, systemic problems, hemoptysis, dyspnea, etc. A diagnosis at the early stages can improve the lifestyle of the patient for a prolonged period (Latimer and Mott 2015). With several reported factors causing cancer including genetic and lifestyle changes, alternative treatment methods with high efficiency are sought after at an alarming demand. Science has been trying all the possible treatment modalities to eradicate cancer since its outburst. However, due to certain limitations and side effects of currently available treatments, science has been developing the idea of treating cancer using plant-based medicine minimizing the side effects and increasing the therapeutic efficiency.

Plants are of common interest among traditional healers since ancient times and are being studied by researchers across the world for their medicinal properties. Plants are a well chosen alternative mode of treatment against cancer in many countries, with around 3000 plants already reported to have an effective potential to fight against cancer (Alves-Silva et al. 2017; Tariq et al. 2017). Natural products, especially those from plants, have gained interest due to their biologically active compounds and their chemotherapeutic effects against cancer. They are biologically friendly and less or non-toxic to normal cells (Mishra and Tiwari 2011). The ability to induce alternative modes of cell death makes these naturally derived anti-cancer drugs a promising candidate for anti-cancer research. Moreover, there is evidence that natural product-derived anti-cancer drugs can induce alternative modes of cell death pathways (Khalid et al. 2016; Gali-Muhtasib et al. 2015).

The genus *Syzygium* is being well studied for its medicinal properties by natural scientists. One of the authors (Chandran) has also performed several biochemical and pharmacological studies on species of *Syzygium* (Chandran et al. 2015, 2016,

2017a, 2017b, 2017c, 2018; Sathyanarayana et al. 2018). All of these studies have proven the tremendous therapeutic potential of *Syzygium calophyllifolium* Walp. The plant has also been proven to be a good remedy for toothaches and inflammation (Rajan et al. 2005; Sathyavathi and Janardhanan 2011). The present study examines the effect of the *Syzygium calophyllifolium* bark methanol (SCBM) extract on A549 lung cancer cells to facilitate an alternative treatment modality.

26.2 MATERIALS AND METHODS

26.2.1 Collection of Plant Materials

The bark was collected during the month of February 2012 from Ooty, India. The collected plant material was identified, and its authenticity was confirmed by comparing the voucher specimen at the herbarium of the Botanical Survey of India, Southern Circle Coimbatore, Tamil Nadu (BSI/SRC/5/23/2012-13/Tech.453). The freshly collected plant material was cleaned to remove adhering dust and then dried under shade. The dried sample was powdered and used for further studies.

26.2.2 Successive Solvent Extraction

The air-dried, powdered tree bark (100 g) was extracted in a Soxhlet extractor using methanol (300 mL), vacuum dried in a rotary vacuum evaporator, and used for further studies.

26.2.3 Cell Culture and Treatment

A commercially purchased A549 lung cancer cell line (ATCC CCL-185) was used for the anti-cancer study. Approximately 5×10^5 cells were seeded in 3.4 cm^2 diameter culture dishes and cultured in a Roswell Park Memorial Institute 1640 Medium (Sigma-Aldrich, R8758), supplemented with 10% fetal bovine serum (Gibco, 306.00301) and 1% antibacterial (penicillin) agents. The culture was maintained at 37°C with 5% carbon dioxide (CO_2) and 85% humidity for 4 hours to allow the cells to attach. Culture dishes with more than 90% confluence were used for these experiments. The experimental groups were divided into untreated controls and the cells treated with SCBM at three different doses (5, 10, and 20 µg/mL).

26.2.4 Cellular Morphology-Inverted Microscopy

The morphology of the SCBM extract (5, 10, and 20 µg/mL)-treated cells was analyzed after 24 hours of incubation using an inverted light microscope (Wirsam, Olympus CKX41). Once the digital images were recorded, the cells were trypsinized using 1 mL/25 cm^2 of TrypLE Express (Invitrogen, 12605-028) and resuspended in Hank's Balanced Salt Solution to perform further assays.

26.2.5 Cellular Proliferation-Adenosine Triphosphate Luminescent Assay

The CellTiter-Glol luminescent assay (Promega, G7571, Anatech Analytical Technology, Bellville, South Africa) is a homogeneous method for determination of cellular proliferation and quantification of the adenosine triphosphate (ATP) present in the metabolically active cells. An equal volume (50 µL) of the reconstituted ATP reagent and the cell suspension was mixed on a shaker for 2 minutes to induce cell lysis, followed by incubation at room temperature for 10 minutes in the dark to stabilize the luminescent signal. The luminescent signal was read using the 1420 Multilabel Counter Victor3 (PerkinElmer, Separation Scientific).

26.2.6 Cytotoxicity-Lactate Dehydrogenase Assay

The membrane integrity was assessed by estimating the amount of lactate dehydrogenase (LDH) present in the culture media. The cytosolic enzyme LDH will be released into the media due to membrane damage. The CytoTox 96® X assay (Anatech, Promega G 400) was used to measure the LDH released. An equal volume (50 µL) of reconstituted LDH reagent and cell culture medium was mixed and incubated in the dark at room temperature for 30 minutes. The colorimetric compound was measured spectrophotometrically at 490 nm (PerkinElmer, Victor3™).

26.2.7 Hoechst Stain

The damage caused by the SCBM extract was analyzed using Hoechst nuclear stain. The MCF-7 cells were cultured in 3.4 cm diameter culture dishes over sterile cover slips and allowed to reach a confluence of 80%. The cells were then treated with different concentrations of SCBM (5, 10, and 20 µg). After 24 hours of incubation, the cells were stained with 1 µg/mL Hoechst stain (Hoechst 33258, H21491) for 30 minutes. Thereafter, the cells were rinsed with Phosphate Buffer Saline (PBS), and the blue fluorescent signal was examined using the Olympus fluorescent microscope.

26.2.8 Statistical Analysis

All the results were expressed as mean ± SEM. The statistical significance was determined by using the one-way Analysis of Variance (ANOVA) followed by Dunnett's multiple comparison tests. $p < 0.05$ was considered statistically significant.

26.3 RESULTS AND DISCUSSION

This research was an attempt to explore the possibility of using drugs of natural origin against lung cancer. The methanol extract of SCBM was treated and analyzed on A549 cells in vitro. The cytotoxic effects of potentially active drugs can be determined using various qualitative and quantitative biochemical assays. In the present study, the cell morphology, ATP, LDH, and nuclear damage using Hoechst stain were studied.

26.3.1 Cellular Morphology

The change in cellular morphology was observed using inverted microscopy. The A549 cells were treated and incubated with different doses of the SCBM extracts. An evident change in cellular morphology was observed after 24 hours. The loss of membrane integrity, cell to cell contact, rounding, vacuolation, and cell detachment for the culture plate surface were seen as indications of cell death. These signs were more evident in the treated groups as compared to the control (untreated) ones (Figure 26.1). These morphological features depicted the toxicity of the SCBM extract on A549 cells, indicating the signs of apoptosis (Monga et al. 2013). However, this has to be confirmed using further biochemical and molecular study.

26.3.2 ATP Cell Metabolism Assay

Mitochondria generate ATP as an energy molecule for their metabolic activities, and any damage to this organelle will result in the imbalanced ATP production and ATP-dependent mechanisms necessary for cell functions (Maurer and Meyer 2016). ATP is one of the indications of healthy cell metabolism, which can effectively be assessed using a luminescent quantification assay. As observed in the cell morphology, Figure 26.2 expresses higher cellular metabolism in the control group, with the least ATP in 20 μg/mL of the SCBM extract.

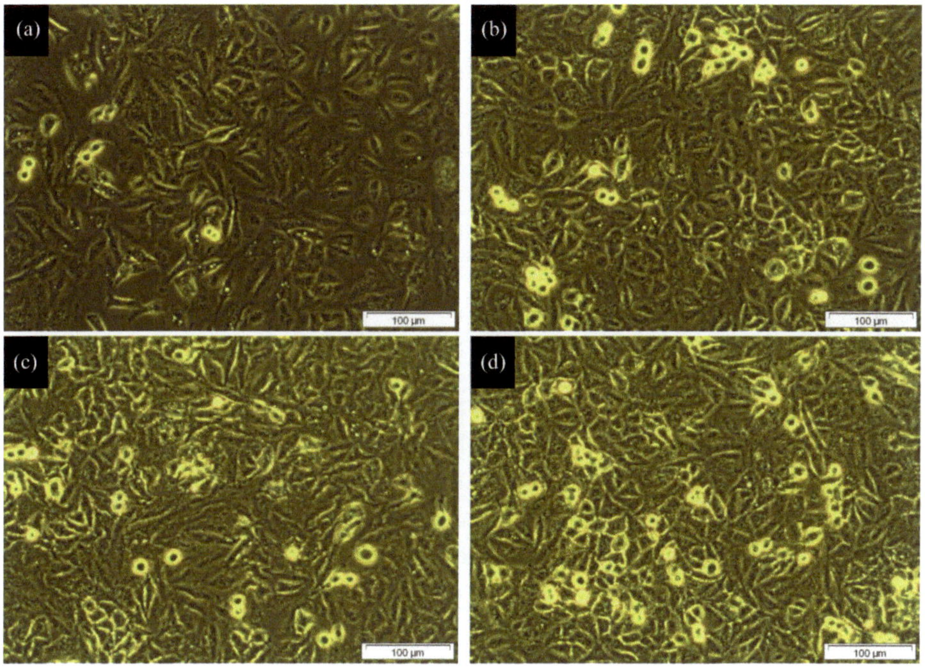

FIGURE 26.1 Morphology of A549 cells after the SCBM extract treatment. (a) Control; (b) SCBM 5 μg; (c) SCBM 10 μg; and (d) SCBM 20 μg (10x objective).

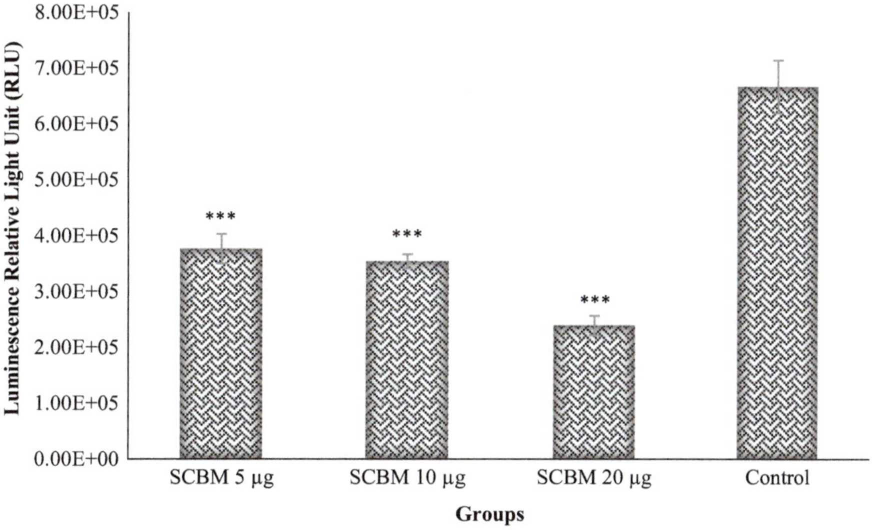

FIGURE 26.2 ATP metabolism in A549 cells treated with the SCBM extract. The data represent the mean ± SEM. Significantly different at ***$p < 0.001$ when compared to the control *S. calophyllifolium* bark methanol (SCBM) extract.

A significant reduction in the cell proliferation was observed in a dose-dependent manner (5, 10, and 20 μg/mL) as compared to the control. As the concentration of the extract was increased, the ATP levels were also decreased. The results of the present study support the findings previously reported by one of the authors (Chandran) in breast cancer (MCF-7) cells (Chandran et al. 2018).

26.3.3 Cytotoxicity-LDH Assay

LDH is a cytoplasmic enzyme which is released into the extracellular space upon plasma membrane damage and is found in nearly all types of cells (Gurunathan et al. 2015). The results of the LDH assay are in agreement with the ATP and morphology studies. The loss of cell membrane integrity released the LDH to the culture media when treated with the SCBM extract. The toxic effect of the extract was found best in the highest dose of the extract (Figure 26.3). The comparable levels of LDH implicate the cytotoxic effects of the extract than the untreated control as shown in Figure 26.3. The results of the LDH in the media were directly proportional to the dose of the extract. Xia et al. (2007) reported a correlation between the LDH upregulation and the induction of apoptosis.

26.3.4 Hoechst Stain

Hoechst stain helps to identify the extent of nuclear damage induced by the SCBM extract. The cells responded variably to the doses of the extract, wherein the highest dose showed visible nuclear damage. The nucleus of the control cells stained uniformly, signifying its intact and dense nature. The nucleus of the extract-treated cells showed nuclear margination, chromatin condensation, and deformation of the nuclear size and integrity as compared to the untreated control (Figure 26.4).

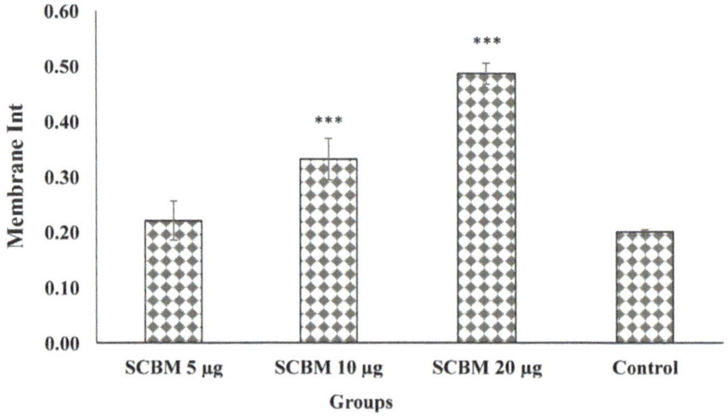

FIGURE 26.3 LDH levels in A549 cells treated with the SCBM extract. The data represent the mean ± SEM. Significantly different at ***$p < 0.001$ when compared to the control *S. calophyllifolium* bark methanol (SCBM) extract.

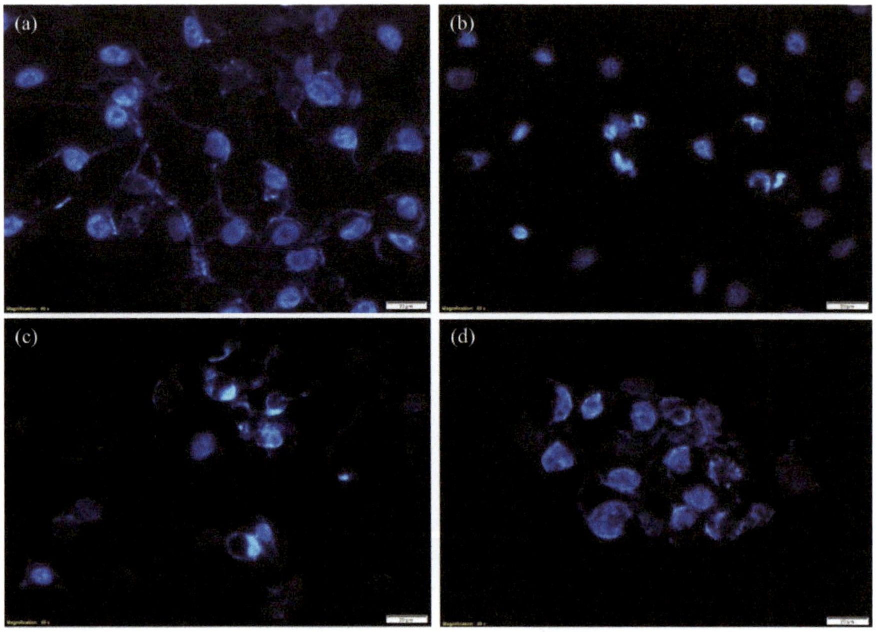

FIGURE 26.4 Nuclear damage study in A549 cells using Hoechst stain. (a) Control; (b) SCBM 5 μg; (c) SCBM 10 μg; and (d) SCBM 20 μg (40x objective).

The fluorescent images of the nucleus also supported the fact that the extract promoted DNA damage and cell death.

26.4 CONCLUSION

Natural drugs have gained interest among a majority of the population around the globe. To complement this, the present study was designed to formulate the cytotoxic activity of the SCBM extract against lung cancer cells in vitro. The morphology of the treated cells explained the degree of toxicity induced by the SCBM extract. The reduced ATP and increased LDH levels also implicated the cytotoxic effect of the extract toward A549 cells. The proposed activity of the extract in this study is never a replacement for drugs popularly used for the treatment for lung cancer in-fact a recommendation on how to utilize such natural products for the treatment with best check on diseases like cancer and lesser side effects.

ACKNOWLEDGMENTS

This work is based on the research supported by the South African Research Chairs Initiative of the Department of Science and Technology and National Research Foundation of South Africa (Grant No 98337), as well as grants received from the University of Johannesburg (URC), the National Research Foundation (NRF), and the Council for Scientific and Industrial Research (CSIR)–National Laser Centre (NLC) Laser Rental Pool Programme.

REFERENCES

Alves-Silva, J. M., Romane, A., Efferth, T., and Salgueiro, L. 2017. North African medicinal plants traditionally used in cancer therapy. *Frontiers in Pharmacology.* 8:1–24.

Bray, F., Ferlay, J., Soerjomataram, I., Siegel, R. L., Torre, L. A., and Jemal, A. 2018. Global cancer statistics 2018: GLOBOCAN estimates of incidence and mortality worldwide for 36 cancers in 185 countries. *CA: A Cancer Journal for Clinicians.* 68(6):394–424.

Chandran, R., Abrahamse, H., and Thangaraj, P. 2018. Cytotoxic, analgesic and anti-inflammatory properties of *Syzygium calophyllifolium* bark. *Biomedicine and Pharmacotherapy.* 103:1079–1085.

Chandran, R., Abrahamse, H., Parimelazhagan, T., and Gowtham, D. 2017a. *Syzygium mundagam* bark methanol extract restores skin to normal in diabetic wounded rats. *Biomedicine and Pharmacotherapy.* 94:781–786.

Chandran, R., George, B. P., Abrahamse, H., and Parimelazhagan, T. 2017b. Therapeutic effects of *Syzygium mundagam* bark methanol extracts on Type-2 diabetic complications in rats. *Biomedicine and Pharmacotherapy.* 95:167–174.

Chandran, R., Parimelazhagan, T., and George, B. P. 2017c. Antihyperglycemic activity of the bark methanolic extract of *Syzygium mundagam* in diabetic rats. *Alexandria Journal of Medicine.* 53(4):317–324.

Chandran, R., Saravanan, S., Sajeesh, T., and Parimelazhagan, T. 2016. Antidiabetic activity of *Syzygium calophyllifolium* in Streptozotocin-Nicotinamide induced Type-2 diabetic rats. *Biomedicine and Pharmacotherapy.* 82:547–554.

Chandran, R., Sathyanarayanan, S., Rajan, M., Kasipandi, M., and Parimelazhagan, T. 2015. Antioxidant, hypoglycemic and anti-hyperglycemic properties of *Syzygium calophyllifolium* Walp. *Bangladesh Journal of Pharmacology.* 10:672–680.

Gali-Muhtasib, H., Hmadi, R., Kareh, M., Tohme, R., and Darwiche, N. 2015. Cell death mechanisms of plant-derived anticancer drugs: Beyond apoptosis. *Apoptosis.* 20:1531–1562.

Gurunathan, S., Han, J. W., Park, J. H., Kim, E., Choi, Y. J., Kwon, D. N., and Kim, J. H. 2015. Reduced graphene oxide-silver nanoparticle nanocomposite: A potential anticancer nanotherapy. *International Journal of Nanomedicine.* 100:6257–6276.

Khalid, E. B., Ayman, E. E., Rahman, H., Abdelkarim, G., and Najda, A. 2016. Natural products against cancer angiogenesis. *Tumor Biology.* 37:14513–14536.

Latimer, K. M., and Mott, T. F. 2015. Naval Hospital Pensacola, Pensacola, Florida. *American Family Physician.* 91(4): 250–256.

Maurer, L. L., and Meyer, J. N. 2016. A systematic review of evidence for silver nanoparticle-induced mitochondrial toxicity. *Environmental Science: Nano.* 3:311–322.

Mishra, B. B., and Tiwari, V. K. 2011. Natural products: An evolving role in future drug discovery. *European Journal of Medicinal Chemistry.* 46:4769–4807.

Monga, J., Pandit, S., Chauhan, C. S., and Sharma, M. 2013. Cytotoxicity and apoptosis induction in human breast adenocarcinoma MCF-7 cells by (+)-cyanidan-3-ol. *Experimental Toxicololgy and Pathology.* 65(7–8):1091–1100.

Rajan, S., Jayendran, M., and Sethuraman, M. 2005. Folk herbal practices among Toda tribe of the Nilgiri Hills in Tamil Nadu, India. *Journal of Natural Remedies.* 5(1):52–58.

Roy, A., Ahuja, S., and Bharadvaja, N. 2017. A review on medicinal plants against cancer. *Journal of Plant Sciences and Agricultural Research.* 2(1):008.

Sathyanarayanan, S., Chandran, R., Thankarajan, S., Abrahamse, H., and Parimelazhagan, T. 2018. Phytochemical composition, antioxidant and antibacterial activity of wild edible fruit *Syzygium calophyllifolium* Walp. *Journal of Food Science and Technology.* 55(1):341–350.

Sathyavathi, R., and Janardhanan K. 2014. Wild edible fruits used by Badagas of Nilgiri District, Western Ghats, Tamilnadu, India. *Journal of Medicinal and Plants Research.* 8(2):128–132.

Tariq, A., Sadia, S., Pan, K., Ullah, I., Mussarat, S., Sun, F., Abiodun,O. O., Batbaatar, A., Li, Z., Song, D., Xiong, Q., Ullah, R., Khan, S., Basnet, B. B., Kumar, B., Islam, R., and Adnan, M. 2017. A systematic review on ethnomedicines of anticancer plants. *Phytotherapy Research.* 31:202–264.

Xia, G. H., Chen, B. A., Shao, Z. Y., Lu, H. X., Konstanze, D., and Hartmut, D. 2007. Mechanism of 2-methoxyestradiol-induced apoptosis in myelodysplastic syndrome MUTZ-1 cell line. *Zhongguo Shi Yan Xue Ye XueZaZhi.* 15(2):296–301.

Index